Aquatic
and
Surface
Photochemistry

Aquatic
and
Surface
Photochemistry

Edited by

George R. Helz, Ph.D.
Director, Maryland Water Resources Research Center
Professor, Department of Chemistry and Biochemistry
University of Maryland
College Park, Maryland

Richard G. Zepp, Ph.D.
Research Scientist
Environmental Research Laboratory
U.S. Environmental Protection Agency
Athens, Georgia

Donald G. Crosby, Ph.D.
Professor
Department of Environmental Toxicology
University of California Davis
Davis, California

LEWIS PUBLISHERS
Boca Raton   Ann Arbor   London   Tokyo

Library of Congress Cataloging-in-Publication Data

Helz, G. R.
    Aquatic and surface photochemistry / George R. Helz, Richard G. Zepp,
    Donald G. Crosby.
       p. cm.
    Includes bibliographical references and index.
    ISBN 0-87371-871-2
    1. Water—Pollution.   2. Photochemistry.   3. Water—Purification—Photocatalysis.
I. Zepp, Richard G.   II. Crosby, Donald G.   III. Title.
TD423.H45   1994
628.1′68—dc20   93-22832
                                                              CIP

© 1994 by CRC Press, Inc.
Lewis Publishers is an imprint of CRC Press

No claim to original U.S. Government works
International Standard Book Number 0-87371-871-2
Library of Congress Card Number 93-22832
Printed in the United States of America 1 2 3 4 5 6 7 8 9 0
Printed on acid-free paper

Preface

Interest in environmental photochemistry was kindled in the 1960s and early 1970s through the emergence of various environmental problems. First, it was recognized that certain atmospheric problems, e.g., smog formation and stratospheric ozone depletion, are closely linked to photoreactions of pollutants in the troposphere and stratosphere. As a result of the need to better understand the causes of these problems, atmospheric photochemistry, especially of gas-phase processes, has become an intensely studied, highly developed subdiscipline.

Recognition of the importance of surface and aquatic photochemistry in the environment was slower to develop. Only in the 1980s has this field of study begun to blossom. Concerns about the causes of acid rain and the role of clouds in tropospheric chemistry have stimulated research on photochemistry in atmospheric water droplets. Other studies have revealed that natural photoreactions play an important role in cleansing the aquatic and terrestrial environment of biologically active substances derived from human activities, e.g., agrochemical residues and various other pollutants. A variety of recent studies have further shown that photoreactions can be utilized to remediate contaminated water supplies. Going hand-in-hand with these pollution-related efforts, other recent studies indicate that natural photoreactions have a broad impact on the functioning of ecosystems, with significant effects on the cycling of carbon, oxygen, sulfur, and various trace metals that are biologically important.

The purpose of this book is to provide the interested reader with a broad overview of current research in the emerging field of environmental aquatic and surface photochemistry. To accomplish this purpose, selected reviews and current research articles are blended together to provide an in-depth treatment of various aspects of this research area. For convenience of the reader, the chapters have been subdivided into two main topics: photochemistry in the environment and photochemistry in water treatment.

The section on photochemistry in the environment covers recent research on topics such as aquatic photochemistry of organic pollutants and agrochemicals, photochemical cycling of carbon and transition metals (especially iron), photochemical formation of reactive oxygen species in natural waters, photoreactions in cloud and rain droplets, and photoreactions on environmental surfaces (soil, ash; metal oxides).

The water-treatment section provides discussions and data on both heterogeneous photocatalytic and homogeneous processes. Topics range from applications to mechanistic studies. These chapters illustrate the wide diversity of pollutant classes that are degradable by photochemical techniques and the effects of various reaction conditions on the rates and efficiency of the techniques. Current kinetic studies are presented, which provide new information about the role of adsorption and the nature of the reactive oxidizing species that mediate these photoremediation processes.

The chapters in this book are based on presentations at a symposium that was held at the 203rd National Meeting of American Chemical Society in San Francisco, CA. The meeting brought together leading scientists who are focusing on environmental aspects of aquatic and surface photochemistry.

Acknowledgments

The editors thank the Environmental and Geochemistry Divisions of the American Chemical Society; the donors of the Petroleum Research Fund, which is administered by the American Chemical Society; the U.S. Environmental Protection Agency; and the Maryland Water Resources Research Center for financial support of the symposium upon which this book is based. The editorial assistance of Mary Ellen Atkinson, of the Maryland Water Resources Research Center, is particularly appreciated.

The Editors

George R. Helz   Director, Maryland Water Resources Research Center
Professor, Department of Chemistry and Biochemistry
University of Maryland
College Park, MD 20742
Tel: 301-405-6829   Fax: 301-314-9121

George R. Helz is Professor of Chemistry and Director of the Water Resources Research Center at the University of Maryland, College Park. He holds an A.B. in geology from Princeton University and a Ph.D. in geochemistry from Pennsylvania State University. His research deals with the nature and rates of geochemical processes affecting anthropogenic contaminants. The chemical reactions of transition metals in sulfidic waters, the fate of chloramines in dechlorinated effluents, and the use of accelerator mass spectrometry to study groundwater transport patterns and rates are among his current interests. He has served as Chairman of the Geochemistry Division of American Chemical Society. In 1988, he was an Environmental Science and Engineering Fellow at AAAS.

Richard G. Zepp   Research Scientist
Environmental Research Laboratory
U.S. Environmental Protection Agency
Athens, GA 30613
Tel: 706-546-3428   Fax: 706-546-3636

Richard G. Zepp is a Research Chemist at the Environmental Research Laboratory, U.S. Environmental Protection Agency, Athens, GA and an Adjunct Professor at the Rosenstiel School of Marine and Atmospheric Sciences, University of Miami, FL. He received his B.S. degree from Furman University and a Ph.D. in physical organic chemistry from Florida State University, Tallahassee. His current research interests include processes that provide sources and sinks of atmospheric trace gases in the biosphere, environmental reactions of humic substances and transition metals, pollutant photoreactions in terrestrial and aquatic environments, and photochemical remediation of contaminated waters. He is currently on the Editorial Advisory board of *Environmental Science and Technology* and is a member of the Executive Committee of the American Chemical Society Environmental Chemistry Division.

Donald G. Crosby   Professor, Department of Environmental Toxicology
University of California at Davis
Davis, CA 95616
Tel: 916-752-4529   Fax: 916-752-3394

Donald G. Crosby is Professor Emeritus in the Department of Environmental Toxicology at the University of California, Davis. He received his A.B. degree in chemistry from Pomona College and a Ph.D. in chemistry/biology from California Institute of Technology, Pasadena. After 8 years at Union Carbide, he came to The University of California at Davis in 1961. He is the author of over 200 research papers and 2 books in the areas of environmental photochemistry, marine chemistry, and pesticide chemistry and was founding chairman of the American Chemical Society Division of Pesticide Chemistry (Agrochemicals).

Contributors

John M. Allen
School of the Environment
Environmental Chemistry Laboratory
Duke University
Durham, NC

Ghassan Al-Sayyed
Chemistry Department
University College Cork
Cork, Ireland

Laurence Amalric
URA au CNRS n° 1385
Photocatalyse, Catalyse
  et Environement
Ecole Centrale de Lyon
BP 163, 69131 Ecully Cédex
France

Cort Anastasio
School of the Environment
Environmental Chemistry Laboratory
Duke University
Durham, NC

Detlef W. Bahnemann
Institut für Solarenergieforschung
  GmbH
Sokelantstr. 5
D-30165 Hannover, Germany

Dirk Bockelmann
Institut für Solarenergieforschung
  GmbH
Sokelantstr. 5
D-30165 Hannover, Germany

James R. Bolton
Solarchem Environmental Systems
130 Royal Crest Court
Markham, Ontario, Canada; and
Photochemistry Unit
Department of Chemistry
University of Western Ontario
London, Ontario, Canada

J. R. Brock
Department of Chemical Engineering
University of Texas at Austin
Austin, TX

Stephen R. Cater
Solarchem Environmental Systems
130 Royal Crest Court
Markham, Ontario, Canada

M. Chin
Physical Sciences Laboratory
Georgia Tech Research Institute
Georgia Tech School of Earth and
  Atmospheric Sciences
Georgia Institute of Technology
Atlanta, GA; and
Department of Earth and Planetary
  Sciences
Harvard University
Cambridge, MA

William J. Cooper
Drinking Water Research Center
Florida International University
Miami, FL

Donald G. Crosby
Department of Environmental
  Toxicology
University of California
Davis, CA

Joseph Cunningham
Department of Chemistry
University College Cork
Cork, Ireland

Thomas A. Dahl
Department of Pharmacology and
  Experimental Therapeutics
Tufts University Veterinary, Medical,
  and Dental Schools
136 Harrison Ave.
Boston, MA

Hervé Delprat
URA au CNRS n° 1385
Photocatalyse, Catalyse
  et Environement
Ecole Centrale de Lyon
BP 163, 69131 Ecully Cédex
France

M. Dieckmann
Department of Civil Engineering and
  Geological Sciences
University of Notre Dame
Notre Dame, IN

Jean-Christophe D'Oliveira
URA au CNRS n° 1385
Photocatalyse
Catalyse et Environement
Ecole Centrale de Lyon
BP 163, 69131 Ecully Cédex
France

Susan G. Donaldson
Department of Environmental and
  Resource Science
University of Nevada
Reno, NV

Marifusa Eto
Laboratory of Pesticide Chemistry
Department of Agricultural Chemistry
Kyushu University
Higashi-ku, Fukuoka 812, Japan

Bruce C. Faust
School of the Environment
Environmental Chemistry Laboratory
Duke University
Durham, NC

M. A. Fox
Department of Chemistry
University of Texas
Austin, TX

Peter K. Freeman
Department of Chemistry
Oregon State University
Corvallis, OR

F. H. Frimmel
Engler-Bunte-Institut
Universität Karlsruhe
D-76128 Karlsruhe, Germany

Roland Goslich
Institut für Solarenergieforschung
  GmbH
Sokelantstr. 5
D-30165 Hannover, Germany

K. A. Gray
Department of Civil Engineering and
  Geological Sciences
University of Notre Dame
Notre Dame, IN

Chantal Guillard
URA au CNRS n° 1385
Photocatalyse, Catalyse
  et Environement
Ecole Centrale de Lyon
BP 163, 69131 Ecully Cédex
France

Werner R. Haag
Purus, Inc.
2713 N. First St.
San Jose, CA

Susan A. Hatlevig
Department of Chemistry
Oregon State University
Corvallis, OR

Adam Heller
Department of Chemical Engineering
University of Texas at Austin
Austin, TX

D. P. Hessler
Engler-Bunte-Institut
Universität Karlsruhe
D-76128 Karlsruhe, Germany

Marcus Hilgendorff
Institut für Solarenergieforschung
  GmbH
Sokelantstr. 5
D-30165 Hannover, Germany

Can Hoang-Van
URA au CNRS n° 1385
Photocatalyse, Catalyse
  et Environement
Ecole Centrale de Lyon
BP 163, 69131 Ecully Cédex
France

J. Hoigné
Swiss Federal Institute for Water
  Resources and Water Pollution
  Control (EAWAG)
CH-8600 Dübendorf, Switzerland

Aitken R. Hoy
Photochemistry Unit
Department of Chemistry
The University of Western Ontario
London, Ontario, Canada

Guilherme L. Indig
Department of Chemistry
Boston University
Boston, MA

Kai-Kong Iu
Department of Chemistry and
  Biochemistry
University of Notre Dame
Notre Dame, IN

Guilford Jones, II
Department of Chemistry
Boston University
Boston, MA

Prashant V. Kamat
Radiation Laboratory
University of Notre Dame
Notre Dame, IN

Georgios Karametaxas
Swiss Federal Institute for
  Environmental Science and
  Technology (EAWAG)
CH-8600 Dübendorf, Switzerland

Hansulrich Laubscher
Swiss Federal Institute for
  Environmental Science and
  Technology (EAWAG)
CH-8600 Dübendorf, Switzerland

Darren Lawless
Laboratory of Pure and Applied
  Studies in Catalysis, Environment
  and Materials
Department of Chemistry and
  Biochemistry
Concordia University
Montreal, Canada

David R. S. Lean
National Water Research Institute
Environment Canada
P. O. Box 5050
Burlington, Ontario, Canada

Xinsheng Liu
Department of Chemistry and
  Biochemistry
University of Notre Dame
Notre Dame, IN

Scott A. Mabury
Department of Environmental
  Toxicology
University of California
Davis, CA

Gleb Mamantov
Department of Chemistry
University of Tennessee
Knoxville, TN

Yun Mao
Department of Chemistry and
  Biochemistry
University of Notre Dame
Notre Dame, IN

Theodore Mill
Chemistry Laboratory
SRI International
Menlo Park, CA

Glenn C. Miller
Department of Environmental and
  Resource Sciences
University of Nevada
Reno, NV

William L. Miller
U.S. Environmental Protection Agency
National Research Council
960 College Station Rd.
Athens, GA

Claudio Minero
Dipartimento di Chimica Analitica
Università di Torino
10125 Torino, Italy

L. Nowell
U. S. Geological Survey
Water Resources Division
2800 Cottage Way
Sacaramento, CA

Churl Oh
Department of Chemistry
Boston University
Boston, MA

David F. Ollis
Department of Chemical Engineering
North Carolina State University
Raleigh, NC

Leonardo Palmisano
Dipartimento di Ingegneria
Chimica dei Processi e dei Materiali
Università di Palermo
Palermo, Italy

Ezio Pelizzetti
Dipartimento di Chimica Analitica
Università di Torino
10125 Torino, Italy

Anne-Marie Pelletier
Laboratory of Pure and Applied
  Studies in Catalysis, Environment
  and Materials
Department of Chemistry and
  Biochemistry
Concordia University
Montreal, Canada

Pierre Pichat
URA au CNRS n° 1385
Photocatalyse, Catalyse
  et Environement
Ecole Centrale de Lyon
BP 163, 69131 Ecully Cédex
France

Frances R. Pick
University of Ottawa
Department of Biology
Ottawa, Ontario, Canada

Antonino Sclafani
Dipartimento di Ingegneria
Chimica dei Processi e dei Materiali
Università di Palermo
Palermo, Italy

K.-Michael Schindler
Photochemistry Unit
Department of Chemistry
University of Western Ontario
London, Ontario, Canada

Mario Schiavello
Dipartimento di Ingegneria
Chimica dei Processi e dei Materiali
Università di Palermo
Palermo, Italy

Nick Serpone
Laboratory of Pure and Applied
  Studies in Catalysis, Environment
  and Materials
Department of Chemistry and
  Biochemistry
Concordia U niversity
Montreal, Canada

Chihwen Shao
Drinking Water Research Center
Florida International University
Miami, FL

Michael E. Sigman
Chemistry Division
Oak Ridge National Laboratory
P.O. Box 2008
Oak Ridge, TN

Somkiat Srijarani
Chemistry Department
University College Cork
Cork, Ireland

Ulick Stafford
Department of Chemical Engineering
University of Notre Dame
Notre Dame, IN

Barbara Sulzberger
Swiss Federal Institute for
  Environmental Science and
  Technology (EAWAG)
CH-8600 Dübendorf, Switzerland

Lizhong Sun
Photochemistry Unit
Department of Chemistry
University of Western Ontario
London, Ontario, Canada

R. Szymczak
Australian Nuclear Science and
  Technology Organization
Private Mail Bag No. 1
Menai, NSW 2234, Australia

Rita Terzian
Laboratory of Pure and Applied
  Studies in Catalysis, Environment
  and Materials
Department of Chemistry and
  Biochemistry
Concordia University
Montreal, Canada

J. Kerry Thomas
Department of Chemistry and
  Biochemistry
University of Notre Dame
Notre Dame, IN

Rong Tsao
Pesticide Toxicology Laboratory
Department of Entomology
115 Insectary Building
Iowa State University
Ames, IA

K. Vinodgopal
Department of Chemistry
Indiana University Northwest
Gary, IN

T. D. Waite
Department of Water Engineering
University of New South Wales
Kensington, NSW, Australia

E. L. Wehry
Department of Chemistry
University of Tennessee
Knoxville, TN

P. H. Wine
Physical Sciences Laboratory
School of Chemistry & Biochemistry
School of Earth and Atmospheric
  Sciences
Electro-Optics & Physical Sciences
  Laboratory
Georgia Tech Research Institute
Georgia Institute of Technology
Atlanta, GA

Edward J. Wolfrum
Department of Chemical Engineering
North Carolina State University
Raleigh, NC

C. C. David Yao
Chemistry Laboratory
SRI International
Menlo Park, CA

S. P. Zingg
Chemistry Division
Oak Ridge National Laboratory
P.O. Box 2008
Oak Ridge, TN

Y. Zuo
Drinking Water Research Center
Florida International University
Miami, FL

Contents

PART I
PHOTOCHEMISTRY IN THE ENVIRONMENT

PART II
PHOTOCHEMISTRY IN WATER TREATMENT

PART I

Photochemistry in the Environment

CHAPTER 1

A Review of the Photochemical Redox Reactions of Iron(III) Species in Atmospheric, Oceanic, and Surface Waters: Influences on Geochemical Cycles and Oxidant Formation

Bruce C. Faust

INTRODUCTION

Iron (Fe) is the fourth most abundant element in the Earth's crust and the most abundant transition metal. Sunlight-mediated photochemical reactions of Fe(III) species in natural waters are a driving force that perturbs the chemical cycles of iron and numerous other elements away from their possible equilibrium states. In particular, the photochemical redox (photoredox) reactions of Fe(III) species are sufficiently rapid so as to affect the geochemical cycles of carbon, sulfur, phosphorus, oxygen, and, of course, iron. These effects are manifested in several ways, both directly and indirectly.

Photoredox reactions of Fe(III) species (dissolved complexes, dimers, polymers, and precipitates) reduce Fe(III) to Fe(II).[1-25] In the case of photoredox reactions involving Fe(III) (hydrous) oxides, this can result in increased concentrations of dissolved iron species in solution, due to a pH-dependent release of the photoformed Fe(II) into bulk aqueous solution. Thus, the concentrations of Fe(II) and total dissolved iron in natural waters are increased by photoredox reactions involving Fe(III). This is important because numerous chemical and biological processes are critically dependent upon and linked to the speciation of iron in atmospheric and surface waters.

Photoredox reactions of Fe(III) species have been postulated as atmospheric water drop sources of oxidants, such as $\cdot OH$,[14,15,19,20,23,24] $SO_x^{-\cdot}$,[9,18] $HO_2^{\cdot}/O_2^{-\cdot}$,[23,25] and

H_2O_2,[23,25] all of which are known to play important roles in the oxidation of SO_2 to H_2SO_4 in the atmosphere. Iron(III) catalyzes the aqueous-phase oxidation of S(IV) species (e.g., HSO_3^- and SO_3^{2-}) by oxygen in cloud and fog drops, which under certain conditions can be an important pathway for the oxidation of SO_2 to H_2SO_4 in the atmosphere.[9,13-15,18,26]

It is estimated that anthropogenic SO_2 emissions represent 71% of the global sulfur emissions to the troposphere and 81% of the sulfur emissions to the Northern hemisphere troposphere.[27] Thus, processes responsible for the oxidation of SO_2 to H_2SO_4 in the troposphere will have a significant impact on the atmospheric sulfur cycle. Sulfuric acid formation in clouds contributes substantially to regional-scale acid deposition[28] and, upon evaporation of the cloud drops, to the formation of respirable acidic sulfate aerosols.[29] Furthermore, sulfate aerosols scatter sunlight, which causes visibility degradation. Due to their light-scattering properties, anthropogenic tropospheric sulfate aerosols have also caused a major perturbation to the radiative-transfer characteristics of the atmosphere; this sulfate-aerosol perturbation is opposite in sign and comparable in magnitude to current greenhouse gas forcing and has partially masked global warming trends caused by greenhouse gases.[30] It is ultimately a coupling of the chemical time scale for conversion of SO_2 to H_2SO_4 with the time scales for transport and deposition of SO_2 and sulfate that determines the spatial extent and magnitude of tropospheric-sulfate perturbations to the aforementioned problems. Photochemical reactions that produce oxidants are the driving force for the oxidation of SO_2 to H_2SO_4 in the troposphere.

Figure 1 demonstrates that the wet deposition of sulfate on a regional scale (the northeastern United States) is correlated with solar actinic flux, which initiates the formation of all photooxidants that are responsible for the conversion of SO_2 to sulfate. The lifetime of SO_2 in the troposphere is on the order of several days.[28] Thus, SO_2 emissions from, for example, the Ohio River Valley can affect the deposition of sulfuric acid on a regional basis and can significantly impact the northeastern United States and eastern Canada.[28] However, not all of the SO_2 emissions from the Ohio River Valley are deposited on North America as SO_2 and sulfate (after oxidation). The coupling of the chemical time scales for oxidation of SO_2 to H_2SO_4 with the typical physical time scales for transport and deposition is such that some of the SO_2 is carried out over the Atlantic Ocean before it is converted to H_2SO_4. The question is, "What is controlling the rate of H_2SO_4 deposition to the northeastern United States and eastern Canada?" The correlation observed in Figure 1 is consistent with the hypothesis that the deposition of sulfuric acid over the northeastern United States is limited by the availability (within the time allowed for by transport of the SO_2) of one or more photooxidants which control the oxidation of SO_2 to H_2SO_4 in the troposphere. Thus, photoredox processes responsible for the oxidation of SO_2 to H_2SO_4 can affect the amount of sulfate that is ultimately present in the atmosphere on local, regional, and global scales.

The oxidation of SO_2 to H_2SO_4 occurs by many pathways, some of which involve the intermediacy of water drop Fe(III) photoredox reactions that can form

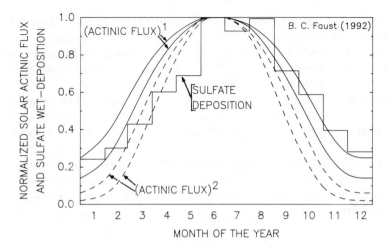

Figure 1. Normalized monthly regional mean wet deposition of sulfate for the northeastern United States for the period 1976–1979 and normalized solar actinic flux for 40°N. Only sites with extensive wet-deposition data were used here:[89] (1) White-face Mountain, NY, (2) Ithaca, NY, (3) Penn State, PA, and (4) Charlottesville, VA. For each of the 12 months, the 3-year mean wet deposition (moles sulfate per square meter) was individually calculated for each of the four sites, and these four monthly site mean wet-deposition values were then summed to obtain a measure of the monthly mean regional wet deposition. This was repeated for each month, and the resulting 12 monthly 3-year mean regional wet-deposition values were summed to obtain the total 3-year mean regional wet deposition for 1 year, which was used to normalize each 3-year mean monthly regional wet-deposition value. Normalized solar actinic flux is reported for two discrete wavelength bands: (1) 305–310 nm (lower curve of each pair) and (2) 310–315 nm (upper curve of each pair).[90] (From Faust, B. C., unpublished results.)

oxidants[9,13-15,18-20,23-25] and can modify the Fe(III) speciation. Furthermore, the oxidation of SO_2 in atmospheric water drops forms H_2SO_4, which lowers the pH and increases the solubility of Fe(III) precipitates. This also affects the Fe(III) speciation, which in turn affects the Fe(III)-catalyzed oxidation of S(IV) by O_2 (see Reference 26 and references cited therein), and the forms of Fe(III) available to microorganims upon deposition of the atmospheric hydrometeors to the ocean.

Iron is a critical, and in some cases possibly limiting, micronutrient for the growth of phytoplankton in the ocean,[10,31-36] which affects the global photosynthetic uptake of CO_2. However, the uptake of iron by marine phytoplankton is limited by the rate of reaction between monomeric Fe(III)-hydroxy species with the iron-transport agent.[10,33] Consequently, processes that affect the speciation of iron deposited to and present in surface ocean waters will affect the rate of phytoplankton and bacteria growth and, therefore, the rates of photosynthetic CO_2 uptake and dimethylsulfide (DMS) emissions by microorganisms in the oceans. Both of these processes, CO_2 uptake and DMS emissions, exert a cooling effect on global climate because CO_2 is a greenhouse gas and DMS is a precursor to marine sulfate aerosols that scatter solar radiation and serve as cloud condensation nuclei.[30]

Independent investigations have concluded that 95%[31] and greater than 99%[37] of the iron in the surface waters of the remote Pacific Ocean is derived from atmospheric deposition. In the atmosphere, iron is nominally present in an aqueous environment, whether in a cloud drop or in an aquated aerosol particle. Thus, aqueous-phase photochemical processes in atmospheric water drops[9,13-15,18-20,23-25,38] undoubtedly affect the speciation of atmospheric iron that is deposited to the ocean. In the ocean, Fe(III) species also undergo photoredox reactions that affect the iron speciation[10,22,39,40] and, therefore, the bioavailability of iron to phytoplankton and bacteria.

From this brief introduction, it is apparent that the photochemical redox reactions of Fe(III) species influence chemical cycles in the atmosphere and in oceanic and surface waters. For this reason, it is useful to review our present understanding of these processes.

BACKGROUND AND THEORY

Fe(III) Speciation in Atmospheric, Oceanic, and Surface Waters

In order to interpret the photochemistry of Fe(III) species in natural waters, it is important to understand the likely speciation of Fe(III) in the various systems because each Fe(III) species exhibits a different intrinsic photoreactivity. Consequently, it is useful to discuss the speciation of Fe(III) in natural waters, although it should be noted that corresponding information on the photoreactivities of the various known natural Fe(III) species is currently not always available. Iron(III) has been reported to occur in various forms in soils and sediments,[41,42] including goethite (α-FeOOH), hematite (α-Fe$_2$O$_3$), lepidocrocite (γ-FeOOH), maghemite (γ-Fe$_2$O$_3$), ferrihydrite [amorphous Fe(OH)$_3$], in layered silicates, and associated with organic matter. However, little direct information is available on the identities of Fe(III) solids present in authentic atmospheric, oceanic, or surface waters. This is probably due to the lower concentrations of Fe(III) present in natural waters (relative to soils, sediments, and aerosols), to the small and variable particle sizes of iron-bearing particles, and to the amorphous nature of many of the precipitates; all of these factors hinder characterization of the Fe(III) solids by the classical techniques of X-ray diffraction and Mössbauer spectroscopy. Thus, much of our understanding of the solid-phase Fe(III) species in natural waters has been inferred from soil, sediment, and aerosol studies; from indirect wet-chemical and physical-separation techniques; and from experiments on the formation of Fe(III) phases under conditions intended to mimic those of the natural water of interest.

Iron(III) hydrous oxide particles have been identified in Esthwaite Lake.[43] The Fe(III) particles were negatively charged (due to the adsorption of humic substances) and exhibited the following composition: 30–40% iron, 4–7% humic carbon, and 1.1–2.8% phosphorus. As can be seen from Figure 2, the particles were typically 0.05–0.4 μm in diameter. X-ray diffraction studies of the particles

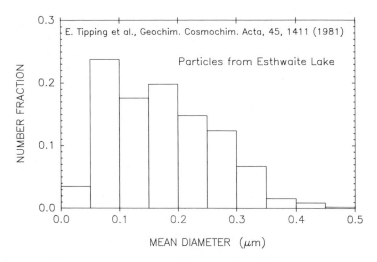

Figure 2. Number size distribution of 252 particles [mostly Fe(III) hydrous oxides] concentrated by ultrafiltration from Esthwaite Lake water, as determined from scanning electron micrographs. (From Tipping, E., Woof, C., and Cooke, D., *Geochim. Cosmochim. Acta*, 45, 1411, 1981. With permission.)

revealed no discernable peaks for air-dried samples, but one broad peak appeared upon heating (393 K) that was consistent with ferrihydrite. An Fe(III)-Fe(II)-orthophosphate-calcium precipitate was found to form at the oxic-anoxic boundary of a eutrophic lake.[44] Kinetically labile forms of Fe(III) have been identified in freshwaters[45] and in seawater;[22] this probably represents dissolved species and some amorphous precipitates. A review of the colloidal and macromolecular forms of Fe(III) is available elsewhere.[46]

Although it has never actually been identified in authentic seawater, akaganéite (β-FeOOH) has historically been considered to be a likely form of Fe(III) in the ocean because it is the form of iron that precipitates from aqueous chloride solutions of Fe(III) that are undergoing hydrolysis.[41] However, lepidocrocite (γ-FeOOH) was found to be a product of the oxidation of Fe(II) by O_2 in aqueous solutions from pH 7.2 to 9.0, and it was formed in ClO_4^-, Cl^-, and SO_4^{2-} media.[47]

Typically, the concentration of "dissolved" Fe(III) increases with depth in the ocean until a depth of about 0.5 km.[32,34] The pH of ocean water decreases from approximately 8.1 at the surface to about 7.5 at a 0.5-km depth.[48] Several processes, including biological uptake of iron in the surface waters, affect the profile of dissolved Fe(III) concentrations in the ocean. However, the solubility of Fe(III) hydrous oxides and the desorption and/or dissociation of Fe(III) from particles and macromolecules might be expected to increase with decreasing pH from the surface to about a 0.5-km depth in the typical oceanic profile. This could partially explain the observed oceanic dissolved iron profiles.

Iron(III) has been detected in atmospheric particulate matter as hematite, maghemite, magnetite (Fe_3O_4), and in the lattice of weathered clays.[49-52] Iron can typically represent 0.5 to 5% of the total aerosol mass. In some cases, a sizable

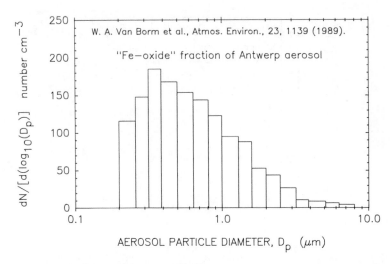

Figure 3. Number size distribution of the "Fe-oxide" fraction of the Antwerp aerosol, as
determined by electon microscope analyses of individual aerosol particles collected by filtration on membrane filters. Note that the actual distribution might not
peak at ca. 0.2–0.3 µm, as implied in this figure, because the authors found
decreased collection efficiency and detection ability for particles with diameters
≤0.4 µm. N = number concentration of particles (number per cubic centimeter).
(From van Borm, W. A., Adams, F. C., and Maenhaut, W., *Atmos. Environ.*, 23,
1139, 1989. With permission.)

percentage (10–90%) of aerosol iron has been reported to be present as Fe(II).[38,52]
From electron microprobe analyses of individual aerosol particles, it was found
that 5 to 32% of Antwerp aerosol particles could be classified as "iron oxides" and
that an additional 5.2% (mean value) of aerosol particles were classified as iron-
rich soil dust.[53] Thus, a substantial fraction of the Antwerp aerosol particles were
found to contain appreciable amounts of iron. Based on manual morphological
analyses, these iron oxide aerosol particles were classified into two distinct
groups.[53] One group of particles was spherical, with no other distinctive characteristics, and was attributed to formation at high temperatures in smelters or
incinerators. The other group contained particles with sharp edges, which could
have originated from soil dust and/or other sources (e.g., rust, coal fly ash). The
diameters of the Antwerp iron oxide aerosol particles were generally less than 3
µm, as illustrated in Figure 3.[53]

The diameters of iron-bearing aerosol particles are sufficiently small (Figure 3)
to allow for their long-range transport into more rural and remote areas of the
atmosphere and the ocean. It is noteworthy that iron has been detected in remote
marine aerosols.[38] Undoubtedly, the speciation of atmospheric iron is altered by
aqueous-phase (photo)chemical reactions in the hydrated aerosol and in cloud drops,
as well as upon deposition into the ocean. It is well known that iron species in
atmospheric aerosols dissolve in acidified aqueous solutions and in seawater.[37,54,55]
A substantial fraction of the dissolution occurs within 10 min. Thus, significant

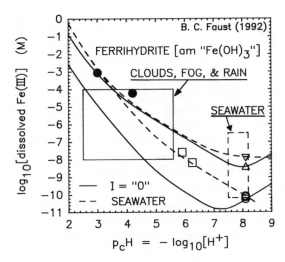

Figure 4. Information related to Fe(III) speciation in natural waters: (1) solubiity of ferrihydrite [amorphous "Fe(OH)$_3$" precipitate] as a function of $p_cH = -\log_{10}[H^+]$ (curves), (2) domains of typical p_cH and total iron concentrations in atmospheric and oceanic waters (boxes), and (3) equilibrium positions of ambient seawater and seawater with added atmospheric aerosol particulate matter (individual symbols). Two solubility curves are shown for each of two sets of conditions: "zero" ionic strength and seawater conditions. Each set of solubility curves represents the likely upper and lower bounds of ferrihydrite solubility under the specified conditions at 298 K. Thermodynamic data[91-96] and the MINTEQ computer program[97] were used to calculate the solubility curves. ○: Ambient dissolved Fe(III) concentration, determined by extraction with organic solvent, in the surface waters of the northeast Pacific Ocean at three locations (mean value of measurements from 0 and 50 m at each location; Tables 1–3 of Reference 31). ▽, △, and □: Mean value (of 22, 13, and 1 samples, respectively) of dissolved Fe(III) concentration (<0.4 μm, <0.05 μm, and <0.05 μm, respectively) released from atmospheric aerosol particles, after 1 hr of leaching, into seawater from the north Pacific Ocean (Tables 4 and 5 and Figure 7, respectively, of Reference 37) ●: Dissolved Fe(III) (<0.4 μm) released into seawater from acidic oil-combustion ash (10 g/L) after 24 hr of leaching.[98] (From Faust, B. C., unpublished results.)

amounts of iron could dissolve during the lifetime of atmospheric water drops and within the residence time of these particles in the photic zone of the ocean.

The speciation of iron leached from atmospheric particulate matter into aqueous solutions and/or seawater is not known. It was observed that aqueous solutions (pH 3.0, H$_2$SO$_4$) of coal fuel ash became reddish brown after 1 hr,[54] which suggests that ferrihydrite [amorphous Fe(OH)$_3$ solid] had formed.

Figure 4 illustrates the solubility diagram for ferrihydrite [amorphous Fe(OH)$_3$ precipitate] in seawater and in water of "zero" ionic strength, as well as the domains of pH and total iron concentration that are typical of atmospheric and oceanic waters. It is expected that under most conditions of natural waters the majority of Fe(III) will be associated with particles or be present as precipitates. But, as can be seen from Figure 4, at lower pH values a substantial fraction of the total Fe(III) could be present as dissolved species in natural waters.

It is also apparent from Figure 4 that uncertainty exists regarding the solubility of ferrihydrite. This uncertainty is due to the time-dependent and particle-size-dependent solubilities of aging suspensions of ferrihydrite, to the variability in reported equilibrium constants for formation of $Fe(OH)_2^+$, and to the uncertainty associated with the existence of the proposed dissolved species $Fe(OH)_3^0$ (aq).

Superimposed on the solubility diagram of ferrihydrite in Figure 4 are data from several studies on the dissolution of iron from atmospheric particulate matter in seawater and aqueous solutions. Within the experimental uncertainties and given the particle-size and time-dependent variability of ferrihydrite solubility, it is reasonable to conclude from Figure 4 that, as a first approximation, ferrihydrite is important in controlling the solubility of Fe(III) in natural waters for time scales of up to at least 1 day. Other Fe(III) phases are also undoubtedly present in natural waters. But, non-negligible amounts of amorphous Fe(III) precipitates often persist even in preparations designed to specifically synthesize a given crystalline Fe(III) hydrous oxide.[11]

The speciation of dissolved Fe(III) in natural waters is also uncertain. There is no definitive evidence for the existence of dissolved Fe(III) complexes with humic and fulvic acids, although a substantial amount of circumstantial evidence is consistent with the existence of such species.[6,7,45,46,56-58] These studies have proposed various types of interactions between Fe(III) and natural organic matter, including (1) adsorption of organic matter onto Fe(III) hydrous oxides, (2) precipitation of an Fe(III)-organic solid, (3) adsorption of Fe(III) onto organic colloids, and (4) Fe(III)-organic complexes. It is likely that Fe(III) complexes with hydroxide ion and organic ligands dominate the dissolved speciation of Fe(III) in most natural waters, with the relative contributions depending on a number of factors such as pH and dissolved organic carbon concentration. In atmospheric water drops and acidic sulfate aerosols, most of the dissolved Fe(III) will be complexed with ligands such as OH^-, SO_3^{2-}/HSO_3^-, SO_4^{2-}, and a variety of organic compounds including dicarboxylates (e.g., oxalate). In acidified seasalt, Fe(III) complexes with chloride and bromide might also be possible.

Photochemistry of Natural Waters

As illustrated in Figure 5, the shortest wavelength of light present in the terrestrial solar spectrum is approximately 300 nm. Thus, in order to undergo sunlight-initiated photoredox reactions, Fe(III) species must exhibit ligand-to-metal charge-transfer transitions at wavelengths (λ) greater than 300 nm. Many Fe(III) species absorb at wavelengths greater than 300 nm, including species with ligands such as OH^-,[14,19,20,59] H_2O,[5] HO_2^-,[60] HSO_3^-/SO_3^{2-},[9,13,18] Cl^-,[5] carboxylates and polycarboxylates,[1,2,16,23-25,59,61-63] and O^{2-} in Fe(III) (hydrous) oxides,[9,13,18,64] as is illustrated in Figure 6. Thus, the potential exists for a wide variety of natural Fe(III) species to undergo sunlight-mediated photoredox reactions.

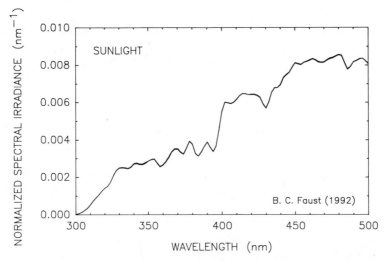

Figure 5. Normalized spectral irradiance of near-equinox mid-day sunlight in North Carolina (100% clear sky, no clouds) with 8-km and ≥24-km visibility on September 15 and October 1 (1987), respectively. Normalized spectral irradiance is $[(E_\lambda)/ \int_{300}^{500} (E_\lambda)(d\lambda)]$, where E_λ = spectral irradiance (watts per square meter per nanometer) of light at wavelength λ. The area under each curve is unity. Solar spectra for the different visibilities are indistinguishable at this level of resolution; their individual data points are not illustrated so as to simplify the figure.

The photoreaction rates of Fe(III) species in sunlit atmospheric and surface waters are often rapid, occurring on time scales of seconds to hours. This is due to several reasons. One, the solar actinic flux in the ultraviolet (UV) region (300–400 nm) is approximately 1 Einstein m^{-2} hr^{-1}, which is sufficiently large to initiate photochemical reactions at fairly rapid rates. Two, most naturally occurring Fe(III) species, including dissolved complexes, dimers, polymers, and crystalline and amorphous (hydrous) oxide precipitates, typically exhibit ligand-to-metal charge-transfer absorption bands that significantly overlap the solar UV spectrum (compare Figures 5 and 6). Three, photoredox reactions of Fe(III) species have sufficiently large quantum efficiencies, although often less than one, to make such reactions feasible in sunlit natural waters.

Solar UV photons are of sufficient energy (300 nm ≈ 400 kJ mol^{-1}, 400 nm ≈ 300 kJ mol^{-1}) to initiate a series of reactions that ultimately transfers one electron from a ligand(L)-centered orbital to an Fe(III)-centered orbital, forming free Fe(II) and a ligand-free radical:

$$Fe(III)\text{-}L + photon \longrightarrow \left[Fe(III)\text{-}L\right]* \tag{1}$$

$$\left[Fe(III)\text{-}L\right]* \longleftrightarrow \left[Fe(II)\text{-}L^{\cdot}\right]_{cage} \tag{2}$$

$$\left[Fe(II)\text{-}L^{\cdot}\right]_{cage} \longleftrightarrow Fe(II) + L^{\cdot} \tag{3}$$

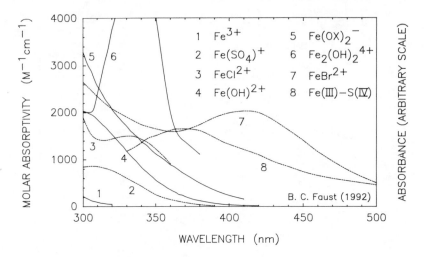

Figure 6. Absorption spectra of various Fe(III) species illustrating the characteristic ligand-to-metal charge-transfer transitions. Left ordinate: Curves 1, 4, 5, and 6 are molar absorptivities. Right ordinate: Curves 2, 3, 7, and 8 are absorption spectra of Fe(III) solutions containing the specified ligand. Various different conditions were used to acquire Spectra 2, 3, 7, and 8. Thus, Spectra 2, 3, and 7, and 8 (1) are not on the same scale, (2) should not be compared with each other, and (3) are included only to show the location of the charge-transfer transition. Literature sources of the data for these spectra are Fe^{3+},[20,99] $Fe(SO_4)^+$,[100] $FeCl^{2+}$,[5] $Fe(OH)^{2+}$,[20] $Fe(OX)_2^-$,[25] $Fe_2(OH)_2^{4+}$,[99] $FeBr^{2+}$,[100] and Fe(III)-S(IV).[9]

The overall quantum yield of Fe(II) and/or free radical formation is dependent upon the competition of complete bond scission and separation of the incipient free radical from the Fe(II) center with a variety of photophysical decay pathways of the various intermediates. Although the reactions in Equations 1 to 3 are illustrated for Fe(III), the mechanism is similar for other metal complexes/species. Several excellent reviews are available that discuss the photochemistry of transition-metal complexes, including iron complexes.[59,61-63]

The rate of direct photolysis of Fe(III) in a natural water, $v_{Fe(III)} = d[Fe(III)]/dt$, can be expressed as

$$v_{Fe(III)} = -\left[Fe(III)\right]_T \sum_i \left(f_i\right)\left(j_i\right) = -j\left[Fe(III)\right]_T \qquad (4)$$

where f_i represents the fraction of total Fe(III) concentration, $[Fe(III)]_T$, present as the ith Fe(III) species, which can be a dissolved or surficial Fe(III) chromophore. The parameter "j" is the mean apparent first-order rate constant (sec^{-1}) for direct photolysis of all Fe(III) species and is a function of f_i (speciation) and j_i, where j_i is the apparent first-order rate constant for direct photolysis of the ith Fe(III) species. The rate constant j_i is a function of several parameters:

Table 1. Spectral and Photochemical Properties of Fe(III)-Hydroxy Species

Fe(III) Species	Wavelength Maxima of Charge-Transfer Transitions (nm)	λ (nm)	Φ_λ
Fe(OH)$^{2+}$	295	313	0.14
		360	0.017
Fe$_2$(OH)$_2^{4+}$	335	350	0.007
α-FeOOH (goethite)	295,[a] 240–270[b]	300–400	(0.19–2.9) $\times$ 10^{-4}

Note: λ = wavelength, Φ_λ = quantum yield at wavelength λ. Data are from References 5, 12, 19, 20, and 64.

[a] Assumed, by formal analogy to the dissolved Fe(OH)$^{2+}$ complex,[19,20] to represent the charge-transfer bands of surface Fe(III)-hydroxyl species.

[b] Range of the lowest energy ligand-to-metal charge-transfer transitions in the polymorphs of Fe$_2$O$_3$ (hematite and maghemite) and FeOOH (goethite and lepidocrocite).[64] There is not unanimous agreement in the literature regarding the assignments of ligand-to-metal charge-transfer transitions of Fe(III) (hydrous) oxides; such transitions might also occur at wavelengths longer than those reported here.

$$j_i \equiv 2.303 \int \left(\Phi_{\lambda,i} \right) \left(\varepsilon_{\lambda,i} \right) \left(I_\lambda \right) d\lambda \tag{5}$$

where $\phi_{\lambda,i}$, and $\varepsilon_{\lambda,i}$ are the quantum efficiency and decadic (base 10) molar absorptivity (absorption cross section, $M^{-1}cm^{-1}$) of the ith Fe(III) species, respectively, and I_λ is the solar actinic flux (spherically integrated spectral irradiance, einstein L^{-1} sec^{-1}), all at wavelength λ. It is seen from Equations 4 and 5 that the rate of Fe(III) photolysis in sunlight depends on the Fe(III) speciation (f_i), on the quantum efficiency of each Fe(III) species ($\phi_{\lambda,i}$), and also on the intrinsic rate of sunlight absorption by each Fe(III) species ($\varepsilon_{\lambda,i} \cdot I_\lambda$).

A measure of the rate of sunlight absorption is the quantity ($\varepsilon_{\lambda,i} \cdot I_\lambda$). It is tempting to make statements regarding the environmental significance of a given photochemical process based solely on knowledge of the quantum efficiency for the process. However, the chemical speciation and each species' rate of sunlight absorption must also be known before any such assessment can be made. For example, the lower quantum yields of Fe$_2$(OH)$_2^{4+}$ vs. those of Fe(OH)$^{2+}$ (see Table 1) are partially compensated for by the larger molar absorptivities and greater spectral overlap of Fe$_2$(OH)$_2^{4+}$ with the solar spectrum[20] (Figures 5 and 6).

Several factors affect the photoreactivity of an Fe(III) species. One important factor is the number of strong bonds that must be broken to liberate the photoformed Fe(II) center. As shown in Table 1, the quantum yields decrease in the order Fe(OH)$^{2+}$ > Fe$_2$(OH)$_2^{4+}$ > α-FeOOH (goethite). This can be explained by considering that with increasing polymerization of the Fe(III) species there is a corresponding increase in the mean number of stronger bonds that must be broken to release each Fe(II) when it is formed. The limited quantum yield data of Table 1 is consistent with this rationalization. However, in general, this analysis is too

simplified for Fe(III) (hydrous) oxides because the crystal structure and resulting electronic configuration, as well as surface and lattice defects/states, will also exert large effects on the photoreactivity of a given Fe(III) species. Quantum yields for higher order polymeric structural units, with more than two Fe(III) centers, are not currently available. Based on this analysis, the quantum yields of Fe(III) polymeric species and of ferrihydrite will probably be between that of $Fe_2(OH)_2^{4+}$ and the Fe(III) hydrous oxides. As can be seen from this discussion, the speciation of Fe(III) plays an important role in the photoreactivity of Fe(III) in natural waters.

GEOCHEMICAL AND REDOX CYCLES

Iron

Based solely on thermodynamic considerations, Fe(II) species are not expected to be present in air-saturated natural waters because molecular oxygen should oxidize all Fe(II) to Fe(III). Yet numerous field studies[4,8,17,38,39,65,66] of atmospheric and surface waters have demonstrated that a large percentage of the total iron, probably often ranging from 10% to 90%, is present as Fe(II); this is especially true as the pH of the natural water decreases. Natural organic matter can reduce Fe(III) in the dark via thermal reactions,[58,67] but these reactions are generally much slower than the corresponding photochemical reactions. A review of the heterogeneous Fe(III) redox reactions is given elsewhere.[68] In some of the field studies, positive correlations have been found between concentrations of Fe(II) and solar actinic flux on a diurnal and seasonal basis.

Laboratory studies have demonstrated that solar UV light initiates photoreduction of Fe(III) species to Fe(II),[1,2,5-7,9-13,16,18-25] via processes that can be represented, in a simplified manner, by the reactions in Equations 1 to 3. In all of these studies, the rates of Fe(II) formation in photochemical reactions were much larger than those in the corresponding thermal (dark) controls. The combination of laboratory and field observations provides very strong support for the hypothesis that photoredox reactions of Fe(III) species are an important, and probably dominant, daytime source of Fe(II) in atmospheric, oceanic, and surface waters.

Dissolved Fe(III) complexes undergo very rapid photoreactions to form Fe(II) under conditions that are typical of atmospheric and surface waters.[1,2,5,7,14,15,19,20,23-25] The mechanism of these reactions is represented in the reactions in Equations 1–3. Table 2 illustrates that the half-lives in sunlight of Fe(III) complexes with hydroxide ion and with dicarboxylates are on the order of minutes. The photolysis of $Fe(OH)^{2+}$ alone[18,19] (half-life = 20 min, Table 2) could easily account for the Fe(II) formation in acid mine drainage streams.[17] The rapid photoreduction of Fe(III) coupled with high rates of Fe(II) oxidation (*vide infra*) strongly suggests that the Fe(III)-Fe(II) redox dynamics in natural waters is probably in a rapid photochemically driven steady state.

Table 2.   Half-Lives of Some Dissolved Fe(III) Species in Sunlight

Species	Half-Life of Fe(III) in Sunlight (min)
$FeOH^{2+}$	20
$Fe(ox)^+/Fe(ox)_2^-$	0.2
$Fe(mal)^+/Fe(mal)_2^-$	5
$Fe(OH)(cit)^-$	0.9

ox = oxalate^{2-}, mal = malonate^{2-}, cit = citrate^{3-}.
Note: Data are from References 19, 20, 23, and 25.

$$Fe(III)\text{-}L \quad \begin{cases} \xrightarrow{\text{light}} & Fe(II) \qquad (6) \\[2ex] \xrightarrow{\text{dark}} & Fe(II) \qquad (7) \end{cases}$$

$$Fe(II) + oxidant(s) \longrightarrow Fe(III) \qquad (8)$$

Here, L represents a ligand such as hydroxide ion, a polycarboxylate anion, etc. In general, the rate of the photoreaction (Equation 6) is greater than that of the thermal reaction (Equation 7) for many Fe(III) species.

A possible general mechanism[23] for the photolysis of Fe(III) polycarboxylates (e.g., oxalate, malonate, citrate) is shown in Figure 7. As can be seen from Figure 7, the fate of the polycarboxylate radical is determined by several competing processes: (1) back reaction to reform Fe(III) (not shown); (2) reaction with O_2, which represents an important sink for O_2 and source of $^.O_2^-/HO_2^.$ in some natural waters (*vide infra*); (3) reduction of another Fe(III) species; and (4) decarboxylation to form CO_2 and a carbon-centered radical (e.g., CO_2^- in the case of oxalate), which can also react with O_2 or another Fe(III).

Decarboxylation occurs rapidly for $C_2O_4^-{}^.$ (half-life ≈ 0.3 µsec[69]); decarboxylation rate constants for other carboxylate radicals are not available. Oxalate radical ($C_2O_4^-{}^.$) and the $CO_2^-{}^.$ derived therefrom react at near diffusion-controlled rates with Fe(III) oxalates;[70,71] corresponding rate constants for the citrate and malonate radicals are not available, although it is likely that their analogous reactions are also rapid. The reactions of Fe(III) with the carbon-centered radicals derived from polycarboxylate radicals are also expected to be rapid, because this is the case for $CO_2^-{}^.$[72,73] and because numerous other carbon-centered radicals react rapidly with Fe(III) species.[74] Carbon-centered radicals (including $CO_2^-{}^.$) also react rapidly with O_2.[74]

Thus, in the mechanism of Figure 7 the fate of the reducing radicals (polycarboxylate radical and the carbon-centered radical derived therefrom) is largely determined by their competitive reactions with Fe(III) and O_2. The relative rates of the reactions of the reducing radical with O_2 or Fe(III) are very important for determining the steady-state concentration of Fe(II) and $HO_2^./^.O_2^-$. If most of the reducing radical reacts with Fe(III), additional Fe(II) is formed. In contrast, if most of the reducing radical reacts with O_2, additional Fe(II) is not formed, and

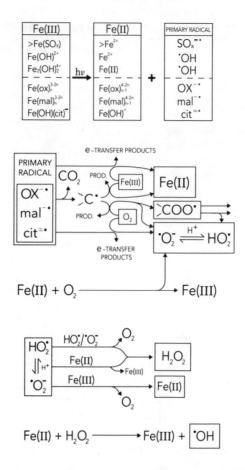

Figure 7. Possible mechanism for the photolysis of Fe(III) complexes with polycarboxylates (ox = oxalate^{2-}, $^-$O(O)C-C(O)O$^-$; mal = malonate^{2-}, $^-$O(O)C-CH$_2$-C(O)O$^-$; cit = citrate^{3-}) and of other ligands.[23] The symbol >C$^\cdot$ represents the carbon-centered radical derived from the decarboxylation of the polycarboxylate radical, or CO$_2^-$ in the case of oxalate. Note that not all polycarboxylate-derived >C$^\cdot$ will necessarily form $^\cdot$O$_2^-$ upon reaction with O$_2$; reaction of O$_2$ with some >C$^\cdot$ will form organic peroxyl radicals (>COO$^\cdot$, or more generally, RO$_2^\cdot$) that could form organic peroxides rather than H$_2$O$_2$.

various oxidants (e.g., $^\cdot$O$_2^-$, HO$_2^\cdot$, H$_2$O$_2$, $^\cdot$OH) are generated sequentially, all of which can reoxidize Fe(II) to Fe(III).[23] In the latter case, the net effect is to lower the apparent quantum yield for Fe(II) formation.[23]

The sources of Fe(II) in atmospheric waters are primarily (1) photoreduction of Fe(III) species, (2) thermal reduction of Fe(III) species by $^\cdot$O$_2^-$, and (3) thermal reduction of Fe(III) by organic compounds and S(IV) species (HSO$_3^-$, SO$_3^{2-}$). By analogy to previous arguments, rates of photoreduction of Fe(III)-organic species are likely to greatly exceed their corresponding thermal (dark) reduction rates. In the remote troposphere, the thermal reduction rate of Fe(III) by S(IV) species (in

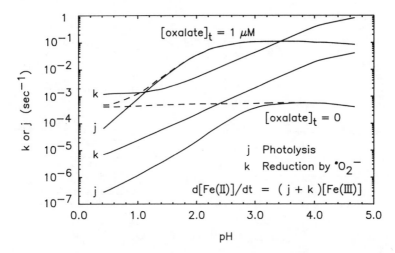

Figure 8. Apparent first-order rate constant (Equation 4) for direct photolysis of all Fe(III) species (j) and for thermal reduction of all Fe(III) species by $^{\cdot}O_2^-$ (k) in atmospheric water drops as a function of pH at 298 K. Two sets of conditions are used for direct photolysis (j): (1) no oxalate and (2) 1 μM total oxalate. Two sets of conditions are used for thermal reduction by $^{\cdot}O_2^-$ (k); these are ($[HO_2^{\cdot}] = [^{\cdot}O_2^-]$) = 1 or 20 nM.[76,79] Solid line (——): j_i = 0 for Fe(SO$_4$)$^+$. Dashed line (– – –): j_i for Fe(SO$_4$)$^+$ = j_i for FeOH^{2+} Calculations are for clear-sky, solar-noon, June sunlight.[20] j_i (Equations 4 and 5, sec^{-1}): 8.2×10^{-8} (Fe^{3+}), 6.3×10^{-4} (FeOH^{2+}), 0.15 [Fe(ox)$_n^{3-2n}$], 0 [assumed for Fe(OH)$_2^+$, Fe(SO$_4$)$_2^-$]. Equilibrium constants: references 24 and 101 (and references therein). Rate constants [see Reference 101, or assumed (A)] for $^{\cdot}O_2^-$ (M^{-1} s^{-1}): 1.5×10^8 [FeOH^{2+}, Fe(SO$_4$)$^+$, Fe(ox)$_n^{3-2n}$ (A)], 0 [Fe(OH)$_2^+$ (A), Fe(SO$_4$)$_2^-$ (A)]. Rate constants for HO$_2^{\cdot}$ (M^{-1} s^{-1}): 0 to 1×10^3 [FeOH^{2+}, Fe(SO$_4$)$^+$], 0 [Fe(ox)$_n^{3-2n}$ (A), Fe(SO$_4$)$_2^-$ (A), Fe(OH)$_2^+$ (A)], and $(0.066 - 3.5) \times 10^5$ (Fe^{3+}). Ionic strength (H$_2$SO$_4$) varies from 3.5×10^{-5} M at pH equal to 4.67 to 0.57 M at pH equal to 0.42. These calculations do not include sources of Fe(II) from the photoreactions of Fe(III)-S(IV) species. (From Faust, B. C., unpublished results.)

equilibrium with 0.1 ppbv SO$_2$) is slow by comparison with the rates of other processes, based on the reaction rate of FeOH^{2+} with SO$_3^{2-}$.[75] In polluted atmospheres, with regional SO$_2$ concentrations of about 5 ppbv (typical of the eastern United States), this process could be a significant source of Fe(II) in atmospheric water drops with pH greater than 4. In the absence of large sources of acid neutralizing capacity (i.e., wind-blown carbonate mineral dust, seasalt, or NH$_3$), these conditions (high SO$_2$, pH >4) would seem to be rather uncommon. Thus, except for environments with exceptionally high SO$_2$ concentrations (e.g., aerosols, fogs, or low-altitude clouds in the inversion layer of urban areas), thermal reduction of Fe(III) by S(IV) species is not likely to be a significant daytime source of Fe(II) in atmospheric water drops. By contrast, thermal reduction of Fe(III) by S(IV) species and/or organic compounds is likely to be an important nighttime source of Fe(II) in atmospheric water drops throughout the troposphere.

Figure 8 illustrates a comparison between the two most likely daytime sources of Fe(II) to atmospheric water drops: (1) photoreduction of Fe(III) species and

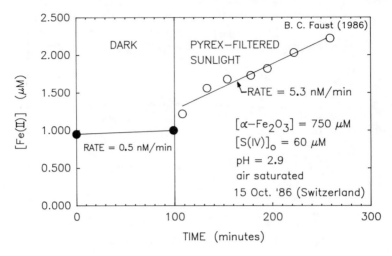

Figure 9. Thermal (dark) dissolution and sunlight-induced photodissolution of hematite (α-Fe_2O_3) by HSO_3^- in air-saturated aqueous solution. (From Faust, B. C., unpublished results.)

(2) thermal reactions of Fe(III) with $\cdot O_2^-$, which is also formed from aqueous-phase photochemical reactions in atmospheric waters.[76-79] As can be seen from Figure 8, both processes exhibit a strong pH dependence. It should be emphasized that uncertainties exist in these calculations and that Figure 8 only represents the formation rate of Fe(II) and does not treat the oxidation of Fe(II) to Fe(III). The oxidation rate of Fe(II) by various oxidants will also increase with pH. Nevertheless, the calculations (Figure 8) indicate that the time scales for daytime formation of Fe(II) in atmospheric water drops range from approximately 1 sec to 1 day. The rates of these reactions are sufficiently rapid to easily account for the formation of Fe(II) in fogs[65,66] and remote marine aerosols.[38] A major uncertainty in these calculations are the quantum yields for the photolysis of other Fe(III) species that are present in acidic fogs, clouds, and aquated aerosols. Furthermore, the calculations of Figure 8 ignore the photoreduction of dissolved and surficial Fe(III)-S(IV) species as a source of Fe(II) to atmospheric water drops, although these processes are known to form Fe(II) rapidly (Figure 9).[9,13]

Some of the daytime sinks for Fe(II) in atmospheric waters are illustrated in Table 3. The most likely pathways for oxidation of Fe(II) are by reactions with $HO_2^{\cdot}$ and H_2O_2 (Table 3). Based on information in Table 3, Fe(II) should be oxidized to Fe(III) on the time scale of minutes in sunlit atmospheric water drops. It has already been demonstrated that dissolved Fe(III) species undergo sunlight photoredox reactions on time scales of minutes. Thus, during the daytime, the dissolved iron in atmospheric water drops is rapidly cycling between Fe(III) and Fe(II), very probably on time scales of minutes, due to rapid (photo)redox reactions occurring in the water drops. It is probable that similar mechanisms could be contributing to the iron redox cycling in acidic surface waters and perhaps even at the higher pH values encountered in oceanic waters.

Table 3. Daytime Sinks for Fe(II) in Atmospheric Water Drops (pH ≤5)

Oxidant	$[Oxidant]_{ss}$ [a] (M)	$k_{OX,Fe2+}$ [b] (M^{-1}/s^{-1})	$k_{OX,Fe2+}[Oxidant]_{ss}$ [c] (sec^{-1})
$\cdot O_2^-$	$1 \times 10^{-10} - 2 \times 10^{-9}$ [d]	1.0×10^7	$(1 - 20) \times 10^{-3}$ [d]
HO_2^-/RO_2^-	$1 \times 10^{-9} - 2 \times 10^{-8}$	$(1.2 - 1.7) \times 10^6$	$(1 - 30) \times 10^{-3}$
H_2O_2	$10^{-6} - 10^{-4}$	5.1×10^1	$(0.05 - 5) \times 10^{-3}$
O_3	1×10^{-9} [e]	8.2×10^5	$(0.8) \times 10^{-3}$
$\cdot OH$	$10^{-14} - 10^{-12}$	4×10^8	$(0.004 - 0.4) \times 10^{-3}$
$O_2(^1\Delta_g)$	$10^{-14} - 10^{-12}$	NA [f]	$(0.004 - 0.4) \times 10^{-3}$
O_2	2.7×10^{-4}	NA	$(0.000003) \times 10^{-3}$ [g]

Note: Effects of S(IV) species, organic compounds, and of other redox-active metals (e.g., copper, manganese) are not included, but could be very significant under some conditions. NA = information is not available.

[a] Typical steady-state concentration of oxidant in sunlit atmospheric water drops.

[b] Bimolecular rate constant for the reaction Fe^{2+} + OXIDANT $----> Fe(III)$.[101–105] Rate constants for reactions of Fe(II) complexes with H_2O_2 could be much larger than the value used here for Fe^{2+}.[106]

[c] Apparent first-order rate constant for the oxidation of Fe(II) in atmospheric water drops by the specified oxidant for a typical daytime steady-state concentration of the oxidant.

[d] Assumes $[\cdot O_2^-]_{ss} = 0.1[HO_2^-]_{ss}$ (pH = 3.8); pKa (HO_2^-) = 4.8.

[e] Based on Henry's law equilibrium with 100 ppbv gas-phase O_3.

[f] Rate constant is unknown, and therefore, a value identical to that for $\cdot OH$ was used.

[g] Asymptotic value of apparent first-order rate constant from Figure 7.15 of Reference 107 for pH less than 5.0.

Numerous studies have shown that Fe(III) (hydrous) oxides also undergo photoreduction reactions[9-13,16,18,21,22,68] and that these reactions are an important natural pathway for dissolving Fe(III) solids and, of course, as a source of dissolved iron. The mechanisms for photoredox reactions occurring at the interface of Fe(III) solids are complex, but involve a combination of excitation of charge-transfer bands in surficial complexes and in the solid itself.[9,11,13,21] Figure 9 illustrates the photoreductive dissolution of hematite by S(IV) species under conditions typical of acidified atmospheric waters. Figure 10 illustrates the photodissolution of ferrihydrite in seawater. From these and other studies, it is evident that photoredox reactions of crystalline and amorphous Fe(III) solids can dissolve the particles and can serve as an important source of dissolved iron in sunlit atmospheric, oceanic, and surface waters. The photoformed Fe(II) can be reoxidized by various oxidants [e.g., O_2, H_2O_2, HO_2^-, RO_2^-, $\cdot OH$, $O_2(^1\Delta_g)$] to form dissolved Fe(III). Depending on the precipitation rate, steady-state concentrations of dissolved Fe(III) could be established that are significantly higher than those predicted by equilibrium models. However, even if precipitation of Fe(III) is rapid, the newly formed Fe(III) is not likely (at least initially) to be crystalline solids, but is more likely to form amorphous precipitates (e.g., ferrihydrite) that have a much higher solubility than do crystalline Fe(III) (hydrous) oxides. Thus, photoredox reactions are sources of dissolved iron: (1) directly through photodissolution reactions and (2) indirectly through the resultant formation of amorphous Fe(III) precipitates that exhibit higher solubilities than corresponding crystalline

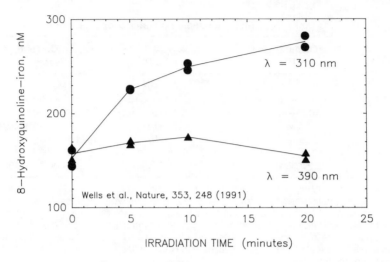

Figure 10. Photoinduced increase in the "dissolved" Fe(III) concentration, as measured by 8-hydroxyquinoline, in irradiated (310 and 390 nm) seawater (from near the Gironde River estuary) to which 4 μM of freshly prepared ferrihydrite was added. (From Wells, M. L., et al., *Nature (London)*, 353, 248, 1991. With permission.)

solids. This is significant because, as noted in the introduction, many chemical and biological processes are dependent upon the availability of dissolved iron. A conceptual picture (oversimplified) of these processes might be the following:

$$\text{crystalline Fe(III)} + \text{light} \longrightarrow \text{Fe(II)} \tag{9}$$

$$\text{amorphous Fe(III)} + \text{light} \longrightarrow \text{Fe(II)} \tag{10}$$

$$\text{dissolved Fe(III)} + \text{light} \longrightarrow \text{Fe(II)} \tag{11}$$

$$\text{Fe(II)} + \text{oxidant(s)} \longrightarrow \text{dissolved Fe(III)} \tag{12}$$

$$\text{dissolved Fe(III)} \longrightarrow \text{amorphous Fe(III)} \tag{13}$$

$$\text{amorphous Fe(III)} \longrightarrow \text{crystalline Fe(III)} \tag{14}$$

The crystallization of Fe(III) precipitates (the reaction in Equation 14) would probably occur on time scales longer than those of the other reactions and of the residence time of iron in the atmosphere or in the photic zone of marine waters and freshwaters.[20,41] Figure 11 presents data[80] that is consistent with this scheme; the solubility of ambient ferrihydrite in seawater, as measured by 8-hydroxyquinoline, decreases only by a factor of 2 over a 4-day period in the dark. Figure 12 illustrates that the net accumulation rate of ^{59}Fe in particles greater than 0.2 μm

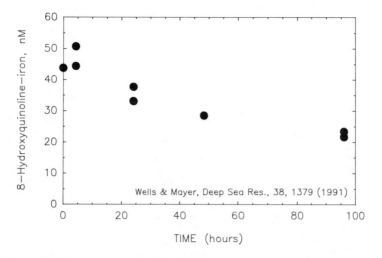

Figure 11. Decrease in the ambient "dissolved" iron concentration, as measured by 8-hydroxyquinoline, in surface seawater upon storage in the dark. The water was collected from the Damariscotta River estuary at mid-day under a clear sky. (From Wells, M. L. and Mayer, L. M., *Deep Sea Res.*, 38, 1379, 1991. With permission.)

is decreased in light relative to dark controls.[40] This is also consistent with the aforementioned mechanism, because photoreduction of Fe(III) is expected to lower the steady-state concentration of particulate iron in natural waters. Thus, in many situations, the "pool" of amorphous Fe(III) could undergo rapid internal cycling, in comparison to the slower reactions that form or dissolve crystalline solids.

It has been reported that the species Fe(OH)$^+$ exhibits an absorption spectrum that significantly overlaps the solar UV and visible spectrum and that it undergoes direct photooxidation to Fe(III):[81]

$$FeOH^+ + light \xrightarrow{\text{H}_2\text{O}} Fe(OH)_2^+ + H^\cdot \tag{15}$$

It was suggested that in the ancient Earth's atmosphere, with low concentrations of molecular oxygen, this process alone (and in the absence of molecular oxygen) would have been sufficiently rapid to form Fe(III) present in the Precambrian-banded iron formations.[81] A similar argument has been made for other Fe(II)-hydroxy species.[82] It has long been argued that the appearance of the Precambrian banded iron formations on Earth indicated the first occurrence of molecular oxygen in the ancient atmosphere. It should be noted that photochemical and thermal reductive dissolution reactions of Fe(III) precipitates would have recycled some of the Fe(III) precipitates back to Fe(II). In the present atmosphere, with 20.9% O_2, the fate of most H$^\cdot$ would be to react with aqueous O_2 to form $HO_2^\cdot$, which is a precursor to H_2O_2. Combination of two H$^\cdot$ forms H_2, which probably represents a minor source of H_2 to the atmosphere.

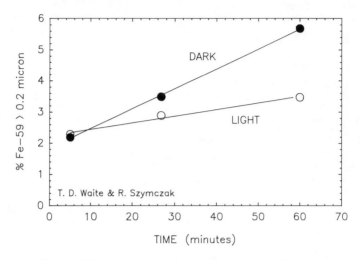

Figure 12. Effect of light on the net accumulation rate in surface seawater of ^{59}Fe-containing particles greater than 0.22 μm from the addition of [59]Fe(III).[40] Similar results were obtained for the addition of [59]Fe(II). (After Waite, T. D. and Szymczak, R., *American Chemical Society Meeting*, 1992.)

Other Elements

Carbon

The presence of Fe(III) in natural waters should increase the rate of photomineralization of natural organic matter by several possible mechanisms. Photolysis of Fe(III)-hydroxy species is known to form ·OH radicals[5,11,12,14-16,19,20,23,24] (*vide infra*), which oxidize numerous organic compounds at near-diffusion-controlled rates. The photolysis of 1 μM FeOH^{2+} alone produces 2.3 μM hr^{-1} of hydroxyl radical.[19,20] It is reasonable to assume that conditions are favorable for formation of FeOH^{2+} and other Fe(III)-hydroxy species in acidified atmospheric and surface waters.[19,20] Furthermore, the pH range of acidified atmospheric and surface waters is sufficiently low (pH = 2.5–5.5, Figure 4) that HCO$_3^-$ and CO$_3^{2-}$ will be present at concentrations that are too low [pK_{a1}(H$_2$CO$_3^*$) = 6.3] to be a significant sink for ·OH in these systems. Thus, a substantial fraction of the photogenerated ·OH could be expected to oxidize organic compounds present in acidified atmospheric and surface waters.

As discussed previously, sufficient circumstantial evidence exists to argue for the formation of Fe(III)-organic complexes in natural waters, and for coordination of the Fe^{3+} center through carboxylate functional groups. Some of these Fe(III) species will undoubtedly undergo photoinduced decarboxylation reactions analogous to those illustrated in Figure 7 for Fe(III) polycarboxylates. This could represent an additional pathway for mineralization of organic compounds and an abiotic source of CO$_2$ to the atmosphere. Iron(III) was demonstrated to

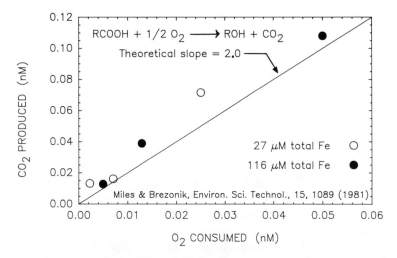

Figure 13. CO_2 produced and O_2 consumed over a 6-hr period in the iron-mediated photomineralization of organic matter in Lake Mize water: pH = 3.8, [dissolved organic carbon] = 3080 μM (37 mg/L), ambient [total iron] = 27 μM; glass- and plexiglass-filtered light from a tungsten lamp, 298 K. (After Miles, C. J. and Brezonik, P. L., *Environ. Sci. Technol.*, 15, 1089, 1981.)

accelerate the photochemical consumption of O_2 and the corresponding production of CO_2 in surface waters from Florida; these waters had dissolved organic carbon concentrations of 1080–5660 μM (13–68 mg/L), pH values of 3.6–7.4, and total iron concentrations of 1.8–36 μM.[6] Figure 13 shows that the observed ratio of CO_2 produced to O_2 consumed in the iron-mediated photoreaction was close to the theoretical ratio of 2.0 demanded by the following stoichiometry:[6]

$$RCOOH + 1/2\ O_2 \xrightarrow{\text{Fe(III) + light}} ROH + CO_2 \qquad (16)$$

This illustrates how the photoredox reactions of Fe(III) species can link the cycles of two major elements (carbon and oxygen).

Sulfur

Photoredox reactions of Fe(III) can affect the atmospheric water drop oxidation of S(IV) species (SO_3^{2-}, HSO_3^-) in at least two ways: (1) by forming oxidants that oxidize S(IV) species and (2) by controlling the speciation of iron, which in turn affects the Fe(III)-catalyzed oxidation of S(IV) by O_2.

Photolysis of $Fe(OH)^{2+}$ is an important source of ·OH in atmospheric water drops (*vide infra*).[14,15,19,20] Photolysis of Fe(III) species forms Fe(II), which can react with H_2O_2 to also form ·OH. The hydroxyl radical is known to initiate the free radical chain oxidation of the S(IV) species.[83] Photolysis of Fe(III) polycarboxylates[1,2,23-25] is a source of HO_2^- and, therefore, of H_2O_2 (*vide infra*). The aqueous-phase oxidation of S(IV) species by H_2O_2 in cloud, fog, and rain drops is the dominant mechanism for the oxidation of SO_2 to H_2SO_4 in the atmosphere.[84] Current computer models of cloud drop chemistry do not incorporate aqueous-phase photochemical sources of peroxides[77,78] and thus underestimate daytime rates of SO_2 oxidation in polluted cloudy boundary layers.

The Fe(III)-catalyzed thermal oxidation of S(IV) species by O_2 has been extensively studied and, under certain conditions, is considered to be an important mechanism for the formation of H_2SO_4 in the atmosphere.[9,13,18,26,75] The reaction is catalyzed by dissolved Fe(III) species. Thus, any process that affects the speciation of iron in atmospheric water drops will affect the rate of this reaction. As shown in Figure 9, sunlight initiates the photoreductive dissolution of hematite by S(IV) in air-saturated solutions.[9,13,18] Increasing concentrations of dissolved Fe(III) species, caused by the photodissolution of hematite, will increase rates of the thermal Fe(III)-catalyzed oxidation of S(IV) by O_2 in solution.[9,13,18,26,75] Figures 7 and 8 illustrate that the photoredox reactions of Fe(III) and the thermal reactions of photochemically formed $\cdot O_2^-$ with Fe(III) can have a pronounced effect on the speciation of iron and, therefore, on oxidant formation in atmospheric water drops, which will affect the oxidation of S(IV).

Phosphorus

Several reports have indicated that Fe(III), natural organic matter, and phosphorus are all present together in particulate and, possibly, dissolved species in natural waters.[7,43,44] Several types of interaction are possible, but the exact nature of the reported interactions has not been elucidated, although one study[44] indicated a remarkably constant stoichimetric composition of a precipitate. The effects of photoredox reactions on these entities have not been extensively studied.

It has been observed that irradiation of acidic bog waters (pH = 5.2–5.8) results in the simultaneous appearance of Fe(II) and "soluble reactive phosphorus" (SRP) (i.e., dissolved orthophosphate species $H_nPO_4^{n-3}$)[7] as shown in Figure 14. During the course of the irradiation, the net release of SRP vs. Fe(II) was Δ[SRP] to Δ[Fe(II)] = 1:90.2 on a molar basis (Figure 14), which is nearly identical to the independently determined ratio of the total dissolved concentrations of SRP to iron of 1:89.5. The large difference (a factor of 90) between the net amount of photoreleased phosphorus and Fe(II) suggests that photoreduction of Fe(III) particulates is followed by the simultaneous release of Fe(II) and adsorbed-coprecipitated orthophosphate. It is conceivable that some of the released Fe(II) is present as Fe(II) in the original unphotolyzed species. The photodependent release of SRP and Fe(II) in sunlight was also reported to occur on time scales of approximately 10 min.[7]

The photorelease of orthophosphate species is potentially an important, and heretofore minimally investigated, process that might have widespread ecological ramifications.[7] It is generally acknowledged that dissolved orthophosphates are the preferred forms of phosphorus for assimilation by algae and microorganisms. The dynamics of the abiotic photoinitiated release of orthophosphate parallel the algal

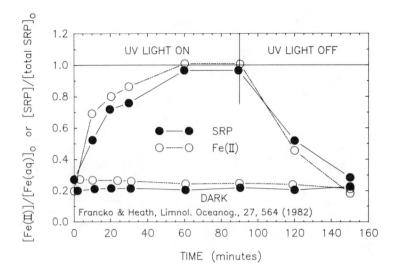

Figure 14.   Relationship between the dynamics of Fe(II) and "soluble reactive phosphorus" (SRP) concentrations in filtered (0.45 μm) Crazy Eddy Bog water in the dark and during UV irradiation (wavelength less than 400 nm; lower λ bound not specified). SRP is determined by the molybdate blue procedure and presumably represents $H_nPO_4^{-3}$ species. $[Fe(aq)]_o$ = 89.5 μM and [total SRP]$_o$ = 1.16 μM are the total dissolved (<0.45 μm) iron and SRP concentrations, respectively, at time zero. (From Francko, D. A. and Heath, R. T., *Limnol. Oceanogr.*, 27, 564, 1982. With permission.)

requirements for phosphorus that are driven by photosynthesis and, therefore, might possibly modulate algal growth dynamics (and CO_2 uptake) in some surface waters.

Chromium

The rapid photoredox reactions of Fe(III) species could drastically affect the geochemical cycling of other important elements. The photoreduction of Fe(III) as a source of Fe(II), followed by rapid reaction of Fe(II) with Cr(VI), has been invoked as the mechanism for the reduction of dissolved Cr(VI) species to highly insoluble Cr(III) species in sunlit surface waters.[85] It was argued[85] that this process accelerates the removal of chromium from estuaries and from the surface waters of the ocean.

Oxygen

Figure 15 shows that the photochemical consumption rate of O_2 by swamp water is critically dependent upon the iron concentration.[6] The O_2 photoconsumption rate also exhibited a linear dependence on the dissolved organic carbon concentration. Oxygen consumption rates were found to be approximately three times greater in sunlight than with the type of irradiation used in the laboratory.[6] Methylation of the carboxyl groups resulted in a 50%

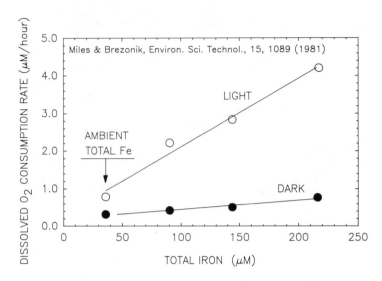

Figure 15. Effect of added iron on the mean rate of O_2 consumption over a 6-hr period by Austin Cary cypress swamp water: pH = 3.6, [dissolved organic carbon] = 5660 μM (68 mg/L), ambient [total iron] = 36 μM ; glass- and plexiglass-filtered light from a tungsten lamp, 298 K. (From Miles, C. J. and Brezonik, P. L., *Environ. Sci. Technol.*, 15, 1089, 1981. With permission.)

decrease in the O_2 photoconsumption rate relative to the untreated control. Based on this and on the correlation found between CO_2 formation and O_2 consumption (Figure 13), the authors proposed the reaction in Equation 16 as the stoichiometry for this process.[6]

Figure 7 illustrates that oxygen can be consumed at various steps in the photolysis of Fe(III) polycarboxylates through the oxidation of reducing radical intermediates and Fe(II). Other pathways are probably also operative. For example, the photolysis of Fe(III)-hydroxy species forms an ·OH radical,[5,11,12,14-16,19,20] which readily abstracts hydrogen atoms from carbon atoms, forming a carbon-centered radical that reacts rapidly with O_2 to form the corresponding peroxyl radical. Molecular oxygen is the terminal electron acceptor in the oxidation of organic matter in natural waters, and there are numerous pathways by which Fe(III) photochemical reactions can mediate the oxidation of organic matter by O_2. This provides a clear example of an Fe(III) photochemical link between the elemental cycles of carbon and oxygen.

FORMATION OF OXIDANTS IN ATMOSPHERIC, OCEANIC, AND SURFACE WATERS

Peroxyl Radicals (HO$_2$ AND RO$_2$) and Superoxide Radical (·O$_2^-$)

Figure 7 illustrates the mechanism for the photochemical formation of $HO_2^-/·O_2^-$ from the photolysis of Fe(III)-polycarboxylate complexes.[1,2,23-25,79] A critical step

Table 4.   Sinks for $HO_2^{\cdot}/^{\cdot}O_2^-$ in Atmospheric Water Drops, Some of Which Form H_2O_2

Reaction (A + B → Products)	$k_{A,B}$[a] $(M^{-1}\,s^{-1})$	Rate = $k_{A,B}$[A][B] $(nM\,s^{-1})$
H_2O_2-forming reactions		
$HO_2^{\cdot} + HO_2^{\cdot} \rightarrow H_2O_2 + O_2$	8.3×10^5	0.057
$HO_2^{\cdot} + {}^{\cdot}O_2^- (+H^+) \rightarrow H_2O_2 + O_2$	9.7×10^7	0.66
$HO_2^{\cdot} + Fe(II) (+H^+) \rightarrow H_2O_2 + Fe(III)$	1.2×10^6	5.6
$HO_2^{\cdot} + Cu(I) (+H^+) \rightarrow H_2O_2 + Cu(II)$	1×10^9	460
${}^{\cdot}O_2^- + Fe(II) (+H^+) \rightarrow H_2O_2 + Fe(III)$	1.0×10^7	3.7
${}^{\cdot}O_2^- + Cu(I) (+H^+) \rightarrow H_2O_2 + Cu(II)$	1×10^{10}	370
O_2-forming reactions		
$HO_2^{\cdot} + Fe(III) \rightarrow O_2 + Fe(II) + H^+$	$< 10^4$	<0.05
$HO_2^{\cdot} + Cu(II) \rightarrow O_2 + Cu(I) + H^+$	10^8	46
${}^{\cdot}O_2^- + Fe(III) \rightarrow O_2 + Fe(II)$	1.5×10^8	56
${}^{\cdot}O_2^- + Cu(II) \rightarrow O_2 + Cu(I)$	1×10^{10}	370
${}^{\cdot}O_2^- + O_3(aq) (+H^+) \rightarrow 2\,O_2 + {}^{\cdot}OH$	1.5×10^9	1.1

Note: These calculations contain numerous uncertainties, such as the effects of dissolved organic compounds and S(IV) species (which are not treated here), and are provided only for purposes of internal relative comparisons, for atmospheric water drops of the following composition: pH = 3.7, $[HO_2^{\cdot}]$ = 9.26 nM, $[{}^{\cdot}O_2^-]$ = 0.74 nM, [Fe(III)] = [Fe(II)] = 0.5 μM, [Cu(II)] = [Cu(I)] = 50 nM, $[O_3(aq)]$ = 1 nM (based on Henry's law equilibrium with 100 ppbv gas-phase O_3).
[a] Bimolecular rate constants are from Reference 101.

in this mechanism is the competition between Fe(III) and O_2 for the reducing radicals (the polycarboxylate radical and the carbon-centered radical derived therefrom). If most of the reducing radicals are intercepted by O_2, then $HO_2^{\cdot}/^{\cdot}O_2^-$ and organic peroxyl radicals $(RO_2^{\cdot})$ could be generated at fairly rapid rates because Fe(III) polycarboxylates photolyze rapidly in sunlight (Table 2). The yield of H_2O_2 will also depend critically upon competition between H_2O_2-producing reactions and O_2-producing reactions, as is illustrated in Table 4. In turn, this competition will depend of the steady-state redox speciation of Fe(III)/Fe(II) and Cu(II)/Cu(I) in the atmospheric water. From this discussion and from the previous discussion regarding the sources and sinks for Fe(II), it seems very likely that the fates of $^{\cdot}O_2^-/HO_2^{\cdot}$ and Fe(III)/Fe(II) in atmospheric water drops are intertwined. It is seen from Table 4 that copper complexes will also play very important roles in determining the lifetime of $HO_2^{\cdot}/^{\cdot}O_2^-$ and yield of H_2O_2 in atmospheric water drops.

Other photochemical reactions of Fe(III) species could give rise to peroxyl radicals. As shown in Figure 7, the photolysis of $FeOH^{2+}$ and other Fe(III)-hydroxy species forms an $^{\cdot}OH$ radical,[5,11,12,14-16,19,20] which will readily attack C–H bonds in organic compounds (RH) forming carbon-centered radicals that will react with diffusion-controlled rates which dissolve O_2, ultimately giving rise to peroxyl radicals:[79]

$${}^{\cdot}OH + RH \longrightarrow H_2O + R^{\cdot} \tag{17}$$

$$R^{\cdot} + O_2 \longrightarrow RO_2^{\cdot} \tag{18}$$

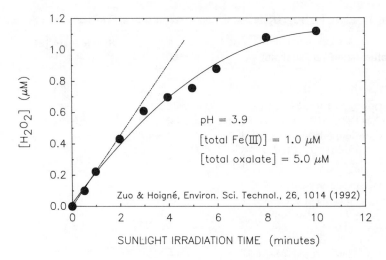

Figure 16.   Hydrogen peroxide formation in a sunlit (mid-day, September 14, 1990) aqueous solution of Fe(III) and oxalate. Ionic strength equals 0.03 M, 283–300 K. (From Zuo, Y. and Hoigné, J., *Environ. Sci. Technol.*, 26, 1014, 1992. With permission.)

$$R^{\cdot} + O_2 \longrightarrow R^+ + {}^{\cdot}O_2^-/HO_2^{\cdot} \tag{19}$$

For some organic peroxyl radicals ($RO_2^{\cdot}$), for example α-hydroxylalkylperoxyl radicals, unimolecular elimination of $HO_2^{\cdot}$ is also possible:[79]

$$R_i(R_j)C(OH)O_2^{\cdot} \longrightarrow R_i(R_j)C{=}O + HO_2^{\cdot} \tag{20}$$

Hydrogen Peroxide (H_2O_2)

Figure 16 illustrates that the photolysis of Fe(III)-oxalate solutions can result in rapid rates of H_2O_2 formation under conditions where most of the Fe(III) is coordinated to oxalate.[23,25] It is likely that this process is a significant source of H_2O_2 to fog and low clouds trapped within the inversion layer of an urban area where dicarboxylic acids are formed from the gas-phase reactions of ozone with alkenes and aromatic compounds. The lifetimes of alkenes and aromatic compounds in the troposphere are short, due to rapid reactions with gas-phase ${}^{\cdot}OH$ and O_3. The depletion of oxalate is sufficiently rapid in such solutions[25] that some other process, such as formation of oxalate or transport of oxalate to the cloud/fog, could possibly be rate limiting for this process in the atmosphere.

The extent to which this and similar processes contribute to the aqueous-phase photochemical formation of H_2O_2 in clouds on regional and continental scales remains to be demonstrated. Three pieces of evidence from studies on authentic atmospheric waters[77,78] suggest that other processes are also important for the aqueous-phase photochemical formation of peroxides in clouds and fogs. One, only a very weak correlation was found between the aqueous-phase peroxide photoformation rate and the iron concentration (filtered, 0.5 µm) of authentic cloud waters. Two, the quantum yields for peroxide formation, albeit based on the total absorbance of the atmospheric water sample, vary by approximately a factor of 2.3 from 313 to 334 nm for authentic atmospheric waters, whereas quantum yields for the Fe(III)-oxalate system are invariant over this wavelength region.[86] Three, the aqueous-phase photochemical formation of peroxides is correlated with the normalized fluorescence and the dissolved organic carbon concentration of cloud waters from rural regions of the United States and Canada.[77,78]

If such processes are generally important, then Fe(III) photochemistry provides an important link between the atmospheric chemical cycles of carbon, sulfur, and oxygen because water drop oxidation of bisulfite by H_2O_2 is the dominant source of H_2SO_4 in the atmosphere.[84]

Hydroxyl Radical ($^{\cdot}$OH)

Photolysis of dissolved and surficial Fe(III)-hydroxy species forms Fe(II), as discussed earlier, and also is a direct source of $^{\cdot}$OH radical via the following reactions:[5,11,12,14–16,19,20]

$$FeOH^{2+} + light \longrightarrow Fe^{2+} + {}^{\cdot}OH \qquad (21)$$

$$Fe_2(OH)_2^{4+} + light \longrightarrow Fe^{2+} + Fe(III) + {}^{\cdot}OH \qquad (22)$$

$$>FeOH + light \longrightarrow >Fe(II) + {}^{\cdot}OH \qquad (23)$$

where >FeOH indicates a hydroxylated Fe(III) group at the surface of any amorphous or crystalline Fe(III) (hydrous) oxide. The quantum yield of $^{\cdot}$OH varies considerably between the different Fe(III)-hydroxy species (Table 1). The half-life for photolysis of $FeOH^{2+}$ in sunlight is approximately 20 min (Table 2) and represents an important source of $^{\cdot}$OH to atmospheric water drops.[19,20] Field studies on acid mine drainage systems (pH = 3–4, H_2SO_4) demonstrated that dissolved and particulate forms of Fe(III) photolyze to produce Fe(II); it is highly likely that $^{\cdot}$OH was also formed in these systems. By analogy, such reactions to form Fe(II) and $^{\cdot}$OH are also likely to occur in atmospheric water drops (pH = 3–4, H_2SO_4). These reactions could also explain, in part, the occurrence of Fe(II) in urban fog and remote marine aerosols.[38,65,66]

In cases where the speciation of Fe(III) in atmospheric and surface waters is dominated by Fe(III) complexes with S(IV) species, organic ligands, or other ligands, but not by Fe(III)-hydroxy species, the photochemical reactions of Fe(III) will produce Fe(II), but not necessarily $^{\cdot}$OH directly [note however that mixed ligand complexes, e.g., Fe(OH)(L), are a possible direct photochemical source of

·OH]. In these cases, the hydroxyl radical can also be formed indirectly, from reaction of photoformed Fe(II) with H_2O_2:[14,15,20,24]

$$Fe(II) + H_2O_2 \longrightarrow Fe(III) + \cdot OH \qquad (24)$$

and this also constitutes an important source of ·OH to atmospheric water drops.[14,15,20] In the case of Fe(III) polycarboxylates (Figure 7), they could be important photochemical sources of both Fe(II) and H_2O_2, and the reactions of the "self-generated" Fe(II) and H_2O_2 could lead to rapid ·OH formation rates in these systems.[23,24] In this model, polycarboxylates have three important roles. First, polycarboxylates increase the concentration of dissolved Fe(III) because they form strong complexes with Fe^{3+}. Second, Fe(III) polycarboxylates undergo rapid photolytic reactions that reduce Fe(III) to Fe(II) and that simultaneously reduce O_2 to H_2O_2 through dismutation of HO_2^-/O_2^- derived from the reaction of O_2 with the polycarboxylate radicals, carbon-centered radicals, and Fe(II) (Figure 7). Third, H_2O_2 reacts much faster with Fe(II) in the presence of polycarboxylates,[106] presumably because Fe(II) polycarboxylates are more reactive than Fe^{2+} toward H_2O_2, and therefore, polycarboxylates apparently increase the rate of ·OH formation in the Fenton reaction. However, a major uncertainty regarding this reaction is the yield of hydroxyl radicals from reaction of H_2O_2 with various Fe(II) species. As shown in Figure 17, the yield of ·OH per Fe(II) oxidized by H_2O_2 is approximately one for various Fe(II) species[24] in the presence of different polycarboxylates that are representative of the types of coordination environments experienced by Fe(II) in atmospheric, oceanic, and surface waters. Thus, the photolysis of Fe(III) species could even form ·OH in atmospheric and surface waters where the Fe(III)

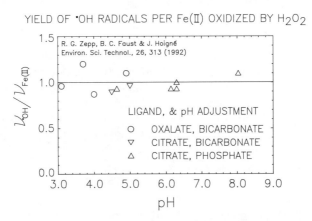

Figure 17.   Rate of ·OH radical photoformation (v_{OH}) in air-saturated solutions of Fe(III) and a ligand (ligand = oxalate or citrate) containing 100 μM added hydrogen peroxide, divided by the rate of Fe(II) photoformation ($v_{Fe(II)}$) in the same solution (without added hydrogen peroxide) that was purged with argon. [Fe(III)] equals 50–200 μM; 295–298 K. Data are from Table I of Reference 24.

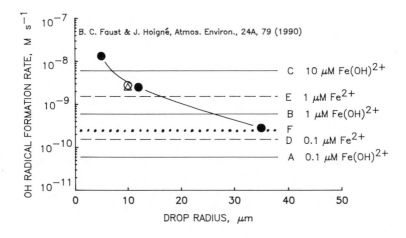

Figure 18. Calculated formation rates of ·OH radical, from different sources, in cloud and fog drops. All results are for clear sky at mid-day on June 30, 1987 for 47.4°N (solar zenith angle = 24.3°). Lines A, B, and C (—): photolysis of Fe(OH)²⁺ (rxn. 21, j_i = 6.3 × 10⁻⁴ sec⁻¹) present at a steady-state concentration of 0.1 μM (A), 1.0 μM (B), and 10 μM (C). Lines D and E (– – – –): Fenton reaction (rxn. 24) of 30 μM H₂O₂ with Fe²⁺ present at a steady-state concentration of 0.1 μM (D) and 1 μM (E), using a rate constant of 51 M^{-1} s⁻¹.[103] Line F (· · · · ·): photolysis of 30 μM H₂O₂, using 2j_i = 8.25 × 10⁻⁶ sec⁻¹, obtained by multiplying the rate constant of Reference 108 by 0.75 and 1.1 to correct for altitude and latitude-seasonal differences, respectively. (●—●,○): scavenging of gas-phase ·OH present at 2.4 × 10⁶ molecules per cubic centimeter[108] (●—●) and 4.0 × 10⁶ molecules per cubic centimeter[109] (○). (●—●): values from Figure 4 of Reference 108 (sticking coefficient = 1) multiplied by 0.75, 1.1, and 2.0 to correct for altitude and latitude-seasonal differences and for sunlight attenuation by clouds, respectively. (○): value from Table 3 of Reference 109, multiplied by 1.1 to correct for latitude-seasonal differences. (△): scavenging of HO₂⁻ from the gas phase, followed by reaction of ·O₂⁻ with aqueous ozone (at pH equal to 4.16) to yield ·OH (from Table 3 of Reference 109 multiplied by 1.1 to correct for latitude-seasonal differences). (From Faust, B. C. and Hoigné, J., *Atmos. Environ.*, 24A, 79, 1990. With permission.)

speciation is dominated by Fe(III) polycarboxylates and not by Fe(OH)²⁺ and other Fe(III)-hydroxy species that are known direct photochemical sources of ·OH.

Figure 18 summarizes much of the information regarding possible sources of ·OH radical to cloud and fog drops.[20] It is seen that the photolysis of micromolar concentrations of Fe(OH)²⁺ alone constitutes an important source of ·OH to atmospheric water drops (Figure 18). Moreover, reaction of Fe(II) at micromolar concentrations with H₂O₂ (30 μM) is also an important source of ·OH radical to atmospheric water drops (Figure 18). Both of these processes compete effectively with gas-to-drop partitioning of gas-phase hydroxyl radical (Figure 18).

The diagonal line in Figure 18 illustrates that gas-to-drop partitioning of an oxidant (·OH), which depends on the ratio of surface area to volume for the drop, increases with decreasing drop radius. Based on this, it is commonly argued that the relative importance of gas-to-drop partitioning vs aqueous-phase photoformation as sources of a given oxidant to atmospheric water drops increases with decreasing drop radius. However, such arguments neglect the fact that measured solute concentrations (and, therefore, chromophore concentrations) also increase with decreasing radius of the atmospheric water drop.[88] Thus aqueous-phase photoreactions may also proceed at higher rates with decreasing drop radius.

Other Oxidants

Other oxidants (e.g., SO_4^-, SO_3^-) could be formed from aqueous-phase photochemical reactions in atmospheric water drops (Figure 7). It has even been speculated that photolysis of seasalt Fe(III)-bromide complexes could represent a pathway for the conversion of Br^- to Br_2 and might constitute a source of ozone-depleting bromine to the arctic troposphere.[87]

CONCLUSIONS

Photoredox reactions are an important, if not dominant, source of Fe(II) and dissolved iron to atmospheric, oceanic, and surface waters. The photoredox reactions of Fe(III) species also play important roles in the geochemical cycling of iron, carbon, sulfur, phosphorus, oxygen, and probably other elements (e.g., chromium). Moreover, the photochemical reactions of Fe(III) species often provide important links between the elemental cycles of other major elements, such as between carbon and oxygen and between carbon and sulfur.

In natural waters, the photochemistry and chemistry of Fe(III), and the speciation of iron in general, is tightly coupled to oxidant (HO_2^-/O_2^-, H_2O_2, OH, etc.) formation in atmospheric and surface waters. Photochemical reactions of Fe(III) species can directly and indirectly represent significant, and in some cases dominant, sources of these oxidants to atmospheric and surface waters.

Given the rapid (photo)chemical redox reactions of Fe(III) and Fe(II) species, which occur on the time scales of minutes, it is essential that techniques be developed to rapidly sample and characterize natural waters and atmospheric aerosols for their iron speciation, ideally within minutes (or less) of sampling. More studies are needed on the photochemistry of many Fe(III) species that are common constituents in atmospheric, surface, and oceanic waters because major gaps still exist in this information base.

ACKNOWLEDGMENTS

This work was supported by the Andrew W. Mellon Foundation and the Atmospheric Chemistry Program of the U.S. National Science Foundation.

REFERENCES

1. Parker, C. A., Induced autoxidation of oxalate in relation to the photolysis of potassium ferrioxalate, *Trans. Faraday Soc.*, 50, 1213, 1954.

2. Parker, C. A. and Hatchard, C. G., Photodecomposition of complex oxalates. Some preliminary experiments by flash photolysis, *J. Phys. Chem.*, 63, 22, 1959.

3. Behar, B. and Stein, G., Photochemical evolution of oxygen from certain aqueous solutions, *Science*, 154, 1012, 1966.

4. McMahon, J. W., The annual and diurnal variation in the vertical distribution of acid-soluble ferrous and total iron in a small dimictic lake, *Limnol. Oceanogr.*, 14, 357, 1969.

5. Langford, C. H. and Carey, J. H., The charge transfer photochemistry of the hexaaquoiron(III) ion, the chloropentaaquoiron(III) ion, and the μ-dihydroxo dimer explored with *tert*-butyl alcohol scavenging, *Can. J. Chem.*, 53, 2430, 1975.

6. Miles, C. J. and Brezonik, P. L., Oxygen consumption in humic-colored waters by a photochemical ferrous-ferric catalytic cycle, *Environ. Sci. Technol.*, 15, 1089, 1981.

7. Francko, D. A. and Heath, R. T., UV-sensitive complex phosphorus: association with dissolved humic material and iron in a bog lake, *Limnol. Oceanogr.*, 27, 564, 1982.

8. Collienne, R. H., Photoreduction of iron in the epilimnion of acidic lakes, *Limnol. Oceanogr.*, 28, 83, 1983.

9. Faust, B. C., Photo-Induced Reductive Dissolution of Hematite (α-Fe$_2$O$_3$) by S(IV) Oxyanions, Ph.D. dissertation, California Institute of Technology, Pasadena, 1984.

10. Finden, D. A. S., Tipping, E., Jaworski, G. H. M., and Reynolds, C. S., Light-induced reduction of natural iron(III) oxide and its relevance to phytoplankton, *Nature (London)*, 309, 783, 1984.

11. Waite, T. D. and Morel, F. M. M., Photoreductive dissolution of colloidal iron oxides in natural waters, *Environ. Sci. Technol.*, 18, 860, 1984.

12. Cunningham, K. M., Goldberg, M. C., and Weiner, E. R., The aqueous photolysis of ethylene glycol adsorbed on goethite, *Photochem. Photobiol.*, 41, 409, 1985.

13. Faust, B. C. and Hoffmann, M. R., Photoinduced reductive dissolution of α-Fe$_2$O$_3$ by bisulfite, *Environ. Sci. Technol.*, 20, 943, 1986.

14. Weschler, C. J., Mandich, M. L., and Graedel, T. E., Speciation, photosensitivity, and reactions of transition metal ions in atmospheric droplets, *J. Geophys. Res.*, 91, 5189, 1986.

15. Graedel, T. E., Mandich, M. L., and Weschler, C. J., Kinetic model studies of atmospheric droplet chemistry 2. Homogeneous transition metal chemistry in raindrops, *J. Geophys. Res.*, D4, 5205, 1986.

16. Cunningham, K. M., Goldberg, M. C., and Weiner, E. R., Mechanisms for aqueous photolysis of adsorbed benzoate, oxalate, and succinate on iron oxyhydroxide (goethite) surfaces, *Environ. Sci. Technol.*, 22, 1090, 1988.

17. McKnight, D. M., Kimball, B. A., and Bencala, K. E., Iron photoreduction and oxidation in an acidic mountain stream, *Science*, 240, 637, 1988.

18. Faust, B. C., Hoffmann, M. R., and Bahnemann, D. W., Photocatalytic oxidation of sulfur dioxide in aqueous suspensions of α-Fe$_2$O$_3$, *J. Phys. Chem.*, 93, 6371, 1989.

19. Benkelberg, H. J., Deister, U., and Warneck, P., OH quantum yields for the photodecomposition of Fe(III) hydroxo complexes in aqueous solution and the reaction with hydroxymethanesulfonate, in *Physico-Chemical Behavior of Atmospheric Pollutants*, Restellia, G. and Angeletti, G., Eds., Kluwer, Dordrecht, 1990, 263.

20. Faust, B. C. and Hoigné, J., Photolysis of Fe(III)-hydroxy complexes as sources of OH radicals in clouds, fog and rain, *Atmos. Environ.*, 24A, 79, 1990.

21. Siffert, C. and Sulzberger, B., Light-induced dissolution of hematite in the presence of oxalate: a case study, *Langmuir*, 7, 1627, 1990.

22. Wells, M. L., Mayer, L. M., Donard, O. F. X., de Souza Sierra, M. M., and Ackelson, S. G., The photolysis of colloidal iron in the oceans, *Nature (London)*, 353, 248, 1991.

23. Faust, B. C. and Zepp, R. G., Photochemistry of aqueous iron(III)-polycarboxylate complexes: roles in the chemistry of atmospheric and surface waters, *Environ. Sci. Technol.,* in press, 1993.

24. Zepp, R. G., Faust, B. C., and Hoigné, J., Hydroxyl radical formation in aqueous reactions (pH 3–8) of iron(II) with hydrogen peroxide: the photo-Fenton reaction, *Environ. Sci. Technol.,* 26, 313, 1992.

25. Zuo, Y. and Hoigné, J., Formation of hydrogen peroxide and depletion of oxalic acid in atmospheric water by photolysis of iron(III)-oxalato complexes, *Environ. Sci. Technol.,* 26, 1014, 1992.

26. Martin, L. R., Hill, M. W., Tai, A. F., and Good, T. W., The iron-catalyzed oxidation of sulfur(IV) in aqueous solution: differing effects of organics at high and low pH, *J. Geophys. Res.,* 96, 3085, 1991.

27. Langner, J. and Rodhe, H., A global three-dimensional model of the tropospheric sulfur cycle, *J. Atmos. Chem.,* 13, 225, 1991.

28. Schwartz, S. E., Acid deposition: unraveling a regional phenomenon, *Science,* 243, 753, 1989.

29. Koutrakis, P., Wolfson, J. M., Spengler, J. D., Stern, B., and Franklin, C. A., Equilibrium size of atmospheric aerosol sulfates as a function of the relative humidity, *J. Geophys. Res.,* 94, 6442, 1989.

30. Charlson, R. J., Schwartz, S. E., Hales, J. M., Cess, R. D., Coakley, J. A., Jr., Hansen, J. E., and Hofmann, D. J., Climate forcing by anthropogenic aerosols, *Science,* 255, 423, 1992.

31. Martin, J. H., and Gordon, R. M., Northeast Pacific iron distributions in relation to phytoplankton productivity, *Deep Sea Res.,* 35, 177, 1988.

32. Martin, J. H., Gordon, R. M., Fitzwater, S., and Broenkow, W. W., VERTEX: phytoplankton/iron studies in the Gulf of Alaska, *Deep Sea Res.,* 36, 649, 1989.

33. Hudson, R. J. M. and Morel, F. M. M., Iron transport in marine phytoplankton: kinetics of cellular and medium coordination reactions, *Limnol. Oceanogr.,* 35, 1002, 1990.

34. Martin, J. H., Gordon, R. M., and Fitzwater, S. E., Iron in Antartic waters, *Nature (London),* 345, 156, 1990.

35. Price, N. M., Andersen, L. F., and Morel, F. M. M., Iron and nitrogen nutrition of equatorial Pacific plankton, *Deep Sea Res.,* 38, 1361, 1991.

36. Sunda, W. G., Swift, D. G., and Huntsman, S. A., Low iron requirement for growth in oceanic phytoplankton, *Nature (London),* 351, 55, 1991.

37. Zhuang, G., Duce, R. A., and Kester, D. R., The dissolution of atmospheric iron in surface seawater of the open ocean, *J. Geophys. Res.,* 95, 16207, 1990.

38. Zhuang, G., Yi, Z., Duce, R. A., and Brown, P. R., Link between iron and sulfur cycles suggested by detection of Fe(II) in remote marine aerosols, *Nature (London),* 355, 537, 1992.

39. Hong, H. and Kester, D. R., Redox state of iron in the offshore waters of Peru, *Limnol. Oceanogr.,* 31, 512, 1986.

40. Waite, T. D. and Szymczak, R., Redox transformations of iron in marine systems: effects of organics and light, extended abstract, American Chemical Society meeting, San Francisco, 1992.

41. Murray, J. W., Iron oxides, in *Marine Minerals,* Burns, R. G., Ed., *Rev. Mineral.,* Vol. 6, chap. 2, Mineralogical Society, American Bookcrafters Inc., Chelsea, MI, 1979, 47.

42. Stucki, J. W., Goodman, B. H., and Schwertmann, U., *Iron in Soils and Clay Minerals,* Nato Adv. Studies Inst. Ser. C, Vol. 217, D. Reidel, Boston, 1985.

43. Tipping, E., Woof, C., and Cooke, D., Iron oxide from a seasonally anoxic lake, *Geochim. Cosmochim. Acta,* 45, 1411, 1981.

44. Buffle, J., DeVitre, R. R., Perret, D., and Leppard, G. G., Physico-chemical characteristics of a colloidal iron phosphate species formed at the oxic-anoxic interface of a eutrophic lake, *Geochim. Cosmochim. Acta,* 53, 399, 1989.

45. Tipping, E., Woof, C., and Ohnstad, M., Forms of iron in the oxygenated waters of Esthwaite Water, U.K., *Hydrobiologica,* 92, 383, 1982.

46. Mill, A. J. B., Colloidal and macromolecular forms of iron in natural waters 1: a review, *Environ. Technol. Lett.,* 1, 97, 1980.

47. Sung, W. and Morgan, J. J., Kinetics and product of ferrous iron oxygenation in aqueous systems, *Environ. Sci. Technol.*, 14, 561, 1980.
48. Morel, F. M. M., *Principles of Aquatic Chemistry*, Wiley, New York, 1983, 226.
49. Kopcewicz, M. and Dzienis, B., Mössbauer study of iron in atmospheric air, *Tellus*, 23, 176, 1971.
50. Fukasawa, T., Iwatsuki, M., Kawakubo, S., and Miyazaki, K., Heavy-liquid separation and X-ray diffraction analysis of airborne particulates, *Anal. Chem.*, 52, 1784, 1980.
51. Hansen, L. D., Silberman, D., and Fisher, G. L., Crystalline components of stack-collected, size-fractionated coal fly ash, *Environ. Sci. Technol.*, 15, 1057, 1981.
52. Nigam, A. N., Tripathi, R. P., Ramasheshu, P., and Chopra, R. K., Characterization of the iron-bearing component present in aerosol over Jodphur, the capital city of the Thar Desert (India), *Atmos. Environ.*, 22, 425, 1988.
53. van Borm, W. A., Adams, F. C., and Maenhaut, W., Characterization of individual particles in the Antwerp aerosol, *Atmos. Environ.*, 23, 1139, 1989.
54. Brimblecombe, P. and Spedding, D. J., The dissolution of iron from ferric oxide and pulverized fuel ash, *Atmos. Environ.*, 9, 835, 1975.
55. Williams, P. T., Radojevic, M., and Clarke, A. G., Dissolution of trace metals from particles of industrial origin and its influence on the composition of rainwater, *Atmos. Environ.*, 22, 1433, 1988.
56. Koenings, J. P., In situ experiments on the dissolved and colloidal state of iron in an acid bog lake, *Limol. Oceanogr.*, 21, 674, 1976.
57. Perdue, E. M., Beck, K. C., and Reuter, J. H., Organic complexes of iron and aluminum in natural waters, *Nature (London)*, 260, 418, 1976.
58. Langford, C. H., Kay, R., Quance, G. W., and Khan, T. R., Kinetic analysis applied to iron in a natural water containing model ions, organic complexes, colloids, and particles, *Anal. Lett.*, 10, 1249, 1977.
59. Balzani, V. and Carassitti, V., *Photochemistry of Coordination Compounds*, Academic Press, New York, 1970, chap. 10.
60. Evans, M. G., George, P., and Uri, N., The $[Fe(OH)]^{2+}$ and $[Fe(O_2H)]^{2+}$ complexes, *Trans. Faraday Soc.*, 34, 230, 1949.
61. Lever, A. B. P., Charge transfer spectra of transition metal complexes, *J. Chem. Educ.*, 51, 612, 1974.
62. Endicott, J. F., Charge-transfer photochemistry, in *Concepts in Inorganic Photochemistry*, Adamson, A. W. and Fleischauer, P. D., Eds., Wiley, New York, 1975, chap. 3.
63. Stasicka, Z. and Marchaj, A., Flash photolysis of coordination compounds, *Coord. Chem. Rev.*, 23, 131, 1977.
64. Sherman, D. M. and Waite, T. D., Electronic spectra of Fe^{3+} oxides and oxide hydroxides in the near IR to near UV, *Am. Mineral.*, 70, 1262, 1985.
65. Behra, P. and Sigg, L., Evidence for redox cycling of iron in atmospheric water droplets, *Nature (London)*, 344, 419, 1990.
66. Pehkonen, S. O., Erel, Y., and Hoffmann, M. R., Simultaneous spectrophotometric measurement of Fe(II) and Fe(III) in atmospheric water, *Environ. Sci. Technol.*, 26, 1731, 1992.
67. Theis, T. L. and Singer, P. C., Complexation of iron(II) by organic matter and its effect on iron(II) oxygenation, *Environ. Sci. Technol.*, 8, 569, 1974.
68. Stumm, W. and Sulzberger, B., The cycling of iron in natural environments: considerations based on laboratory studies of heterogeneous redox processes, *Geochim. Cosmochim. Acta*, 56, 3233, 1992.
69. Mulazzani, Q. G., d'Angelantonio, M., Venturi, M., Hoffman, M. Z., and Rodgers, M. A. J., Interaction of formate and oxalate ions with radiation-generated radicals in aqueous solution. Methylviologen as a mechanistic probe, *J. Phys. Chem.*, 90, 5347, 1986.
70. Cooper, G. D. and DeGraff, B. A., On the photochemistry of the ferrioxalate system, *J. Phys. Chem.*, 75, 2897, 1971.

71. Cooper, G. D. and DeGraff, B. A., The photochemistry of the monoxalatoiron(III) ion, *J. Phys. Chem.*, 76, 2618, 1972.

72. Goldstein, S., Czapski, G., Cohen, H., and Meyerstein, D., Formation and decomposition of iron-carbon σ-bonds in the reaction of iron(II)-poly(amino carboxylate) complexes with CO_2^- free radicals. A pulse radiolysis study, *J. Am. Chem. Soc.*, 110, 3903, 1988.

73. Neta, P., Huie, R. E., and Ross, A. B., Rate constants for reactions of inorganic radicals in aqueous solution, *J. Phys. Chem. Ref. Data*, 17, 1027, 1988.

74. Ross, A. B. and Neta, P., Rate constants for reactions of aliphatic carbon-centered radicals in aqueous solution, *Natl. Stand. Ref. Data Ser.*, NSRDS-NBS-70, U.S. National Bureau of Standards, 1982.

75. Conklin, M. H. and Hoffmann, M. R., Metal ion-sulfur(IV) chemistry. 3. Thermodynamics and kinetics of transient iron(III)-sulfur(IV) complexes, *Environ. Sci. Technol.*, 22, 899, 1988.

76. Allen, J. M. and Faust, B. C., Aqueous-phase photochemical formation of peroxyl radicals and singlet molecular oxygen in cloud water samples from across the United States, in *Aquatic and Surface Photochemistry*, Helz, G., Zepp, R., and Crosby, D., Eds., Lewis Publishers, Boca Raton, FL, 1993, chap. 19.

77. Anastasio, C., Allen, J. M., and Faust, B. C., Aqueous-phase photochemical formation of peroxides in authentic cloud waters, in *Aquatic and Surface Photochemistry*, Helz, G., Zepp, R., and Crosby, D., Eds., Lewis Publishers, Boca Raton, FL, 1993, chap. 18.

78. Faust, B. C., Anastasio, C., Allen, J. M., and Arakaki, T., Aqueous- phase photochemical formation of peroxides in authentic cloud and fog waters, *Science*, 260, 73, 1993.

79. Faust, B. C. and Allen, J. M., Aqueous-phase photochemical sources of peroxyl radicals and singlet molecular oxygen in clouds and fog, *J. Geophys. Res.*, 97, 12913, 1992.

80. Wells, M. L. and Mayer, L. M., The photoconversion of colloidal iron oxyhydroxides in seawater, *Deep Sea Res.*, 38, 1379, 1991.

81. Braterman, P. S., Cairns-Smith, A. G., Sloper, R. W., Truscott, T. G., and Craw, M., Photo-oxidation of iron(II) in water between pH 7.5 and 4.0, *J. Chem. Soc. Dalton Trans.*, 1441, 1984.

82. Mauzerall, D. C., The photochemical origins of life and photoreaction of ferrous ion in the archaen oceans, *Origins Life Evol. Biosphere*, 20, 293, 1990.

83. Huie, R. E., and Neta, P., Rate constants for some oxidations of S(IV) by radicals in aqueous solutions, *Atmos. Environ.*, 21, 1743, 1987.

84. Gunz, D. W. and Hoffmann, M. R., Atmospheric chemistry of peroxides: a review, *Atmos. Environ.*, 24A, 1601, 1990.

85. Kieber, R. J., and Helz, G. R., Indirect photoreduction of aqueous chromium(VI), *Environ. Sci. Technol.*, 26, 307, 1992.

86. Hatchard, C. G. and Parker, C. A., A new sensitive chemical actinometer II. Potassium ferrioxalate as a standard chemical actinometer, *R. Soc. London Proc.*, A235, 518, 1956.

87. McConnell, J. C., Henderson, G. S., Barrie, L., Bottenheim, J., Niki, H., Langford, C. H., and Templeton, E. M. J., Photochemical bromine production implicated in Arctic boundary-layer ozone depletion, *Nature (London)*, 355, 150, 1992.

88. Bächmann, K., Röder, A., and Haag, I., Determination of anions and cations in individual raindrops, *Atmos. Environ.*, 26A, 1795, 1992.

89. The MAP3S/RAINE Research Community, The MAP3S/RAINE precipitation chemistry network: statistical overview for the period 1976–1980, *Atmos. Environ.*, 16, 1603, 1982.

90. Peterson, J. T., Calculated Actinic Fluxes (290–700 nm) for Air Pollution Photochemistry Applications, Report EPA-600/4–76–025, U.S. Environmental Protection Agency, Washington, D.C., 1976.

91. Milburn, R. M. and Vosburgh, W. C., A spectrophotometric study of the hydrolysis of iron(III) ion. II. Polynuclear species, *J. Am. Chem. Soc.*, 77, 1352, 1955.

92. Langmuir, D., The Gibbs free energies of substances in the system $Fe-O_2-H_2O-CO_2$ at 25°C, *U.S. Geol. Surv. Prof. Pap.*, 650-B, B180, 1969.

93. Byrne, R. H. and Kester, D. R., Solubility of hydrous ferric oxide and iron speciation in seawater, *Mar. Chem.*, 4, 255, 1976.

94. Byrne, R. H. and Kester, D. R., A potentiometric study of ferric ion complexes in synthetic media and seawater, *Mar. Chem.*, 4, 275, 1976.

95. Smith, R. M. and Martell, R. M., *Critical Stability Constants, Vol. 4: Inorganic Complexes*, Plenum Press, New York, 1976.

96. Wagman, D. D., Evans, W. H., Parker, V. B., Schumm, R. H., Halow, I., Bailey, S. M., Churney, K. L., and Nuttall, R. L., The NBS tables of chemical thermodynamic properties: selected values for inorganic and C_1 and C_2 organic substances in SI units, *J. Phys. Chem. Ref. Data*, 11 (Suppl. 2), 1982.

97. Felmy, A. R., Girvin, D. C., and Jenne, E. A., MINTEQ — A Computer Program for Calculating Aqueous Geochemical Equilibria, EPA-600/3-84-032, U.S. Environmental Protection Agency, Athens, GA, 1985.

98. Breslin, V. T. and Duedall, I. W., Metal release from particulate oil ash in seawater, *Mar. Chem.*, 22, 31, 1987.

99. Knight, R. J. and Sylva, R. N., Spectrophotometric investigation of iron(III) hydrolysis in light and heavy water at 25°C, *J. Inorg. Nucl. Chem.*, 37, 779, 1975.

100. Maslowska, J., Mieszane kompkeksy zelaza(III). I. Spektrofotometryczne badanie ukladu $Fe(ClO_4)_3$-NaBr-Na_2SO_4-H_2O, *Rocz. Chem.*, 41, 1857, 1967.

101. Bielski, B. H., Cabelli, D. E., and Arudi, R. L., Reactivity of HO_2/O_2^- radicals in aqueous solution, *J. Phys. Chem. Ref. Data*, 14 (Suppl.), 1041, 1985.

102. Løgager, T., Holcman, J., Sehested, K., and Pedersen, T., Oxidation of ferrous ions by ozone in acidic solutions, *Inorg. Chem.*, 31, 3523, 1992.

103. Wells, C. F. and Salam, M. A., Complex formation between Fe(II) and inorganic anions. Part 1. Effect of simple and complex halide ions on the Fe(II) + H_2O_2 reaction, *Trans. Faraday Soc.*, 63, 620, 1967.

104. Buxton, G. V., Greenstock, C. L., Helman, W. P., and Ross, A. B., Critical review of rate constants for reactions of hydrated electrons, hydrogen atoms and hydroxyl radicals ($\cdot$OH/$\cdot$O$^-$) in aqueous solution, *J. Phys. Chem. Ref. Data*, 17, 513, 1988.

105. Neta, P., Huie, R. E., and Ross, A. B., Rate constants for reactions of peroxyl radicals in fluid solutions, *J. Phys. Chem. Ref. Data*, 19, 413, 1990.

106. Rush, J. D., Maskos, Z., and Koppenol, W. H., Distinction between hydroxyl radical and ferryl species, *Methods Enzymol.*, 186, 148, 1990.

107. Stumm, W. and Morgan, J. J., *Aquatic Chemistry*, Wiley, New York, 1981, 467.

108. Chameides, W. L. and Davis, D. D., The free radical chemistry of cloud droplets and its impact upon the composition of rain, *J. Geophys. Res.*, 87, 4863, 1982.

109. Jacob, D. J., Chemistry of OH in remote clouds and its role in the production of formic acid and peroxymonosulfate, *J. Geophys. Res.*, 91, 9807, 1986.

CHAPTER 2

Photoredox Transformations of Iron and Manganese in Marine Systems: Review of Recent Field Investigations

T. D. Waite and R. Szymczak

INTRODUCTION

Thermodynamically, both iron (Fe) and manganese (Mn) are expected to predominate in oxidized form in seawater with iron present as Fe(III) oxyhydroxides and manganese present as Mn(III) and/or Mn(IV) oxides. However, there is now convincing evidence that light may induce significant departures from these thermodynamically expected forms.

Both the rates of oxidation of Mn(II) and the reduction of Mn(III,IV) oxides are now recognized to be influenced by light. Thus, the rate of oxidation, which is mediated by bacteria in marine waters, appears to be significantly suppressed by solar radiation.[1,2] In addition, light may also enhance the rate of reductive dissolution of particulate manganese oxides.[1,3,4]

Light has also been shown to solubilize iron oxides under acidic conditions,[5-7] but the tendency of photoproduced Fe(II) to adsorb to particulate surfaces, and to undergo rapid surface-assisted reoxidation to Fe(III), appears to limit the extent of light-assisted dissolution at seawater pH. While there is some evidence that light may induce the formation of measurable concentrations of Fe(II) in seawater,[8-10] it is more likely that a more significant effect may be the formation of a relatively soluble coating of amorphous iron oxides on suspended particulate matter.[11]

In this chapter, we first present the results of recent studies of specific (photo)redox processes involving iron and manganese in surface seawaters. Then in an attempt to assess the conjunctive effects of these processes in seawaters, we present and discuss the results of diurnal studies of iron and manganese speciation and partitioning between filterable and nonfilterable phases in a variety of marine systems.

EXPERIMENTAL

Process Investigations

Investigations of iron and manganese redox processes were undertaken aboard the R. V. Columbus Iselin in eastern Caribbean waters in the vicinity of the Orinoco River outflow region in late 1988. Oxidation processes were studied by addition of radiolabeled iron or manganese salts to surface (2 m) seawater samples (obtained from a Teflon™* bow line), while dissolution processes were investigated using preformed radiolabeled iron and manganese oxides.

The effects of light on both particle formation and dissolution processes were investigated using 365-nm illumination provided by an Osram XBO 100-W mercury lamp mounted in a Rofin arc lamp and power supply system. The 365-nm mercury line was isolated by passing the beam through a narrow bandpass filter. The intensity of the light entering the reaction vessel was determined by potassium ferrioxalate actinometry to be 96 μEinsteins/cm²/min.

The change in partitioning between solid and solution phases in these studies was determined using filtration followed by γ-counting of the iron or manganese activity in filtrate or filtrand as appropriate. Complete descriptions of the procedures used in these studies are given elsewhere.[12,13]

Diurnal Studies

In an attempt to assess the importance of light-driven processes in influencing the speciation of iron and manganese in seawater, we have investigated the variation in speciation and/or filterability of naturally occurring iron and manganese in a number of marine systems over a diurnal cycle.

The marine systems chosen for study are such that investigations are conducted on a unique water mass with mixing between water masses minimized as much as possible. The four systems chosen for study are as follows:

- System 1 — 50-L tanks of surface water from locations within the Great Barrier Reef along Australia's northeastern coastline placed on the deck of the vessel used for sampling. Three different waters have been used, each with different suspended sediment load.
- System 2 — A 15,000-m³ aquarium of the Great Barrier Reef Marine Park Authority in Townsville, Australia, with natural light input and designed to closely simulate conditions on the Great Barrier Reef.

---

*Teflon™ is a registered trademark of E. I. du Pont Nemours and Company, Inc., Wilmington, DE.

- System 3 — The One Tree Island coral reef located in the southern region of the Great Barrier Reef. This reef is relatively unique in that waters are enclosed within the lagoon for almost a complete tidal cycle. Studies have been conducted on the "pie crust" near the outer rim of the reef where the water depth over the reef is particularly shallow.

Iron and manganese analyses were performed on samples using a method similar to that described by Sturgeon et al.[14] with modifications as outlined below. Surface seawater samples were collected using acid-cleaned 500-mL polypropylene bottles attached to a metal-free sampling rod and, immediately upon return to the laboratory (a delay of typically 5–10 min), passed directly through duplicate 0.22-μm Millipore filters arranged in parallel using a fully enclosed acid-washed filtration system incorporating Teflon tubing and polypropylene (Millipore, Swinnex) filter holders. The initial 50 mL passing through each filter was discarded, then subsequent 100-mL portions were passed at 15 mL min^{-1} through C18-bonded silica SEP-PAK cartridges (Waters Associates) containing preadsorbed 8-hydroxyquinoline (5 mL, 0.05 M in methanol) and placed immediately after the Millipore membrane filters. The cartridges were placed in acid-washed polypropylene vials for return to the ANSTO laboratories, where they were eluted with 5 mL of 5% HNO_3 (Merck Suprapur) and the iron and manganese concentrations in the eluates were determined by graphite furnace atomic absorption spectrometry (GFAAS) (Varian AA975). All data were corrected for total process blanks (90 $\pm$ 20 pmol for iron and 63 $\pm$ 9 pmol for manganese) and for a recovery efficiency of 96%.

Hydrogen peroxide concentrations were also determined in some instances using a modified version of the scopoletin-horseradish peroxidase method.[15]

RESULTS AND DISCUSSION

Process Studies

Manganese

Mn(II) Oxidation. Particle generation following addition of approximately 1.5 nM $MnCl_2$ to surface seawater samples from eastern Caribbean waters typically exhibited linear time dependence. While added insight into the processes involved in particulate manganese formation is needed, the most likely mechanism involves initial uptake of Mn(II) on particle surfaces, followed by heterogeneous catalysis of the adsorbed Mn(II) to Mn oxyhydroxides.[12] Bacteria are known to catalyze the oxidation of Mn(II), and the particulates to which Mn(II) are initially adsorbed may be bacteria or other surfaces with a resident bacterial population.

Linearity of particulate manganese formation suggests that adequate surface sites are available for continuing Mn(II) uptake. New surface sites may obviously be generated through the formation of fresh manganese oxyhydroxides and,

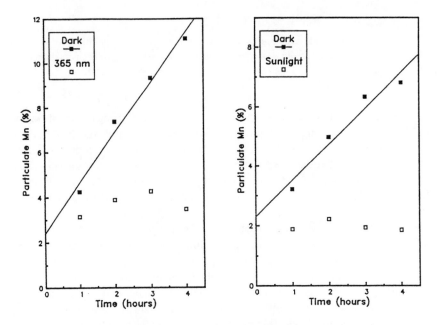

Figure 1. Effect of 365-nm and sunlight illumination on the percentage of radiolabeled
Mn(II) added to samples from eastern Caribbean Station 41 that is retained on
a 0.22-μm pore size Millipore filter. (From Waite, T. D. and Szymczak, R., *J.
Geophys. Res.*, 98, 2361,1992. With permission.)

indeed, is the most likely scenario accounting for the observed constancy in
particulate manganese formation rates. Manganese(II) uptake on manganese
oxyhydroxides (or indeed on any surface) is expected to occur at rates significantly higher than rates of particulate manganese formation,[4] suggesting that the
rate-limiting step in these studies is that of heterogeneous (most likely bacterially
mediated) Mn(II) oxidation.

As previously reported by Sunda and Huntsman,[2] illumination induces
nonlinearity in particle generation, presumably by inhibiting bacterial production
of fresh manganese oxyhydroxides. As shown in Figure 1, nonlinearity is induced
both by sunlight and by 365-nm radiation on photolysis of samples from Station
41 (see Waite and Szymczak[12] for station positions) with maxima in the proportion of particulate manganese present after 2–3 hr. Thereafter, the proportion of
particulate manganese decreases on continued illumination.

MnO$_x$ Photodissolution.   As shown for samples from selected eastern Caribbean stations in Figure 2, essentially constant rates of 365-nm-induced dissolution
of approximately 2 nM of preformed manganese oxides added to the samples are
observed. (Zero or very low dissolution rates were observed in the dark.)

The dissolution process is believed to occur as a result of excitation of natural
organic substances adsorbed to the surfaces of manganese oxides. Such excitation
contributes to an enhanced rate of intramolecular electron transfer, resulting in

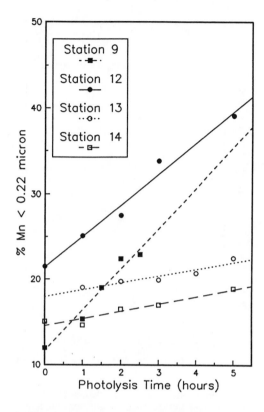

Figure 2. Effect of 365-nm photolysis on extent of dissolution of radiolabeled particulate manganese oxide added to seawater samples from Stations 9, 12, 13, and 14. (From Waite, T. D. and Szymczak, R., *J. Geophys. Res.*, 98, 2361,1992. With permission.)

Mn(II) formation at the oxide surface. Depending on the affinity of the oxide surface for manganous ions, a portion of the Mn(II) produced will be rapidly released to solution, resulting in dissolution of the oxide.[4]

Under the conditions of these experiments, a direct proportionality between the rate of oxide dissolution and total concentration of organic reducing agent is to be expected. While the data set is not large, such a proportionality is observed (Figure 3; $R^2 = 0.94$) (assuming a proportionality between organic concentration and absorbance at 365 nm).

A comparison of the absolute rates of Mn(II) oxidation and Mn(III,IV) oxide photodissolution for waters from various eastern Caribbean stations is shown in Table 1 and indicates that particulate manganese dissolution rates are generally significantly greater than particulate manganese formation rates. As such, factors influencing the dissolution of particulate manganese oxides (concentration, surface affinity and photoreactivity of organic matter, light regime, etc.) are expected to have a significant impact on the steady-state partitioning of manganese between the dissolved and particulate phase.

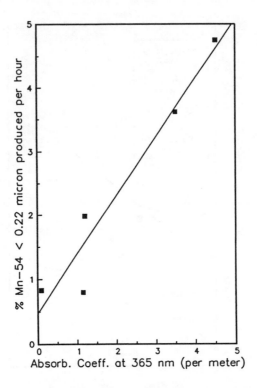

Figure 3. Correspondence between 365-nm absorbance and rate of photodissolution of added particulate manganese oxide for samples from Stations 9, 12, 13, 14, and 16. (From Waite, T. D. and Szymczak, R., *J. Geophys. Res.*, 98, 2361,1992. With permission.)

Iron

Fe(II) Oxidation and Fe(III) Hydrolysis.   Addition of both $^{59}FeCl_3$ and $^{59}FeCl_2$ to eastern Caribbean surface seawater samples (each to a concentration of approximately 8.5 n*M*) resulted in particle generation as indicated by an increasing percentage of ^{59}Fe captured on 0.2-μm Nuclepore filters (see Figure 4) and presumably occurred as a result of oxide formation followed by aggregation of these oxides to a filterable size.

Undertaking these oxidation-particle growth studies under 365-nm illumination resulted in a significant reduction in extent of particle capture (0.06 min^{-1} in the light compared to 0.02 min^{-1} in the dark).

A number of hypotheses can be developed to account for such an effect including:

1. lowered oxidation potential of seawater medium on illumination
2. increased reducing ability of naturally occurring humic substances
3. modification in surface charge properties of oxides on illumination
4. continual light-induced dissolution of freshly formed Fe(III) oxides

Table 1. Comparison of Absolute Rates of Particulate Manganese Formation and Dissolution at Selected Eastern Caribbean Stations

Station No.	Particulate Mn Formation Rate (pmol/L/hr)	Particulate Mn Dissolution Rate (pmol/L/hr)
7	0.76	180[a]
9	0.26	190
12	45	1340
13	11	21.6
14	5.8	4.4
16	7.7	148
16[b]	—	<7.5
30	36	109
30[b]	—	32.2
39	12	44.7
39[c]	0.07	49.1
41	16	—
41[c]	—	18.1
44	1.8	—
44[c]	—	21.2

Adapted from Waite, T. D. and Szymczak, R., *J. Geophys. Res.*, 98, 2361,1992.

[a] Assuming "natural" manganese oxides react as similar rates to radiolabeled preformed oxide.

[b] No illumination.

[c] Sample prefiltered through a 0.22-μm membrane.

Some support for the latter two hypotheses is offered by the results presented in Figure 4A, which indicate that light also retards Fe(III) hydrolysis and particle growth on addition of 8.5 nM $^{59}FeCl_3$ to the same seawater sample.

FeO$_x$ Photodissolution. No measurable increase in 0.2-μm filterable iron was observed on photolysis of approximately 11 nM of preformed FeO$_x$ added to surface seawater samples. Any net release of Fe(II) from particle surfaces as a result of photoreductive dissolution may be expected to be small and, if occurring in these systems, is presumably below the detection limit of this tracer technique (0.1 nM).

Indeed, recent investigations of the diurnal variability of ferrozine-reactive iron in northern Australian surface waters[16] indicates very low concentrations of "reactive" iron in oligotrophic waters (≤0.05 nM) and somewhat higher light-dependent concentrations in more productive coastal waters. For example, in the Gulf of Carpentaria, ferrozine-reactive iron concentrations increased from 0.37 ± 0.10 nM at 6:00 a.m. to 0.86 ± 0.14 nM at 12 noon (zenith) and returned to 0.14 ± 0.10 nM at 6:00 p.m. Essentially, all the increase in ferrozine-reactive iron was due to an increase in soluble iron with the concentration of particle-bound ferrozine-reactive iron remaining approximately constant.

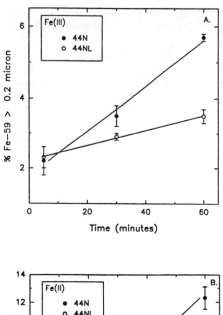

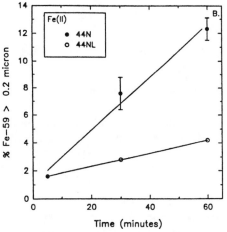

Figure 4.   Percentage of ^{59}Fe that is retained on a 0.20-μm Nuclepore membrane filter as a function of time after addition of (A) ^{59}Fe(III) and (B) ^{59}Fe(II) to surface seawater from Station 44 either maintained in darkness (44) or illuminated with 365-nm light (44 L). The ^{59}Fe addition in each case represents an addition of 8.5 nM total iron. (From Waite, T. D. and Szymczak, R., *J. Geophys. Res.*, 98, 2371,1992. With permission.)

Diurnal Studies

System 1 — 50 L Tanks

The results from studies of iron and manganese in 50-L tanks exposed to natural sunlight on the deck of the research vessel, the R. V. Lady Basten, are given in

Figures 5A, 5B and 5C. In addition to the concentrations of iron and manganese that pass through a 0.22-μm Millipore filter at various times of the day (and night), we also report observed concentrations of hydrogen peroxide.

For each tank study, there is a distinct increase in concentration of both iron and manganese in the less than 0.22-μm fraction throughout the day. Hydrogen peroxide concentrations are observed to also increase throughout the day at rates ranging from 12.5 nM hr^{-1} in Tank 1 to 27 and 23 nM hr^{-1} in Tanks 2 and 3, respectively.

Concentrations of manganese in the less than 0.22 μm fraction increase from nighttime values on the order of 0.2 nM in each case at rates ranging from 50 pM hr^{-1} (Tank 1) to 170 pM hr^{-1} (Tank 2) and reach peak concentrations of approximately 1 nM at 4:00 p.m. in Tank 1, 1.5 nM at 8:00 p.m. in Tank 2, and 1.7 nM at 6:00 p.m. in Tank 3.

In the case of manganese, where little homogeneous chemical oxidation of Mn(II) is expected, it is reasonable to conclude that we are observing changes from a dissolved pool of manganous manganese to bacterially produced manganese-rich particles. Thus, it is apparent that light is producing relatively dramatic changes in the distribution of manganese between dissolved and particulate forms on a daily basis.

Once a maximum in the dissolved manganese concentration is reached, the concentration of manganese in the filtrate then falls slowly through the night, presumably as the ability of bacteria to catalyze the oxidation of Mn(II) returns to full capacity.

Filterable iron concentrations are somewhat more erratic than those of manganese and (from our earlier process discussions) may reflect the generation of less than 0.22-μm colloidal particles either as a result of a continuous formation of fresh iron oxyhydroxides or as a result of light-induced disaggregation processes. The possibility of increase in the steady-state concentration of soluble Fe(II) as a result of photoreductive processes should also not be ignored. Interestingly, the apparent light-induced increase in concentration of filterable (<0.22 μm) iron and manganese is independent of total iron and manganese concentrations.

In comparison to the relatively slow increase in proportion of particulate manganese present on removal of light, the concentration of iron in the less than 0.22-μm fraction drops rapidly once a maximum has been achieved. Many scenarios can be proposed at this stage to account for such changes in iron partitioning. For example, does light alter slightly the affinity of Fe(II) for particle surfaces at seawater pH or affect the extent of particle-particle interactions (as proposed earlier)? Alternatively, does the light-induced production of H_2O_2 influence the rate of oxidation of Fe(II) and, thence, the steady-state concentration of Fe(II)?

Systems 2 and 3 — Aquarium and Coral Reef

Results from these systems are given in Figures 6A and 6B and, while a little more erratic, show similar trends to those reported for the 50-L tank studies.

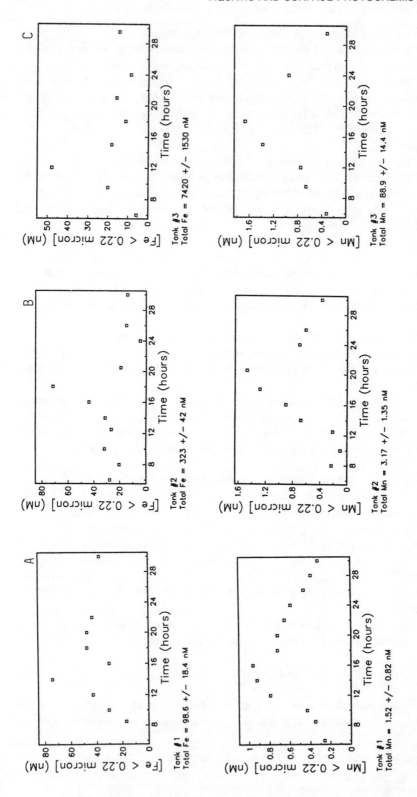

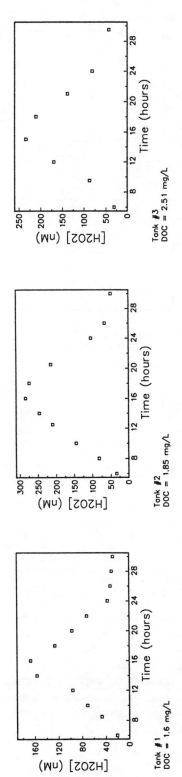

Figure 5. Variation in 0.22-μm filterable iron and manganese and hydrogen peroxide concentrations in 50-L tanks exposed to natural light and containing surface seawaters from selected locations of the Great Barrier Reef. Significant differences in suspended sediment concentrations were evident between the three samples as indicated by **(A)**. Tank 1 containing approximately 100 nM total iron, **(B)** Tank 2 containing approximately 320 nM total iron, and **(C)** Tank 3 containing over 7000 nM total iron.

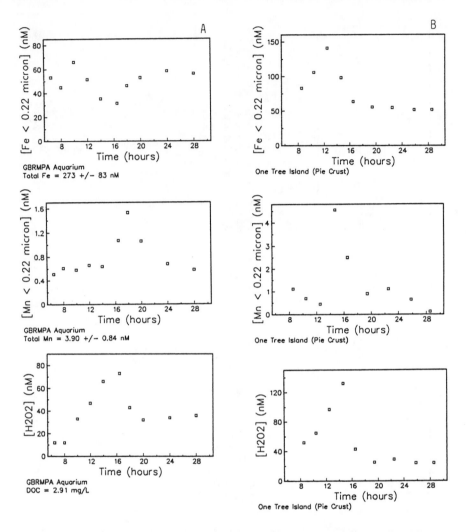

Figure 6. Variation in 0.22-μm filterable iron and manganese and hydrogen peroxide
concentrations in **(A)** Great Barrier Reef Marine Park Authority aquarium "reef
tank" in Townsville and **(B)** in surface waters from the "pie crust" of One Tree
Island, a semienclosed coral reef in the southern portion of the Great Barrier Reef
in northern Australia.

In the case of the aquarium of the Great Barrier Reef Marine Park Authority,
distinct peaks in filterable manganese and hydrogen peroxide concentrations are
observed, but no clear diurnal variations in iron partitioning between less than
0.22 and greater than 0.22 μm size classes can be documented.

In the One Tree Island case, a coral reef in the southern portion of the Great
Barrier Reef, sharp maxima in concentrations of hydrogen peroxide and filterable
(<0.22 μm) iron and manganese are evident with relatively high concentrations

(at least compared to the tank studies) in the less than 0.22 μm fraction. In this case, light certainly reaches the bottom sediments with the possibility of photoassisted elemental release.

SUMMARY

Some clarification of the relative importance of various photoredox processes has been achieved through batch studies using surface seawaters. Thus, light is observed to lower the rate of oxidation of Mn(II) by inhibiting bacteria catalyzing this heterogeneous process. In addition, light increases the rate of reductive dissolution of particulate manganese oxides with the rate of reduction exhibiting a direct proportionality to the organic content of the medium (or more particularly, the near-UV absorbing ability of surface located charge-transfer complexes). A comparison of the absolute rates of Mn(II) oxidation and Mn(III,IV) oxide photodissolution for waters from various eastern Caribbean stations is shown in Table 1 and indicates that particulate manganese dissolution rates are generally significantly greater than particulate manganese formation rates. As such, factors influencing the dissolution of particulate manganese oxides (concentration, surface affinity and photoreactivity of organic matter, light regime, etc.) are expected to have a significant impact on the steady-state partitioning of manganese between dissolved and particulate phase.

Light is observed to retard the hydrolysis-aggregation processes leading to particle formation and growth on addition of soluble ferric ions to seawater medium. Since similar phenomena are observed on addition of both Fe(II) and Fe(III) to seawater, it appears likely that either (1) illumination increases the surface charge properties of iron oxides, leading to a retardation in rate of aggregation, and/or (2) a continual light-induced dissolution of freshly formed Fe(III) oxides occurs retarding the growth of ferric oxyhydroxide particles.

The extent of photoreductive dissolution of iron oxides in seawater appears to be limited by the tendency of Fe(II) to be retained at the particle surface. However, in some waters, and particularly those that appear relatively productive, measurable increases in the steady-state concentration of "ferrozine-reactive" iron are observed, while small, distinct increases in concentration of ferrozine-reactive iron are observed at times of elevated solar irradiance.

Diurnal studies of the partitioning of manganese between 0.22-μm filtrand and filtrates indicates a strong dependency on light intensity. This is to be expected given the ability of light to retard both Mm(II) oxidation and to induce the reductive dissolution of Mn(III,IV) oxides. Since the vast majority of particulate manganese occurs in relatively large "clumps" around the oxidizing bacteria, it is reasonable to conclude that we are observing a change from a dissolved pool of manganous manganese to bacterially produced manganese-rich particles.

Such a conclusion is not possible for iron for which nucleation-aggregation processes are considerably more important. The rather erratic (though apparently light dependent) variation in filtrand-filtrate partitioning through the course of a day most likely reflects the sensitivity of the redox, nucleation, and aggregation-disaggregation processes to physicochemical conditions in the seawater medium. Thus, concentrations of oxidants (such as hydrogen peroxide) and reductants (such as organic acids) in the seawater medium may vary with (light dependent) biological activity. In addition, organic acids and photoproduced Fe(II) may alter the surface properties (and hence the interactive behavior) of particulate iron oxides.

Particularly in the case of iron, more extensive studies in which a wide range of parameters are monitored under any given set of conditions are needed in order to elucidate the importance of the various proposed processes.

REFERENCES

1. Sunda, W. G. and Huntsman, S. A., Effect of sunlight on redox cycles of manganese in the southwestern Sargasso Sea, *Deep Sea Res.*, 35, 1297, 1988.
2. Sunda, W. G. and Huntsman, S. A., Diel cycles in microbial manganese oxidation and manganese redox speciation in coastal waters of the Bahama Islands, *Limnol. Oceanogr.*, 35, 325, 1990.
3. Sunda, W. G., Huntsman, S. A., and Harvey, G. R., Photoreduction of manganese oxides and the supply of manganese to marine plants, *Nature (London)*, 301, 234, 1983.
4. Waite, T. D., Wrigley, I. C., and Szymczak, R., Photoassisted dissolution of a colloidal Mn oxide in the presence of fulvic acid, *Environ. Sci. Technol.*, 22, 778, 1988.
5. Waite, T. D. and Morel, F. M. M., Photoreductive dissolution of colloidal iron oxides in natural waters, *Environ. Sci. Technol.*, 18, 860, 1984.
6. McKnight, D. M., Kimball, B. A., and Bencala, K. E., Iron photoreduction and oxidation in an acidic mountain stream, *Science*, 240, 637, 1988.
7. Sulzberger, B. D., Suter, C., Siffert, S., Banwart, S., and Stumm, W., Dissolution of Fe(III) (hydr)oxides in natural waters: laboratory assessment on the kinetics controlled by surface chemistry, *Mar. Chem.*, 28, 127, 1989.
8. Hong, H. and Kester, D. R., Redox state of iron in the offshore waters of Peru, *Limnol. Oceanogr.*, 31, 512, 1986.
9. O'Sullivan, D., Hanson, A. K., Miller, W. L., and Kester, D. R., Measurement of Fe(II) in surface water of the equatorial Pacific, *Limnol. Oceanogr.*, 36, 1727, 1991.
10. Kuma, K., Nakabayashi, S., Suzuki, Y., Kudo, I., and Matsunaga, K., Photoreduction of Fe(II) by dissolved organic substances and existence of Fe(II) in seawater during spring blooms, *Mar. Chem.*, 37, 15, 1992.
11. Wells, M. L. and Mayer, L. M., The photoconversion of colloidal iron oxyhydroxides in seawater, *Deep Sea Res.*, 1993, in press.
12. Waite, T. D. and Szymczak, R., Manganese dynamics in surface waters of the eastern Caribbean, *J. Geophys. Res.*, 98, 2361, 1992.
13. Waite, T. D. and Szymczak, R., Particulate iron formation dynamics in surface waters of the eastern Caribbean, *J. Geophys. Res.*, 98, 2371, 1992.
14. Sturgeon, R. E., Berman, S. S., Willie, S. N., and Desaulniers, J. A. H., Preconcentration of trace elements from seawater with silica-immobilized 8-hydroxyquinoline, *Anal. Chem.*, 53, 2337, 1981.
15. Szymczak, R. and Waite, T. D., Photochemical activity in waters of the Great Barrier Reef, *Estuarine Coastal Shelf Sci.*, 33, 6 05, 1991.
16. Espey, Q. I., Szymczak, R., and Waite, T. D., Bioavailable forms of iron in surface wateats from shelf and coral reef waters of northern Australia, *Mar. Chem.*, submitted.

CHAPTER 3

Photoredox Reactions at the Surface of Iron(III) (Hydr)Oxides

Barbara Sulzberger, Hansulrich Laubscher, and Georgios Karametaxas

SIGNIFICANCE OF PHOTOREDOX REACTIONS AT THE SURFACE OF IRON(III) (HYDR)OXIDES FOR NATURAL WATER SYSTEMS

Iron is one of the most abundant elements in the environment. In oxic aquatic systems, iron is predominantly present in particulate form. These Fe(III) (hydr)oxides have large surface areas, up to several hundred square meters per gram, and may specifically adsorb organic and inorganic compounds. In the presence of light, these adsorbed compounds may become oxidized. In many cases, the photochemical oxidation of an adsorbed compound is accompanied by the photochemical reductive dissolution of the solid phase, yielding dissolved Fe(II). In the presence of oxygen, hydrogen peroxide, or other suitable oxidants, Fe(II) undergoes oxidative precipitation. The reoxidation of Fe(II) may produce an $Fe(OH)_3$ that is less polymeric and less crystalline than aged Fe(III) hydroxides. The product may be more soluble and quicker to approach equilibrium with monomeric Fe(III) species and may control iron uptake by phytoplankton.[1,2] Complexation by organic ligands, the extent of which is largely unknown in seawater,[3] may stabilize ferrous iron or control the reactivity of dissolved ferric iron. Commercial humic acid was found to have sufficient affinity for iron to compete with Fe(III) hydrolysis in seawater, and limited evidence was obtained for an interaction with dissolved organic matter in coastal seawater.[4] The formation of bioavailable iron is an important issue in seawater; iron is essential for the growth of phytoplankton because it is required for the synthesis of chlorophyll and nitrate reductase. Iron deficiency is postulated in offshore areas ranging from the tropical equatorial Pacific to the polar Antarctic, where major nutrients occur in

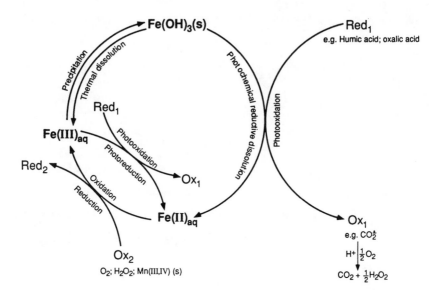

Figure 1.   Schematic representation of the photoredox cycling of iron. The photochemical formation of Fe(II) may proceed through different pathways: (1) through the photochemical reductive dissolution of Fe(III) (hydr)oxides and (2) through photolysis of dissolved Fe(III) coordination compounds that are formed via oxidation of Fe(II) or via thermal, nonreductive dissolution of particulate iron, a reaction that is catalyzed by Fe(II).

excess.[5] Overall, the iron photoredox cycle catalyzes the oxidation of reduced compounds, i.e., of organic carbon (Figure 1). Thus, in the photic zone of surface waters, the geochemical iron cycle is linked to that of carbon. Mopper et al.[6] have suggested that the photochemical degradation pathway is the rate-limiting process for the removal of a large fraction of oceanic dissolved organic carbon. The involvement of iron in this photochemical degradation appears likely, although no data are available yet. The iron-mediated oxidation of organic compounds is a source of H_2O_2 through the reaction of the primary photoproduct, which is mostly a radical, with oxygen, yielding $\cdot O_2^-$, followed by protonation and disproportionation reactions.[7] In atmospheric water, the photoredox cycle of iron is not only linked to the carbon cycle, but also to the sulphur cycle.[8] It has been proposed,[9-11] that the oxidation of S(IV) by O_2 in atmospheric water is catalyzed by Fe(III) (hydr)oxide particles. Due to comparatively low pH-values in atmospheric water, a high fraction of total iron is present as dissolved Fe(II). Maximal concentrations of 0.2 mM of Fe(II) were reported in fog water, with the concentration of Fe(II) increasing both with decreasing pH and with the light intensity.[12] It is assumed that the atmospheric transport of continental weathering products is responsible for much of the mineral material and iron entering the open ocean and that it is probably the dominant source of the nutrient iron in the photic zone.[13] Thus, photoredox reactions of iron are important in global iron, electron, and carbon cycling in atmospheric and surface waters.

GENERAL RATE EXPRESSIONS FOR HETEROGENEOUS PHOTOREDOX REACTIONS OCCURRING AT MINERAL SURFACES

The photochemical oxidation of compounds that are adsorbed at mineral surfaces may proceed through different pathways: (1) ligand-to-metal charge-transfer process within a surface complex formed between a surface metal ion of a metal (hydr)oxide and a ligand that is specifically adsorbed at the surface of the solid phase; (2) photooxidation of adsorbed species by valence-band holes of a semiconducting mineral; the photoelectrons in the conduction band are either transferred to adsorbed molecular oxygen or scavenged by surface metal centers, resulting eventually in detachment of the reduced surface centers; (3) oxidation of adsorbed species by OH· radicals, resulting from the scavenging of the photoholes by surface hydroxyl groups. Which of these pathways prevails may depend strongly on whether a specific or a nonspecific adsorption of the reductant takes place, i.e., a chemical interaction at the solid-water interface in terms of surface complexation may favor pathway (a) and (b).[14-16] Pathway (a) may be looked at as a unimolecular process, where electron-transfer occurs within the primary chromophore, whereas pathways (b) and (c) are bimolecular processes, where electron transfer occurs between an electronically excited state of a metal (hydr)oxide and a quencher of the electronically excited state, i.e., the electron donor. General schemes of a unimolecular and of a bimolecular heterogeneous photoredox process are shown in Figures 2 and 3, respectively. (The notation "primary chromophore" is used for the chromophore responsible for the photo-chemical reaction, i.e., the photochemical reaction occurs from one electronically excited state of the primary chromophore.) The rate of formation of primary photoproducts, which in the case of reducible metal (hydr)oxides is a reduced surface metal center and an oxidized reductant, is the product of the rate of light absorption by the primary chromophore, $I_{A\lambda}$, and the quantum yield, Φ_λ, at a given wavelength: λ, $R_\lambda = I_{A\lambda} \cdot \Phi_\lambda$. The primary chromophore is either a surface complex, >MR, or the metal (hydr)oxide bulk phase, **M**.

If we consider the kinetics of the photochemical reductive dissolution of a metal (hydr)oxide, then the rate depends in addition on the efficiency of detach-ment of the reduced surface metal centers from the crystal lattice, η_{DET}, since detachment of the reduced surface metal centers may be in competition with their oxidation by a suitable oxidant. The rate, R_λ, of formation of dissolved reduced metal species, M^{-1}, then for both pathways is

$$R_\lambda = I_{A\lambda} \cdot \Phi_\lambda \cdot \eta_{DET} \tag{1}$$

According to Beer-Lambert's law, the rate of light absorption by the primary chromophore, $I_{A\lambda}$, is given by Equations 2 and 3 for the uni- and bimolecular pathway respectively.

$$I_{A\lambda} = 2.303 \cdot I_{0\lambda} \cdot [>MR] \cdot \varepsilon_\lambda \cdot l_\lambda \qquad \text{if } 2.303 \cdot [>MR] \cdot \varepsilon_\lambda \cdot l_\lambda \ll 1 \tag{2}$$

$$I_{A\lambda} = 2.303 \cdot I_{0\lambda} \cdot [\mathbf{M}] \cdot \varepsilon_\lambda \cdot l_\lambda \qquad \text{if } 2.303 \cdot [\mathbf{M}] \cdot \varepsilon_\lambda \cdot l_\lambda \ll 1 \tag{3}$$

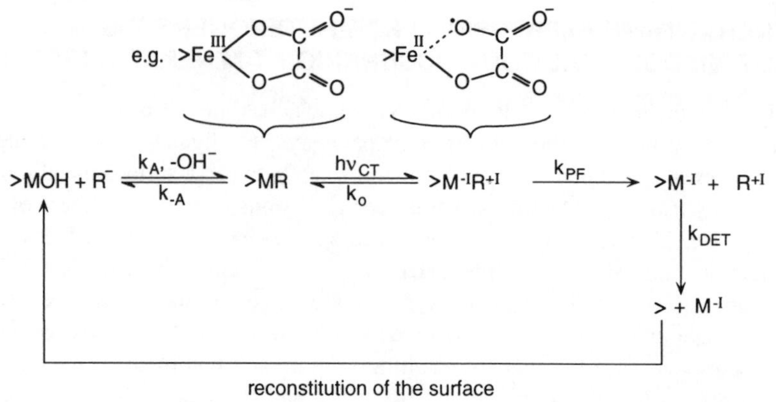

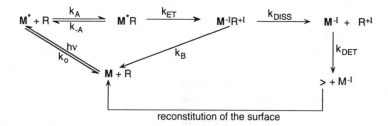

reconstitution of the surface

Figure 2. Scheme of a unimolecular heterogeneous photoredox process. >MOH stands for a surface metal center of a metal (hydr)oxide; R for a reductant; >MR for a surface complex, formed between a surface metal center and a reductant; and > for the remaining lattice surface of the metal (hydr)oxide after detachment of the reduced surface metal center. k_A is the rate constant of adsorption of the reductant at the metal (hydr)oxide surface, i.e., the rate constant of surface complex formation, e.g., of formation of an Fe(III) oxalato surface complex. The other symbols are explained throughout the text.

where $I_{0\lambda}$ in moles of photons per liter per hour is the volume-averaged light intensity that is available to the primary chromophore; ε_λ in liter per mole per centimeter is the molar decadic extinction coefficient of the primary chromophore, and l_λ is the average light path length in centimeters.

At steady state of the electronically excited state, the quantum yield can be expressed in terms of rate constants. In **unimolecular** processes, the quantum yield is the rate constant of primary **photoproduct** formation k_{PF}, i.e., of formation

reconstitution of the surface

Figure 3. Scheme of a bimolecular heterogeneous photoredox process. **M** stands for a metal (hydr)oxide; **M*R** for a reductant that is (specifically or nonspecifically) adsorbed at the surface of a metal (hydr)oxide in its electronically excited state (indicated with an asterisk); **M⁻ᴵ** for a reduced metal (hydr)oxide, i.e., the photo-electron is trapped at a surface metal center of the (hydr)oxide; and > for the remaining lattice surface of the metal (hydr)oxide after detachment of the re-duced surface metal center. The other symbols are explained throughout the text.

of a reduced surface metal center, $>M^{-I}$, and the oxidized reductant, R^{+I}, divided by the sum of all rate constants that deactivate the electronically excited state, i.e., the charge-transfer state, through chemical and physical processes:

$$\Phi_\lambda = \frac{k_{PF}}{k_{PF} + k_o} \tag{4}$$

where k_o stands for the sum of all the rate constants that deactivate the electronically excited state via physical processes. (The primary photoproduct is defined as the product that occurs in one step from an electronically excited state of the primary chromophore.) In **bimolecular** photoredox reactions occurring at mineral surfaces, the quantum yield of primary photoproduct formation is not merely a fraction of rate constants, but depends in addition on the concentration of the quencher of the electronically excited state, i.e., on the surface concentration of the adsorbed electron donor, which, for small solution concentrations, is directly proportional to the solution concentration of the reductant, [R]. The quantum yield of formation of the primary photoproduct, $M^{-I}R^{+I}$, is then

$$\Phi_{\lambda ET} = \frac{k_{ET} \cdot K_A [R]}{k_{ET} \cdot K_A [R] + k_o} \tag{5}$$

where k_{ET} is the rate constant of electron transfer from the adsorbed reductant to the metal (hydr)oxide in its electronically excited state, and K_A is the equilibrium constant of adsorption of the reductant at the mineral surface, $K_A = k_A/k_{-A}$. Furthermore, in a bimolecular heterogeneous photoredox reaction, the formation of the dissociated photoproducts, e.g., of a reduced metal (hydr)oxide, M^{-I}, and the oxidized reductant, R^{+I}, occurs in two steps: (1) electron transfer from the adsorbed electron donor to the metal (hydr)oxide in its electronically excited state, i.e., the photoholes are trapped by the adsorbed electron donor; and (2) dissociation of the primary photoproduct, $M^{-I}R^{+I}$. Thus, via a bimolecular pathway, the overall quantum yield of formation of an oxidized reductant, R^{+I}, is the product of the quantum yield of electron transfer and of the efficiency of dissociation of the oxidized reductant from the surface, η_{DISS}:

$$\Phi_\lambda = \Phi_{\lambda ET} \cdot \eta_{DISS} \tag{6}$$

For steady-state conditions of the primary photoproduct, $M^{-I}R^{+I}$, the efficiency of dissociation of the oxidized reductant from the surface is equal to the rate constant of dissociation, k_{DISS}, divided by the sum of the rate constants of disappearance of the primary photoproduct:

$$\eta_{DISS} = \frac{k_{DISS}}{k_{DISS} + k_B} \tag{7}$$

where k_B is the rate constant of back electron transfer to the ground state reactants.

As discussed, the rate expression for the photochemical reductive dissolution of a metal (hydr)oxide also includes the efficiency of detachment of the reduced surface metal center from the crystal lattice, η_{DET}. At steady state of the reduced surface metal centers, the efficiency of detachment is given by the following expression:

$$\eta_{DET} = \frac{k_{DET}}{k_{DET} + k_{Ox}[Ox]} \tag{8}$$

where k_{DET} is the rate constant of detachment of the reduced surface metal centers, k_{Ox} is the rate constant of their reoxidation, and [Ox] is the concentration of a suitable oxidant for the oxidation of the reduced surface metal centers. The rate expressions for the photochemical reductive dissolution of a metal (hydr)oxide via the uni- and the bimolecular pathway are then given by Equations 9 and 10, respectively:

$$R_\lambda = 2.303 \cdot I_{0\lambda} \cdot [>MR] \cdot \varepsilon_\lambda \cdot l_\lambda \cdot \frac{k_{PF}}{k_{PF} + k_o} \cdot \frac{k_{DET}}{k_{DET} + k_{Ox}[Ox]} \tag{9}$$

$$R_\lambda = 2.303 \cdot I_{0\lambda} \cdot [M] \cdot \varepsilon_\lambda \cdot l_\lambda \cdot \frac{k_{ET}K_A[R]}{k_{ET}K_A[R] + k_o}$$

$$\cdot \frac{k_{DISS}}{k_{DISS} + k_B} \frac{k_{DET}}{k_{DET} + k_{Ox}[Ox]} \tag{10}$$

As can be seen from these two rate expressions, irrespective of whether a uni- or a bimolecular heterogeneous photoredox reaction takes place, the rate of heterogeneous photoredox reactions where a reductant is oxidized at a mineral surface depends on the surface concentration of the electron donor.

If instead of being detached from the surface the reduced surface metal center is reoxidized, then the metal (hydr)oxide can be considered as a photocatalyst, PC, i.e., a light-absorbing species that enables a surface photoredox reaction, but remains unchanged after the overall process:

$$PC \longrightarrow PC* \tag{11}$$

$$PC* + D + A \longrightarrow PC + D^+ + A^{\cdot -} \tag{12}$$

Specific Adsorption of a Reductant at a Mineral Surface

Not only with regard to the potential functioning of a surface complex as a primary chromophore is the extent and type of adsorption of a reductant at a

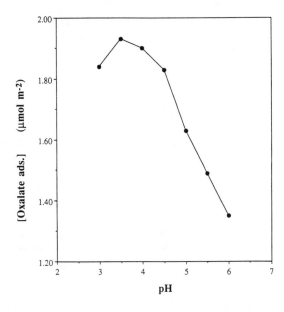

Figure 4. pH dependence of the extent of adsorption of oxalic acid at the surface of lepidocrocite (γ-FeOOH). Ionic strength: 5 mM (NaClO$_4$); temperature: 25°C.[54]

mineral surface of importance, but also with regard to the efficiency of charge separation. Using methylviologen (MV^{2+}) as the electron acceptor, Frei et al.[17] have reported that the yield of photochemical MV$^+$ formation with TiO$_2$ as photocatalyst was much higher in the presence of phenylfluorone as an electron donor that forms a bidentate surface complex at the surface of TiO$_2$ than in its absence where surface hydroxyl groups act as hole trappers.

The surface complex formation model[18,19] interprets specific adsorption in terms of the coordination chemistry at the solid-water interface, e.g., the chemical interaction of a central metal ion of the surface lattice with a ligand. Thereby, chemical bonds are formed. The stability of surface complexes, as with complexes in solution, can be quantified by mass action law constants. Spectroscopic investigations with magnetic resonance methods (EPR, ENDOR, ESEEM),[20] with EXAFS,[21,22] with FTIR,[23,24] and with STM and AFM[25] have confirmed the inner-sphere structure of different surface complexes. The surface complex formation model finds wide application.[26,27] The model permits the prediction of the distribution of reactive elements (e.g., reductants) between particle surfaces and the solution as a function of pH and other solution variables. The pH dependence of adsorption of weak acids and anions can be predicted with the help of the ligand exchange model from the acid-base equilibria of both the anion and the (hydr)oxide. Figure 4 shows the experimentally determined pH dependence of the adsorption of oxalic acid at the surface of lepidocrocite (γ-FeOOH). From this strong pH dependence of the extent of adsorption of a reductant at a mineral surface, we will expect a strong pH dependence of the rate of a photoredox reaction occurring at

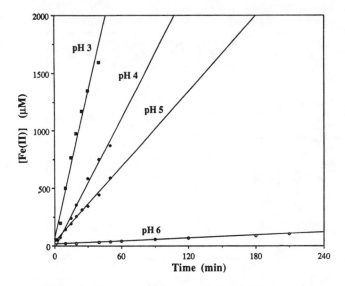

Figure 5. pH dependence of the rate of the photochemical reductive dissolution of γ-FeOOH with oxalate as the reductant in deaerated suspensions. Experimental conditions: 0.5 g L^{-1} γ-FeOOH; initial oxalate concentration: 1 mM; ionic strength: 5 mM (NaClO$_4$); temperature: 25°C; light source: white light from a 1000-W high-pressure xenon lamp that was filtered by the bottom window of the Pyrex glass vessel, which acts as cutoff filter ($\lambda_{1/2}$ = 350 nm); incident light intensity, I$_o$ ≈ 0.5 kW m^{-2}.[54]

a mineral surface. This is shown in Figure 5 for the photochemical reductive dissolution of lepidocrocite with oxalic acid as the reductant.

The specific adsorption of an electron donor at a mineral surface can often be described with a Langmuir isotherm. A Langmuir-type adsorption is consistent with the formation of a surface complex by surface ligand exchange:[28]

$$>\!MOH + HL^- \rightleftarrows \; >\!ML^- + H_2O; \qquad K^S = \frac{\{>\!ML^-\}}{\{>\!MOH\} \cdot [HL^-]} \qquad (13)$$

$$\{>\!ML^-\} = \frac{K^S S_T [HL^-]}{1 + K^S [HL^-]} \qquad (14)$$

where K^s is the conditional microscopic equilibrium constant of the adsorption equilibrium, and S$_T$ is the concentration of the total surface sites that are available for the adsorption of the anion HL$^-$, S$_T$ = {>MOH} + {>ML$^-$}. Equation 14 thus follows directly from Equation 13. Different authors have reported that the rate of the photocatalytic oxidation of an electron donor at the TiO$_2$ surface varied as a function of the dissolved concentration of the electron donor according to a Langmuir-type function.[29,30] It has also been shown experimentally with various

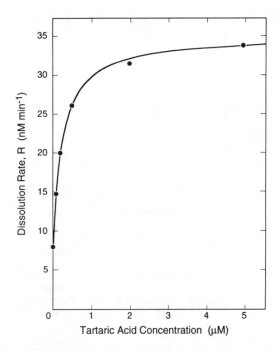

Figure 6. Dependence of the rate of dissolution of 5 μM γ-FeOOH on the solution concentration of tartaric acid. pH equals 4.0; ionic strength 0.01 M (NaCl).[31]

electron donors that the rate of dissolved Fe(II) formation through the photochemical reductive dissolution of Fe(III) (hydr)oxides is a Langmuir-type function of the dissolved electron donor concentration (Figure 6).[31]

WAVELENGTH DEPENDENCE OF THE RATE OF PHOTOREDOX REACTIONS AT IRON (HYDR)OXIDE SURFACES

Studies of the wavelength dependence of the rate of the photochemical reductive dissolution of Fe(III) (hydr)oxides have been carried out in order to identify the primary chromophore and the electronically excited state of the primary chromophore involved in this heterogeneous photoredox reaction.[10,32,33] Apparently, only light in the near-UV region ($\lambda < 400$ nm) leads to an enhancement of the dissolution of Fe(III) (hydr)oxides, brought about by a surface redox process (Figure 7). From these studies, it appears likely that a surface complex, formed from the specific adsorption of a reductant at the surface of Fe(III) (hydr)oxide particles, acts as the primary chromophore. However, the involvement of the bulk solid phase, i.e., a band-to-band transition within the semiconducting mineral, may not be excluded. Hematite (α-Fe$_2$O$_3$) has a complicated valence-band structure, due to the stable $3d^5$ electronic configuration of the Fe^{3+} cations in α-Fe$_2$O$_3$,

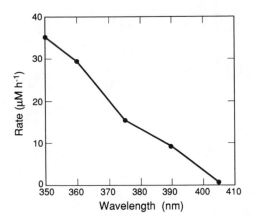

Figure 7.   Rate of the photochemical reductive dissolution of hematite, $R_\lambda = dFe(II)/dt$, in the presence of oxalate as a function of the wavelength at constant incident light intensity ($I_o = 1000 \mu E\ L^{-1}h^{-1}$). The hematite suspensions were deaerated; initial oxalate concentration equals 3.3 mM; pH equals 3. (In order to keep the rate of the thermal dissolution constant, a high enough concentration of Fe(II), $[Fe^{2+}] = 0.15$ mM, was added to the suspensions from the beginning. Thus, the rates correspond to dissolution rates due to the surface photoredox process.)[33]

which lowers the energy of the 3d orbitals into the region of the O 2p band.[34] These authors report changes that occur in the valence-band structure as the surface stoichiometry is varied by sputtering and annealing, which suggest that the O 2p contribution to the valence band lies primarily in the upper portion of the band with the cation contribution concentrated in the lower portion of the band. An electronic transition from the valence band to the conduction band of α-Fe_2O_3 thus may be interpreted in terms of a ligand-to-metal charge-transfer transition. Although α-Fe_2O_3 absorbs light at wavelengths higher than 400 nm, this is probably absorption, due not to ligand-to-metal charge-transfer transitions, but probably to the spin-state transition of iron.

If a surface complex acts as the primary chromophore, then the oxidation of the adsorbed reductant and the reduction of the surface Fe(III) occurs through a ligand-to-metal charge-transfer transition within the surface complex.[33,35] The observed wavelength dependence of the photochemical reductive dissolution of Fe(III) (hydr)oxides may have consequences for the steady-state concentration of Fe(II) at different depths of the photic zone, since longer wavelength light can penetrate to a greater depth as compared to shorter wavelength light. Wells et al.[36] have reported that light increases the lability of colloidal iron in seawater of pH 8 and that generally only wavelengths below 400 nm lead to an increase in the lability of iron due to light. From model calculations combining the wavelength dependence of the attenuation of the light by dissolved organic carbon (DOC) and the wavelength dependence of the labilization of colloidal iron, the authors predict that the depth-integrated formation of labile iron will be maximal at 390 nm.

Case Studies

A Comparative Study of the Kinetics of the Photochemical Reductive Dissolution of Various Iron(III) (Hydr)oxides

Reduction of surface metal centers leads to an enhancement of the dissolution of oxide minerals over nonreductive dissolution. This may be explained by the decrease in the bond strength between the reduced surface metal centers and the neighboring oxygen ions in terms of a decrease in the Madelung energy. As discussed in the previous section, in reductive dissolution of oxide minerals, instead of being detached from the surface, the reduced surface metal centers may be reoxidized by a suitable oxidant. Since different Fe(III) (hydr)oxides exhibit different coordination geometries within the surface lattice, different reactivities of reduced surface iron centers may be expected. In aerated lepidocrocite (γ-FeOOH) suspensions, the photochemical reductive dissolution with oxalate as the reductant occurs according to the following overall stoichiometry:

$$\gamma\text{-FeOOH} + C_2O_4^{2-} + \frac{1}{4}O_2 + 4\,H^+ \xrightarrow{\text{hv}} Fe^{2+} + 2\,CO_2 + \frac{5}{2}H_2O \quad (15)$$

Thus, one dissolved Fe(II) is formed per oxidized oxalate (Figure 8). Note, that because of the relatively slow oxidation of Fe^{2+} by oxygen at pH 3,[37] a constant concentration of dissolved Fe(II) is maintained over a time scale of hours once oxalate has completely disappeared from the solution due to oxidation. Hence, with lepidocrocite as the solid phase, detachment of the reduced surface iron centers outcompetes their oxidation.

Unlike lepidocrocite, in aerated goethite (α-FeOOH) and hematite (α-Fe_2O_3) suspensions, the photochemical oxidation of oxalate is not accompanied by the formation of equivalent concentrations of dissolved Fe(II) (Figure 9). The lack of appreciable amounts of dissolved Fe(II) in an irradiated aerated goethite or hematite (not shown) suspension may be explained by the reoxidation of surface Fe(II) by oxygen or another suitable oxidant such as H_2O_2. Unlike Fe_{aq}^{2+}, surface Fe(II) may be readily oxidized even at low pH. The oxidation of Fe(II) by oxygen is favored thermodynamically and kinetically by hydrolyis and by specific adsorption to hydrous oxide surfaces (Figure 10). This is because Fe(II) may more readily associate (probably outer-spherically) with O_2 if it is present as complexes with OH^- or as complexes with oxo surface groups of hydrous oxides.[38-40] As explained by Luther,[40] the OH^- ligands donate electron density to the Fe(II) through both the σ and π systems, which results in metal basicity and increased reducing power. Fe(II) bound to silicates is more readily oxidized by O_2 than Fe_{aq}^{2+}. This has been demonstrated for hornblende.[41] From these considerations, we conclude that the redox couple surface Fe(II)/surface Fe(III) has a much lower redox potential as compared to the redox couple $Fe_{aq}^{2+}/Fe_{aq}^{3+}$ and, therefore, that surface Fe(II) is more readily oxidized than dissolved Fe(II).

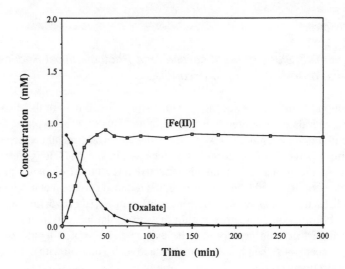

Figure 8.   Concentration of dissolved Fe(II) and oxalate as a function of time upon photo-chemical reductive dissolution of γ-FeOOH in an aerated suspension at pH 3. Experimental conditions: 0.5 g L^{-1} γ-FeOOH; initial oxalate concentration: 1 mM; ionic strength: 5 mM (NaClO$_4$); temperature: 25°C; light source: white light from a 1000-W high-pressure xenon lamp that was filtered by the bottom window of the Pyrex glass vessel, which acts as cutoff filter ($\lambda_{1/2}$ = 350 nm); incident light intensity, I$_o$ ≈ 0.5 kW m^{-2}.

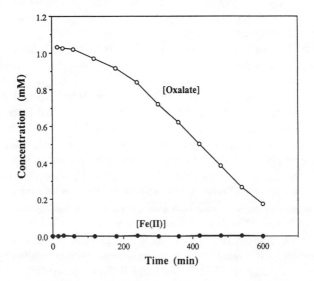

Figure 9.   Concentration of dissolved Fe(II) and of oxalate as a function of time upon photochemical reductive dissolution of α-FeOOH in an aerated suspension at pH 3. The experimental conditions are the same as those described in Figure 8.

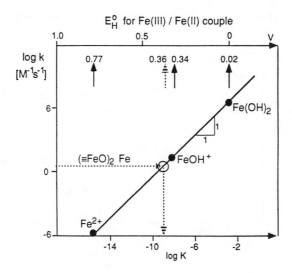

Figure 10. Log of the reaction rate constants of the oxidation of inorganic Fe(II) species by O_2 as a function of the log of the equilibrium constant. The redox potentials, $E_H°$, for the corresponding Fe^{III}/Fe^{II} redox couples are also shown. From the observed rate of the oxygenation of Fe(II) inner spherically adsorbed on a goethite surface (data from Tamura et al.[55]), the equilibrium constant and the redox potential can be estimated. Data for Fe^{2+} and $Fe(OH)^+$ are from Singer and Stumm,[56] for $Fe(OH)_2$ from Millero et al.,[57] and for $(>FeO)_2Fe$ goethite surface from Tamura et al.[55] (Modified from Wehrli, B., *Aquatic Chemical Kinetics*, Stumm, W., Ed., Wiley-Interscience, New York, 1990, 311.)

Although reductive dissolution of hematite and goethite is very slow in oxic environments even at low pH, the rate of the photocatalyzed oxidation of oxalate is higher in aerated suspensions as compared to deaerated suspensions (Figure 11). Plausibly, this is due to the trapping of the oxalate radical, $\cdot C_2O_4^-$, by oxygen:

$$\cdot C_2O_4^- + O_2 \longrightarrow 2\,CO_2 + \cdot O_2^- \tag{16}$$

and thus preventing back electron transfer. Thus, aged Fe(III) (hydr)oxides may act as efficient photocatalysts for the oxidation of organic compounds in the photic zone of surface waters:

$$\text{organic compound} + O_2 + 2\,H^+ \xrightarrow[\text{Fe(III)\,(s)}]{h\nu} \text{oxidized organic compound} + H_2O_2 \tag{17}$$

In the absence of oxygen, the oxalate radical undergoes a fast decarboxylation reaction, yielding CO_2 and the $\cdot CO_2^-$ radical,[42] which is a strong reductant and can, in a thermal redox reaction, reduce a second surface Fe(III), so that overall two dissolved Fe(II) are formed per oxidized oxalate in deaerated Fe(III) (hydr)oxide suspensions (Figure 12):

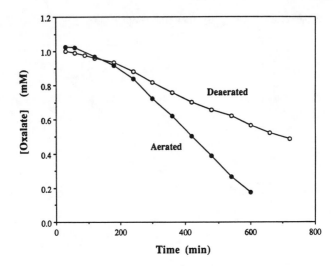

Figure 11. Concentration of oxalate in solution as a function of time upon irradiation of a deaerated and an aerated α-FeOOH suspension at pH 3. The experimental conditions are the same as those described in Figure 8.

$$\gamma\text{-FeOOH} + \frac{1}{2}\,C_2O_4^{2-} + 3\,H^+ \xrightarrow{\ \text{hv}\ } Fe^{2+} + CO_2 + 2\,H_2O \qquad (18)$$

This 2:1 stoichiometry is also observed in deaerated goethite and hematite suspensions.

Photochemical Oxidation of EDTA in the Presence of Lepidocrocite

Ethylenediaminetetraacetate (EDTA) is a powerful chelating agent being widely used for the removal of metal ions interfering with industrial processes, e.g., photographic developing, paper production, and textile dyeing. Since its biodegradation is very slow, little EDTA is removed by wastewater treatment. The resulting release into the aquatic environment potentially causes the mobilization of heavy metals bound to solid surfaces. It is well known that Fe(III)-EDTA complexes are photolyzed with relatively high quantum yields.[43] EDTA adsorbs strongly to mineral surfaces, e.g., to Fe(III) (hydr)oxides.[44,45] The photodissolution of iron oxides in the presence of EDTA has been studied by Litter and collaborators.[32,46,47]

The photochemical oxidation of an organic compound that is adsorbed at the surface of an Fe(III) (hydr)oxide may occur as an interplay of photochemical and thermal redox processes (Figure 1): the photochemical oxidation of the reductant occurring at the surface is — in case of nonaged Fe(III) (hydr)oxides — accompanied by the formation of dissolved Fe(II). Fe(II) catalyzes the thermal dissolution of the solid phase in the presence of suitable ligands. Thereby, dissolved

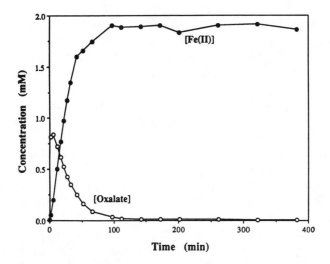

Figure 12.   Concentration of dissolved Fe(II) and oxalate as a function of time upon photochemical reductive dissolution of γ-FeOOH in a deaerated suspension at pH 3. The experimental conditions are otherwise the same as those described in Figure 8.

Fe(III) coordinated with the ligand is formed. These Fe(III) complexes may then be readily photolyzed in solution, as is the case for Fe(III)-EDTA complexes. The Fe(II)-catalyzed thermal dissolution of iron oxides has been studied by several investigators.[48-50] The catalytic effect of Fe(II) is explained in terms of electron transfer from the adsorbed Fe(II) through a bridging ligand to the surface Fe(III). Electron transfer results in the formation of a reduced surface iron that is readily removed from the crystal lattice. This dissolution pathway leads to an increase in the concentration of dissolved Fe(III), while the concentration of Fe(II) does not change; Fe(II) acts as a catalyst for the dissolution of Fe(III) (hydr)oxides. In case of oxalate as the ligand, the rate of dissolved Fe(III) formation through this Fe(II)-catalyzed dissolution is[49,50]

$$R = k\left[C_2O_4^{2-}\right]^n \cdot \frac{K^S S_T\left[Fe^{2+}\right]}{1 + K^S\left[Fe^{2+}\right]} \tag{19}$$

where K^s is the conditional microscopic equilibrium constant of the adsorption of Fe(II) at the Fe(III) (hydr)oxide surface through a bridging ligand, and S_T is the concentration of total surface sites for formation of the ternary surface complex: $S_T = \{>Fe^{III}C_2O_4Fe^{II\,+}\} + \{>Fe^{III}C_2O_4^-\}$. Due to the catalytic effect of Fe(II), the photochemical formation of Fe(II) may proceed as an autocatalytic reaction.[33,46,51] This is the case if the photochemical Fe(II) formation occurs predominantly through photolysis of dissolved Fe(III) complexes that are formed via the Fe(II)-catalyzed dissolution of the solid phase. If, on the other hand, the photochemical

Fe(II) formation occurs through the photochemical reductive dissolution of the solid phase, then we will expect the rate to depend on the concentration of the surface complex, which, for steady-state conditions of the surface species is constant; thus, the dissolution rate will be constant. Photolysis of dissolved Fe(III) complexes on the other hand will follow first-order kinetics. Which pathway predominates depends among other factors on the type of the ligand/reductant present. EDTA seems to be an efficient bridging ligand in the Fe(II)-catalyzed dissolution of iron oxides.[48]

One of the oxidation products of EDTA is formaldehyde. Figure 13 shows the concentration of CH_2O as a function of time upon irradiation of an aerated lepidocrocite suspension at pH 3. Interestingly, the formaldehyde concentration increases with time according to a sigmoidal function, indicating first autocatalytic and then first-order kinetics. The initial autocatalytic formaldehyde formation is observed because the photochemical EDTA oxidation does not only occur at the surface of lepidocrocite, but predominantly through photolysis of dissolved Fe(III)-EDTA complexes whose concentration increases with time. Within 60 min after irradiation, about 80% of the initially added EDTA is present as dissolved Fe(III)-EDTA, due to the Fe(II)-catalyzed dissolution of lepidocrocite (Figure 13). After this initial phase, the Fe(III)-EDTA concentration decreases according to first-order kinetics, due to photolysis. Per photolyzed Fe(III)-EDTA complex, one formaldehye and one Fe(II) is formed under these experimental conditions. At pH 7, the formaldehyde formation occurs at a similar rate in an aerated irradiated lepidocrocite suspension (Figure 14). Under these conditions, however, the photochemical formaldehyde formation is not followed by the formation of equivalent concentrations of Fe(II).

CONCLUDING REMARKS

Environmentally important processes occur to a large extent at the solid-water interface, and thus, it is obvious that heterogeneous photoredox reactions play a key role in the transformation of organic and inorganic compounds and in the biogeochemical cycling of elements such as transition metals and carbon in natural aquatic systems.[27] Aquatic particles may be organisms, particulate organic matter, or minerals. Most of the oxide minerals occurring in the natural environment have semiconducting properties and thus can act as photocatalysts for the photochemical transformation of organic and inorganic compounds that are specifically adsorbed at the surface of these metal (hydr)oxides. With reducible oxide minerals such as Fe(III) (hydr)oxides, the photochemical oxidation of an adsorbed compound is often accompanied by the reductive dissolution of the solid phase, yielding dissolved Fe(II). Dissolved ferric iron, resulting from oxidation of Fe(II), may be stabilized by complexation with organic ligands. The photochemical reductive dissolution of particulate iron and the

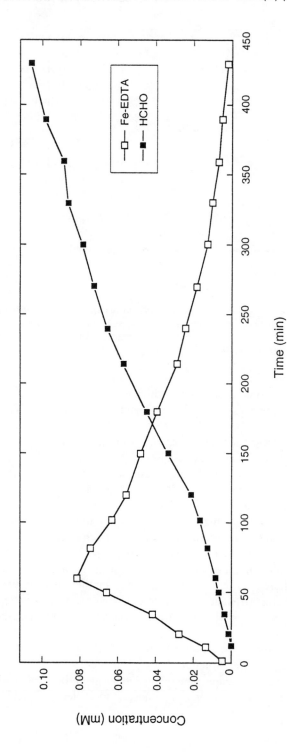

Figure 13. Concentration of formaldehyde and of the dissolved Fe(III)–EDTA complex as a function of time upon the photochemical oxidation of EDTA in an irradiated aerated γ-FeOOH suspension at pH 3. Experimental conditions: 0.05 g L^{-1} γ-FeOOH; initial EDTA concentration: 0.1 mM; ionic strength: 1 mM (NaClO$_4$); temperature: 20°C; light source: white light from a 1000-W high-pressure xenon lamp; incident light intensity, $I_o \approx 0.5$ kW m^{-2}.^{58}

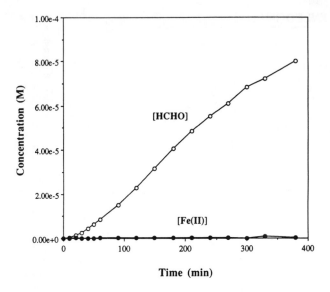

Figure 14.   Concentration of formaldehyde and of dissolved Fe(II) as a function of time
upon the photochemical oxidation of EDTA in an irradiated aerated γ-FeOOH
suspension at pH 7. The experimental conditions are otherwise the same as
those described in Figure 13.[58]

subsequent reoxidation of $Fe(II)_{aq}$ may cause departure from the control of
marine iron chemistry by mineral solubility, which is of great importance for the
bioavailability of iron, since iron can only be taken up by the biota in dissolved
form.[2]

The kinetics of photochemical processes occurring at mineral surfaces depends primarily on the extent and type of adsorption of reactive species at the
mineral surface. Surface complexes formed from the specific adsorption of a
ligand at the surface of a metal (hydr)oxide may act as primary chromophore.
With oxide minerals that exhibit large surface areas, the concentration of such
surface complexes can become high enough to be of importance as light-
absorbing species in natural environments. The absorption spectrum due to a
surface complex may occur at longer wavelengths as compared to the absorption
spectrum of a semiconducting mineral,[15] and thus, a surface complex acting as
the primary chromophore may make better use of the sunlight that reaches the
Earth's surface. Also, the efficiency of a heterogeneous photoredox reaction
may be higher if a ligand-to-metal charge-transfer transition of a surface complex is involved.[15,33]

The efficiency of photoredox reactions occurring at mineral surfaces is determined to a large extent by the structure, both geometrical and electronic, of
surface species. Surface spectroscopic methods and molecular orbital calculations
of surface species[52,53] are needed in order to elucidate their electronic and geometrical properties.

ACKNOWLEDGMENTS

We thank Carrick Eggleston of the Swiss Federal Institute for Environmental Science and Technology (EAWAG), Dübendorf, Switzerland, for having prereviewed this manuscript; and we thank Bettina Bartschat, Stephan Hug, David Sedlak, and Werner Stumm, EAWAG, for stimulating discussions.

REFERENCES

1. Rich, H. W. and Morel, F. M. M., Availability of well-defined iron colloids to the marine diatom *Thalassiosira* weissflogii, *Limnol. Oceanogr.*, 35, 348, 1990.
2. Morel, F. M. M., Hudson, R. J. M., and Price, N. M., Limitation of productivity by trace metals in the sea, *Limnol. Oceanogr.*, 36, 1742, 1991.
3. Landing, W. M., Chemical kinetic/thermodynamic models for iron in the ocean, *Appl. Geochem.*, 3, 89, 1988.
4. Hudson, R. J. M., Covault, D. T., and Morel, F. M. M., Investigations of iron coordination and redox reactions in seawater using ^{59}Fe radiometry and ion-pair solvent extraction of amphiphilic iron complexes, *Mar. Chem.*, 38, 209, 1992.
5 . Martin, J. H., Gordon, R. M., and Fitzwater, S. E., Iron limitation?, *Limnol. Oceanogr.*, 36, 1793, 1991.
6. Mopper, K., Zhou, X., Kieber, R. J., Kieber, D. J., Sikorski, R. J., and Jones, R. D., Photochemical degradation of dissolved organic carbon and its impact on the oceanic carbon cycle, *Nature (London)*, 353, 60, 1991.
7. Zuo, Y. and Hoigné, J., Formation of hydrogen peroxide and depletion of oxalic acid in atmospheric water by photolysis, *Environ. Sci Technol.*, 26, 1014, 1992.
8. Zhuang, G., Yi, Z., Duce, R. A., and Brown, P. R., Link between iron and sulphur cycles suggested by detection of Fe(II) in remote marine aerosols, *Nature (London)*, 355, 537, 1992.
9. Jacob, D., Gottlieb, E. W., and Prather, M. J., Chemistry of polluted cloudy boundary layer, *J. Geophys. Res.*, 94, 12975, 1989.
10. Faust, B. C. and Hoffmann, M. R., Photoinduced reductive dissolution of α-Fe$_2$O$_3$ by bisulfite, *Environ. Sci. Technol.*, 20, 943, 1986.
11. Hoffmann, M. R., Catalysis in aquatic environments, in *Aquatic Chemical Kinetics: Reaction Rates of Processes in Natural Waters*, Stumm, W., Ed., Wiley-Interscience, New York, 1990.
12. Behra, P. and Sigg, L., Evidence for redox cycling of iron in atmospheric water, *Nature (London)*, 344, 419, 1990.
13. Duce, R. A. and Tindale, N. W., Chemistry and biology of iron and other trace metals, *Limnol. Oceanogr.*, 36, 1715, 1991.
14. Bahnemann, D., Bockelmann, D., and Goslich, R., Mechanistic studies of water detoxification in illuminated TiO$_2$ suspensions, *Sol. Energy Mater.*, 24, 564, 1991.
15. Moser, J., Punchihewa, S., Infelta, P. P., and Grätzel, M., Surface complexation of colloidal semiconductors strongly enhances interfacial electron-transfer rates, *Langmuir*, 7, 3012, 1991.
16. Tunesi, S. and Anderson, M. A., Influence of chemisorption on the photodecomposition of salicylic acid and related compounds using suspended TiO$_2$ ceramic membranes, *J. Phys. Chem.*, 95, 3399, 1991.
17. Frei, H., Fitzmaurice, D. J., and Grätzel, M., Surface chelation of semiconductors and interfacial electron transfer, *Langmuir*, 6, 198, 1990.
18. Stumm, W., Kummert, R., and Sigg, L., A ligand exchange for the adsorption of inorganic and organic ligands at hydrous oxide interfaces, *Croat. Chem. Acta*, 53, 291, 1980.

19. Schindler, P. W. and Stumm, W., The surface chemistry of oxides, hydroxides and oxide minerals, in *Aquatic Surface Chemistry*, Stumm, W., Ed., Wiley-Interscience, New York, 1987.

20. Motschi, H., Aspects of the molecular structure in surface complexes: spectroscopic investigations, in *Aquatic Surface Chemistry*, Stumm, W., Ed., Wiley-Interscience, New York, 1987.

21. Hayes, K. F., Roe, A. L., Brown, G. E., Jr., Hodgson, K. O., Leckie, J. O., and Parks, G. A., In situ X-ray absorption study of surface complexes: selenium oxyanions on α-FeOOH, *Science*, 238, 783, 1987.

22. Wehrli, B., Friedl, G., and Manceau, A., Role of surface reactions in manganese cycling at the sediment-water interface, in *Aquatic Chemistry*, Huang, C. P. and Morgan, J. J., Eds., Advances in Chemistry Series, American Chemical Society, Washington, D. C., in press.

23. Tunesi, S. and Anderson, M. A., Surface effects in photochemistry: an in situ cylindrical internal reflection Fourier-transform infrared investigation of the effect of ring substituents on chemisorption onto TiO_2 ceramic membranes, *Langmuir*, 8, 487, 1992.

24. Tejedor-Tejedor, M. I., Yost, E. C., and Anderson, M. A., Characterization of benzoic and phenolic complexes at the goethite/aqueous solution interface using cylindrical internal reflection Fourier transform infrared spectroscopy. 2. Bonding structures, *Langmuir*, 8, 525, 1992.

25. Hochella, M. F., Jr., Atomic structure, microtopography, composition, and reactivity of mineral surfaces, in *Mineral-Water Interface Geochemistry*, Hochella, M. F., Jr. and White, A. F., Eds., *Rev. Mineral.*, Mineralogical Society of America, Bookcrafters Inc., Chelsea, MI, 1990, 23.

26. Dzombak, D. and Morel, F. M. M., *Aquatic Sorption: Stability Constants for Hydrous Ferric Oxide*, Wiley-Interscience, New York, 1990.

27. Stumm, W., *Chemistry of the Solid-Water Interface*, Wiley-Interscience, New York, 1992.

28. Stumm, W. and Furrer, G., The dissolution of oxides and aluminium silicates: examples of surface-coordination-controlled kinetics, in *Aquatic Surface Chemistry*, Stumm, W., Ed., Wiley-Interscience, New York, 1987.

29. Herrmann, J. M., Mozzanega, M. N., and Pichat, P., Oxidation of oxalic acid in aqueous suspensions of semiconductors illuminated with UV or visible light, *J. Photochem.*, 22, 333, 1983.

30. Al-Ekabi, H., Serpone, N., Pelizzetti, E., Minero, C., Fox, M. A., and Draper, R. B., Kinetic studies in heterogeneous photocatalysis. 2. TiO_2-mediated degradation of 4-chlorophenol alone and in a 2,4-dichlorophenol, and 2,4,5-trichlorophenol in air-equilibrated aqueous media, *Langmuir*, 2, 250, 1989.

31. Waite, T. D., Photoredox chemistry of colloidal metal oxides, in *Geochemical Processes at Mineral Surfaces*, Davis, J. A. and Hayes, K. F., Eds., ACS Sump. Ser. No. 323, Washington, D.C., 1986.

32. Litter, M. I. and Blesa, M. A., Photodissolution of iron oxides I. Maghemite in EDTA solutions, *J. Colloid Interface Sci.*, 125, 679, 1988.

33. Siffert, C. and Sulzberger, B., Light-induced dissolution of hematite in the presence of oxalate: a case study, *Langmuir*, 7, 1627, 1991.

34. Kurtz, R. L. and Henrich, V. E., Surface electronic structure and chemisorption on corundum transition-metal oxides: α-Fe_2O_3, *Phys. Rev. B*, 36, 3413, 1987.

35. Waite, T. D. and Morel, F. M. M., Photoreductive dissolution of colloidal iron oxide: effect of citrate, *J. Colloid Interface Sci.*, 102, 121, 1984.

36. Wells, M. L., Mayer, L. M., Donard, O. F. X., de Souza Sierra, M. M., and Ackelson, S. G., The photolysis of colloidal iron in the oceans, *Nature (London)*, 353, 248, 1991.

37. Stumm, W. and Lee, G. F., Oxygenation of ferrous iron, *Ind. Eng. Chem.*, 53, 143, 1961.

38. Wehrli, B. and Stumm, W., Oxygenation of vanadyl(IV): effect of coordinated surface-hydroxyl groups and OH^-, *Langmuir*, 4, 753, 1989.

39. Wehrli, B., Redox reactions of metal ions at mineral surfaces, in *Aquatic Chemical Kinetics*, Stumm, W., Ed., Wiley-Interscience, New York, 1990, 311.

40. Luther, G. W., III, The frontier-molecular-orbital theory approach in geochemical processes, in *Aquatic Chemical Kinetics*, Stumm, W., Ed., Wiley-Interscience, New York, 1990, 173.

41. White, A. F. and Yee, A., Aqueous oxidation-reduction kinetics associated with coupled electron-cation transfer from iron-containing silicates at 25°C, *Geochim. Cosmochim. Acta*, 49, 1263, 1985.

42. Prasad, D. R. and Hoffman, M. Z., Pulsed-laser flash and continuous photolysis of ion-pair complexes of methyl viologen and ethylenediamine tetra-acetic acid in aqueous solution, *J. Chem. Soc. Faraday Trans. 2*, 82, 2275, 1986.

43. Lockhard, H. B. Jr. and Blakeley, R. V., Aerobic photodegradation of Fe(III)-(ethylenedinitrilo)tetraacetate, *Environ. Sci. Technol.*, 9, 1035, 1975.

44. Chang, H. C., Healy, T. W., and Matijevic, E., Interactions of metal hydrous oxides with chelating agents, *J. Colloid Interface Sci.*, 92, 469, 1983.

45. Bondietti, G. L., Sinniger, J., and Stumm, W., The reactivity of Fe(III) (hydr)oxides; effects of ligands in inhibiting the dissolution, *Colloids Surf.*, in print.

46. Litter, M. I., Baumgartner, E. C., Urrutia, G. A., and Blesa, M. A., Photodissolution of iron oxides III: the interplay of photochemical and thermal processes in maghemite/carboxylic acid systems, *Environ. Sci. Technol.*, 25, 1907, 1991.

47. Litter, M. I. and Blesa, M. A., Photodissolution of iron oxides IV: a comparative study on the photodissolution of hematite, magnetite and maghemite in EDTA media, *Can. J. Chem.*, 70, 2502, 1992.

48. Borghi, E. B., Regazzoni, A. E., Maroto, A. J. G., and Blesa, M., Reductive dissolution of magnetite by solutions containing EDTA and Fe(II), *J. Colloid Interface Sci.*, 130, 299, 1988.

49. Suter, D., Siffert, C., Sulzberger, B., and Stumm, W., Catalytic dissolution of iron(III) (hydr)oxides by oxalic acid in the presence of Fe(II), *Naturwissenschaften*, 75, 571, 1988.

50. Suter, D., Banwart, S., and Stumm, W., Dissolution of hydrous iron(III) oxides by reductive mechanisms, *Langmuir*, 7, 809, 1991.

51. Cornell, R. M. and Schindler, P. W., Photochemical dissolution of goethite in acid/oxalate solution, *Clays Clay Miner.*, 35, 347, 1987.

52. Hoffmann, R., *Solids and Surfaces: A Chemist's View of Bonding in Extended Structures*, VCH Publishers, New York, 1988.

53. Calzaferri, G. and Hoffmann, R., The symmetrical octasilasesquioxanes $X_8Si_8O_{12}$: electronic structure and reactivity, *J. Chem. Soc. Dalton Trans.*, 917, 1991.

54. Sulzberger, B. and Laubscher, H. U., Photochemical reductive dissolution of lepidocrocite: the effect of pH, in *Aquatic Chemistry*, Huang, C. P. and Morgan, J. J., Eds., ACS, Adv. Chem. Ser., in press.

55. Tamura, H., Goto, K., and Nagayama, M., The effect of ferric hydroxide on the oxygenation of ferrous ions in neutral solutions, *Corros. Sci.*, 26, 197, 1976.

56. Singer, Ph. C. and Stumm, W., Acidic mine drainage — the rate determining step, *Science*, 167, 1121, 1970.

57. Millero, F., Sotolongo, S., and Izaguirre, M., The oxidation kinetics of Fe(II) in seawater, *Geochim. Cosmochim. Acta*, 51, 793, 1987.

58. Karametaxas, G. and Sulzberger, B., Photodegradation of EDTA in the presence of particulate iron, in preparation.

CHAPTER 4

Photochemical Reactions in Atmospheric Waters: Role of Dissolved Iron Species

J. Hoigné, Y. Zuo, and L. Nowell

INTRODUCTION

In most regions of the globe, about 15% of the volume of the lower troposphere is composed of clouds. A parcel of air, therefore, spends about one seventh of its time in a cloud system. In addition, the distribution patterns of clouds cause air to remain typically for less than 12 hr between clouds. Within clouds, the high dispersion of the liquid phase, forming a few hundred droplets per milliliter, allows for an efficient mass transfer between the gaseous phase and the droplets. Therefore, for many atmospheric compounds, the liquid phase acts as a very efficient reactor, despite its limited volume, only accounting for a fraction of about 50 ppbv of the lower troposphere. Comprehensive critical reviews describing and interpreting the role of cloud systems have recently been published by Lelieveld and Crutzen.[1,2]

The effectiveness of the atmospheric liquid-phase reactor is due to the following:

1. Some reactants, such as H_2O_2 and $HO_2^-/\cdot O_2^-$, reach in the equilibrated aqueous-phase concentrations that are nearly 1 million times higher in concentration than in the gaseous phase.
2. Reactants of high water solubility (e.g., HO_2^-) become separated from those exhibiting low solubility (e.g., NO).
3. Compounds dissolved in water may form new species by dissolution, hydration, dissociation, and complex formation (Figure 1). These allow for further types of reactions, as illustrated in Figure 2 for the case of iron species.
4. Within water droplets, solar radiation is enhanced because of light-scattering and in-droplet reflections.[1,2]

0-87371-871-2/94/$0.00+$.50

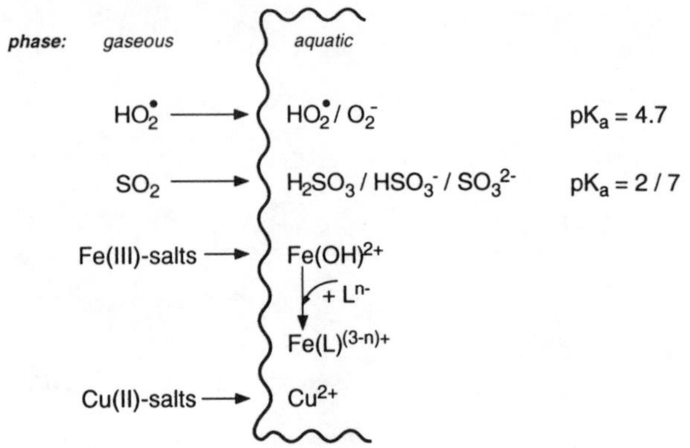

phase: *gaseous* | *aquatic*

$HO_2^\bullet \longrightarrow$ $HO_2^\bullet / O_2^-$ $pK_a = 4.7$

$SO_2 \longrightarrow$ $H_2SO_3 / HSO_3^- / SO_3^{2-}$ $pK_a = 2 / 7$

Fe(III)-salts $\longrightarrow$ $Fe(OH)^{2+}$
                                          $+ L^{n-}$
                                       $Fe(L)^{(3-n)+}$

Cu(II)-salts $\longrightarrow$ Cu^{2+}

Figure 1.   Examples of new chemical species formed when gaseous solutes dissolve in atmospheric water.

With this background, we can examine the role of transition metals occurring in atmospheric water. Figure 2 depicts a few different primary reactions that might occur when Fe(III) is dissolved in water in presence of ligand-forming solutes, dissolved photooxidants, and sunlight. Comparing the estimated rate constants for the different reactions, we conclude that Fe(III) may undergo very fast photolysis, mainly when complexed by oxalate or a similar ligand. It has, therefore, become of interest to study the chemical processes that occur in such systems, including measurements of the rate of reoxidation of photolytically produced Fe(II) and comparisons of the role of dissolved iron with that of copper species.

The speciation of Fe(III) (10 μM) in the presence of varying concentrations of oxalate, Ox, is depicted in Figure 3. A significant fraction of the dissolved Fe(III)

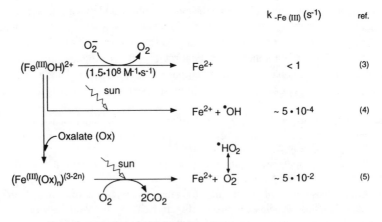

Figure 2.   Main species of Fe(III) formed in the aqueous phase of the troposphere and its reactivities. Estimated kinetic data for pH equal to 4.7 and irradiation by average in cloud daytime sun elevation.

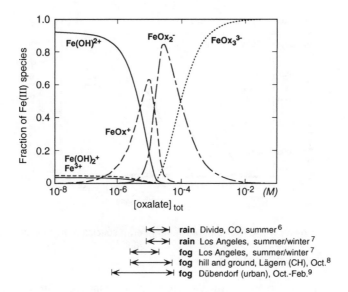

Figure 3.   Calculated speciations of dissolved Fe(III) in water at pH 4 vs concentration of oxalate. Comparision with ranges of oxalate concentrations reported for different atmopsheric waters. $[Fe(III)]_{tot} = 10 \ \mu M$. (Adapted from Zuo, Y. and Hoigné, J., *Environ. Sci. Technol.*, 26, 1014, 1992.)

is present as $FeOx^+$, even when as little as 1 μM oxalate is present. The $FeOx_2^-$ and FOx_3^{3-} species are formed when about 10 μM or 100 μM oxalate is present. The bars below the speciation diagram indicate that oxalate concentrations for many types of rain and fog water assume values in these critical ranges of 1–100 μM. Reviews show that reported concentrations of iron in such atmospheric waters are of comparable magnitudes as those measured for oxalic acid.[8-11] Therefore, at least a fraction of dissolved iron is potentially present as an oxalate complex. Also interactions by similar ligands, such as pyruvic acid, may occur in atmospheric waters. Compounds such as acetic acid and formaldehyde may be present at concentrations that are two orders of magnitude higher than that of oxalate. However, their stability constants for complexation with iron are very low and therefore have not been considered here.

EXPERIMENTAL RESULTS

Experimental methods have been described elsewhere.[4,5,11] The subsequent finding that Fe^{2+} may reduce DPD^+ and therefore stoichiometrically interfere with the DPD-based determination of hydrogen peroxide[23] does not lead to significant corrections for the data presented in this chapter.

Figure 4 shows that the ultraviolet (UV) absorption spectrum of dissolved $Fe(OH)^{2+}$, which is the dominating aquo complex of Fe(III) in pure water of pH

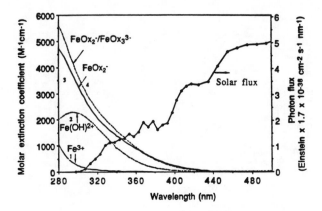

Figure 4. Comparision of UV spectra of different Fe(III) species with the noon September solar spectrum for 47.4°N. (From Zuo, Y. and Hoigné, *J., Environ. Sci. Technol.,* 26, 1014, 1992. With permission.)

4–5, exhibits some overlap with the tropospheric solar spectrum in the UV-A region.[4] The overlap becomes much higher when the iron forms an oxalate complex. Also, quantum efficiencies for the photolysis of the iron oxalate species are much higher than those reported for the Fe(III)-hydroxo complexes.[11] Correspondingly, the presence of oxalate highly sensitizes the photoprocesses.

Figure 5A depicts the rate of reduction of Fe(III) in samples present in tubes exposed to sunlight under anaerobic conditions in the presence of oxalate. Already, after 17 sec of August noon sunlight, half of the Fe(III) was photolytically reduced. In aerated solutions (Figure 5B), reoxidations seem also to occur, and a photostationary-state condition is established within about 2 min. The ratio of [Fe(II)]/[Fe(III)] became about 2:1.

Figure 6 shows how the rate of photolytic degradation of oxalate and the corresponding formation of hydrogen peroxide varies with the concentration of Fe(III). In September noon sunlight, 5–10 μM of oxalate were photolytically degraded within 14 min, even when only 1 μM of Fe(III) was present. The fact that much more oxalate was degraded than iron was present suggests the involvement of a chain reaction in which iron acts as a chain carrier. The rate of photolytic conversions, however, decreased with lower levels of iron concentrations.

INTERPRETATION OF RESULTS

Figure 7 shows the proposed reaction pathways following the primary photoexcitation of an Fe(III) oxalate complex. The primary ligand-to-metal charge-transfer process occurring in the photoexcited state is able to transfer an electron to oxygen. This leads to the formation of $\cdot O_2^-$. Oxalic acid is thereby oxidized to CO_2. Fe(II) is released. The situation is rather different from that occurring at high concentrations of iron oxalate typically applied for actinometry; in such solutions,

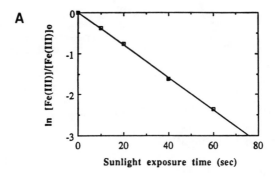

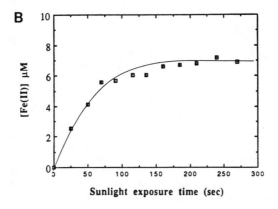

Figure 5. **(A)** Reduction of Fe(III) ($[iron]_o = 10$ μM) in presence of 30 μM oxalate in deaerated solutions at pH 4.0 *vs.* August noon (CH-Dübendorf) sunshine exposure time. **(B)** Measured formation of Fe(II) in aerated solution under conditions described in **(A)**. (From Zuo, Y., Ph.D. thesis, ETHZ No 9727, Swiss Federal Institute of Technology, Zurich, 1992.)

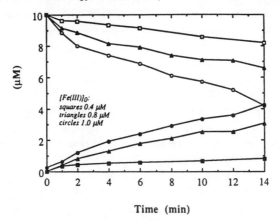

Figure 6. Photodepletion of oxalic acid (declining lines) and formation of H_2O_2 (rising lines) at low concentrations of oxalic acid and Fe(III) in aerated solutions. $[Oxalate]_o = 10$ μM, pH = 4, solar noon August sunshine.[11]

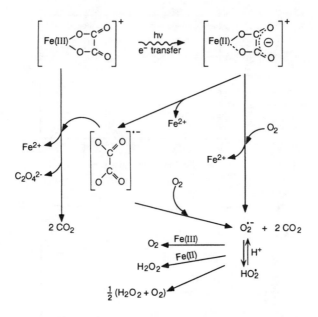

Figure 7. Proposed scheme of reactions for the photoinduced reduction of Fe(III) and oxidation of oxalate for aerated solutions containing Fe(III) oxalate. For simplicity, only one oxalate ligand is shown. The scheme also includes the pathway of reactions occurring at low Fe(III) concentrations (right side) where the oxidation of the oxalate radical predominantly occurs by oxygen and not by Fe(III). (Adapted from Zuo, Y. and Hoigné, J., *Environ. Sci. Technol.*, 26, 1014, 1992.)

the oxalate radical is oxidized by Fe(III) oxalate. This prevents the reduction of O_2 and the formation of H_2O_2.

Figure 8 summarizes possible fates of $HO_2^{\cdot}/\cdot O_2^{-}$ in atmospheric waters. Because such waters generally have a pH ≤ 5, a significant fraction of $\cdot O_2^{-}$ is protonated to $HO_2^{\cdot}$ ($pK_a = 4.7$). The aqueous $HO_2^{\cdot}$ will equilibrate with the surrounding gas phase within milliseconds.[2] The fraction remaining in the aqueous phase reacts through several possible pathways. Depending on conditions, some of those pathways yield kinetically comparable sinks for $HO_2^{\cdot}/\cdot O_2^{-}$. Considering tropospheric conditions, the summer daytime (47°N) concentration of $HO_2^{\cdot}$ is predicted to be on the order of $10^8 \, \text{cm}^{-3}$, and the concentration in the equilibrated aqueous phase at pH 4.7 becomes about 10 nM for both $HO_2^{\cdot}$ and $\cdot O_2^{-}$. Under such daylight conditions, $\cdot O_2^{-}$ reacts with $HO_2^{\cdot}$ within about 1 sec. Under conditions of lower light intensities, the concentration of $HO_2^{\cdot}$ becomes much lower and this reaction loses relative importance.

Even when the concentration of aqueous O_3 is larger than $^1/_{15}$ that of $HO_2^{\cdot}$, the reaction of $\cdot O_2^{-}$ with O_3 becomes dominant. The typical day and night concentration of 40 ppbv of O_3 in the gaseous phase equilibrates to an aqueous concentration of dissolved ozone of about 1 nM. For most situations, this concentration is sufficient to make the reaction of $\cdot O_2^{-}$ with O_3 more important than the bimolecular reaction of $HO_2^{\cdot}/\cdot O_2^{-}$. By reacting with ozone, $HO_2^{\cdot}/\cdot O_2^{-}$ is converted to OH· radicals

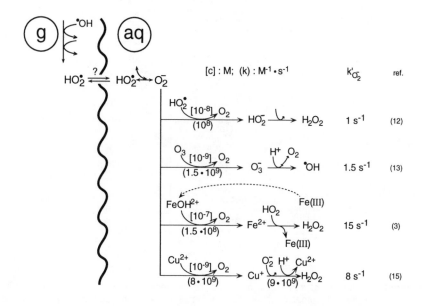

Figure 8. Summary of main reactions by which aqueous $\cdot O_2^-$ becomes transformed in atmospheric waters. Assumed concentrations of reactants (numbers above arrows in brackets given in molar units) approximate those considered for summer daytime by Lelieveld and Crutzen[1] or Chameides and Davis.[20] The given second-order rate constants (numbers given below arrows in parenthesis) are rounded values ($M^{-1} s^{-1}$) from cited literature. The calculated pseudo-first-order rate constants for the depletion of $\cdot O_2^-$ (k'_{O-2}) are rather different from those deduced in Figure 6 of Reference 16. There we have based all calculations on the depletion of the sum of $HO_2 + \cdot O_2^-$ and on the overall effect of Fe_{tot} and Cu_{tot} due to the calculated steady-state concentrations of all species occurring in the system.

and not to H_2O_2. However, OH radicals may then be efficiently converted back to $HO_2^-/\cdot O_2^-$ through reaction with solutes such as formaldehyde or formic acid.[1,2]

Already in the presence of only 100 nM Fe(OH)$^{2+}$ or of only 1 nM dissolved Cu^{2+}, most $\cdot O_2^-$ can react with the reduced transition-metal species and produce H_2O_2. Based on many reported observations and reviews, we can conclude that traces of copper are ubiquitous in all atmospheric waters.[14,15] Therefore, von Piechowski et al. redetermined rate constants for $\cdot O_2^-$ reacting with Cu^{2+} and Cu^+ by producing $\cdot O_2^-$ by pulse radiolysis of aqueous solutions containing dissolved Cu(II) and following the UV spectra by kinetic spectroscopy in the microsecond range.[15,16] Comparisons of these reaction rates (Figure 8) show that copper becomes the dominant oxidant for $\cdot O_2^-$ even when its concentration only assumes 2% of that of dissolved Fe(III) and one fifth of that of dissolved ozone. That means that copper will become a dominant radical converter even when its concentration is only a few percent of that of Fe(III). The Cu(I) formed will convert further $\cdot O_2^-$ into H_2O_2 or reduce Fe(III).[16]

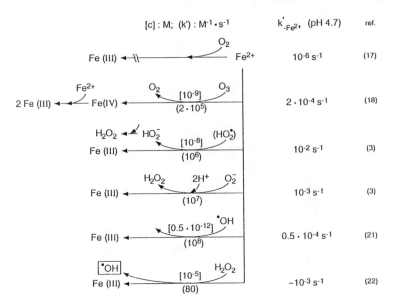

Figure 9. Summary of main back reactions for the reoxidation of Fe(II) in atmospheric waters. Assumed concentrations of reactants (numbers above arrows in brackets given in molar units) approximate those considered for summer daytime by Lelieveld[1] and Crutzen[1] or Chameides and Davis.[20] The given second-order rate constants (numbers given below arrows in parenthesis) are rounded values (M^{-1} s^{-1}) from cited literature.

Figure 9 summarizes the fate of the photochemically produced Fe(II). In the low pH range of atmospheric waters, dissolved oxygen will reoxidize Fe(II) only through very slow reactions.[17] Instead, reoxidations will occur via reactions of atmospheric photooxidants. We have determined a rate constant of $2 \cdot 10^5\,M^{-1}\,s^{-1}$ for the reaction of aqueous ozone with Fe^{2+} by following the rate of depletion of of ozone in presence of an excess of Fe(II), when both were injected as solutions into a capillary tube and the concentration of the ozone residual was determined at the exit of the capillary by adding the indigo reagent.[18] Recently, Logager et al. have determined a rate constant for this reaction of $8 \cdot 10^5\,M^{-1}\,s^{-1}$ by using a stop-flow procedure.[19] Both groups have concluded that O_3 oxidizes Fe(II) by an oxygen-transfer reaction, which leads to Fe(IV) as an intermediate. Thus, no electron-transfer reaction occurs that would produce O_3^- and finally lead to OH radicals. Because the reaction of ozone with Fe(II) proceeds relatively slowly, HO_2^- becomes the dominant oxidant already when the aqueous concentration of HO_2^- is only one tenth that of ozone. Therefore, at least during the daytime, HO_2^- could become a much more efficient oxidant for Fe(II). The aqueous-phase OH· radical concentrations are much lower, and therefore, OH· radicals do not significantly contribute to the reoxidation of Fe(II). Hydrogen peroxide exhibits only a very low reaction rate constant, and it therefore becomes a significant oxidant only when its concentration assumes a value that is 10^4 times that of HO_2^- and 10^3 times that of O_3. Such a scenario is to be expected only during the nighttime.

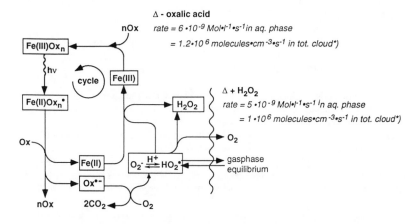

Figure 10. Scheme of photochemical cycling of iron in presence of oxalate (Ox) showing the iron-catalyzed photooxidation of oxalate and the formation of hydrogen peroxide. The rates for the transformations occurring within the aqueous phase have been used for the estimate of the rate of transformation occurring in the total cloud system by accounting for an in-cloud liquid water content of 0.3 g/m³. (Adapted from Zuo, Y. and Hoigné, J., *Environ. Sci. Technol.*, 26, 1014, 1992.)

SUMMARY

Summarizing the data given in Figure 10, we conclude that during the daytime, Fe(II) is reoxidized within minutes by photooxidants to form dissolved Fe(III), which will become complexed with residual oxalate. This iron complex will again be photoreduced within less than 1 min to Fe(II). Therefore, an Fe(III)–Fe(II) photochemical-chemical cycle occurs in atmospheric waters. In the presence of approximately 1 μM iron, approximately 6 nM/sec of oxalate will be oxidized and nearly an equivalent amount of hydrogen peroxide will be produced. This means that in a parcel of cloud containing 0.3 ppmv of aqueous phase, about 10^6 molecules per cubic centimeter per second of oxalic acid are transformed to hydrogen peroxide. Therefore, a cloud may act as a source of hydrogen peroxide instead of being a sink. This conclusion is of relevance for quantifying the oxidation of atmospheric SO_2 to sulfate. The role of copper interfering with this cycle will be studied more in detail.

For further generalizations, more information is required on the reaction of $\cdot O_2^-$ with other iron and copper and manganous species occurring in atmospheric waters.[23]

ACKNOWLEDGMENTS

We thank David Sedlak for discussions and corrections. This final review, summarizing a part of a subproject of HALIPP (EUROTRAC), has been supported by the Swiss Federal Office for Education and Science.

REFERENCES

1. Lelieveld, J. and Crutzen, P. J., Influences of cloud photochemical processes on tropospheric ozone, *Nature (London)*, 343, 227,1990.
2. Lelieveld, J. and Crutzen, P. J., The role of clouds in tropospheric photochemistry, *J. Atmos. Chem.*, 12, 229, 1991.
3. Rush, J. D. and Bielski, H. J., Pulse radiolytic studies of the reactions of HO_2/O_2^- with Fe(II)/ Fe(III) ions. The reactivity of HO_2/O_2^- with ferric ions and its implication on the occurrence of the Haber-Weiss reaction, *J. Phys. Chem.*, 89, 5062, 1985.
4. Faust, B. C. and Hoigné, J., Photolysis of Fe(III)-hydroxy complexes as sources of OH radicals in clouds, fog and rain, *Atmos. Environ.*, 24A, 79, 1990.
5. Zuo, Y., and Hoigné, J., Formation of H_2O_2 and depletion of oxalic acid by photolysis of Fe(III)-oxalato complexes in atmospheric waters, *Environ. Sci. Technol.*, 26, 1014, 1992.
6. Norton, R. B., Roberts, J. M., and Huebert, B. J., Tropospheric oxalate, *Geophys. Res. Lett.*, 10, 517, 1983.
7a. Steinberg, S., Kawamura, K., and Kaplan, I. R., The determination of keto acids and oxalic acid in rain, fog and mist by HPLC, *Intern. J. Environ. Anal. Chem.*, 19, 251, 1985.
7b. Kawamura, K., Steinberg, S., and Kaplan, I. R., Capillary GC determination of short-chain dicarboxylic acids in rain, fog, and mist, *Intern. J. Environ. Anal. Chem.*, 19, 175, 1985.
8. Joos, F. and Baltensperger, U., A field study on chemistry, S(IV) oxidation rates and vertical transport during fog conditions, *Atmos. Environ.*, 25A, 217, 1991.
9. Behra, P. and Sigg, L., Evidence for redox cycling of iron in atmospheric water droplets, *Nature (London)*, 344, 419, 1990.
10. Zhuang, G., Zhen, Y., Duce, R. A., and Brown, P. R., Link between iron and sulphur cycles by detection of Fe(II) in remote marine aerosols, *Nature (London)*, 355, 537, 1992.
11. Zuo, Y., Photochemistry of Iron(III)/Iron(II) Complexes in Atmospheric Liquid Phases and Its Environmental Significance, Ph.D. thesis, ETHZ No. 9727, Swiss Federal Institute of Technology, Zurich, 1992.
12. Bielski, B. H. J., Cabelli, D., Arudi, R., and Ross, A., Reactivity of HO_2/O_2^- radicals in aqueous solution, *J. Phys. Chem. Ref. Data*, 14, 1041, 1985.
13. Bühler, R. E., Staehelin, J., and Hoigné, J., Ozone decomposition in water studied by pulse radiolysis. 1. HO_2/O_2 and HO_3/O_3 as intermediates, *J. Phys. Chem.*, 88, 2560, 1984.
14. Xue, H., Gonçalves, M. L., Reutlinger, M., Sigg, L., and Stumm, W., Copper(I) in fogwater: determination and interactions with sulfite, *Environ. Sci. Technol.*, 25, 1716, 1991.
15. von Piechowski, M., Der Einfluss von Kupferionen auf die Redox-chemie des troposphärischen Wassers: Kinetische Untersuchungen, Ph.D. thesis, ETHZ No. 9512, Swiss Federal Institute of Technology, Zurich, 1991.
16. von Piechowski, M., Bühler, R., and Hoigné, J., The kinetics of the reaction of O_2^- with Cu^{2+} and Cu^+ measured by pulse radiolysis combined with kinetic spectroscopy, *Ber.Bunsenges. Phys. Chem.*, 97, 762, 1993.
17. Stumm, W. and Morgan, J. J., *Aquatic Chemistry*, Wiley-Interscience, New York, 1981.
18. Nowell, L. H. and Hoigné, J., Interaction of iron (II) and other transition metals with aqueous ozone, Proc. Int. Ozone Assoc., 8th World Congress on Ozone, Zurich, September 1987, IOA, Zurich, 1987, E 80.
19. Logager, T., Holcman, J., Sehested, K. and Pedersen, Th., Oxidation of ferrous ions by ozone in acidic solution, *J. Inorg. Chem.*, 31, 3523, 1992.
20. Chameides, W. L. and Davis, D. D., The free radical chemistry of cloud droplets and its impact upon the composition of rain, *J. Geophys. Res.*, 87, 4863, 1982.
21. Buxton, G. V., Greenstock, C. L., Helman, W. P., and Ross, A. B., Critical review of rate constants for reactions of hydrated electrons, hydrogen atoms and hydroxyl radicals in aqueous solution, *J. Phys. Chem. Ref. Data*, 17, 513, 1988.
22. Walling, C., Fenton's reagent revisited, *Acc. Res.*, 8, 125, 1975.
23. Sedlak, D. and Hoigné, J., The role of copper and oxalate in the redox cycling of iron in atmospheric waters, *Atmos. Environ.*, 1993, in press.

CHAPTER 5

A Temperature-Dependent Competitive Kinetics Study of the Aqueous-Phase Reactions of OH Radicals with Formate, Formic Acid, Acetate, Acetic Acid, and Hydrated Formaldehyde

M. Chin and P. H. Wine

INTRODUCTION

Organic acids are major contributors to the acidity of precipitation in the remote troposphere.[1] Identification of the atmospheric sources of organic acids has proven to be difficult. Direct emission into the atmosphere from vegetation has been suggested to be an important source,[2] as has emission from formicine ants.[3] Photochemical sources of organic acids are also thought to be important. For example, Jacob and Wofsy have shown that the water-induced isomerization of Criegee biradicals, which are produced via ozone-olefin reactions,[4] can be a major source of organic acids in regions where olefin concentrations are large, i.e., polluted areas or tropical forests.[5,6] In regions where olefin levels are low (marine environments, for example), the reaction of hydrated formaldehyde with OH radicals in cloud water is thought to be an important atmospheric source of formic acid.[7-12]

$$OH + H_2C(OH)_2 \rightarrow H_2O + HC(OH)_2 \quad (R1)$$

$$HC(OH)_2 + O_2 \rightarrow HO_2 + HCOOH \quad (R2)$$

Recent laboratory work has confirmed that the above sequence of reactions produces formic acid (HCOOH) with unit yield[13] [other OH(aq) reactions, of course, compete with R1 in cloud water].

0-87371-871-2/94/$0.00+$.50

Modeling studies suggest that reactions of carboxylic acid anions with OH radicals in cloud droplets represents an important, and possibly dominant, tropospheric removal mechanism for carboxylic acids.[6,7,12]

$$OH + HCOO^- \rightarrow H_2O + CO_2^- \tag{R3}$$

$$OH + CH_3COO^- \rightarrow H_2O + CH_2CO_2^- \tag{R4}$$

The effectiveness of the aqueous-phase sink is strongly pH dependent because the protonated acids are both more volatile and less reactive toward $OH(aq)$[14] than their unprotonated counterparts.

$$OH + HCOOH \rightarrow H_2O + COOH \tag{R5}$$

$$OH + CH_3COOH \rightarrow H_2O + CH_2COOH \tag{R6}$$

Recently, it has become apparent that the reactions of $OH(aq)$ with hydrated formaldehyde, formate, and formic acid in cloud water can play a major role in global tropospheric chemistry.[15-17] In particular, $H_2CO(g)$, an important intermediate in gas-phase hydrocarbon oxidation, can be destroyed by uptake into cloud water and subsequent aqueous-phase oxidation via R1, R2, R3 (or R5 at low pH), and R7.

$$CO_2^-(COOH) + O_2 \rightarrow CO_2 + O_2^-(HO_2) \tag{R7}$$

The above aqueous-phase chemistry reduces the formation rates of $HO_x(g)$ and CO via oxidation of $H_2CO(g)$. Furthermore, the O_2^- radicals that are generated via R7 react with $O_3(aq)$ to recycle OH:

$$O_2^- + O_3 \rightarrow O_3^- + O_2 \tag{R8}$$

$$O_3^- \rightarrow O_2 + O^- \xrightarrow{\;H^+\;} OH \tag{R9}$$

The chain-reaction sequence R1, R2, R3, R7, R8, R9 may represent an important sink for tropospheric O_3.[16,17]

To accurately assess the role of clouds in global tropospheric chemistry, accurate kinetic data for $OH(aq)$ reactions with $H_2C(OH)_2$, $HCOO^-$, HCOOH, CH_3COO^-, and CH_3COOH are needed. Some room temperature kinetic data for R1, R3, R4, R5, and R6 are reported in the literature,[18-27] although the kinetics database is not as good as one would like, particularly for the crucial reaction R1.[18,24] Temperature-dependent kinetic data is available only for R3.[27] In model studies of cloud chemistry, it has often been assumed that activation energies for fast, aqueous-phase free radical reactions are equal to 3 kcal mol^{-1},[12,17] a value which is close to but apparently a little smaller than ΔE_{diff}, the effective "activation energy" for aqueous-phase diffusion.[28,29] However, it has recently been demonstrated[29] that the temperature dependence of a variety of aqueous-phase reactions over a wide temperature range can be rationalized in terms of the relationship

$$k_{obs}^{-1} = k_{diff}^{-1} + k_{react}^{-1} \tag{1}$$

where k_{obs} is the measured bimolecular rate coefficient, k_{diff} is the encounter rate coefficient of the two reacting species, and k_{react} is the rate coefficient that would be measured if diffusion was not rate influencing. According to Equation 1, the activation energy for a reaction where k_{obs} (298 K) $\approx 10^8$ M^{-1} s^{-1} could range anywhere from $\approx \Delta E_{diff}$ to $\ll \Delta E_{diff}$, depending upon the A-factor and activation energy which characterize k_{react} (T). Hence, measurements of temperature-dependent rate coefficients for R1, R3, R4, R5, and R6 are needed. Such measurements are reported in this chapter. A competitive kinetics technique is employed with thiocyanate (SCN$^-$) used as the reference reactant.

EXPERIMENTAL APPROACH

The laser flash photolysis-long path absorption (LFP-LPA) technique was employed in this study. The LFP-LPA apparatus and its adaptation for temperature-dependence studies is described elsewhere.[30–33]

The transient species monitored in this work was (SCN)$_2^-$; it was generated as follows:

$$H_2O_2 + h\nu(248 \text{ nm}) \rightarrow 2\,OH \tag{R10}$$

$$OH + SCN^- \rightarrow SCNOH^- \rightarrow SCN + OH^- \tag{R11}$$

$$SCN + SCN^- \rightleftarrows (SCN)_2^- \tag{R12}$$

The wavelength of peak (SCN)$_2^-$ absorbance is approximately 475 nm, and the peak extinction coefficient is approximately 7600 M^{-1} cm^{-1}.[34,35] In the chemical systems of interest, (SCN)$_2^-$ could be monitored at 475 nm without spectral interference from other absorbing species. In most experiments, the White cell was adjusted for 38 passes, giving an absorption pathlength of 95 cm. With an electronic time constant of 1 µsec and reasonable signal averaging (i.e., 64 shots), the (SCN)$_2^-$ detection limit was below 10^{-9} M.

The H$_2$O$_2$ used in this study was Aldrich semiconductor grade (ACS), 30–32% in H$_2$O. Thiocyanate, formate, and acetate were ACS reagent grade sodium or potassium salts. Water used for making solutions was purified by a Millipore Milli-Q system equipped with filters for removing particulates, ions, and organics. For experiments under acidic conditions, pH was adjusted with ACS reagent grade perchloric acid, 70% in H_2O, while for experiments where pH is greater than 5.7, adjustment of pH involved addition of sodium hydroxide. Liquid monomeric formaldehyde was prepared using the method of Spence and Wild.[36] Concentrations of formaldehyde in aqueous solution were determined using a colorimetric method[37,38] involving quantitative conversion to 3,5-diacetyl-1,4-dihydro-2,6 dimethylpyridine (ε = 8000 M^{-1} cm^{-1} at 412 nm) via reaction with acetylacetone and ammonia.

Solutions were prepared immediately before being used in experiments. Typically, the solution reservoir was placed in a thermostated bath for about 30 min to allow temperature equilibration. Solutions were then flowed through the 50-cm^3 reaction cell at a typical flow rate of 2 cm^3 sec^{-1} and were not recirculated. The laser repetition rate was 0.035 Hz; hence, no aliquot of solution was subjected to more than one laser flash. Most experiments employed unpurged, air-saturated solutions. Purging with UHP argon had no effect on observed kinetics in studies of R3 and R4. Purging was avoided in studies of R1, R5, and R6 to prevent volatilization of $H_2C(OH)_2$, HCOOH, and CH_3COOH from solution into the gas phase.

RESULTS

All experiments were carried out under pseudo-first-order conditions with H_2O_2, SCN$^-$, and X (X = $H_2C(OH)_2$, HCOO$^-$, HCOOH, CH_3COO^-, or CH_3COOH) in large excess over OH. The H_2O_2 concentration was 1.0×10^{-4} M in all experiments, and the SCN$^-$ concentration was varied over the range $(0.5–5) \times 10^{-5}$ M. The concentration of OH immediately after the laser flash was typically 1×10^{-7} M. Experimental conditions were such that virtually all OH produced by the laser flash reacted with either SCN$^-$ or X, i.e. the fraction of OH removed by H_2O_2 or background impurities was negligible. Therefore, under the assumption that the (SCN)$_2^-$ absorption signal per unit OH that reacts with SCN$^-$ is independent of [X] (this assumption is discussed below), the following relationship is obeyed.

$$\frac{A_o}{A_x} = 1 + \frac{k_{OH+X}}{k_{OH+SCN^-}} \frac{[X]}{[SCN^-]} \qquad (2)$$

In Equation 2, A_o is the absorption signal observed with [X] = 0 after OH removal has gone to completion but before significant (SCN)$_2^-$ decay has occurred, and A_x is the analogous signal observed under identical experimental conditions but with X added to the solution. According to Equation 2, a plot of A_o/A_x vs [X]/[SCN$^-$] should be linear with slope = k_{OH+X}/k_{OH+SCN^-}.

Typical 475-nm absorbance temporal profiles observed following laser flash photolysis of H_2O_2-SCN$^-$-X solutions are shown in Figure 1. As typified by the data in Figure 1, experimental conditions were such that absorbance rise times

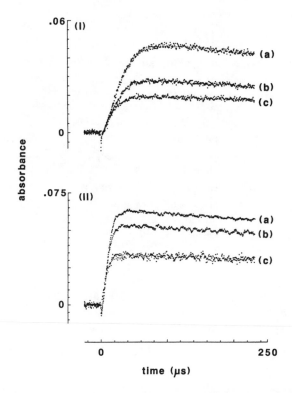

Figure 1. Typical $(SCN)_2^-$ temporal profiles observed following 248-nm laser flash photolysis of H_2O_2-SCN$^-$-X solutions. X = HCOO$^-$; $[H_2O_2]$ = 1 × 10^{-4} M; [SCN$^-$] = (I)
5 × 10^{-6} M, (II) 2 × 10^{-5} M; [HCOO$^-$] = (Ia, IIa) 0, (Ib, IIb) 2 × 10^{-5} M, (Ic) 3 × 10^{-5} M,
(IIc) 8 × 10^{-5} M.

were always much faster than absorbance decay times. Hence, A_o and A_x could be
obtained with good accuracy from the peak absorbance signals. Typical plots of
A_o/A_x vs [X]/[SCN$^-$] are shown in Figure 2; rate coefficient ratios obtained from
the slopes of such plots are summarized in Table 1.

DISCUSSION

Evaluation of Absolute Rate Coefficients

In order to obtain temperature-dependent absolute rate coefficients from the
results in Table 1, the OH + SCN$^-$ rate coefficient must be known as a function
of temperature. Temperature-dependent kinetics studies of R11 have been carried
out by Elliot and Simsons[27] and in our laboratory.[33] Elliot and Simsons employed
a competitive kinetics method where the OH + SCN$^-$ rate coefficient was measured relative to the rate coefficients for OH reactions with formate and t-butanol,

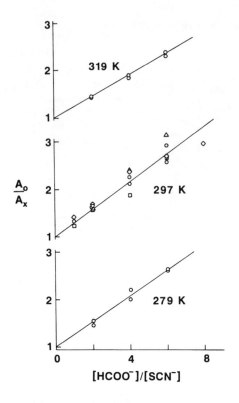

Figure 2. Typical plots of A_0/A_x vs [X]/[SCN$^-$]. X = HCOO$^-$; [H$_2$O$_2$] = 1 × 10^{-4} M; [SCN$^-$] = (◇) 5 × 10^{-6} M, (○) 1 × 10^{-5} M, (□) 2 × 10^{-5} M, (△) 5 × 10^{-5} M. Solid lines are obtained from linear least squares analyses and give the following values for $k_{OH + X}/k_{OH + SCN^-}$ (errors are 1σ and represent precision only): 0.270 ± 0.014 at 279 K, 0.288 ± 0.017 at 297 K, and 0.225 ± 0.006 at 319 K.

the OH + formate, t-butanol rate coefficients were measured relative to the rate coefficient for the OH + ferricyanide (Fe(CN)$_6^{-4}$) reaction, and the OH + ferricyanide rate coefficient was measured directly. Our approach was direct (i.e., no competitors), but involved analysis of double exponential (SCN)$_2^-$ appearance temporal profiles; at low SCN$^-$ concentrations, such experiments are sensitive to the OH + SCN$^-$ rate coefficient.[33,34] At 297 K, the OH + SCN$^-$ rate coefficients obtained from the two studies[27,33] agree very well. However, the activation energy obtained from our data is about 1.0 kcal mol^{-1} larger than the activation energy derived from Elliot and Simsons' results. Since there are advantages and disadvantages to both methods employed to obtain temperature-dependent rate coefficients for R11, we have decided to weigh the two data sets equally. Adopting such an approach leads to the following temperature-dependent rate coefficient:

$$\ln k_{R11} = 28.7655 - 1655/T \ M^{-1} \ s^{-1}$$

Table 1.   Rate Coefficient Ratios Determined in This Study

		$k_{OH + SCN^-}/k_{OH + X}$ [a]		
X	pH	279 K	297 K	319 K
$H_2C(OH)_2$	1.5—5.7	13.6 ± 1.4	15.2 ± 0.9	18.1 ± 1.6
$HCOO^-$	5.7	3.70 ± 0.19	3.47 ± 0.20	4.44 ± 0.11
HCOOH	0.3—1.0	95.3 ± 1.9	117 ± 9	129 ± 12
CH_3COO^-	6.5	163 ± 7	169 ± 11	155 ± 9
CH_3COOH	1.5—2.0	629 ± 28	699 ± 98	730 ± 21

[a] Errors are 1σ and represent precision only.

The above Arrhenius expression has been employed in conjunction with the relative rate data in Table 1 to obtain temperature-dependent absolute rate coefficients for R1, R3, R4, R5, and R6. The results are plotted in Arrhenius form in Figure 3. The solid lines in Figure 3 are obtained from linear least squares analyses of the ln k vs T^{-1} data. These analyses give the following Arrhenius expressions in units of $M^{-1}s^{-1}$:

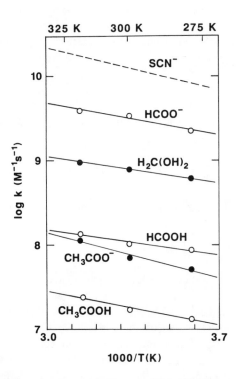

Figure 3.   Arrhenius plots of OH(aq) reactions with $H_2C(OH)_2$, $HCOO^-$, HCOOH, CH_3COO^-, and CH_3COOH. Solid lines are obtained from least squares analyses and give the Arrhenius expressions quoted in the text. Dashed line is the assumed Arrhenius expression for the OH + SCN^- reference reaction.

OH + H$_2$C(OH)$_2$: $\ln k_{R1} = (23.887 \pm 0.290) - (1020 \pm 90)/T$

OH + HCOO$^-$: $\ln k_{R3} = (26.029 \pm 1.326) - (1240 \pm 390)/T$

OH + CH$_3$COO$^-$: $\ln k_{R4} = (24.067 \pm 0.527) - (1770 \pm 160)/T$

OH + HCOOH: $\ln k_{R5} = (21.807 \pm 0.526) - (991 \pm 156)/T$

OH + CH$_3$COOH: $\ln k_{R6} = (21.144 \pm 0.293) - (1330 \pm 90)/T$

Errors in the above expressions are 1σ and represent precision only. It should be noted that the uncertainties in the above activation energies do not include the uncertainty in the activation energy for the OH + SCN$^-$ reaction (estimated to be ± 0.5 kcal mol^{-1}).

Possible Secondary Chemistry Complications

An important experimental parameter in this competitive kinetics investigation is the concentration of SCN$^-$. We have chosen to employ relatively low SCN$^-$ concentrations, ranging from $(0.5-5) \times 10^{-5}M$. Use of low SCN$^-$ concentrations has two important advantages. First, production of reactive transient species via 248-nm laser flash photolysis of SCN$^-$ is kept at a level where such species cannot perturb the chemistry under investigation.[39,40] Second, in air-saturated solutions, the radical products (Y) of R1, R3, R4, R5, and R6 should react with O$_2$ on a time scale which is sufficiently rapid so that Y cannot possibly react with SCN$^-$.

$$OH + X \rightarrow Y + H_2O(OH^-) \qquad (R13)$$

$$Y + O_2 \rightarrow HO_2(O_2^-) + \text{other products} \qquad (R14)$$

$$Y + SCN^- \rightarrow SCN + Y^- \qquad (R15)$$

The occurrence of R15 followed by R12 could potentially lead to underestimation of the rate coefficients of interest. On the other hand, the occurrence of R14 produces HO$_2$(O$_2^-$) radicals that, under our experimental conditions, are unreactive on the time scale for (SCN)$_2^-$ appearance.[41]

The use of low SCN$^-$ concentrations does, in contrast to the advantages discussed above, have one potentially important disadvantage. Over the range of SCN$^-$ concentrations employed in this study, the SCN $\leftrightarrow$ (SCN)$_2^-$ equilibrium is such that 10–50% of the radicals exist as SCN. Hence, the occurrence of the reactions

$$SCN + X \rightarrow SCN^- (HSCN) + Y \qquad \text{(R16)}$$

could lead to overestimation of the rate coefficients of interest; while SCN^- appears to be a viable competitor for a wide range of OH(aq) kinetics studies,[14,42] systematic errors due to R16 have been reported for X equal to ferrocenyl-substituted carboxylic acids.[43] Two pieces of experimental evidence lead us to conclude that R16 was not a problem in this study. First, as typified by the 297 K results for X = $HCOO^-$ shown in Figure 2, no systematic trend in k_{OH+X}/k_{OH+SCN^-} as a function of $[SCN^-]$ is observed for any of the reactants of interest. Second, the occurence of R16 would manifest itself as an increase in the observed $(SCN)_2^-$ decay rate with increasing $[X]$ at constant $[SCN^-]$ and as a decrease in the observed $(SCN)_2^-$ decay rate with increasing $[SCN^-]$ at constant $[X]$; as typified by the $(SCN)_2^-$ temporal profiles shown in Figure 1, no such behavior was observed for any of the reactions studied.

In a recent study of the OH-initiated oxidation of formaldehyde in aqueous solution, McElroy and Waygood obtained evidence that the radical product of R1 reacts with H_2O_2 via two channels, one of which regenerates OH.[13]

$$HC(OH)_2 + H_2O_2 \rightarrow OH + HCOOH + H_2O \qquad \text{(R17a)}$$

$$\rightarrow H_2C(OH)_2 + HO_2 \qquad \text{(R17b)}$$

McElroy and Waygood reported the rate coefficients $(3.5 \pm 1.2) \times 10^6$ and $(7.4 \pm 1.7) \times 10^5 \, M^{-1} s^{-1}$ for channels R17a and R17b, respectively. Under our experimental conditions, $[O_2] \approx 3 \, [H_2O_2]$. Since R2 is very fast, i.e., $k_{R2} \sim 4 \times 10^9 \, M^{-1} s^{-1}$,[13,44] systematic error in our measurement of the OH + $H_2C(OH)_2$ rate coefficient due to the occurence of R17a must be negligible.

Comparison With Previous Work

As pointed out above, some room temperature kinetic data is available for each of the reactions, R1, R3, R4, R5, and R6,[18–27] although the kinetic database is not very good, particularly for crucial cloud reaction[15–17] R1. A critical review has recently been published that considers all previous kinetics studies of R1, R3, R4, R5, and R6.[14] Our results are compared with the recommendations of the critical review authors in Table 2. The agreement of our 297 K results with literature values is quite good, although for R1, R4, and R5 we obtain rate coefficients that are lower than current recommendations by factors of 1.28, 1.29, and 1.21, respectively; these differences are well within the combined uncertainties of the values being compared.

The only temperature-dependent data with which to compare our results is the study of R3 by Elliot and Simsons;[27] they measured the OH + $HCOO^-$ rate coefficient at four temperatures over the range 292–352 K using a pulse

Table 2. Comparison of the 297-K Rate Coefficients Obtained in This Study
with a Recent Critical Evaluation of All Previous Literature[14]

	297-K Rate Coefficient (10^7 M^{-1} s^{-1})	
Reaction	Ref. 14	This Work[a]
$OH + H_2C(OH)_2$	100	78 ± 12[b]
$OH + HCOO^-$	320	340 ± 39
$OH + HCOOH$	13	10.1 ± 1.3
$OH + CH_3COO^-$	8.5	7.0 ± 0.8
$OH + CH_3COOH$	1.6	1.7 ± 0.3

a Errors are 1σ and include an estimated 10% 1σ uncertainty in the 297 K rate
   coefficient for the reference reaction, $OH + SCN^-$.
b Error includes an estimated $\pm 10\%$ uncertainty in the $H_2C(OH)_2$ concentration.

radiolysis-competitive kinetics method with $Fe(CN)_6^{4-}$ as the competitor. Assuming an activation energy of 3.11 kcal mol^{-1} for the reference reaction
(measured as part of their study[27]), Elliot and Simsons reported an activation
energy 2.03 kcal mol^{-1} for R3 — about 0.36 kcal mol^{-1} lower than the activation
energy reported in this study. The agreement is excellent considering that the
combined uncertainties of the two studies, including estimated uncertainties in
the activation energies for the reference reactions, appears to be close to 1.0 kcal
mol^{-1}.

Implications for Atmospheric Chemistry

The most significant result obtained in this work is the relatively slow rate
coefficient for the $OH + H_2C(OH)_2$ reaction, R1. Our 297-K rate coefficient
for R1 is a factor of 2.6 lower than the value assumed by Lelieveld and Crutzen
in recent modeling studies of the role of clouds in tropospheric photochemistry;[16,17] hence, predicted reductions in gas-phase levels of HO_x, CO, and O_3
due to cloud processing of formaldehyde would probably be somewhat dimin-
ished if our values for $k_{R1}(T)$ were incorporated into Lelieveld and Crutzen's
model.

The temperature-dependence studies reported in this chapter demonstrate that
activation energies for fast free radical reactions are often considerably lower than
the "activation energy" for aqueous diffusion. A significant effort will be required
to accurately establish $k(T)$ values for important aqueous-phase atmospheric free
radical reactions.

ACKNOWLEDGMENTS

This research was supported through Grants R814527 and R816559 from the
U.S. Environmental Protection Agency, Office of Exploratory Research.

REFERENCES

1. Galloway, J. N., Likens, G. E., Keene, W. C., and Miller, J. M., The composition of precipitation in remote areas of the world, *J. Geophys. Res.*, 87, 8771, 1982.
2. Keene, W. C. and Galloway, J. N., Considerations regarding sources of formic and acetic acids in the troposphere, *J. Geophys. Res.*, 91, 14466, 1986.
3. Graedel, T. E., Atmospheric formic acid from formicine ants, *EOS Trans. Am. Geophys. Union*, 68, 273, 1987.
4. Atkinson, R. and Lloyd, A. C., Evaluation of kinetic and mechanistic data for modeling of photochemical smog, *J. Phys. Chem. Ref. Data*, 13, 315, 1984.
5. Jacob, D. J. and Wofsy, S. C., Photochemistry of biogenic emissions over the Amazon forest, *J. Geophys. Res.*, 93, 1477, 1988.
6. Jacob, D. J. and Wofsy, S. C., Photochemical production of carboxylic acids in a remote continental atmosphere, in *Proceedings of the NATO Advanced Research Workshop on Acid Deposition Processes at High Elevation Sites*, Unsworth, M. H. and Fowler, D., Eds., D. Reidel, Dordrecht, Netherlands, 1989.
7. Chameides, W. L. and Davis, D. D., Aqueous phase source for formic acid in clouds, *Nature (London)*, 304, 427, 1983.
8. Graedel, T. E. and Goldberg, K. I., Kinetic studies of raindrop chemistry. 1. Inorganic and organic processes, *J. Geophys. Res.*, 88, 10865, 1983.
9. Adewuyi, Y. G., Cho, S.-Y., Tsay, R.-P., and Carmichael, G. R., Importance of formaldehyde in cloud chemistry, *Atmos. Environ.*, 18, 2413, 1984.
10. Chameides, W. L., The photochemistry of a remote marine stratiform cloud, *J. Geophys. Res.*, 89, 4739, 1984.
11. Graedel, T. E., Mandich, M. L., and Weschler, C. J., Jr., Kinetic model studies of atmospheric droplet chemistry. 2. Homogeneous transition metal chemistry in raindrops, *J. Geophys. Res.*, 91, 5205, 1986.
12. Jacob, D. J., Chemistry of OH in remote clouds and its role in the production of formic acid and peroxymonosulfate, *J. Geophys. Res.*, 91, 9807, 1986.
13. McElroy, W. J. and Waygood, S. J., Oxidation of formaldehyde by the hydroxyl radical in aqueous solution, *J. Chem. Soc. Faraday Trans.*, 87, 1513, 1991.
14. Buxton, G. V., Greenstock, C. L., Helman, W. P., and Ross, A. B., Critical review of rate constants for reactions of hydrated electrons, hydrogen atoms, and hydroxyl radicals ($\cdot$OH/$\cdot$O$^-$) in aqueous solution, *J. Phys. Chem. Ref. Data*, 17, 513, 1988.
15. Keene, W. C. and Galloway, J. N., The biogeochemical cycling of formic and acetic acids through the troposphere: an overview of current understanding, *Tellus*, 40B, 322, 1988.
16. Lelieveld, J. and Crutzen, P. J., Influences of cloud photochemical processes on tropospheric ozone, *Nature (London)*, 343, 227, 1990.
17. Lelieveld, J. and Crutzen, P. J., The role of clouds in tropospheric photochemistry, *J. Atmos. Chem.*, 12, 229, 1991.
18. Hart, E. J., Thomas, J. K., and Gordon S., A review of the radiation chemistry of single-carbon compounds and some reactions of the hydrated electron in aqueous solution, *Radiat. Res. Suppl.*, 4, 74, 1964.
19. Thomas, J. K., Rates of reaction of the hydroxyl radical, *Trans. Faraday Soc.*, 61, 1965.
20. Adams, G. E., Boag, J. W., Currant, J., and Michael, B. D., in *Pulse Radiolysis*, Ebert, M., Keene, J. P., Swallow, A. J., and Baxendale, J. H., Eds., Academic Press, New York, 1965, chap. 10.
21. Baxendale, K. H. and Khan, A. A., The pulse radiolysis of p-nitrosodimethylaniline in aqueous solution, *Int. J. Radiat. Phys. Chem.*, 1, 11, 1969.
22. Buxton, G. V., Pulse radiolysis of aqueous solutions. Some rates of reaction of OH and O$^-$ and pH dependence of the yield of O$_3^-$, *Trans. Faraday Soc.*, 65, 2150, 1969.

23. Willson, R. L., Greenstock, G. L., Adams, G. E., Wageman, R., and Dorfman, L. M., The standardization of hydroxyl radical rate data from radiation chemistry, *Int. J. Radiat. Phys. Chem.*, 3, 211, 1971.

24. Markovic, V. and Sehested, K., Radiolysis of aqueous solutions of some simple compounds containing aldehyde groups. Part I: formaldehyde, in *Proceedings of the Third Tihany Symposium on Radiation Chemistry*, Dobo, J. and Hedvig, P., Eds., Akademiai Kiado, Budapest, Hungary, 1972, 1281.

25. Fisher, M. M. and Hamill, W. H., Electronic processes in pulse irradiated aqueous and alcoholic systems, *J. Phys. Chem.*, 77, 171, 1973.

26. Wolfenden, B. S. and Willson, R. L., Radical-cations as reference chromogens in kinetic studies of one-electron transfer reactions: pulse radiolysis studies of 2,2'-azinobis-(3-ethylbenzthiazoline-6-sulphonate), *J. Chem. Soc. Perkin Trans.*, 2, 805, 1982.

27. Elliot, A. J. and Simsons, A. S., Rate constants for reactions of hydroxyl radicals as a function of temperature, *Radiat. Phys. Chem.*, 24, 229, 1984.

28. Glasstone, S., Laidler, K. J., and Eyring, H., *The Theory of Rate Processes*, McGraw-Hill, New York, 1941, 523.

29. Elliot, A. J., McCracken, D. R., Buxton, G. V., and Wood, N. D., Estimation of rate constants for near-diffusion controlled reactions in water at high temperatures, *J. Chem. Soc. Faraday Trans.*, 86, 1539, 1990, and references therein.

30. Wine, P. H., Mauldin, R. L., III, and Thorn, R. P., Kinetics and spectroscopy of the NO_3 radical in aqueous ceric nitrate-nitric acid solutions, *J. Phys. Chem.*, 92, 1156, 1988.

31. Tang, Y., Thorn, R. P., Mauldin, R. L., III, and Wine, P. H., Kinetics and spectroscopy of the SO_4^- radical in aqueous solution, *J. Photochem. Photobiol. A: Chem.*, 44, 243, 1988.

32. Wine, P. H., Tang, Y., Thorn, R. P., Wells, J. R., and Davis, D. D., Kinetics of aqueous phase reactions of the SO_4^- radical with potential importance in cloud chemistry, *J. Geophys. Res.*, 94, 1015, 1989.

33. Chin, M. and Wine, P. H., A temperature dependent kinetics study of the aqueous phase reactions $OH + SCN^- \rightarrow SCNOH^-$ and $SCN + SCN^- \leftrightarrow (SCN)_2^-$, *J. Photochem. Photobiol. A: Chem.*, 1993, 69, 17, 1992.

34. Baxendale, J. H., Bevan, P. L. T., and Stott, D. A., Pulse radiolysis of aqueous thiocyanate and iodide solutions, *Trans. Faraday Soc.*, 64, 2389, 1968.

35. Schuler, R. H., Patterson, L. K., and Janata, E., Yield for the scavenging of OH radicals in the radiolysis of N_2O-saturated solutions, *J. Phys. Chem.*, 84, 2088, 1980.

36. Spence, R. and Wild, W., The preparation of liquid monomeric formaldehyde, *J. Chem. Soc.*, 338, 1935.

37. Nash, T., The colorimetric estimation of formaldehyde by means of the Hantzsch Reaction, *Biochem. J.*, 55, 416, 1953.

38. Clifton, C. L. and Huie, R. E., Rate constants for hydrogen abstraction reactions of the sulfate radical, SO_4^-. Alcohols, *Int. J. Chem. Kinet.*, 21, 677, 1989.

39. Luria, M. and Treinin, A., The photochemistry of NCS^- in solution, *J. Phys. Chem.*, 72, 305, 1968.

40. Dogliotti, L. and Hayon, E., Flash photolysis study of sulfite, thiocyanate, and thiosulfate ions in solution, *J. Phys. Chem.*, 72, 1800, 1968.

41. Bielski, B. H. J., Cabelli, D. E., and Arudi, R. L., Reactivity of HO_2/O_2^- radicals in aqueous solution, *J. Phys. Chem. Ref. Data*, 14, 1041, 1985.

42. Dorfman, L. M. and Adams, G. E., The reactivity of hydroxyl radical in aqueous solutions, NSRDS-NBS-46, U.S. Department of Commerce, National Bureau of Standards, 1973, chap. 3.

43. Logan, S. R. and Salmon, G. A., Discrepancies between the rate constants for the reactions of hydroxyl radicals with ferrocenyl-substituted carboxylic acids determined by direct measurement and by competition with thiocyanate-ion, *Radiat. Phys. Chem.*, 24, 593, 1984.

44. Bothe, E. and Schulte-Frohlinde, D., Reaction of dihydroxymethyl radicals with molecular oxygen in aqueous solution, *Z. Naturforsch. Teil B*, 35, 1035, 1980.

CHAPTER 6

Factors Affecting Photolysis of Organic Compounds on Soils

Glenn C. Miller and Susan G. Donaldson

INTRODUCTION

Soil surfaces receive large quantities of organic contaminants, including combustion products, pesticides, and naturally occurring substances. Due to the presence of strong chromophores and a variety of indigenous reactants in soils, photochemical processes can alter both soil surfaces and the chemicals sorbed to those surfaces. The heterogeneity of surfaces, however, has not allowed successful modeling of photolysis processes as compared to water or air, which offer greater homogeneity. Recent efforts have sought to understand at least qualitatively how various factors affect photochemical processes. Several factors must be examined and understood before an assessment of the importance of photochemical processes on soil surfaces can be made. The three primary factors which affect photolysis include the following:

1. Depth of photolysis — In order to model photolysis, an estimate of the depth dependence of photolysis in the solid phase must be made. Only that portion of applied compounds that exists within this zone will be available for phototransformation reactions.
2. Photochemical quenching-sensitization reactions — Naturally occurring organic and inorganic substances can potentially affect the fundamental photochemical processes occurring at soil-atmosphere interface, either by generating reactive species or by serving to alter the photoproduct distribution and rates of direct photolysis.

0-87371-871-2/94/$0.00+$.50
© 1994 by CRC Press, Inc.

3. Transport processes — Photolysis at the soil-atmosphere interface will be a function of transport to and away from the sunlight-irradiated surface. Water will transport contaminants into and out of the photic zone, and can also result in redistribution of particles at the soil surface.

The following discussion will review efforts to better understand the three factors listed above. We will also describe new work in our laboratory coupling the transport of contaminants to the soil surface with photochemical reactions.

DEPTH OF PHOTOLYSIS

Soil contains both organic and inorganic chromophores that substantially limit the penetration of light into soils. The brown color of soils is indicative of broad absorption of light, which is greatest in the short wavelengths and trails off at the longer wavelengths. Humic substances, iron and manganese oxides, and reflective surfaces all contribute toward limiting light penetration. Competitive light absorption is greatest in the ultraviolet (UV) region. This is also the region of the sunlight spectrum that is most active in photochemical transformations. Except for a variety of dyes, most compounds absorb sunlight most strongly in the UV portion of the spectrum. The close association of organic contaminants with the humic fraction of soil also will tend to enhance light screening of the contaminants. Due to both competitive absorption of light and also physical shading by soil constituents, it is expected that the depth of direct photolysis in soils will be relatively shallow.

A series of experiments were conducted to estimate the depth of both direct and indirect photolysis.[1] These experiments were based on the expectation that when soils containing a uniformly incorporated, photolabile compound were irradiated, photolysis would proceed until all of that percentage of the chemical that was exposed to light was transformed. The average depth of photolysis is calculated as the fractional loss of a uniformly incorporated chemical multiplied by the depth of the soil examined. For example, 25% loss of a photolabile chemical from a 1.0-mm deep soil would indicate an average photolysis depth of 0.25 mm. Table 1 presents the characteristics of two of the soils used in these studies, and Tables 2 and 3 provide a range of photolysis depths which were determined for these soils. The chemicals used to establish photolysis depth were flumetralin and disulfoton. Flumetralin undergoes rapid direct photolysis in sunlight in solution ($t_{1/2}$ = 20 min

Table 1. Soil Properties of Designated Agricultural Areas

Soil Designation	% Organic Matter	pH	% Clay	% Silt	% Sand	Bulk Density
Kracaws	0.8	7.5	8	47	45	1.28
Main Station Farm	2.0	7.2	14	32	54	1.32
Montana Grain	2.2	7.6	22	28	50	1.60

Table 2. Estimated Mean Photolysis Depths in Main Station Farm Soil

	Soil Depth (mm)	Outdoor Estimated Photolysis Depth[a] (mm)	Indoor Estimated Photolysis Depth (mm)
A. Disulfoton	0.5	0.2	0.1
(indirect	1.0	0.4	0.3
photolysis)	1.9	0.7	0.2
	3.8	0.7	0.4
B. Flumetralin	0.5	0.1	0.1
(direct	1.0	0.1	0.2
photolysis)	1.9	0.2	0.2
	3.8	0.4	0.2

[a] Estimates of photolysis depth were derived by multiplying the percent recovery of starting material by soil depth when photolysis rates approached zero.

in mid-day mid-summer sunlight) and also has a low vapor pressure, thus limiting loss by volatilization. Disulfoton reacts with singlet oxygen generated in soils by an indirect process. In both cases, the average photolysis depth is shallow. In well-controlled conditions using artificial lights in the laboratory, the photolysis depths ranged from 0.1 to 0.2 mm for flumetralin, while outdoors the measured average photolysis depths were larger, up to 0.4 mm. The deeper photolysis depths outdoors were probably the result of wind shaking the soil containers, which would have mixed the soils and exposed new surfaces to sunlight. The photolysis depth of disulfoton was determined to be greater, probably due to the ability of singlet oxygen to diffuse to lower depths in soils.[2] Under laboratory conditions, the photolysis depths for disulfoton were 0.1–0.4 cm, while outdoors the measured photolysis depths ranged from 0.2 to 0.8 mm. Vapor transport of these compounds would also result in larger measured average photolysis depths. However, using a homologous series of aryl ketones with identical photolysis rates but varying vapor pressures, vapor transport in dry soil was not observed for compounds with vapor pressures less than 10^{-4} mm mercury, which includes most pesticides.[3]

Table 3.   Estimated Mean Photolysis Depths in Montana Grain Soil

	Soil Depth (mm)	Outdoor Estimated Photolysis Depth[a] (mm)	Indoor Estimated Photolysis Depth (mm)
A. Disulfoton	0.4	0.3	0.2
(indirect	0.8	0.4	0.3
photolysis)	1.6	0.6	0.2
	3.1	0.8	0.4
B. Flumetralin	0.4	0.1	0.1
(direct	0.8	0.1	0.1
photolysis)	1.6	0.2	0.2
	3.1	0.1	0.2

[a] Estimates of photolysis depth were derived by multiplying the percent recovery of starting material by soil depth when photolysis rates approached zero.

Photolysis depths are unlikely to be uniform. Soils are highly heterogeneous media, and both reflection and refraction of light is likely to allow light penetration, in some cases, which is much deeper than the average depth. Alternatively, it is probable that some particles may restrict depths of photolysis to less than 10 µm. Soil textural analysis may be helpful in predicting approximate photolysis depths. Under any circumstances, however, the lack of light penetration into soils will restrict photolysis to only very shallow depths, on the order of 0.5 mm or less, except in relatively transparent sands.

PHOTOCHEMICAL SENSITIZATION-QUENCHING PROCESSES

Humic substances in natural waters are known to serve as energy-transfer agents to oxygen, resulting in the formation of singlet oxygen.[4] Although soil organic matter is not identical in composition to dissolved organic matter, the basic structural characteristics are sufficiently similar that the same type of energy-transfer reactions are likely to occur. Indeed, singlet oxygen was shown to be generated on soil surfaces using chemical traps, deuterated tetramethylethylene, and 1,2-dimethylcyclohexene.[5] The rate of production was sufficiently rapid that half of the initial concentration of these compounds was lost in less than 4 hr of mid-summer, mid-day sunlight. Singlet oxygen reactions are likely to be important for sulfides[6] and other compounds that are susceptible to reactions with this oxidant.

When the photolysis rates of singlet oxygen traps were measured as a function of organic content of the soil, no clear correlation was found.[7] In fact, some soils with the lowest organic matter content showed the greatest ability to generate singlet oxygen. Additional work using silica gel, alumina, and very low-organic desert sands indicated that these materials also were capable of generating singlet oxygen.[6] The mechanism of this inorganic sensitization was not determined, although the involvement of lattice defects is likely since oxygen quenched the phosphorescence of illuminated nontransition-metal surfaces.[8]

If singlet oxygen (and other potential oxidants) is produced on irradiated soil surfaces, it is reasonable to predict that the soil organic matter would be oxidized to the depth of singlet oxygen penetration. However, experiments that exposed bulk soils of varying organic content to 30 days of summer sunlight in thin layers did not result in an appreciable loss of organic matter from the soils (unpublished data). The soil organic matter appears to be surprisingly stable to both direct photolysis and reactions with singlet oxygen. Additional work is warranted to understand how soil organic matter is affected at the irradiated atmosphere-soil interface.

The photochemistry of compounds sorbed on soils has been shown to vary from that in solution.[9] Octachlorodibenzo-p-dioxin (OCDD) undergoes photolysis on soil surfaces resulting predominantly in reduction at the 1,4,6,9 positions and producing, in part, 2,3,7,8-tetrachlorodibenzo-p-dioxin (TCDD). In solution, photolysis occurs predominantly on the 2,3,7,8 positions. Production of 2,3,7,8-TCDD was not observed. The same general photochemical pathway was confirmed in later work on soils contaminated with OCDD,[10] but this photoreductive behavior was not observed for octachlorodibenzofuran; preferential photoreduction on soils resulting in the toxic 2,3,7,8-substituted dibenzofurans did not occur.

Nilles and Zabik[11] also observed substantially different photoproducts from the herbicide basalin on soils relative to solution photolysis. On soils, a nitroso photoproduct was observed that was not detected in irradiated solutions. These photoproduct differences are likely due to physical constraints on compounds in the sorbed state. Sorption on soils likely affects the vibrational and rotational properties of molecules, thus having a significant effect on the photochemical fate of those compounds. Interpretation of photoproduct differences on soils relative to solutions is also complicated by the light-screening effects of soils. Photochemically reactive photoproducts can be stabilized on soils by light screening, while in solution these same photoproducts would be further photochemically transformed.

TRANSPORT PROCESSES

Because the depth of photolysis on soil surfaces is shallow, any process that moves compounds into or out of the sunlight-exposed zone will affect the rate of photolysis. This can involve either physical mixing or erosion of soil or transport of chemicals through the bulk soil. Although water is the most common medium available for transport in soils, solvents may also exert an effect on the transport of compounds. For example, the photolysis rate of 2,3,7,8-TCDD on thin soil layers was shown to be substantially enhanced when the organic solvent hexadecane was added to the soil layers. When these soils were irradiated with FS40 lamps (λ_{max} = 310 nm), there was a substantial decrease in concentrations of TCDD in the presence of organic solvent (Figures 1 and 2).[12] Hexadecane appeared to act as a hydrocarbon film on the soil particles, which allowed mobilization of TCDD and migration to the irradiated soil surface. Rates of photolysis decreased as the organic matter content in soils (0.8% in Kracaws, 2.2% in Montana Grain) increased. The Montana Grain soil also had a much higher clay fraction (22 vs 8%), providing more sorptive area and greater screening of the light.

Although hydrocarbon transport will be significant for hydrophobic contaminants where oils are present, transport of contaminants in water is of more general environmental significance, particularly for those materials that exhibit lower sorption to soils. Sorption of organic molecules will vary according to the organic matter content of soils, moisture content, particle size (surface area to volume ratios), mineralogy, pH, and the presence of other organic compounds or cosolvents. For many low-solubility organic compounds (i.e., pesticides), sorption is generally controlled by the organic carbon fraction of soils and sediments.[13,14] The organic fraction of the soil will dominate sorption of nonionic organic compounds

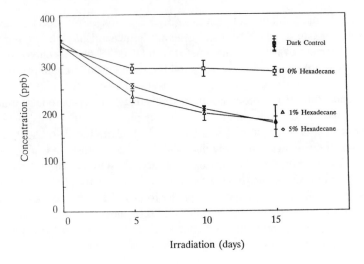

Figure 1. Sunlamp photolysis of TCDD on Montana Grain soil.

when the total organic carbon exceeds 0.1%.[15] Correlations have also been made to the water solubility of the solute and its octanol-water partition coefficient, although Ainsworth et al.[16] found that the organic carbon content of the particular soil horizon was best correlated with degree of sorption of carbazole. Changes in pH can result in changes in the surface charge on clay minerals or may affect the ionization state in which a compound exists, thus altering the degree of sorption.[17] Sorbed hydrophobic compounds may be displaced from soil surfaces by water molecules[18] or by changes in solubility that result from mixtures of chemicals or

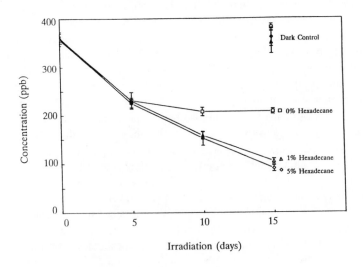

Figure 2. Sunlamp photolysis of TCDD on Kracaws soil.

solvents.[19] Displacement from particle surfaces increases the rate of transport relative to the rate of water flow and results in more rapid translocation.

The movement of water-soluble chemicals carried in capillary rise soil water to the soil-atmosphere interface has been well documented.[20-22] Mahnken and Weber[23] studied the effect of capillary rise on the upward transport of the herbicides triasulfuron and chlorsulfuron. Using a simple soil column and five different water and evaporation treatments, they found the greatest enhancement came in soil columns fed with a continuous supply of water at the base. This redistribution was found to be the inverse of leaching experiments, suggesting that similar mechanisms are in effect for both unsaturated flow leaching and capillary rise.

In related work, Spencer[24] examined the volatilization of pesticides from soils as affected by water evaporation from the bulk soil. The rate of evaporation is dependent on moisture content, the hydraulic conductivity of soil (which is a function of volumetric water content), boundary layer conditions such as temperature and wind speed, and the depth to water table. Spencer classified chemicals as Category I, II, or III, based on Henry's law constants. Category I compounds have high Henry's law constants and move to the surface, evaporating rapidly. Category III compounds have low Henry's law constants and relatively high water solubilities. These compounds move to the surface, but do not volatilize. Category II compounds are intermediate in their behavior.

The extent of phototransformation processes at the atmosphere-soil interface may have been underestimated, particularly for these chemicals that move with evaporating water, and have low vapor pressure (Category III compounds). Although photolysis occurs only in the top 100–500 μm of soil, chemicals that are transported in water can move to the surface and thus be exposed to both direct and indirect photochemical processes.

For water-soluble, photolabile compounds, photolysis in soils is therefore likely to be a significant transformation pathway. In the presence of transport in evaporating water, weakly sorbed compounds will move to the soil surface and into the photic zone. Rates of transport are expected to be the most rapid in saturated soils with high hydraulic conductivities, such as sands. The degree to which transport will result in measurable photolysis rates will depend on the degree of sorption/solubilization of the compound, the rate of evaporation from the soil, and rates of photolysis in aqueous solution and at the soil surface. Any decrease in the amount of light that reaches the soil surface, whether seasonal or due to shading by plant cover, will also be important.

We have examined the effect of water transport on the photolysis of two moderately water-soluble pesticides, napropamide and imazaquin, which have low vapor pressures. Microcosms of loamy sand soil (6 cm in diameter and 6 cm in depth) were supplied with water from the bottom, with the water table maintained in the soil at the 4.5 cm depth. Placed atop 3 cm of clean soil was 3 cm of uniformly spiked soil, and the samples were exposed to artificial light, again at a maximum wavelength of 310 nm. Water was supplied to the soils on a constant basis to approximate saturated flow. Other soils were maintained in an air-dry

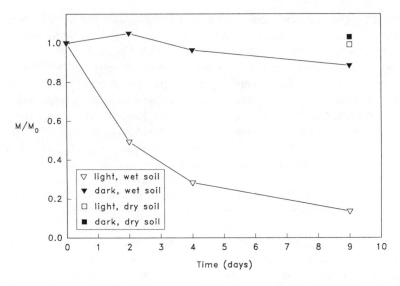

Figure 3. Sunlamp photolysis of napropamide in loamy sand soil subjected to saturated flow vs air-dry conditions.

state. Photolysis of napropamide was rapid (Figure 3), with less than 15% remaining after 9 days of continuous irradiation. No loss was seen in the dry soils. The decrease in concentration in the wet, dark controls may be due to irreversible sorption or biodegradation, although the soils were autoclaved prior to incorporation of the pesticides.

The rate of photolysis was slower for imazaquin (Figure 4), with the same behavior seen in the dry and the dark controls. Differences in the rate of photolysis may be due, in part, to differences in degree of light absorbance. Napropamide has a half-life on the order of 7 min in water solution when exposed to the artificial lights, as compared to imazaquin ($t_{1/2}$ = 120 min).

Profiles of the distribution of each chemical with depth in the soil (Figures 5 and 6) show that the movement of napropamide to the soil surface was much more rapid, presumably due to increased water solubility. Sorption was minimized by the low-organic matter content of the soil (0.3%). These results suggest that the rate of transport is the limiting factor on the rate of photochemical loss, and processes that increase rates of transport can be expected to have a significant effect on the degree to which photolysis occurs in soils.

SUMMARY

Photolysis on solid surfaces depends on the distribution of the compound relative to the photic zone. Photolysis will only occur within a shallow surface

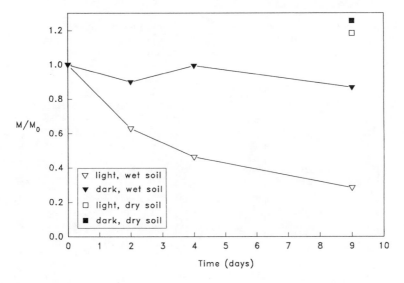

Figure 4.   Sunlamp photolysis of imazaquin in loamy sand soil subjected to saturated flow vs air-dry conditions.

zone, the depth of which depends on soil characteristics and the mechanism of photodegradation. Light absorption and photolysis of organic contaminants will be influenced by sorption reactions, which are related to the soil organic matter content, and by singlet oxygen formation. Transport of soluble organic compounds to the soil surface in evaporating water or in soil films may provide a substantial increase in the rate of photolysis at soil surfaces. Assessing all these factors is necessary before an accurate estimate of the relative importance of photolysis as transformation process in soils can be made.

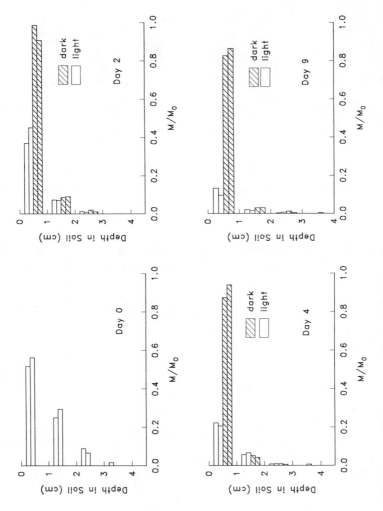

Figure 5. Distribution of napropamide with depth in loamy sand soil undergoing constant evaporation while irradiated by sunlamps.

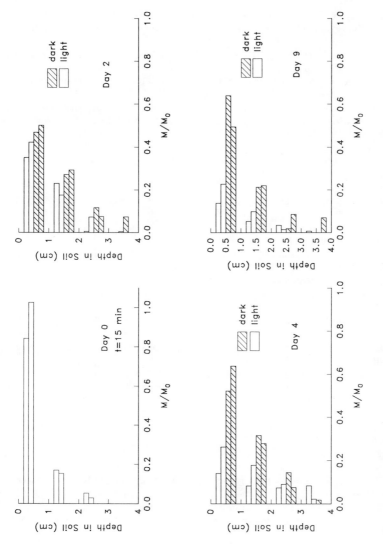

Figure 6. Distribution of imazaquin with depth in loamy sand soil undergoing constant evaporation while irradiated by sunlamps.

REFERENCES

1. Hebert, V. R. and Miller, G. C., Depth dependence of direct and indirect photolysis on soil surfaces, *J. Agric. Food Chem.*, 38, 913, 1990.
2. Miller, G. C., Hebert, V. R., and Miller, W. W., Effects of sunlight on organic contaminants at the atmosphere-soil interface, in *Reactions and Movement of Organic Chemicals in Soils*, Sawhney, B., Ed., Soil Science Society of America, Madison, WI, 1989a, 99–110.
3. Kieatiwong, S. and Miller, G. C., Photolysis of aryl ketones with varying vapor pressures on soil, *Environ. Toxicol. Chem.*, 11, 173, 1992.
4. Zepp, R. G., Wolfe, N. L., Baughman, G. L., and Hollis, R. C., Singlet oxygen in natural waters, *Nature (London)*, 267, 421, 1977.
5. Gohre, K. and Miller, G. C., Photooxidation of thioether pesticides on soil surfaces, *J. Agric. Food Chem.*, 34, 709, 1986.
6. Gohre, K. and Miller, G. C., Photochemical generation of singlet oxygen on nontransition-metal oxide surfaces, *J. Chem. Soc. Faraday Trans.*, 81, 793, 1985.
7. Gohre, K., Scholl, R., and Miller, G. C., Singlet oxygen reactions on irradiated soil surfaces, *Environ. Sci. Technol.*, 20, 934, 1986.
8. Tench, A. J. and Pott, G. T., Surface states in some alkaline earth oxides, *Chem. Phys. Lett.*, 26, 590, 1974.
9. Miller, G. C., Hebert, V. R., Miille, M. J., Mitzel, R., and Zepp, R. G., Photolysis of octachlorodibenzo-p-dioxin on soils: production of 2,3,7,8-TCDD, *Chemosphere*, 18(1–6), 1265, 1989b.
10. Tysklind, M., Carey, A. E., Rappe, C., and Miller, G. C., Photolysis of OCDF and OCDD on soil, *Chemosphere*, 1992, submitted
11. Nilles, G. P. and Zabik, M. J., Photochemistry of bioactive compounds — multiphase photodegradation of basalin, *J. Agric. Food Chem.*, 22, 684, 1974.
12. Kieatiwong, S., Nguyen, L. V., Hebert, V. R., Hackett, M., Miller, G. C., Miille, M. J., and Mitzel, R., Photolysis of chlorinated dioxins in organic solvents and on soils, *Environ. Sci. Technol.*, 24, 1575, 1990.
13. Karickoff, S. W., Organic pollutant sorption in aqueous systems, *J. Hydraul. Eng.*, 110, 707, 1984.
14. Gerstl, Z. and Yaron, B., Behavior of bromacil and napropamide in soils: I. Adsorption and degradation, *Soil Sci. Soc. Am. J.*, 47, 474, 1983.
15. Pignatello, J. J., Sorption dynamics of organic compounds in soils and sediments, in *Reactions and Movement of Organic Chemicals in Soils*, Sahwney, B., Ed., Soil Science Society of America, Madison, WI, 1989, 45–97.
16. Ainsworth, C. C., Zachara, J. M., and Smith, S. C., Carbazole sorption by surface and subsurface material: influence of sorbent and solvent properties, *Soil Sci. Soc. Am. J.*, 53, 1391, 1989.
17. Renner, K. A., Meggitt, W. F., and Penner, D., Effect of soil pH on imazaquin and imazethapyr adsorption to soil and phytotoxicity to corn (*Zea mays*), *Weed Sci.*, 36, 78, 1988.
18. Boesten, J. J. T. I. and van der Linden, A. M. A., Modeling the influence of sorption and transformation on pesticide leaching, *J. Environ. Qual.*, 20, 425, 1991.
19. Brusseau, M. L., Wood, A. L., and Rao, P. S. C., Influence of organic cosolvents on the sorption kinetics of hydrophobic organic chemicals, *Environ. Sci. Technol.*, 25, 903, 1991.
20. Sharma, P. K., Sinha, A. K., and Chaudhary, T. N., Upward flux of water and deep-placed P in relation to soil texture, water table depth and evaporation rate, *J. Agric. Sci.*, 104, 303, 1985.
21. Sheppard, M. I., Thibault, D. H., and Mitchell, J. H., Element leaching and capillary rise in sandy soil cores: experimental results, *J. Environ. Qual.*, 16, 273, 1987.

22. Spencer, W. F., Cliath, M. M., Jury, W. A., and Zhang, L.-Z., Volatilization of organic chemicals from soil as related to their Henry's Law constants, *J. Environ. Qual.*, 17, 504, 1988.

23. Mahnken, G. E. and Weber, J. B., Capillary movement of triasulfuron and chlorsulfuron in Rion sandy loam soil, *Proc. South. Weed Sci. Soc.*, 41, 332, 1988.

24. Spencer, W. F., Volatilization of pesticide residues, in *Fate of Pesticides in the Environment*, Biggar, J. W. and Seiber, J. N., Eds., Division of Agriculture and Natural Resources, University of California, 1987, Pub. #3320.

CHAPTER 7

Recent Advances in the Photochemistry of Natural Dissolved Organic Matter

William L. Miller

INTRODUCTION

Natural organic matter present in surface waters provides a principal pathway for the absorbance of solar energy in aquatic systems. The consequent cascade of photophysical and photochemical processes have received increasing scientific attention as their relevance to various related environmental fields has become more apparent. In applications ranging from remotely sensed pigment estimates to the photochemical treatment of wastewater, the absorbance of light by dissolved organic matter has proven to be an integral part of the processes in question. Some very good reviews of environmental photochemistry have been presented in recent years,[1-6] as well as special journal issues and symposium volumes with a number of papers on the subject.[7-10] In the last several years, exciting new advances have been made in our understanding of photochemical mechanisms involving natural organic matter and the importance of photochemical transformations in global biogeochemical cycles. Consequently, this chapter will focus on recent research in the photochemistry of naturally occurring organic matter. Some of the work reviewed here introduces new concepts and techniques, while some reinforces previously published results. The reader is directed to the reviews cited above for a more comprehensive view of progress made in the field prior to 1989. While recognizing the importance of heterogeneous photochemical processes in natural waters, my emphasis will be on studies of dissolved organic carbon. Distinction between isolated and unisolated organic matter will be made with the terms dissolved humic substances (DHS) and dissolved organic matter

0-87371-871-2/94/$0.00+$.50

111

(DOM), respectively. This review includes selected publications from about 1989 to the present, covering primary photophysical changes, photochemical transients, stable photoproducts, and proposed biogeochemical consequences of these photochemical processes.

PHOTOPHYSICS AND PRIMARY RADICALS

Light Absorbance

Absorbance of light by DOM is an essential event triggering photophysical and photochemical processes in natural waters. Consequently, DOM absorption spectra are a critical consideration for prediction of photochemical reactivity in natural waters. An inherent hurdle to any general interpretation of absorbance spectra is the myriad of organic chromophores available for incorporation into natural DOM. Most natural water samples exhibit a featureless exponential decrease in absorbance from the ultraviolet (UV) into the visible portion of the spectrum. At first glance, this spectrum appears to contain little information. Variations in the intensity and exponential nature of these spectra have been used, however, to investigate source material and structural features of natural organic material.

Recent field observations have reported calculated best fit slopes (S*) for log-linearized absorbance spectra that provide information related to DOM source material and possible organic transformations. Measured spectra in a tropical estuary[11] and in the Gulf of Mexico and adjacent coastal waters[12,13] give values for S* ranging from 0.007 to 0.0363 nm^{-1}; a result more variable than reported previously.[14,15] In general, S* values for whole water samples were higher upstream than downstream, higher in open ocean water than in near-coastal waters, and higher for isolated marine fulvic acid (FA) than for marine humic acids (HA).

The causes of these spectral variations are unknown. Carder et al.[13] point out that the resulting spectral characteristics for whole water samples can be modeled using varied mixing ratios of FA and HA, each fraction with different S* values. A simple spectral summation may not be strictly valid in all cases, however, since mixtures of high- and low-molecular weight (MW) fractions of a soil FA have been shown to result in absorbance values greater than those observed in the unfractionated sample, especially in the red end of the spectrum.[16] Extensive reversed-phase HPLC fractionation of Suwannee River FA[17] shows numerous components with the isolated hydrophobic fractions showing higher S* values than those for the hydrophilic portion isolated from the same sample. Clearly, variations in the relative proportions of these various components could explain at least part of the field variations of S*.

Dark incubations of estuarine water caused increases in S* values and were proposed to reflect changes in the mean MW of DOM with lower MW compounds exhibiting the highest S* values.[11] Conversely, dark incubations of radiolabeled glucose and leucine added to natural seawater have resulted in microbial alteration

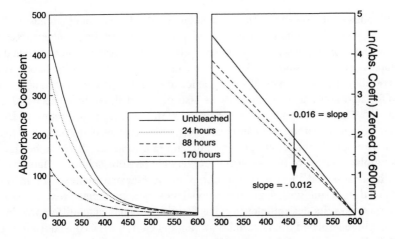

Figure 1. Change in spectral qualities for irradiated Suwannee River water. Samples were exposed to simulated solar light (1000-W xenon lamp, approximately 65 mW/cm²) in sealed quartz tubes for 24, 88, and 170 hr. The left panel shows changes in absorbance coefficients. The right panel compares linear regression lines for ln(Abs) vs wavelength as a function of exposure (lines normalized at 600 nm).

to produce higher MW dissolved materials.[18] Figure 1 shows that irradiation of terrestrial DOM from the Suwannee River results in a decrease in S* with increased time of irradiation,[19] a result similar to that of Wang et al.[20] for 2-hr irradiations of a model compound, salicylic acid, in aquatic systems. This argues against photochemical processing of terrestrial DOM as a mechanism explaining increased S* for marine samples. It is clear, however, from light-induced photobleaching of absorbance and concurrent changes in spectral quality that solar radiation causes significant alteration of DOM in natural waters.

Observed S* differences between marine and freshwater samples probably reflect different source material for incorporation into humic material. This is supported by recent stable carbon isotope data and ^{13}C-NMR studies that show open ocean DOM to be predominantly of marine origin.[21] Distinct compositional differences between riverine and marine DHS[21,22] have led to conclusions that freshwater DHS is rapidly removed from seawater. A complete understanding of the relation between the chemical structure of DOM and observed spectral characteristics of natural waters will require additional study.

Fluorescence

A portion of the light absorbed by aquatic DOM is reirradiated in the form of fluorescence. Even though the quantum efficiency of this process appears to be low,[14] fluorescence spectroscopy is suitably sensitive for direct characterization of DOM without potential artifacts introduced by separation, concentration, and/or derivitization. The reader is directed to an overview published by Senesi[23] for discussion of the mechanistic aspects of fluorescence and the applications and limitations of its use in evaluating aqueous DOM. Recent field studies[12,24-31] of fluorescence distributions have focused primarily on marine systems. Results show two to three times lower fluorescent efficiency in surface waters than in deep samples. This is presumably due to photochemical bleaching in the photic zone. A correlation between natural fluorescence and both apparent oxygen utilization (AOU) and nutrient distributions was also observed throughout the water column.[24,26] These observations are consistent with oxidative regeneration of DOM from particulate carbon and consequent formation or release of fluorescent materials. Biological productivity, river input, and rain were not shown to be significant global sources of fluorescence.[26] Reversible photobleaching of fluorescence has been observed[27] and may also contribute to elevated fluorescence in water removed from sunlight. Three-dimensional and contour mapping (excitation λ vs emission λ vs fluorescent intensity) has identified fundamental differences in the fluorophores isolated from coastal seawater, the Gulf of Mexico, and from different depths in the Black Sea.[12,25] Differences between estuarine and "blue" water samples indicate a shift to higher excitation and emission energies for the open ocean samples. Black Sea samples showed excitation-emission maxima which provide evidence for three distinct fluorophores with different vertical distributions. This technique shows promise as yet another tool for the investigation of organic sources and processes relative to DOM distributions in natural waters.

Measured rates of fluorescence photobleaching in marine samples have been used in conjunction with a shallow mixing model that incorporates calculated incident radiation and measured light attenuation and wind speeds to predict fluorescence vertical profiles.[27,28] There is general agreement between model predictions and fluorescence observations in the upper ocean (0–100 m), with the major model limitation being the ability of the physical mixing algorithms to simulate the mixed layer depth. This study demonstrates progress in the development of photochemical process models for natural waters. It also confirms that close collaboration between modelers of physical and photochemical processes will be required to obtain a comprehensive view of the distribution of photochemical processes and products in natural waters.

The chemical nature of DOM and DHS fluorophores has been examined using molecules that quench fluorescence.[32,33] Recent studies using charged and uncharged nitroxides provided information on dynamic (collisional) fluorescence quenching,[33] and additional work with methylviologen and copper has shown static (associational) quenching characteristics.[32] The extent of steady-state quenching by cationic probes is inhibited by increasing ionic strength and decreasing pH, a result effectively modeled using Columbic interactions at a negatively charged fluorophore. Additional information from these studies shows that fluorescent lifetimes for HA and FA ranged from less than 1 to about 7 nsec. Size fractionation of Suwannee River HA revealed fluorescence in all size classes with high-MW fractions being less fluorescent and more susceptible to cationic quenchers than low-MW fractions. All these data suggest the existence of multiple fluorophores with a possible correlation between fluorescent characteristics and carboxylic acid content in DHS.

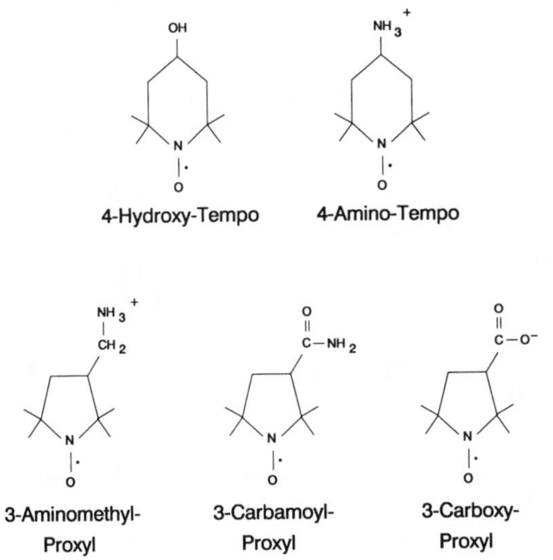

Figure 2.   Nitroxide probes used in the studies reviewed.

Recently, synchronous fluorescence scanning techniques have shown additional evidence for multiple fluorophores associated with Suwannee River FA.[34] The difference between FA fluorescent scans with and without added metals shows evidence for multiple binding sites for divalent metals at pH 7.5 with no such multiple sites present at pH 5.0. There appears to be some specificity for metal binding, since Al(III), but not Mg(II), competes with Cu(II) for sites on FA. It seems that chemical interactions which influence the electrostatic environment of DOM fluorophores in solution will also influence observed fluorescent spectra. Therefore, the fluorescent differences observed between natural water samples should be carefully interpreted.

Recent Work with Molecular Probes

In addition to energy released as fluorescence, the absorbance of sunlight by naturally occurring DOM generates a variety of photochemical transients.[35,36] These include excited triplet state DOM, solvated electrons, organic radical cations, hydroxyl radicals, and peroxy radicals. The use of various molecular probes (Figure 2) has been an effective technique for identifying and quantifying these reactive transients. Zafiriou et al.[36] have published a summary and critical review of the use of molecular probes in natural waters. Their compilation of chemical, temporal, and spatial data on photophysics, photochemistry, and methods for observation of reactive transients provides a good base for future development in this expanding area of natural water photochemistry.

Reactive transient fluxes for coastal marine systems from Maine to Florida and blue water stations throughout the Caribbean have been studied using NO as a free radical probe.[36,37] The measured total free radical fluxes, i.e., change in NO, for noon sun ranged from approximately 0.1 to 1.0 nM/min/sun in blue water, from 0.7 to 4.9 nM/min/sun for coastal water, and as high as 9.7 nM/min/sun at one station in the Gulf of Maine. Vertical distributions of the photochemical potential for radical formation, i.e., radical flux for a given irradiation intensity, appear to be oceanographically consistent when compared to natural fluorescence, salinity, and temperature measurements. Using ^{15}NO as a specific probe for superoxide radical (O_2^-), these same studies showed that O_2^- accounts for about one third of the total radical flux. Another general observation is the existence of regional homogeneity in surface water radical fluxes separated by distinct boundaries between different waters with varying photochemical potential. These may arise from different sources of photoreactive material with varied irradiation histories.

Numerous studies have been presented which use variously substituted nitroxides (Figure 2) as molecular probes for carbon-centered radicals associated with irradiated DOM.[33,36,38-42] It has been shown that nitroxides probe for the same type of transients as molecular oxygen, but do not react irreversibly with O_2^- or with peroxyl radicals, the principal products for reactions between oxygen and carbon radicals. Due to this characteristic, nitroxides can provide a direct measure of reactive centers participating in the photooxidation of DOM. Most of the literature on nitroxides to date has been directed at method confirmation. There is good evidence that fluorescent tagging of the nitroxide radical adducts formed in irradiated samples will allow simultaneous determination of various carbon-centered radical types, and studies on natural DOM should be forthcoming. A new development originating from these nitroxide studies has been a growing body of data showing that radical sites are generated within humic substances. Evidence for this includes a dramatic increase in incorporated fluorescence, which results from radical sites tagged with the nitroxide probe. It seems likely that this method, in conjunction with other molecular probes, should result in an improved ability to evaluate primary photochemical processes in a more quantitative fashion.

Another reactive transient, which has been effectively probed in recent years, is the hydroxyl radical (OH). Because OH is the most reactive photochemically produced free radical in the environment, its steady-state concentrations in natural waters are expected to be extremely low[4,10] and are difficult to measure. New data from Mopper and Zhou[43-45] have confirmed that methanol and benzoic acid are sensitive probes for OH in seawater. Their results show steady-state concentrations of OH in irradiated surface ocean water ($\sim 1 \times 10^{-18}$ M) to be much lower than coastal (~ 11–14×10^{-18} M) and freshwaters (~ 200–850×10^{-18} M).[10,44] These differences reflect both OH source material (nitrate, nitrite, hydrogen peroxide, transition metals,[46] and DOM) and OH sinks. Sunlight-irradiated deep ocean and upwelled waters showed greater rates of OH photoproduction than did nutrient-depleted surface waters. DOM was shown to be the major source for OH in all waters evaluated, and linear correlations (significant at $p < 0.05$) were seen with both light absorbance at 300 nm and DOM fluorescence. Although less well characterized, the sinks for OH in natural waters may also involve DOM either through direct reaction in freshwater systems or possibly via reactions with OH daughter compounds (ex. bromine and carbonate radicals in seawater). This dual role of DOM as a source and sink for OH may prove important for a complete description of DOM reactions in natural waters.

SECONDARY PHOTOPRODUCTS

Hydrogen Peroxide

Primary and secondary reactive transients can be categorized based on reaction time scales[36] using the 1-μsec "oxygen wall" as the point at which most primary radicals encounter O_2 and proceed to secondary reactive transient processes. The resulting longer-lived photoproducts have a greater likelihood of building to trace concentrations that are readily observable in field studies. One secondary reactive transient that has received recent attention in freshwater,[47-49] coastal and estuarine,[50-52] and open ocean[51] systems is hydrogen peroxide (H_2O_2). Diel H_2O_2 variations at field sites and observed accumulation in laboratory irradiations confirm photochemical reactions as a H_2O_2 source. There also appears to be a significant contribution from wet atmospheric deposition,[48,51,52] which can alter observed temporal patterns and overwhelm the photochemical production signal. Hydrogen peroxide exhibits vertical distributions which closely follow density structure and may prove useful as a tracer of photic zone processes or atmospheric input. Data on biological release of H_2O_2 indicates that this source is insignificant in most systems. In fact, surface-bound enzymes have been proposed as the major sink for H_2O_2.[49,50] In estuarine waters, 65–80% of the H_2O_2 decomposes via catalase (resulting in H_2O and O_2) and 20–35% via peroxidase systems (resulting in H_2O and an oxidized product, probably through reaction with aromatic compounds).

The major photochemical source for H_2O_2 in natural waters is DOM. Consequently, H_2O_2 distributions can be used as indicators of DOM photochemical activity in natural waters. Using the typical diel range of H_2O_2 concentrations as a proxy, a general trend of photochemical reactivity progressing from fresh (10–400 nM) > estuarine (40–200 nM) > coastal (15–110 nM) > oceanic waters (100–150 nM) can be seen. This is expected if DOM and biological activity are indeed the processes controlling H_2O_2 concentrations. The generation of hydroxyl radicals via photo-Fenton type processes and the activation of enzymes by H_2O_2 may impact DOM cycling in certain environments, but the importance of these reactions is not yet well known.

Low-Molecular Weight Carbon Compounds

Recent work at the University of Miami[53-58] in Florida has shown that an array of low-molecular weight (LMW) carbon compounds results from the irradiation

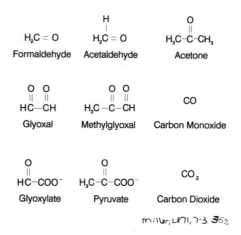

miller, LOT, 7-3 352

Figure 3. Carbon compounds produced by irradiation of natural water identified in the studies reviewed.

of DOM in natural waters (Figure 3). The fragmentation of high MW DOM may have important consequences to subsequent biological and photochemical lability. A strong correlation exists between the photochemical production of LMW carbonyl compounds and the photodegradation of DOM in natural waters. The absorbance of light at 300 nm, the rate at which that absorbance is lost due to photobleaching of DOM chromophores, and the natural water fluorescence all show linear correlations to LMW carbonyl photoproduction.

DOM with a nominal molecular weight greater than 500 contains the photochemical precursor(s) for pyruvate formation in oceanic deep water,[55] but no specific information is published for the size fractions responsible for photoproduction of other LMW carbonyls. Work with radiolabeled glycine bound to soil humic material[56] shows decreases in radioactivity contained in size fractions with a nominal MW greater than 5000 after irradiation ($\lambda < 380$ nm). Although no specific products were measured, there was a simultaneous ^{14}C increase in the MW fraction less than 500 with little change in the fraction between 500 and 5000 MW. This may indicate a DOM fraction of intermediate MW that is resistant to photochemical degradation. Additional wavelength-dependent production studies[57] show that no production of identifiable LMW carbonyls occurs at wavelengths greater than 320 nm, a result in contrast to the wavelength dependence of fluorescence bleaching[57] and H_2O_2 photoproduction[4] which exhibit activity into the visible portion of the spectrum. Much like OH production, the oceanic vertical distribution of the potential for photoproduction of formaldehyde shows that production in deep water is more efficient than in surface water and that profiles correlate well to DOM fluorescence.

Carbon Gases

The photodegradation of DOM in natural waters not only results in optical changes and production of soluble LMW compounds, it also forms trace gases with potential for direct loss to the atmosphere. Carbonyl sulfide (COS) is a potential precursor for production of atmospheric sulfate aerosols which may serve as condensation nuclei for the formation of clouds in remote locations.[59] New data on the photoproduction of COS shows that the UV-B (290–320 nm) portion of the spectra is primarily responsible for its production[60] and that coastal regions are more productive for COS on a global scale than is the open ocean.[61]

Carbon monoxide (CO) is also generated in both fresh[19,62] and marine[19,62-65] waters when exposed to sunlight. DOM appears also to be the source for CO photoproduction, and linear correlations between its production and DOM fluorescence and UV-light absorbance have been noted. Wetland waters, with their high DOM absorbance, produce CO at rates over 100 times greater than those reported for Sargasso Sea water.[62] When normalized to extinction coefficients at 350 nm, however, a variety of wetland, lake, river, and marine waters give approximately equivalent CO production rates under simulated solar radiation. Wavelength-dependent studies show highest CO production in the UV-B with significant production extending to 650 nm.

Marine samples from various depths have also been evaluated for CO and LMW carbonyl production rates using simulated solar radiation.[58] In these samples, CO accounted for 82 (1–20 m), 74 (20–150 m), and 60% (500–4000 m) of the total measured carbon-based product yield. The highest production of CO was found in surface and deep samples (18.6 and 16.1 nM C/hr, respectively). Recent extensive field studies of CO[63,64] have confirmed distributions consistent with photochemical production. Carbon monoxide was supersaturated in surface waters with respect to atmospheric levels, exhibited a diel variation, and showed correlations to light intensity. Based on very low measured rates of dark consumption of CO, the dominant sink for CO in oligotrophic waters appears to be atmospheric venting,[63] a direct loss of DOM carbon.

Direct mineralization of DOM to carbon dioxide (CO_2) by photochemical processes has also recently been reported for sterile filtered Suwannee River samples.[19] The rates of CO_2 photoproduction in these acidic waters (pH 4) were about 20 times greater than CO rates measured in the same samples, and production rates for both gases exhibited a linear correlation to photobleaching and initial absorbance of light at 350 nm. While similar to previous results,[65] it is unclear whether these rates of CO_2 photoproduction can be extrapolated to different systems.

BIOGEOCHEMICAL CONSEQUENCES OF DOM PHOTOCHEMISTRY

Optical Properties of Natural Waters

A recurring observation in many of these studies of natural water DOM photochemistry is the simple linear relationship between many photoprocesses and the absorption of light at a particular wavelength. At first glance, this seems an unlikely result considering the wide variety of environments and potential diversity of the organic precursors available for incorporation into DOM and DHS. However, the absorbance of light is the essential event for photochemical reactions, and perhaps the similarities suggest common chromophores or a common photochemical pathway involved in many of these observed photoreactions. As work progresses on the chemical nature of DOM and the primary photochemical events behind observed photoproduction, understanding of these phenomena may be forthcoming.

While we may have to wait for a complete mechanistic description of correlations between photochemical reactions and simple absorbance measurements, we do not have to wait to use these correlations for a broader approach to natural water photochemistry. There is great incentive for the accurate description of DOM spectral qualities in natural waters. As pointed out by Carder et al.,[13,66] a very small absorbance at 440 nm by DOM can add a significant error to estimates of chlorophyll based on remotely sensed color. One approach to this problem is to use a remotely sensed wavelength in the visible portion of the spectra (ex. 412)[66] for mathematical reconstruction of the entire DOM spectrum. This then allows a wavelength-by-wavelength correction of the interfering DOM absorbance when estimating chlorophyll concentrations. Accordingly, it may then be possible to use this "DOM correction" in conjunction with light intensity values and simple correlations to estimate a number of photochemical production rates over large areas.

Remotely sensed fluorescence may also provide the same applications to natural waters, since similar correlations exist between DOM fluorescence and observed photoproduction. Laser-induced fluorescence represents a sensitive and rapid method for evaluation of natural waters from ships or aircraft.[67,68] Either by relating fluorescence to absorbance spectra or by developing relationships between laser-induced fluorescence and photoproduction, remote sensing of photochemical properties may become viable.

Both remote methods may permit expansion of photochemical prediction to broad areal coverage and even global estimates. In fact, use of both of these techniques in concert may give new insights into regional or global trends in photochemistry. Effort in this area appears to be proceeding rapidly, but much more work is ahead. Accurate descriptions of the link between DOM photochemistry and the optical properties of natural waters presents a dilemma. While the absorbance-fluorescent characteristics of DOM in natural waters might be used to predict photochemical reactivity of a given water mass, those same photochemical processes will result in a fundamental change of the spectral quality and reactivity of DOM. These changes may be quite similar, however, based on the fact that correlations exist between waters with different light histories. Additional insight into photochemical mechanisms is needed to progress with confidence to predictions based on remote sensing data. Numerous corrections will be required to interpret remotely sensed color in terms of real time photochemistry, but this does not seem beyond the capability of present technologies.

Sources for Trace Gases

The substantial production of trace gases by photochemical reactions suggest a significant flux to the atmosphere from natural waters. However, knowledge of production rates alone may provide only minimal information when considering trace gas fluxes since numerous other factors must also be examined. An obvious consideration is variations in light intensity and spectral quality dictated by latitude, cloud cover, time of day, and the atmospheric transparency to wavelengths active in the photoproduction of trace gases. Production in the water column will be dictated by light attenuation and the concentration of photochemical precursors, two factors which are closely related. Waters such as wetlands and black water rivers have potential for high "per volume" production rates due to intense absorbance of light. This strong absorbance quickly attenuates active radiation and allows significant production only very near the surface. Clear waters may display low "per volume" production rates, but allow much deeper penetration of photoactive radiation. Consequently, the vertically integrated photoproduction for these vastly different systems may be similar, depending on their mixing regimes. Well-mixed water masses can alternate exposure of DOM between intense surface light and dimly lit or dark deeper waters. The time lapse between production and venting must also be considered in conjunction with in situ consumption kinetics. It is clear that accurate prediction of gas fluxes will require integration of both spatial and temporal data on photoproduction and in situ consumption.

Some recent estimates have been made for the flux of COS and CO from natural waters.[61,62,69] There is considerable uncertainty for the natural source strengths for both of these reactive gases, and recent estimates are of great interest to atmospheric modelers. Measured concentrations of COS in various marine environments were used to estimate a total flux of COS from the worlds oceans of approximately 0.35 Tg S/year.[61] Coastal/shelf waters are thought to contribute the majority of this flux (0.22 Tg S/year). Also, the coastal ocean and wetland environments potentially can supply a significant flux of CO to the atmosphere.[62] Although some disagreement exists as to the fate of CO in natural waters, the use of wavelength-dependent quantum yield data for CO photoproduction in conjunction with computer models of solar radiation as a function of latitude and season give a good upper bound for CO flux in these waters.[62] A global estimate has been made for CO flux to the atmosphere from the oceans that incorporates models for solar radiation with wind and temperature-controlled transfer velocities.[69] An empirical relation between surface CO concentrations and calculated solar radiation was derived and used for calculation of global CO distributions. This approach may only apply to oligotrophic environments, since regional distinctions in photochemical reactivity have been shown for coastal marine waters.[37] An estimate is given for the global oceanic flux of CO to the atmosphere of 165 ± 80 Tg/year, about 15% of the global flux due to natural sources.[70] If correlations between DOM absorbance and CO production prove universally valid, the coastal ocean should prove a significant, and perhaps dominant, addition to the global CO source from natural waters.

Organic Geochemistry

One of the more important developments in recent years in the field of natural water photochemistry is the demonstration that photochemical transformations of DOM and DHS can significantly influence the cycling of organic carbon. A perplexing question in carbon geochemistry has involved the inconsistencies between riverine carbon input to the ocean, the oceanic mixing cycle, and the measured ^{14}C age of oceanic DHS. Dissolved humic substances are resistant to biological oxidation and appear to pass conservatively through estuarine systems with little chemical alteration. It is estimated from riverine flux data for dissolved organic carbon (DOC) that without a sink other than microbial oxidation the deep oceans would be filled with terrestrial carbon in a time less than the measured age of DOM in the deep sea.[71,72] This is not consistent with evidence that oceanic DOM is largely of marine origin.[21,22] Thus, the question has been "What happens to riverine carbon in the sea?"

It now appears that photochemical processing may control the loss of terrestrially derived DOM and its subsequent rate of remineralization to CO_2. Several recent studies[54-58,71] point out that the LMW photoproducts of DOM are also biological substrates which stimulate production of microbial biomass. Photoproduction of CO represents a loss of organic carbon either through direct transfer to the atmosphere[64] or biological oxidation to CO_2.[58] Several estimates of the importance of photochemical degradation of biologically refractory DOM in the global carbon cycle have been made based on the production rates of LMW carbonyl compounds and CO.[57,58] The resulting calculations show that this pathway may degrade terrestrial carbon in the oceans on the order of approximately 1000–4000 years (~1–4 oceanic mixing cycles).

These preliminary models assume that all of the photochemically degradable portion of terrestrial DOM is transformed into biologically oxidizable or volatile compounds. This may result in an overestimate of carbon loss, since photobleaching of DOM occurs faster than the rate of carbon transformation to identifiable products,[57] suggesting production of soluble, colorless compounds that may exhibit no change in bioavailability or volatility. On the other hand, as the authors of these studies point out, not all photoproducts are included in these calculations, and accounting for additional products such as CO_2 and organic acids will increase the rate of carbon loss and make these residence time estimates conservative. Admittedly, some crude approximations exist in these calculations (ex. light intensity and penetration, mixing depths, advective losses, and sediment contributions), but they do serve as a good first estimate of the role of natural water photochemistry in the geochemical cycling of terrestrial carbon in the sea. As models are refined, the case supporting photochemical processing as the mechanism required to reconcile the steady-state oceanic DOM concentrations with measured ^{14}C ages may become stronger.

SUMMARY AND CONCLUSIONS

Good progress has been made in recent years in the study of DOM photochemistry. Different experimental approaches and analytical techniques have been brought to bear on a diverse area of research in studies ranging from laboratory photophysics to extensive field expeditions. Studies of spectral characteristics have revealed differences for samples from different environments and implied that the use of remotely sensed data may soon become useful in field surveys and global modeling of natural water photochemistry. Fundamental features of DOM chromophores/fluorophores and their primary reactions have been explored with molecular probes, resulting in new data on both photochemical and chemical reactivity. The longer-lived products of photochemical reactions have been studied in natural waters and appear to be important in redox chemistry, atmospheric exchange of reactive trace gases, and in the biogeochemical cycling of carbon in natural waters.

The field of natural water photochemistry appears poised to proceed to a cohesive view of both the fundamental processes behind photochemical phenomenon and the global impact those processes have on natural water cycles of carbon and associated trace materials. Extensive work is required to broaden our appreciation of both the details and the generalities of photochemical processes. Increased knowledge of primary photochemical reactions will certainly improve our ability to understand field observations. This endeavor is progressing rapidly and no doubt will produce interesting findings in the near future.

ACKNOWLEDGMENTS

The author would like to thank Drs. Neil Blough and Richard Zepp for their comments on the manuscript.

REFERENCES

1. Zepp, R. G., Environmental photoprocesses involving natural organic matter, in *Humic Substances and Their Role in the Environment*, Frimmel, F. H. and Christman, R. F., Eds., John Wiley & Sons, New York, 1988, 193.
2. Zepp, R. G., Photochemical fate of agrochemicals in natural waters, in *Pesticide Chemistry, Advances in International Research, Development, and Legislation*, Frehse, H., Ed., VCH Publishers, Weinheim, 1991, 329.
3. Hoigné, J., Faust, B. C., Haag, W. R., Scully, F. E., Jr., and Zepp, R. G., Aquatic humic substances as sources and sinks of photochemically produced transient reactants, in *Aquatic Humic Substances: Influence on Fate and Treatment of Pollutants*, Suffit, I. H. and MacCarthy, P., Eds., Advances in Chemistry Series No. 219, American Chemical Society, Washington, D. C., 1989, 363.

4. Cooper, W. J., Sunlight-induced photochemistry of humic substances in natural waters: major reactive species, in *Aquatic Humic Substances: Influence on Fate And Treatment of Pollutants*, Suffit, I. H. and MacCarthy, P., Eds., Advances in Chemistry Series No. 219, American Chemical Society, Washington, D.C., 1989, 333.

5. Mopper, K., Zika, R., and Fischer, A., Photochemistry and photophysics of marine humic substances, in *Humic Substances: Vol. 4*, MacCarthy, P., Gjessing, E. T., Mantoura, R. S. C., and Sequi, P., Eds., Wiley-Interscience, 1993, in press.

6. Zepp, R. G., Sunlight-induced oxidation and reduction of organic xenobiotics in water, in *Fate of Pesticides and Chemicals in the Environment*, Schnoor, J. L., Ed., John Wiley & Sons, New York, 1992, 127.

7. Zika, R. G. and Cooper, W. J., Eds., *Photochemistry of Environmental Aquatic Systems*, ACS Symposium Ser. No. 327, American Chemical Society, Washington, D.C., 1987.

8. Frimmel, F. H. and Christman, R. F., Eds., *Humic Substances and Their Role in the Environment*, John Wiley & Sons, New York, 1988.

9. Special issue on humic and fulvic compounds, *Anal. Chim. Acta*, 232(1), 1990.

10. Blough, N. V. and Zepp, R. G., *Effects of Solar Radiation on Biogeochemical Dynamics in Aquatic Environments*, Woods Hole Oceanographic Institute Technical Report, WHOI-90, Woods Hole, MA, 1990.

11. Pages, J. and Gadel, F., Dissolved organic matter and UV absorption in a tropical hyperhaline estuary, *Sci. Total Environ.*, 99, 173, 1990.

12. Green, S. A. and Blough, N. V., Adsorption and fluorescence spectra of waters from the Gulf of Mexico and western coastal Florida, in ACS, Division of Environmental Chemistry, Preprints of Papers Presented at the 203rd National Meeting, San Francisco, CA, April 5–10, 1992, Vol. 32, No. 1, Paper Nos. 247, 226, 1992.

13. Carder, K. L., Steward, R. G., Harvey, G. R., and Ortner, P. B., Marine humic and fulvic acids: their effects on remote sensing of ocean chlorophyll, *Limnol. Oceanogr.*, 34(1), 68, 1989.

14. Zepp, R. G. and Schlotzhauer, P. F., Comparison of photochemical behavior of various humic substances in water: III. Spectroscopic properties of humic substances, *Chemosphere*, 10(5), 479, 1981.

15. Bricaud, A., Morel, A., and Prieur, L., Absorption by dissolved organic matter of the sea (yellow substance) in the UV and visible domains, *Limnol. Oceanogr.*, 26(1), 43, 1981.

16. Wang, Z., Pant, B. C., and Langford, C. H., Spectroscopic and structural characterization of a Laurentian fulvic acid: notes on the origin of the color, *Anal. Chim. Acta*, 232, 43, 1990.

17. Saleh, F. Y., Ong, W. A., and Chang, D. Y., Structural features of aquatic fulvic acids. Analytical and preparative reversed-phase high-performance liquid chromatography separation with photodiode array detection, *Anal. Chem.*, 61, 2792, 1989.

18. Brophy, J. E. and Carlson, D. J., Production of biologically refractory dissolved organic carbon by natural seawater microbial populations, *Deep Sea Res.*, 36(4), 497, 1989.

19. Miller, W. L. and Zepp, R. G., Photochemical carbon cycling in aquatic environments: formation of atmospheric carbon dioxide and carbon monoxide, in ACS, Division of Environmental Chemistry, Preprints of Papers Presented at the 203rd National Meeting, San Francisco, CA, April 5–10, 1992, Vol. 32, No. 1, Paper No. 85, 158, 1992.

20. Wang, W. H., Beyerle-Pfnür, R., and Lay, J. P., Photoreaction of salicylic acid in aquatic systems, *Chemosphere*, 17(6), 1197, 1988.

21. Hedges, J. I., Hatcher, P. G., Ertel, J. R., and Meyers-Schulte, K. J., A comparison of dissolved humic substances from seawater with Amazon River counterparts by ^{13}C-NMR spectrometry, *Geochim. Cosmochim. Acta*, 56, 1753, 1992.

22. Malcolm, R. L., The uniqueness of humic substances in each of soil, stream and marine environments, *Anal. Chim. Acta*, 232, 19, 1990.

23. Senesi, N., Molecular and quantitative aspects of the chemistry of fulvic acid and its interactions with metal ions and organic chemicals. Part II. The fluorescence spectroscopy approach, *Anal. Chim. Acta*, 232, 77, 1990.

24. Hayase, K., Tsubota, H., and Sunada, I., Relationships of fluorescence and AOU in three north Pacific water samples, *Sci. Total Environ.*, 81/82, 315, 1989.

25. Coble, P. G., Green, S. A., Blough, N. V., and Gagosian, R. B., Characterization of dissolved organic matter in the Black Sea by fluorescence spectroscopy, *Nature (London)*, 348, 432, 1990.

26. Che, R. F. and Bada, J. L., The fluorescence of dissolved organic matter in seawater, *Mar. Chem.*, 37, 191, 1992.

27. Kouassi, A. M. and Zika, R. G., Light-induced alteration of the photophysical properties of dissolved organic matter in seawater. Part I. Photoreversible properties of natural water fluorescene, *Neth. J. Sea Res.*, 27(1), 25, 1990.

28. Kouassi, A. M., Zika, R. G., and Plane, J. M. C., Light-induce alteration of the photophysical properties of dissolved organic matter in seawater. Part II. Estimates of the environmental rates of the natural water fluorescence, *Neth. J. Sea Res.*, 27(1), 33, 1990.

29. Hayase, K., Yamamoto, M., Nakazawa, I., and Tsubota, H., Behavior of natural fluorescence in Sagami Bay and Tokyo Bay, Japan — vertical and lateral distributions, *Mar. Chem.*, 20, 265, 1987.

30. Hayase, K., Tsubota, H., Sunada, I., Goda, S., and Yamazaki, H., Vertical distribution of fluorescent organic matter in the North Pacific, *Mar. Chem.*, 25, 373, 1988.

31. Laane, R. W. P. M. and Kramer, K. J. M., Natural fluorescence in the North Sea and its major estuaries, *Neth. J. Sea Res.*, 26, 1, 1990.

32. Milne, P. J. and Zika, R. G., Luminescence quenching of dissolved organic matter in seawater, *Mar. Chem.*, 27, 147, 1989.

33. Green, S. A., Morel, F. M. M., and Blough, N. V., Investigation of the electrostatic properties of humic substances by fluorescence quenching, *Environ. Sci. Technol.*, 26, 294, 1992.

34. Cabaniss, S. E., Synchronous fluorescence spectra of metal-fulvic acid complexes, *Environ. Sci. Technol.*, 26(6), 1133, 1992.

35. Haag, W. R. and Mill, T., Survey of sunlight-induced transient reactants in surface waters, in *Effects of Solar Radiation on Biogeochemical Dynamics in Aquatic Environments*, Blough, N. V. and Zepp, R. G., Eds., Woods Hole Oceanographic Institute Technical Report, WHOI-90-09, Woods Hole, MA, 1990, 82.

36. Zafiriou, O. C., Blough, N. V., Micinski, E., Dister, B., Kieber, D., and Moffett, J., Molecular probe systems for reactive transients in natural waters, *Mar. Chem.*, 30, 45, 1990.

37. Zafiriou, O. C. and Dister, B., Photochemical free radical production rates: Gulf of Maine and Woods Hole — Miami transect, *J. Geophys. Res.*, 96(C3), 4939, 1991.

38. Blough, N. V., Electron paramagnetic resonance measurements of photochemical radical production in humic substances. 1. Effects of O_2 and charge on radical scavenging by nitroxides, *Environ. Sci. Technol.*, 22(1), 77, 1987.

39. Green, S. A., Simpson, D. J., Zhou, G., and Blough, N. V., Intramolecular quenching of excited singlet states by stable nitroxyl radicals, *J. Am. Chem. Soc.*, 112, 7337, 1990.

40. Kieber, D. J. and Blough, N. V., Fluorescence detection of carbon-centered radicals in aqueous solution, *Free Radical Res. Commun.*, 10(1–2), 109, 1990.

41. Kieber, D. J. and Blough, N. V., Determination of carbon-centered radicals in aqueous solution by liquid chromatography with fluorescence detection, *Anal. Chem.*, 62, 2275, 1990.

42. Kieber, D. J., Johnson, C. G., and Blough, N. V., Mass spectrometric identification of the radical adducts of a fluorescamine-derivatized nitroxide, *Free Radical Res. Commun.*, 16(1), 35, 1992.

43. Zhou, X. and Mopper, K., Determination of photochemically produced hydroxyl radicals in seawater and freshwater, *Mar. Chem.*, 30, 71, 1990.

44. Mopper, K. and Zhou, X., Hydroxyl radical photoproduction in the sea and its potential impact on marine processes, *Science*, 250, 661, 1990.

45. Mopper, K. and Zhou, X., Photoproduction of hydroxyl radicals at the sea surface and its potential impact on marine processes, *Effects of Solar Radiation on Biogeochemical Dynamics in Aquatic Environments*, Woods Hole Oceanographic Institute Technical Report, WHOI-90-09, Woods Hole, MA, 1990, 151.

46. Zepp, R. G., Faust, B. C., and Hoigné, J., Hydroxyl radical formation in aqueous reactions (pH 3–8) of iron(II) with hydrogen peroxide: the photo-Fenton reaction, *Environ. Sci. Technol.*, 26, 313, 1992.

47. Cooper, W. J., Lean, D. R. S., and Carey, J., Spatial and temporal patterns of hydrogen peroxide in lake waters, *Can. J. Fish. Aquat. Sci.*, 46(7), 1227, 1989.

48. Cooper, W. J. and Lean, D. R. S., Hydrogen peroxide concentration in a northern lake, Photochemical formation and diel variability, *Environ. Sci. Technol.*, 23(11), 1425, 1989.

49. Cooper, W. J. and Zepp, R. G., Hydrogen peroxide decay in waters and suspended soils: evidence for biologically mediated process, *Can. J. Fish. Aquat. Sci.*, 47(5), 888, 1990.

50. Moffett, J. W. and Zafiriou, O. C., An investigation of hydrogen peroxide chemistry in surface waters of Vineyard Sound with $H_2{}^{18}O_2$ and $^{18}O_2$, *Limnol. Oceanogr.*, 35(6), 1221, 1990.

51. Miller, W. L., An Investigation of Peroxide, Iron, and Iron Bioavailability in Irradiated Marine Waters, Ph.D. thesis, University of Rhode Island, 1990.

52. Szymczak, R. and Waite, T. D., Photochemical activity in waters of the Great Barrier Reef, *Estuarine Coastal Shelf Sci.*, 33, 605, 1991.

53. Mopper, K. and Stahovec, W. L., Sources and sinks of low molecular weight organic carbonyl compounds in seawater, *Mar. Chem.*, 19, 305, 1986.

54. Kieber, D. J. and Mopper, K., Photochemical formation of glyoxylic and pyruvic acids in seawater, *Mar. Chem.*, 21, 135, 1987.

55. Kieber, D. J., McDaniel, J. A., and Mopper, K., Photochemical source of biological substrates in seawater: implications for geochemical carbon cycling, *Nature (London)*, 341, 637, 1989.

56. Amador, J. A., Alexander, M., and Zika, R. G., Sequential photochemical and microbial degradation of organic molecules bound to humic acid, *Appl. Environ. Microbiol.*, 55(11), 2843, 1989.

57. Kieber, R. J., Zhou, X., and Mopper, K., Formation of carbonyl compounds from UV-induced photodegradation of humic substances in natural waters: fate of riverine carbon in the sea, *Limnol. Oceanogr.*, 35(7), 1503, 1990.

58. Mopper, K., Zhou, X., Kieber, R. J., Kieber, D. J., Sikorski, R. J., and Jones, R. D., Photochemical degradation of dissolved organic carbon and its impact on the oceanic carbon cycle, *Nature (London)*, 353, 60, 1991.

59. Hofmann, D. J., Increase in the stratospheric background sulfuric acid aerosol mass in the past 10 years, *Science*, 249, 996, 1990.

60. Zepp, R. G. and Andreae, M. O., Photosensitized formation of carbonyl sulfide in sea water, in *Effects of Solar Radiation on Biogeochemical Dynamics in Aquatic Environments*, Woods Hole Oceanographic Institute Technical Report, WHOI-90-09, Woods Hole, MA, 1990, 180.

61. Andreae, M. O., Photochemical production of carbonyl sulfide in coastal and open ocean waters, in *Effects of Solar Radiation on Biogeochemical Dynamics in Aquatic Environments*, Woods Hole Oceanographic Institute Technical Report, WHOI-90-09, Woods Hole, MA, 1990, 56.

62. Valentine, R. L. and Zepp, R. G., Formation of carbon monoxide from the photodegradation of terrestrial dissolved organic carbon in natural waters, *Environ. Sci. Technol.*, 27, 409, 1993.

63. Jones, R. D., Carbon monoxide and methane distribution and consumption in the photic zone of the Sargasso Sea, *Deep Sea Res.*, 38(6), 625, 1991.

64. Gammon, R. H. and Kelly, K. C., Photochemical production of carbon monoxide in surface waters of the Pacific and Indian oceans, in *Effects of Solar Radiation on Biogeochemical Dynamics in Aquatic Environments*, Woods Hole Oceanographic Institute Technical Report, WHOI-90-09, Woods Hole, MA, 1990, 58.

65. Chen, Y., Khan, S. U., and Schnitzer, M., Ultraviolet irradiation of dilute fulvic acid solutions, *Soil Sci. Soc. Am. J.*, 42, 292, 1978.

66. Carder, K. L., Hawes, S. K., Baker, K. A., Smith, R. C., Steward, R. G., and Mitchell, B. G., Reflectance model for quantifying chlorophyll a in the presence of productivity degradation products, *J. Geophys. Res.*, 96(C11), 20599, 1991.

67. Donard, O. F. X., Lamotte, M., Belin, C., and Ewald, M., High sensitivity fluorescence spectroscopy of Mediterranean waters using a conventional or a pulsed laser excitation source, *Mar. Chem.*, 27, 117, 1989.

68. Ferrari, G. M. and Tassan, S., On the accuracy of determining light adsorption by "yellow substance" through measurements of induced fluorescence, *Limnol. Oceanogr.*, 36, 777, 1991.

69. Erickson, D. J., III, Ocean to atmosphere carbon monoxide flux: global inventory and climate implications, *Global Biogeochem. Cycles*, 3(4), 305, 1989.

70. Khalil, M. A. K. and Rasmussen, R. A., The global cycle of carbon monoxide: trends and mass balance, *Chemosphere*, 20(1–2), 227, 1990.

71. Ertel, J. R., Photochemistry of dissolved organic matter: an organic geochemical perspective, in *Effects of Solar Radiation on Biogeochemical Dynamics in Aquatic Environments*, Woods Hole Oceanographic Institute Technical Report, WHOI-90-09, Woods Hole, MA, 1990, 79.

72. Deuser, W. G., Whither organic carbon?, *Nature (London)*, 332, 396, 1988.

CHAPTER 8

Organic Dyes Bound to Polyelectrolytes: Photophysical Probes of Binding Domains and Biopolymer Conformation

Guilford Jones, II, Churl Oh, and Guilherme L. Indig

INTRODUCTION

The binding of organic dyes, including textile dyes, in aquatic media is of current interest in terms of the potential for pollution of natural waters near manufacturing sites. The mechanisms of dye binding and the types of experimental techniques that can be confidently used to probe binding phenomena are not well established. We have employed a combination of photophysical methods for investigation of the binding of an organic dye, styryl-7 (S7), to polyelectrolytes having notably different properties. Poly(methacrylic acid) (PMAA) is known to bind organic cations in two modes, one of which involves electrostatic attraction to charged carboxylic acid residues (pH > 6.0) and the other in which the polymer is "hypercoiled" into a globular form (pH < 4.0) and develops a hydrophobic domain for lipophilic species.[1,2] The homopolymer of L-glutamic acid (PLGL) provides a different kind of environment for dye binding and features a helix to random coil transition in the vicinity of pH 6.0 for aqueous media.[3] These studies serve as a prelude to a more detailed investigation of the binding of dyes to humic acids, complex biopolymer electrolytes, that are found in natural aquatic systems.[4] Here, we report the results of deployment of S7 as a probe of polymer microenvironment for these polymers in which spectrometric techniques including absorption, fluorescence emission, and circular dichroism were utilized.

EXPERIMENTAL

Poly(methacrylic acid) was prepared as previously described; a weight-averaged molecular weight of 25 kDa (DP = 290) was determined for the sample by viscosity measurements on aqueous solutions.[2] Styryl-7 purchased from Kodak was recrystallized three times from methanol, and the purity was monitored by thin layer chromatography (TLC) (silica gel, 4:1 DMF:methanol v/v). PLGL (61.2 kDa, DP = 405) was purchased from Sigma and used as received. For circular dichroism (CD) studies, the pH of the samples containing 2 mM phosphate buffer was adjusted by addition of 0.2 N NaOH. The absorption spectra were obtained using a Beckman DU-7 spectrometer; the emission was monitored with an SLM 48000 phase-shift fluorometer. The CD spectra were obtained employing a Model 62 DS CD spectrophotometer from AVIV Associates. All experiments were carried out at room temperature unless otherwise indicated. The commercial sample of humic acid (Aldrich, cat # H1,675-2, lot #01816HH) was treated as follows: a saturated aqueous solution was centrifuged during 20 min (15,000 rpm, Sorval rotor SS36) to deposit particles and decrease ash content. The supernatant was acidified with HCl to pH 2.0 and kept overnight at room temperature. The precipitate was centrifuged to 5000 rpm, the pellet washed with HCl solution pH 2.0, and finally dialyzed against phosphate buffer 0.01 M, pH 7.4, and then against water (MilliQ) using cellulose membrane tubing (Spectrapor; molecular weight cutoff = 3500 Da). The final high-molecular weight fraction was lyophilized and stored at –20°C.

RESULTS AND DISCUSSION

Absorption Properties

Styryl-7 displayed a remarkable shift to the red in its long-wavelength absorption band on decrease of solvent polarity. This property was exhibited by the shift in peak absorption of the dye in water (520 nm) vs t-butyl alcohol (588 nm); the regularity of the effect is illustrated in Figure 1a in which the absorption frequency is plotted against the solvent polarity parameter, π^*. The solvatochromic effect is consistent with a spectral sensitivity exhibited by other merocyanine-type dyes in which the ground state is more polar than the first excited state.[5] A large bathochromic shift in the absorption spectrum for S7 was also observed for an aqueous PMAA solution on adjustment of the pH downward as shown in Figure 1b. In this experiment, a large excess of polyelectrolyte was used (expressed as the ratio of concentrations of polymer residues and dye, P/D) (Figure 1c). The dye, in fact, is acting as a sensitive probe of the titration of PMAA and reports on the alteration of polymer coil dimensions (i.e., a change from extended, rod-like, random coil to the protein-like hypercoiled form).[1,2]

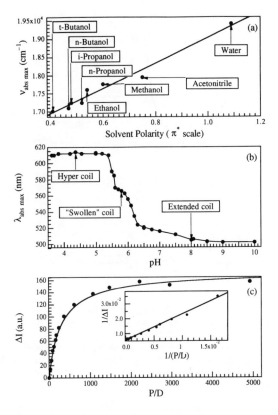

Figure 1. **(a)** The effect of solvent polarity on the absorption frequency maximum of S7. **(b)** The effect of pH on the absorption maximum wavelength of S7 in aqueous PMAA (P/D 2000). **(c)** The change of S7 fluorescence intensity as a function of the concentration ratio between the monomer unit of PMAA and the dye (P/D) in acetate buffer 0.06 M pH 3.8 and at 20°C. [S7] = 8.2 μM. Inset: Benesi-Hildebrand plot.

S7   PMAA   PLGL

A "metachromatic" property was also prevalent for S7 in that the addition of very dilute polymer binding agent at the appropriate pH (polymer charge) brought about dye aggregation.[2] We have observed the formation of aggregates ("dimers") for S7 and PMAA and for PLGL, with spectra for the latter shown in Figure 2a. In the presence of the negatively charged polyelectrolytes (higher pH regime), a new absorption band that shifted to the blue of the spectrum of free S7 was distinct

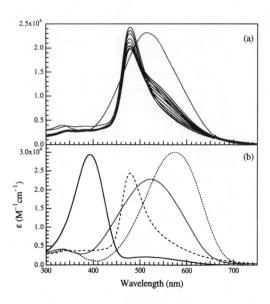

Figure 2. **(a)** The effect of pH on the absorption spectra of S7 in aqueous PLGL solutions. In order of decreasing absorption maximum (solid lines), the pH values were 5.90, 6.30, 6.50, 6.75, 6.95, 7.15, 7.35, 7.45, 7.55, 7.70, and 8.60. The dashed line is S7 in water (in the absence of PLGL). **(b)** From the left, in order of increasing absorption maximum wavelength, absorption spectra for S7 in the protonated form at pH 2.0, as a dimer bound to PLGL, in water, and in ethylene glycol.

when the P/D ratio is quite low, indicative of a very heavy loading of dye counterions. There is some reversal of the aggregation phenomenon on elevation of pH beyond 6.0 which produces the maximum number of PLGL charge sites. The hyposochromically shifted absorption band (H-band) is characteristic of the formation of sandwich-type dye aggregates.[6]

The environmental influences on the absorption properties of S7 are summarized in Figure 2b. Besides the monomer band for pure water and for the less polar ethylene glycol solvent, the aggregate band on binding to PLGL and another blue-shifted band that results at lower pH (formation of an S7 conjugate acid form) are revealed.

Emission Properties

The wavelength maximum in the fluorescence emission for S7 near 700 nm was much less sensitive to environmental perturbation (i.e., the emitting state less polar than the dye ground state).[5] However, the emission intensity was greatly affected by solvent properties with a particular sensitivity shown for bulk or microviscosity. For example, the S7 emission intensity was increased more than 20-fold on increase in the percentage of ethylene glycol in water (up to 100%) for mixed alcohol-water solutions. Similar changes in the emission intensity were observed upon addition of PMAA for buffered media at low pH for which the polymer was expected to remain in the hypercoiled form. The alteration of fluorescence yield on interaction of dye with a large excess of polymer (high P/D) is shown in Figure 1c, with an accompanying double-reciprocal plot from which an association constant could be obtained (Benesi-Hildebrand treatment[7]). For the latter, the "concentration" employed is supposed to reflect the molarity of binding sites within the polymer domain. For purposes of comparison, these sites were considered (somewhat arbitrarily) to be uniform; a phenomenological binding constant can then be computed on a "per residue" basis (i.e., the binding constant is expressed in terms of the monomeric concentration instead of macromolecular concentration of the polymer). Therefore, the plot of the data yields an association constant, $K = 470\ M^{-1}$ per residue. Even though the PMAA used was polydisperse, the double-reciprocal plot was quite linear. The emission of S7 decreased significantly under the conditions of dye aggregation (P/D less than 20 and a high concentration or "stacking" of counterions for the charged polyelectrolytes, PMAA or PLGL). The quenching of fluorescence has been commonly observed for sandwich-type aggregates.[2,6]

Circular Dichroism

In the presence of the negatively charged biopolymer, PLGL, bound S7 exhibited induced optical activity (ICD) as revealed in CD spectra. This phenomenon, which has been observed for a number of dye-biopolyelectrolyte complexes,[8] was principally associated with the absorption band assigned to S7 aggregates (480-nm band, Figure 2a). The pH dependence of the ICD was examined in some detail with respect to changes in both the S7 dimer visible band and the CD that could also be monitored in the 200–240 nm UV region (and associated with peptide transitions for native PLGL).[3] The CD spectra for 75 μM S7 with PLGL at P/D 2 at various pHs are shown in Figure 3a (right). The pH dependence of CD transitions with maxima at 208 and 222 nm, bands characteristic of the α-helix conformation of PLGL,[3] is illustrated in Figure 3a (left). The remarkable feature of these data is the extent to which the helix content remains high at pH greater than 6.0, a point at which PLGL normally achieves the random coil conformation in the absence of dye aggregate.

This ability of S7, interpreted as a loading on the peptide of species that are both charge compensating and also relatively hydrophobic, is also demonstrated in Figure 3b, in which the dependence of peptide CD in the UV region on dye concentration is shown. Another indication of the exceptionally high sensitivity of ICD to the conditions of medium and binding was revealed in the pH profile of Figure 3a (right). Thus, within a very narrow pH interval, the optical activity induced in the dye dimer transition is reversed. This feature would be consistent with the existence of two types of dye aggregate (dimer) species, each bound differently to the peptide chain. The model that we tentatively employ envisions

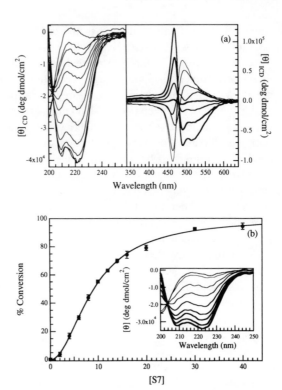

Figure 3.   **(a)** Effect of pH on the CD spectra of S7-PLGL complexes (P/D = 2; [S7] = 75
μ*M*). Left, from the bottom at 222 nm, the pH values were 5.90, 6.30, 6.50, 6.75,
6.95, 7.15, 7.35, 7.70, 7.82, and 8.05; right, in order of decreasing ellipticity at
500 nm, the pH values were 5.90, 6.50, 6.95, 7.45, 8.05, 7.55, 7.82, and 7.70.
(b) The coil-to-helix transition of PLGL induced by S7. The percentage of helix
conversion was calculated based on the ellipticity values at 222 nm; the ellipticity
at 100% conversion was extrapolated from the fit. Inset, from the top at 222 nm,
the S7 concentrations were 0, 2, 4, 6, 8, 10, 12, 14, and 16 μ*M*. [PLGL] = 50 μ*M*.;
pH 6.1; 2 m*M* phosphate buffer.

dye counterions as dimers bound to charged sites on the peptide at regular
intervals that depend upon the degree of charging of the polymer. For example,
at lower pH (pH 6.0), dye species could be site bound at "fourth residues" on the
peptide (approximately at helical turns). Alternatively, a concentration of charge
(to say, third residues on average) for site binding at slightly elevated pH would
result in a significant reorientation of the transition moments associated with the
dye aggregate chromophore and its relationship to the helix axis.[9a-9c]

DYE BINDING TO HUMIC ACID — PRELIMINARY DATA

A commercial sample of humic acid (HA) was subjected to dialysis for
removal of low molecular weight components and examined briefly to determine

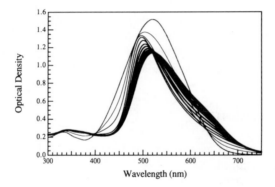

Figure 4. Differential absorption spectra of S7 in water as a function of the HA concentration. From the top, in order of decreasing absorption maximum, the HA concentrations were 0, 13.7, 26.4, 44.3, 60.3, 79.6, 96.4, 114.5, 129.6, 147.2, and 160.7 mg/L. [S7] = 59 μM.; optical pathway = 1 cm; T = 20°C.

effects of S7 binding. Shown in Figure 4 are the result of addition of low (10–100 mg/L) levels of HA to 0.01 mM S7 in water. Two features of the data stand out. Addition of very low levels of HA results in the characteristic blue shift of absorption that we have associated with dye aggregation on the polymer. At higher concentrations of the HA "template," more binding sites are available for occupation by dye monomers and the feature of a red-shifted band (about 560 nm) appears. Appearance of the latter feature would be consistent with the introduction of dye species to HA-binding sites that are less polar than the bulk medium (water). The 560-nm absorption maximum, if taken as the solvatochromic "marker" for these sites, corresponds to a micropolarity similar to that of methanol or acetonitrile (π^* scale, Figure 1a). The 700-nm fluorescence of S7 is strongly quenched by 10–20 mg/L concentrations of our sample of HA. A Stern-Volmer plot of the quenching data was nonlinear, indicating competing mechanisms of dye-polymer interaction (static vs dynamic mechanisms and/or multiple site interaction). The effectiveness of HA in quenching S7 fluorescence is enhanced for neutral solutions (pH 6.7 measured) vs solutions of HA-S7 acetate buffered to pH 3.8. A more thorough study of these trends is currently under way in which a reference HA sample will be employed.

CONCLUSIONS

The laser dye, S7, serves as a remarkably sensitive probe of polymer binding in water as demonstrated for dye complexation with PMAA, PLGL, and a commercial sample of HA. Absorption spectra for S7 reveal large red shifts for binding in polymer domains that are relatively hydrophobic. Aggregate, or dye dimer, bands are displayed for high-polymer loading; these aggregates interact with the asymmetric environment of the peptide, PLGL, in an intimate way that results in high levels of induced circular dichroism (ICD). This induced optical activity is a very sensitive monitor of charge and conformation for the electrolyte biopolymer. The binding effects on the photophysical properties of S7 are demonstrated for a sample of HA for which some indication of dye aggregation and entrapment in hydrophobic microdomains is in evidence.

ACKNOWLEDGMENTS

The support of this research by the Environmental Protection Agency is gratefully acknowledged.

REFERENCES

1. Ghiggino, K. P., *Polymer Photophysics*, Phillips, D., Ed., Chapman and Hall, New York, 1985.
2. Jones, G., II, Oh, C., and Goswami, K., The photochemistry of triarylmethane dyes bound to polyelectrolytes: photoinduced electron transfer involving bound dye monomers and dimers, *J. Photochem. Photobiol. A: Chem.*, 57, 65, 1991.
3. Fasman, G. D., *Poly-Alpha-Amino Acids*, Fasman, G. D., Ed., Academic Press, New York, 1966.
4. Suffet, I. H. and MacCarthy, P., *Aquatic Humic Substances. Influence on Fate and Treatment of Pollutants*, American Chemical Society, Washington, D.C., 1989.
5. Jones, G., II, Photochemistry of laser dyes, in *Dye Laser Principles: with Applications*, Duarte, F. J. and Hillman, L. W., Eds., Academic Press, New York, 1990.
6. Kasha, M., Rawls, H. R., and El-Bayoumi, M. A., The exciton model in molecular spectroscopy, *Pure Appl. Chem.*, 11, 371, 1965.
7. Barra, M., Bohne, C., and Scaiano, J. C., Effect of cyclodextrin complexation on the photochemistry of xanthone: absolute measurement of the kinetics for triplet state exit, *J. Am. Chem. Soc.*, 112, 8075, 1990.
8. Hatano, M., *Induced Circular Dichroism in Biopolymer-Dye Systems*, Springer-Verlag, Berlin, 1986.
9a. Moffitt, W., Optical rotary dispersion of helical polymers, *J. Chem. Phys.*, 25, 467, 1956.
9b. Moffitt, W., Fitts, D. D., and Kirkwood, J. G., Critique of the theory of optical activity of helical polymers, *Proc. Natl. Acad. Sci. U.S.A.*, 43, 723, 1957.
9c. Imae, T. and Ikeda, S., Circular dichroism and structure of the complex of acridine orange with poly(L-glutamic acid), *Biopolymers*, 15, 1655, 1976.

CHAPTER 9

Photochemical Degradation of Triazine and Anilide Pesticides in Natural Waters

F. H. Frimmel and D. P. Hessler

INTRODUCTION

Natural waters may be contaminated with pesticides and other chemicals. Herbicides are especially prone to contaminate water because they are directly applied to soil and may then leach into groundwater, streams, rivers, and lakes. They are a main source of pollution of natural waters, and their concentration is often higher than the limits allowed for drinking water (e.g., a total pesticide concentration of 0.5 µg/L in Germany and other EC countries).

Many pesticides absorb light in the ultraviolet (UV) spectral range and therefore may be degraded photochemically. The use of UV irradiation to replace the classical chemical disinfection of drinking water with chlorine or chlorine dioxide is very promising.[1] Therefore, it is of interest to know to what extent pesticides are degraded in this way and if the humic substances (HS) contained in natural waters influence the photodegradation.

The aquatic photolysis of chemicals can be influenced in several ways by the HS contained in natural waters. On one hand, direct photolysis can be inhibited because competitive light absorption by colored HS occurs. In this case, the HS play the role of a light filter. This leads to a decrease in the photolysis rates in the presence of HS compared to the reactions in pure water.[2] On the other hand, HS can enhance the degradation of pollutants in sunlight by producing radicals or oxidizing species.[3-6] Several studies have indicated that HS are capable of acting as sensitizers for the production of reactive intermediates such as singlet oxygen ($^1O_2$),[7,8] superoxide anion ($^.O_2^-$),[9] hydrogen peroxide (H_2O_2),[10] solvated electron

0-87371-871-2/94/$0.00+$.50

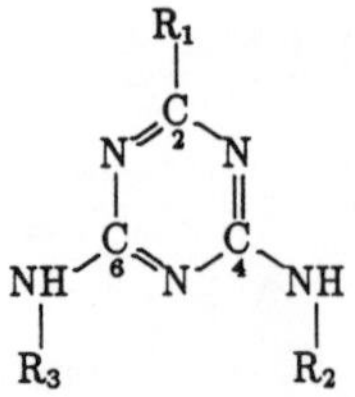

s-Triazine	R_1	R_2	R_3
Atrazine	Cl	$CH(CH_3)_2$	CH_2CH_3
Desethylatrazine	Cl	$CH(CH_3)_2$	H
Simazine	Cl	CH_2CH_3	CH_2CH_3
Terbutylazine	Cl	$C(CH_3)_3$	CH_2CH_3

Figure 1. Structure of s-triazines.

(e^-_{aq}),[11] and HS-derived peroxyradicals (ROO˙).[12,13] The production of these species involves an energy transfer from HS to molecular oxygen or other compounds present in the water under investigation.

The objectives of this study were (1) to determine photoreaction rate constants and quantum yields for the degradation of the s-triazine pesticides atrazine, desethylatrazine, simazine, and terbutylazine (Figure 1) and the anilide pesticides metazachlor and metolachlor (Figure 2) under UV irradiation (λ = 254 nm) in distilled water; (2) to determine the influence of the HS of a natural water on the photolysis; and (3) to compare the results to get an insight on the effect of HS on the photodegradation of the six pesticides investigated.

EXPERIMENTAL METHODS

Materials

Atrazine, desethylatrazine, simazine, terbutylazine, metazachlor, and metolachlor (Ehrensdorfer) were used without further purification. Natural water samples from a bog lake (Hohlohsee, Black Forest, Germany) were filtered through 0.45-μm membranes before use in the photolysis experiments. The samples had a dissolved organic carbon (DOC) of ca. 15 mg/L. Other experiments were conducted in doubly distilled water.

Photolysis

Photolyses were performed in a merry-go-round photoreactor with a 16-W low-pressure mercury lamp (Applied Photophysics, Ltd.) emitting light with the

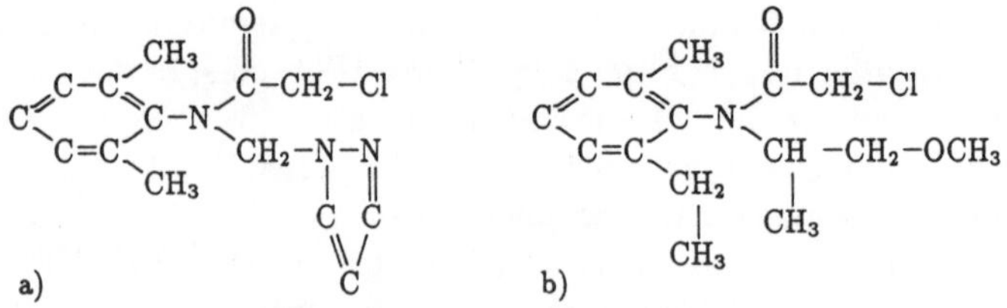

Figure 2. Structure of **(a)** metazachlor and **(b)** metolachlor.

highest intensity at 254 nm. The quartz reaction tubes had an inner diameter of 1 cm. Actinometry was performed using optically thick solutions of potassium ferrioxalate ($A_{254} > 2$).[14] The emitted light intensity P_0 was 7.1 (±0.2) $\times$ 10^{-8} Einstein L^{-1} sec^{-1}.

Analysis

The pesticides and their degradation products were analyzed by HPLC on a Hewlett Packard HP 1090 liquid chromatograph with a diode array detector and a 100×2.1 mm, ODS Hypersil, 5-μm column. A gradient between water (A) and acetonitrile (B) was used as mobile phase (A/B: 79/21 to 0/100).

The dissolved organic carbon (DOC) content of the natural water samples measured using a Beckman TOC-Analyser (Model DC-80) was $DOC_0 = 15$ mg L^{-1} before irradiation. Absorption spectra were obtained on a Perkin-Elmer Lambda 5 UV/VIS spectrophotometer. Fluorescence spectra were measured on a Perkin-Elmer LS-5B luminescence spectrometer at an excitation wavelength of $\lambda_{ex} = 330$ nm.

Kinetic Analysis

The loss of pesticides was followed as a function of irradiation time, and the data were fitted to a first-order rate model

$$\ln\left(C_t/C_0\right) = -k_p \cdot t \qquad (1)$$

where C_0 and C_t are the concentrations of pesticide at times 0 and t, k_p is the first-order rate constant (in seconds^{-1}) and t is the irradiation time in seconds. The photolysis quantum yields, Φ_p, were calculated after determination of k_p, the emitted light intensity P_0, and the absorbance A_{254} of the solution at the irradiation wavelength according to

$$\Phi_p = \frac{k_p \cdot C}{P_0 \cdot \left(1 - 10^{-A_{254}}\right)} \qquad (2)$$

where Φ_p represents the efficiency with which light absorption leads to photodegradation. The nominator of Equation 2 represents the rate of pesticide degradation per unit irradiation time, whereas the denominator represents the absorbed photons per unit time.

RESULTS AND DISCUSSION

Direct Photolysis

Linear regression of $\ln(C_t/C_0)$ vs t yielded photolysis rate constants, k_p, for the pesticides in distilled water. Values for k_p, the pesticide initial concentrations C_0,

Table 1.   Rate Constants k_p and Quantum Yields Φ_p for the Photochemical
            Degradation of Pesticides in Distilled Water

Pesticide	$C_o \cdot 10^5$ (mol L^{-1})	$A_{254} \cdot 10^2$	$k_p \cdot 10^4$ (sec^{-1})	$\Phi_p \cdot 10^2$
Atrazine	2.3	8.0	1.9	3.7 ($\pm$0.3)
Desethylatrazine	2.4	7.8	1.9	3.9 ($\pm$0.3)
Simazine	1.8	5.9	1.9	3.8 ($\pm$0.3)
Terbutylazine	2.3	7.8	1.9	3.7 ($\pm$0.3)
Metazachlor	1.7	0.7	2.5	37 ($\pm$6)
Metolachlor	1.0	0.5	2.8	34 ($\pm$7)

Note: Pesticide initial concentrations C_0 and optical densities A_{254} at $\lambda = 254$ nm.

the optical densities A_{254}, and the calculated quantum yields Φ_p are given in Table 1 for the six investigated pesticides.

The four *s*-triazines showed identical photolysis rate constants and quantum yields. The degradation rates of the two anilide pesticides again were identical, but somewhat higher than for the *s*-triazines; the quantum yields of degradation were about one order of magnitude higher than those for the *s*-triazines. The difference in the rate constants between the two groups was much lower than for the quantum yields. In fact, the much higher molar absorption coefficient at the wavelength of irradiation of the *s*-triazines ($\varepsilon_{254} \approx 3500$ L mol^{-1} cm^{-1}) in comparison to the anilides ($\varepsilon_{254} \approx 500$ L mol^{-1} cm^{-1}) implies that the former absorb light more efficiently than the latter at a given concentration, and therefore, quantum yields of *s*-triazine degradation are lower (Equation 2).

A major degradation product could be observed for the four *s*-triazines in distilled water. This product could be clearly identified as the 2-hydroxy derivative, where the chlorine atom of the *s*-triazine is substituted by an OH group. This derivative has already been identified by other authors.[15] Figure 3 shows the

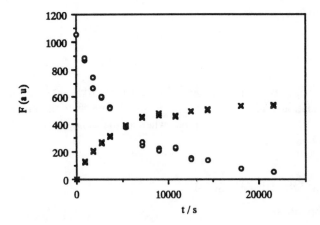

Figure 3.   Degradation of atrazine (o) and formation of the 2-hydroxy derivative (x) as a function of irradiation time t.

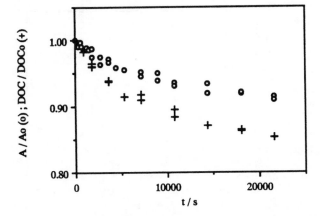

Figure 4. Relative absorbance A/A_0 at $\lambda = 254$ nm (o) and relative DOC value DOC/DOC_0 (+) from natural water as a function of irradiation time t.

degradation rate of atrazine and the production of 2-hydroxyatrazine as a function of irradiation time. In the liquid chromatograms, additional products could be observed at higher retention times. However, they have not been identified. The two anilides show no clearly detectable degradation products.

Influence of Humic Substances

Before irradiation of the pesticides in the natural water, the influence of light on the water itself was investigated through measurements of the optical density at $\lambda = 254$ nm (A_{254}) and of the DOC. The irradiation of the water showed a decrease of less than 10% of the absorbance at $\lambda = 254$ nm and of ca. 15% of the DOC after 6 hr (Figure 4).

Solutions of the six pesticides in the natural water at initial concentrations identical to those in distilled water were irradiated under the same conditions as the pesticide solutions. The degradation followed a pseudo-first-order rate law. The total absorption of the solution was used to calculate the quantum yields of pesticide degradation. To be able to neglect the decrease in absorption of the solution at the wavelength of irradiation due to the degradation of HS, the rate constants and the quantum yields of photodegradation of pesticides were determined during short irradiation periods.

The photodegradation rate of all investigated pesticides was lower in the natural water than in distilled water, as shown in Figure 5 for atrazine. As in distilled water, the pesticides within each of the two groups showed similar behavior (Table 2). The comparison between the rate constants k_p^{HS} in the natural water and k_p in distilled water shows a more pronounced slow-down of the reaction for the anilides than for the s-triazines. This leads to quantum yields of the degradation in the HS-containing water that are 10 and 100 times smaller than in distilled water for the s-triazines and anilides, respectively.

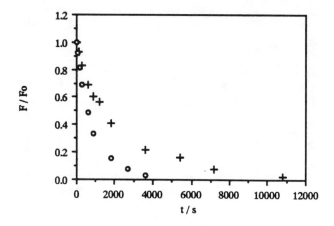

Figure 5. Photochemical degradation of atrazine in distilled water (o) and in natural water (+) containing HS (DOC_0 = 15 mg L^{-1}). Relative concentrations (F/F_0) obtained by HPLC as a function of irradiation time t.

In the HS-containing water, no degradation products in addition to the ones found in distilled water were observed. This does not exclude the possibility of product formation of, e.g., N-dealkylated derivatives, which have the same retention time in the HPLC diagram as HS and therefore could be hidden under the broad HS peak.

The decrease in photolysis rate must be due to the optical filter effect of HS, which shows a strong absorption at the irradiation wavelength. However, a sensitization effect of HS cannot be excluded because it could be hidden by the strong filter effect. To check this possibility, we calculated the quantum yields of degradation of pesticides in the natural water, accounting for the filter effect of HS. The fraction of light absorbed by the pesticides dissolved in the HS-containing water was quantified. This fraction resulted in a correction factor for the quantum yields obtained so far in the natural water. The corrected quantum yields, which gave a value for the efficiency of the pesticide degradation due to their own light absorption and not to the total absorption of the solution, allowed determination of the influence of HS on the photodegradation of the pesticides.

Table 2. Rate Constants for the Photochemical Degradation of Pesticides k_p in Distilled Water and k_p^{HS} in Natural Water and Corresponding Quantum yields Φ_p and Φ_p^{HS}

Pesticide	$k_p \cdot 10^4$ (sec^{-1})	$k_p^{HS} \cdot 10^4$ (sec^{-1})	$\Phi_p \cdot 10^2$	$\Phi_p^{HS} \cdot 10^2$
Atrazine	1.9	1.1	3.7 (±0.3)	0.45 (±0.05)
Desethylatrazine	1.9	1.0	3.9 (±0.3)	0.51 (±0.06)
Simazine	1.9	1.1	3.8 (±0.3)	0.38 (±0.04)
Terbutylazine	1.9	1.2	3.7 (±0.3)	0.33 (±0.04)
Metazachlor	2.5	1.1	37 (±6)	0.47 (±0.05)
Metolachlor	2.8	1.1	34 (±7)	0.36 (±0.04)

To calculate the amount of light absorbed by the pesticide, we assumed that no interaction took place between the HS and the pesticide and that the total absorption of the solution was the addition of the absorption of the separate components.

A possible interaction between HS and the pesticides was studied through absorption and fluorescence measurements. The absorption spectrum (200 nm < λ < 700 nm) of the HS-pesticide solution turned out to be an exact superposition of the absorption of the pesticide in distilled water and the absorption of the natural water. Furthermore, the fluorescence spectrum of the HS-containing water was not influenced by the addition of pesticides. No interaction could be observed by these two methods.

When light of intensity I is absorbed in a solution of thickness dx containing two components at concentrations C_1 and C_2, the intensity decreased is proportional to the concentrations C_1 and C_2, to the thickness dx, and to the intensity I

$$\frac{dI}{dx} = -\left(\alpha_1 C_1 + \alpha_2 C_2\right) \cdot I \tag{3}$$

where α_1 and α_2 represent the absorption coefficients of components 1 and 2, respectively. For a homogeneous solution, the concentration C does not depend on x. After separation of the variables, the integration

$$\int_{I_0}^{I_1} d \ln I = -\int_0^1 \left(\alpha_1 C_1 + \alpha_2 C_2\right) \cdot dx \tag{4}$$

between the incident light intensity I_0 and the light intensity I_1 at the length l and between the solution thickness 0 and l gives

$$I_1 = I_0 \cdot \exp\left\{-\left(\alpha_1 C_1 + \alpha_2 C_2\right) \cdot l\right\} \tag{5}$$

Equation 5 shows that the light intensity decreases exponentially with the solution thickness, the concentration of the absorbing species, and their capacity to absorb light. This equation is equivalent to

$$I_1 = I_0 \cdot 10^{-\left(\varepsilon_1 C_1 + \varepsilon_2 C_2\right) \cdot l} \tag{6}$$

where ε_i represents the molar absorption coefficient of component i ($\varepsilon_i = \alpha_i / \ln 10$). Equation 6 can also be written as

$$\log \frac{I_1}{I_0} = -\left(\varepsilon_1 C_1 + \varepsilon_2 C_2\right) \cdot l \tag{7}$$

where the product $\varepsilon_i C_i l$ represents the absorbance A_i of component i. The intensity change with the path length corresponds to the light absorption through the substances in the solution. Equation 3 can be written as a function of the absorbed light intensity P_a.

$$\frac{dP_a}{dx} = \left(\alpha_1 C_1 + \alpha_2 C_2 \right) \cdot I \tag{8}$$

Under the condition that no interaction is taking place between the components of the solution, the change of light absorption with the path length can be written as

$$\frac{dP_a}{dx} = \frac{d\left(P_{a1} + P_{a2} \right)}{dx} \tag{9}$$

where P_{a1} and P_{a2} represent the absorbed light intensities for the separate components with

$$\frac{dP_{a1}}{dx} = \alpha_1 C_1 I \tag{10}$$

and

$$\frac{dP_{a2}}{dx} = \alpha_2 C_2 I \tag{11}$$

The light absorption P_{a1} of component one in a two-component system is given by

$$P_{a1} = \int_0^1 \frac{dP_{a1}}{dx} \cdot dx \tag{12}$$

Substitution of Equations 5 and 10 in Equation 12 leads to

$$P_{a1} = \alpha_1 C_1 I_0 \int_0^1 \exp\left\{ -\left(\alpha_1 C_1 + \alpha_2 C_2 \right) \right\} \cdot dx \tag{13}$$

and after integration to

$$P_{a1} = \frac{\alpha_1 C_1}{\alpha_1 C_1 + \alpha_2 C_2} \cdot I_0 \left[1 - \exp\left\{ -\left(\alpha_1 C_1 + \alpha_2 C_2 \right) \cdot 1 \right\} \right] \tag{14}$$

After replacement of the absorption coefficient α by the molar absorption coefficient ε, one obtains

$$P_{a1} = \frac{\varepsilon_1 C_1}{\varepsilon_1 C_1 + \varepsilon_2 C_2} \cdot I_0 \left[1 - 10^{-\left(\varepsilon_1 C_1 + \varepsilon_2 C_2 \right) \cdot 1} \right] \tag{15}$$

or

$$P_{a1} = \frac{\varepsilon_1 C_1}{\varepsilon_1 C_1 + \varepsilon_2 C_2} \cdot P_a \qquad (16)$$

where P_a represents the total light absorption of the two-component solution. In other words, the light absorption P_{a1} of component 1 in a two-component solution is smaller by a factor

$$F = \frac{\varepsilon_1 C_1}{\varepsilon_1 C_1 + \varepsilon_2 C_2} \qquad (17)$$

than in a solution of component 1 alone at the same concentration.

To determine the quantum yield of photochemical reaction of one compound in a two-component solution, the absorbed light intensity of the compound of interest has to be expressed by Equation 16. To compare the pesticide solution in distilled water and in the natural water, we define a correction factor f

$$f = \frac{\varepsilon_1 C_1 + \varepsilon_2 C_2}{\varepsilon_1 C_1} \qquad (18)$$

which will be multiplied by the quantum yield Φ_P^{HS} of photochemical degradation of the six pesticides in the HS-containing water to obtain the corrected quantum yields Φ_P^{corr}. The quantum yields Φ_P^{corr} represents the value of the quantum yield of pesticide degradation in the natural water under the hypothesis that HS are not acting as a light filter. The uncorrected (Φ_P^{HS}), the corrected (Φ_P^{corr}) quantum yields, the correction factors f, and the quantum yields Φ_P for the direct photolysis in distilled water are listed in Table 3.

The values obtained for the corrected quantum yields Φ_P^{corr} within the error margins, are identical to the results in distilled water. This result shows that the HS play only the role of a light filter for the degradation of the pesticides and that no sensitization by HS seems to take place. This statement is further supported by measurements in oxygen-free solutions which show no significant changes in the photolysis rates.

Table 3. Quantum Yields Φ_P^{HS} of the Photochemical Degradation of Pesticides in Natural Water, Correction Factors f, and Corrected Quantum Yields Φ_P^{corr} in Comparison to the Quantum Yields Φ_P in Distilled Water

Pesticide	$\Phi_P^{HS} \cdot 10^3$	f	$\Phi_P^{corr} \cdot 10^2$	$\Phi_P \cdot 10^2$
Atrazine	4.5 (±0.5)	8.7 (±0.6)	3.8 (±0.4)	3.7 (±0.3)
Desethylatrazine	5.1 (±0.6)	6.6 (±0.3)	3.6 (±0.5)	3.9 (±0.3)
Simazine	3.8 (±0.4)	9.9 (±0.9)	3.8 (±0.5)	3.8 (±0.3)
Terbutylazine	3.3 (±0.4)	12.4 (±1.4)	3.7 (±0.6)	3.7 (±0.3)
Metazachlor	4.7 (±0.5)	62 (±14)	29 (±7)	37 (±6)
Metolachlor	3.6 (±0.4)	102 (±41)	35 (±15)	34 (±7)

CONCLUSION

The photochemical degradation of the six investigated pesticides is strongly inhibited in natural water containing HS. The decrease of photolysis rate and of the efficiency of the degradation are primarily due to the filter effect of HS, which show a strong absorption at the wavelength of irradiation. The HS show no sensitization of the degradation, which would lead to an enhancement of the degradation. Due to the wavelength of 254 nm used in the experiments, the results do not apply to sunlight conditions or to photoreactions in the aquatic environment. However, for the technical application of short-wavelength irradiation in water treatment, the results are meaningful. They suggest that HS have to be eliminated from the water before UV irradiation in order to obtain an optimum degradation of the pesticides.

ACKNOWLEDGMENTS

We acknowledge the technical assistance of Andrea Schick.

REFERENCES

1. Von Sonntag, C., Disinfection with UV-radiation, in *Process Technologies for Water Treatment*, Stucki, S., Ed., Plenum Press, New York, 1988, 159.
2. Khan, S. U. and Schnitzer, M., UV Radiation of atrazine in aqueous fulvic acid solution, *J. Environ. Sci. Health*, B13, 299, 1978.
3. Mabey, W. R., Tse, D., Baraze, A., and Mill, T., Photolysis of nitroaromatics in aquatic systems. I. 2,4,6-Trinitrotoluene, *Chemosphere*, 1, 3, 1983.
4. Simmons, M. S. and Zepp, R. G., Influence of humic substances on photolysis of nitroaromatic compounds in aqueous systems, *Water. Res.*, 20, 899, 1986.
5. Zepp, R. G., Baughman, G. L., and Schlotzhauer, P. F., Comparison of photochemical behavior of various humic substances in water: I. Sunlight induced reactions of aquatic pollutants photosensitized by humic substances, *Chemosphere*, 10, 109, 1981.
6. Zepp, R. G., Baughman, G. L., and Schlotzhauer, P. F., Comparison of photochemical behavior of various humic substances in water: II. Photosensitized oxygenations, *Chemosphere*, 10, 119, 1981.
7. Haag, W. H. and Hoigné, J., Singlet oxygen in surface waters. 3. Photochemical formation and steady state concentrations in various types of waters, *Environ. Sci. Technol.*, 20, 341, 1986.
8. Zepp, R. G., Wolfe, N. L., Baughman, G. L., and Hollis, R. C., Singlet oxygen in natural waters, *Nature (London)*, 267, 421, 1977.
9. Baxter, R. M. and Carey, J. H., Evidence for photochemical generation of superoxide ion in humic waters, *Nature (London)*, 306, 575, 1983.
10. Cooper, W. J. and Zika, R. G., Photochemical formation of hydrogen peroxide in surface and ground waters exposed to sunlight, *Science*, 220, 711, 1983.
11. Zepp, R. G., Braun, A. M., Hoigné, J., and Leenheer, J. A., Photoproduction of hydrated electrons from natural organic solutes in aquatic environments, *Environ. Sci. Technol.*, 21, 485, 1987.

12. Faust, B. C. and Hoigné, J., Sensitized photooxidation of phenols by fulvic acid and in natural waters, *Environ. Sci. Technol.*, 21, 957, 1987.
13. Mill, T., Hendry, D. G., and Richardson, H., Free-radical oxidants in natural waters, *Science*, 207, 886, 1980.
14. Parker, C. A., A new sensitive chemical actinometer. I. Some trials with potassium ferrioxalate, *Proc. R. Soc. London Ser. A*, 220, 104, 1953.
15. Pape, B. E. and Zabik, M., Photochemistry of bioactive compounds. Photochemistry of selected 2-chloro- and 1-methylthio-4,6-di-(alkylamino)-*s*-triazine herbicides, *J. Agric. Food Chem.*, 18, 202, 1970.

CHAPTER 10

The Relationship of Hydroxyl Reactivity to Pesticide Persistence

Scott A. Mabury and Donald G. Crosby

INTRODUCTION

The importance of hydroxyl in the degradation of pollutants in the aquatic environment has received increased attention as the sources, reaction pathways, and concentrations in natural waters became better known. Hydroxyl is the most reactive oxidant known in the environment, with second-order reaction rate constants for most organic pollutants of 10^9 to $10^{10} M^{-1} s^{-1}$.[1] Numerous constituents of natural waters contribute to its photochemical production; for example, water polluted with nitrate shows an increased ability to generate hydroxyl, thus providing a mechanism for self-cleansing. The extreme reactivity of hydroxyl makes it difficult to monitor directly, requiring the use of surrogate reactions for elucidating reaction pathways and quantifying the concentrations present under various environmental conditions.

Flooded rice fields represent an exceptional opportunity to demonstrate the role of hydroxyl in the degradation of organic pollutants. They are sunlit environments of clear, shallow (10 cm) water to which relatively large amounts of organic pesticides and nitrogenous fertilizers are applied to improve agronomic production. A general observation has been that pesticides are degraded much more rapidly under these conditions than can be attributed solely to the processes of direct photolysis and hydrolysis or the dissipation pathways of volatilization and sorption. The purpose of our research is to relate pesticide reactivity toward hydroxyl with pesticide persistence under field conditions.

This chapter reviews the formation of hydroxyl as relevant to agricultural field water and reaction pathways as they relate to the fate of pesticides and aromatic contaminants. Methods for measuring steady-state hydroxyl concentrations, $[OH]_{ss}$, are also reviewed, and the use of one of these to determine the relative reaction rates of hydroxyl with common pesticides is described.

SOURCES OF HYDROXYL

A variety of reaction pathways are available for the generation of hydroxyl in natural waters. The major sources appear to be the photolysis of nitrate, nitrite, hydrogen peroxide, and dissolved organic carbon (DOC). Nonphotolytic pathways include the Fenton system of iron, manganese, or copper species in conjunction with hydrogen peroxide.[2] The relative importance of any one of these pathways depends on the concentration, quantum yield, and absorption coefficient of the reactive solute.

Many natural waters contain substantial concentrations of nitrate and nitrite ions, which undergo direct photolytic breakdown at wavelengths greater than 290 nm to form hydroxyl. The photolysis of nitrite proceeds as in Equation 1; the conjugate base of hydroxyl, $\cdot O^-$, is quickly protonated ($pK_a = 11.9$, Equation 2).[2] The photolysis of nitrate (NO_3^-) is more complicated (Equations 3a and 3b). Atomic oxygen (O^3P) is generally considered to be much less reactive than hydroxyl, but will react with oxygen to form ozone which then reacts with nitrite or decomposes to hydroxyl.[3] The resulting nitrite radical follows the reaction pathway of Equation 4 to reform nitrate and nitrite.[4] Quantum yields for hydroxyl production from nitrate have been reported to lie in the range of $\Phi = 9.2$–17×10^{-3} at wavelengths greater than 290 nm.[3,4] Nitrite quantum yields are wavelength dependent, decreasing as wavelength increases; typical values range from $\Phi = 0.015$–0.08 between 298 and 371 nm.[4,5]

$$NO_2^- \xrightarrow{\text{hv}} \cdot O^- + NO \qquad (1)$$

$$\cdot O^- + H^+ \longrightarrow \cdot OH \qquad (2)$$

$$NO_3^- \xrightarrow{\text{hv}} NO_2^- + O \qquad (3a)$$

$$NO_3^- \xrightarrow{\text{hv}} \cdot NO_2 + \cdot O^- \qquad (3b)$$

$$2 \cdot NO_2 + H_2O \longrightarrow NO_2^- + NO_3^- + 2H^+ \qquad (4)$$

Hydroxyl generation from nitrate in laboratory systems shows a nonlinear increase as nitrate concentration increases.[3,6,7] This may result from either a quenching of the excited state of nitrate[7] or by scavenging of hydroxyl by nitrite.[3]

Constituents of DOC have been shown to photosensitize the degradation of numerous organic pollutants. Humic substances undergo reactions directly by various mechanisms, including energy transfer from the triplet state and abstraction of electrons or hydrogen atoms.[8] Alternatively, the excited state can produce other reactive intermediates; Takahashi et al.,[9] using electron spin resonance (ESR) spectroscopy, identified hydroxyl, singlet oxygen, aquated electrons, and hydrogen peroxide as products of the photolysis of dissolved humic substances. They concluded that the formation of hydroxyl took place directly through homolytic cleavage of the humic acid, while photolysis of hydrogen peroxide represented a minor pathway. Aquated electrons generally are scavenged by dissolved oxygen ($2 \times 10^{10} \ M^{-1} \ s^{-1}$),[9] forming the superoxide radical anion which rapidly dismutates to form hydrogen peroxide (Equation 5).

$$e^-_{solv} + O_2 \longrightarrow \cdot O_2^- \longrightarrow H_2O_2 \tag{5}$$

Reactions attributable to hydroxyl have been observed upon irradiation of humic substances. Khan and Gamble[10] hypothesized that dealkylation of the triazine herbicide prometryn was mediated by humic-derived hydroxyl, and Kotzias et al.[11] reported the fractionation of humics and the resulting impact on the ability to photoinitiate degradation of cumene: the higher the molecular weight, the higher the ultraviolet (UV) absorbance, and the greater the degradation of cumene. The degree to which humics were degraded under UV radiation affected the production of oxidants and, correspondingly, the ability to degrade organic substances.[12]

The addition of nitrate was shown to increase the production of hydroxyl and alkylperoxides in irradiated solutions of humic acids.[11] It was hypothesized that the additional hydroxyl, produced from nitrate, reacted further with humics to result indirectly in more radicals. Increasing concentrations of nitrate, relative to humic acids, led to a geometrical increase in hydroxyl, contrasting with the quenching seen in simple nitrate photolysis.

Draper and Crosby[13] reported the UV-energized formation of hydrogen peroxide in field water and water amended with certain organic solutes. Tryptophan, hydroquinone, tyrosine, and humic acids derived from soil showed the greatest ability to generate hydrogen peroxide, while the fulvic acid fraction was inactive. Generally, a plateau in hydrogen peroxide concentration was observed after 3 hr of sunlight irradiation. Agricultural waters showed concentrations of hydrogen peroxide as high as 6.8 μM, while highly eutrophic waters had levels up to 32 μM. Hydrogen peroxide was hypothesized to be formed via the solvated electron-superoxide pathway (Equation 5). Cooper and Zika[14] and Cooper et al.[5] showed that hydrogen peroxide generation in water correlated well with total organic carbon content and the amount of light absorbed by the sample; quantum yield was observed to decrease with increasing wavelength, with an $\Phi_{H_2O_2}$ of 7.0×10^{-4} at 297.5 nm and 2.2×10^{-4} at 350 nm.

Hydrogen peroxide produced hydroxyl cleanly upon irradiation by homolysis of the peroxide bond (Equation 6).[1] The quantum yield reportedly was 0.5,[16] although yields generally were low due to the weak absorption in the actinic portion of the sunlight spectrum.[7,17] Production of hydroxyl by nonphotochemical means generally was thought to proceed via Fenton-type reactions with Fe(II) or Cu(I) and H_2O_2 as shown in Equations 7 and 8. Moffett and Zika[18] concluded that in certain seawater microenvironments (cell surfaces, surface microlayers), where concentrations of Fe(II) and Cu(I) could reach a level of 10^{-12} and 10^{-10} M, respectively, the reduction of H_2O_2 would yield a significant amount of hydroxyl. In freshwater, a photoredox cycle would continuously produce hydroxyl by reaction of Fe(II) with DOC-derived H_2O_2, while photoreduction of the Fe(III) replenished the supply of Fe(II).[19]

$$H_2O_2 \xrightarrow{\text{hv}} 2\ ^{\cdot}OH \qquad\qquad (6)$$

$$Fe^{2+} + H_2O_2 \longrightarrow Fe^{3+} + OH^- + ^{\cdot}OH \qquad\qquad (7)$$

$$Cu^+ + H_2O_2 \longrightarrow Cu^{2+} + OH^- + ^{\cdot}OH \qquad\qquad (8)$$

REACTIONS OF PESTICIDES WITH HYDROXYL

Numerous investigators have used photolysis of either hydrogen peroxide or nitrate to produce hydroxyl for reaction studies. In addition, a vast literature is available regarding radiochemical formation of hydroxyl.[1] Hydrogen peroxide produces only hydroxyl without the complicating side reactions present with other methods, although direct photolysis of substrate may complicate matters.

Hydroxyl reaction pathways include hydrogen abstraction, addition to unsaturated systems, electron transfer, and radical interactions. Hydrogen abstraction can occur at any aliphatic position relative to a functional group, but α-abstraction generally predominates[2] to form aldehydes and ketones via peroxyl radical intermediates.[20] Hydroxyl undergoes addition to π electron systems; addition to aromatic rings proceeds through a hydroxycyclohexadienyl radical, which is oxidized by molecular oxygen to phenolic products (Equation 9). Hydroxyl can add to thioethers by S-addition, forming an unstable S-OH radical.[2] Metal ions undergo electron-transfer reactions with hydroxyl, such as the oxidation of Fe(II) to Fe(III) and hydroxide ion. Radical interactions also occur, exemplified by reformation of hydrogen peroxide from two hydroxyls.

$$\cdot OH \ + \ \bigcirc \longrightarrow \bigcirc \overset{H}{\underset{OH}{\cdot}} \overset{O_2}{\longrightarrow} \bigcirc_{OH} \qquad (9)$$

Draper and Crosby[17,20,21] investigated the reactions of numerous pesticides with hydroxyl and compared the degradation products with those formed in irradiated field water. Thiobencarb, a thiolcarbamate herbicide, was shown to undergo the same ring hydroxylations, N-dealkylation, and sulfur oxidation in sterile field water as observed either with hydroxyl generated from photolysis of hydrogen peroxide or by the Fenton system. Singlet oxygen was shown to be unreactive toward this compound. Irradiation of field water solutions produced the same array of photoproducts as in the laboratory system, although larger amounts of thiobencarb sulfoxide and lower concentrations of ring-hydroxylation products were isolated in comparison with products produced with hydroxyl radicals alone. The authors hypothesized that secondary radical reactions could contribute to the difference; quenching of hydroxyl by natural scavengers, such as carbonate, generated other radicals which also could react with thiobencarb.[7] However, the contribution of hydroxyl to the degradation process was clear, given the isolation of the hydroxylated ring products. Investigations with other herbicides showed similar results;[21] singlet oxygen was shown not to degrade phosphorothioates, phosphorodithioates, or other thiolcarbamates.[17]

Hydroxyl generated in the same way has been shown to react readily with other aromatic compounds. Phenols predominated in the reaction with benzene[11,22] and cumene,[11,23] and chlorinated phenols yielded further hydroxylated products such as chlorocatechol and dechlorinated catechols and hydroquinones,[24] while nitrophenols yielded nitrocatechol and nitrohydroquinones.[4] Successive oxidation of 2,4-dinitrotoluene occurred initially at the methyl group and proceeded through the alcohol and aldehyde to the carboxylic acid, which was subsequently decarboxylated. Further oxidation yielded di- and trihydroxynitrobenzenes, followed by ring fission, with isolation of aliphatic acids such as maleic and nitromaleic, glyoxylic, oxalic, and formic.[25] Organophosphorus insecticides engendered greater toxicity by oxidation of the thio group to the oxygen analog; Katagi[26] used experimental and molecular orbital modeling to conclude that hydroxyl was the causative agent.

MEASUREMENT

Efforts to elucidate the importance of hydroxyl in the persistence of organic chemicals in the aquatic environment have included attempts to measure steady-state concentrations, $[OH]_{ss}$. The extreme reactivity and correspondingly short lifetime result in low $[OH]_{ss}$, so hydroxyl has been difficult to monitor directly by such techniques as ESR. There are relatively few reported methods for determining $[OH]_{ss}$, and, generally, these rely on monitoring some particular chemical reaction of hydroxyl. Model reactants have been utilized in both biological and natural water systems;[6,7,23,27,28] the disappearance of the probe chemical was monitored, or, for more specificity, a unique product of the hydroxyl reaction often was measured.

Selection of an appropriate probe requires that the degradation mechanism be specific for hydroxyl or that the products not be appreciably degraded during the course of the measurement. Monitoring of products requires sensitive methods to isolate and quantitate the low levels of hydroxyl-derived products. The use of chemical probes for measuring $[OH]_{ss}$ has followed two approaches. In one, the probe is added to a hydroxyl-generating solution at a low concentration that does not appreciably compete with the natural scavengers of hydroxyl. In the other, it is added at a high enough concentration so that all generated hydroxyl is scavenged. Measured $[OH]_{ss}$ values range from 5×10^{-16} to 1.5×10^{-18} M in natural freshwaters[6,23] and 1.2×10^{-17} to $1.1–12 \times 10^{-18}$ M in coastal and open ocean surface water.[28,29]

Mill, et al.[23] used isopropylbenzene (cumene) at 10^{-4} M as such a probe and quantitated products resulting from hydroxyl attack upon the alkyl side chain (e.g., acetophenone) and the aromatic ring (2- and 4-hydroxyisopropylbenzene). The ratio of ring- to side-chain (R/S) hydroxylation was 2.32; the alkyl side chain also was susceptible to attack by alkylperoxy radicals, and the contribution of this oxidant was subtracted from the final results. When filtered river or lake water (10^{-4} M in cumene) was irradiated for 5 days (>290 nm), the R/S ranged from 1.0 for Boronda Lake to 0.31 for Coyote Creek, indicating that alkylperoxy radicals were contributing to the degradation. Assuming a steady-state concentration of hydroxyl, simple kinetic analysis of the generation of oxidized products showed that an $[OH]_{ss}$ ranged from 1.5×10^{-18} M in lake water to about ten times higher in creek water. No data were presented for the degree to which the probe competed with the natural scavengers for hydroxyl. Russi et al.,[6] in an investigation into the importance of nitrate as a hydroxyl source, reported a higher value of $[OH]_{ss}$ ($5 \times 10^{-16} M$) in river water, using $10^{-5} M$ benzene as probe.

Haag and Hoigné[7] used low micromolar levels of n-butyl chloride (BuCl) to avoid significantly affecting the lifetime of hydroxyl or, correspondingly, the scavenging rate of the water samples. n-Butyl chloride has a reported rate constant with hydroxyl of 3×10^{-9} M^{-1} s^{-1} and is recalcitrant to direct photolysis. The degradation of BuCl was monitored by head-space gas chromatography (GC), and irradiations were conducted in a laboratory merry-go-round photoreactor equipped with a high-pressure mercury lamp and borosilicate glass filter. Hydrogen peroxide and nitrate were added to samples to increase the apparent $[OH]_{ss}$, while the addition of octanol increased the scavenging rate with a concomitant decrease in $[OH]_{ss}$. A plot of added H_2O_2 against $[OH]_{ss}$ showed a linear relationship, with the y-intercept yielding a $[OH]_{ss}$ of $7 \times 10^{-15} M$ for the photoreactor. Extrapolation of these results was required because, under natural sunlight, the loss of BuCl was negligible after only 1 day of irradiation. Assuming 20 to be the factor relating noon sunlight and the photoreactor, a value for $[OH]_{ss}$ of approximately 3×10^{-16} M was obtained. The scavenging rate constant for the lake water was reported as 1×10^{-5} sec^{-1}, which yielded a production rate for hydroxyl of 1×10^{-11} M sec^{-1}. Additional experiments to prove that hydroxyl was the transient oxidant species actually measured were conducted by adding the required amount of octanol in order to double the scavenging rate; the pseudo-first-order loss of BuCl was decreased correspondingly by a factor of 2.

Similar methods were used by Zepp et al.[3] with BuCl, nitrobenzene, and anisole, with disappearance of the probe as the measure of its reaction with hydroxyl. Methylmercuric hydroxide was also used as a probe, monitoring inorganic mercury. Their investigation was concerned with the contribution of nitrate in the production of hydroxyl in natural waters. Laboratory experiments were conducted with octanol as scavenger at concentrations 20-fold that of the probe, while nitrate was generally two orders of magnitude greater than that of the probe. From the quantum yield and the specific light absorption rate, the production rate of hydroxyl was reported as $2.5 \times 10^{-13} M$ sec^{-1} per μM nitrate. These authors calculated maximal $[OH]_{ss}$ for the surface of a lake containing 0.1 mM nitrate (scavenging constant of 10^5 sec^{-1}) to be $2.5 \times 10^{-16} M$, closely approximating the measured value of $3.0 \times 10^{-16} M$ reported for the same water body by Haag and Hoigné.[7]

Mopper and Zhou[28,29] used high concentrations of benzoic acid or methanol to scavenge all of the hydroxyl produced upon irradiation; these probes were chosen because they reacted by different pathways and produced quantifiable, stable products (formaldehyde or p-hydroxybenzoic acid, pHBA). Competition experiments were conducted by irradiating probe solutions at several concentrations to determine the scavenging rate constant for particular water samples and to find the appropriate level of probe; generally, methanol (2–3 mM) or benzoic acid (0.2–5 mM) was added to water samples, irradiated for 4 hr in natural sunlight, and the formaldehyde or pHBA determined. A plot of product formation rate against reciprocal of probe concentration yielded a straight line with a slope equal to the scavenging rate constant and y-intercept representing the photochemical generation rate of hydroxyl. Hydroxyl production rates for seawater ranged from $3.45 \times 10^{-11} M$ sec^{-1} in a coastal surface water to $2.8 \times 10^{-12} M$ sec^{-1} in open ocean surface water, corresponding to $[OH]_{ss}$ of 13.7 and $1.1 \times 10^{-18} M$, respectively. Organic rich freshwater had a production rate of $4.2 \times 10^{-10} M$ sec^{-1}, corresponding to a $[OH]_{ss}$ of $8.4 \times 10^{-16} M$.

HYDROXYL IN AGRICULTURAL FIELD WATER

Agricultural field water is known to contain several of the most important sources of hydroxyl: nitrate, nitrite, DOC, and hydrogen peroxide. However, relatively little has been reported on field water oxidant activity, and no direct measurement of $[OH]_{ss}$ is available. Photooxidants in field water were first reported by Ross and Crosby.[30]

As mentioned previously, nitrate and nitrite have been shown to be photolyzed to $\cdot O^-$, the conjugate base of hydroxyl. At typical California rice field pH of 6 to 9, this species is immediately protonated to provide the hydroxyl radical. Field water commonly contains nitrate, derived largely from fertilizer, and its nitrite reduction product; typical levels in California rice field drainage are 2 to 12 mg/L.[31] Russi et al.[6] reported a positive but nonlinear relationship between natural nitrate concentration (3–45 ppm) and photohydroxylation of benzene, and Haag and Hoigné[7] concluded that nitrate photolysis was the major contribution to $[OH]_{ss}$ in a Swiss lake (4.9 ppm NO_3^-, 66 ppb NO_2^-). The $[OH]_{ss}$, 3×10^{-16} M, was supported by the calculated values of Zepp et al.[3]

Many agricultural waters are especially high in DOC, resulting from intimate contact with the organic matter of soil, animal fertilizers, and plant wastes. A pond may contain as much as 5 mg/L DOC,[6] while typical California rice field water contains from 1 to 10 mg/L. The addition of organic-rich freshwater to humic-deficient seawater resulted in increased hydroxyl formation,[28] and radiation between 280 and 320 nm was correlated with hydroxyl production and simultaneous photobleaching of DOC. However, other authors have reported a negative correlation between DOC and $[OH]_{ss}$, due to radical scavenging and light attenuation.

Relatively high concentrations of hydrogen peroxide have been found in agricultural water. Draper and Crosby[13] reported levels of "oxidant" (as H_2O_2) in irrigation water as high as 6.8 μM (197 $\mu g/L$). Hydrogen peroxide was actually isolated from sunlight-irradiated rice field water and identified,[17] at least part of it a result of the reaction in Equation 5 where e_{solv}^- is formed by the photodegradation of the amino acid tryptophan.[30]

Haag and Hoigné[7] make a strong case for nitrate rather than hydrogen peroxide as the principal photochemical source of hydroxyl based on the low UV absorption and slow formation rate of H_2O_2. They suggest that while DOC might make some contribution, it must be degraded rapidly by hydroxyl. Furthermore, other natural scavengers such as carbonate, bicarbonate, and sulfide could be expected to compete successfully for available ˙OH, leaving relatively little for reaction with pesticides which typically are present at well below 10^{-5} M.

Nonetheless, photooxidation of several pesticides which do not absorb appreciable solar UV radiation, such as molinate and thiobencarb, provided products that clearly were the result of reactions with hydroxyl.[20,21,32] Laboratory irradiation of their aqueous solutions provided the same products and proportions at roughly the same rates, regardless of whether the source of oxidant was filter-sterilized field water, tryptophan in distilled water, or hydrogen peroxide in distilled water. Irradiation of the pesticides in aqueous suspensions of titanium oxide or zinc oxide, clearly a source of ˙OH,[33] likewise produced the same products in both distilled water in the laboratory and in a sunlit rice field. Whatever the source, hydroxyl radicals appear to be important oxidants in agricultural field water.

PESTICIDE REACTIVITY AND PERSISTENCE

Zepp et al.[3] described shallow water bodies high in nitrate and low in natural scavengers as having the highest potential for degradation mediated by hydroxyl. Agricultural rice fields represent a situation in which all the major generators of hydroxyl — nitrate, hydrogen peroxide, and DOC — are known to occur along with a relatively large input of pesticides. The pesticides are dissipated from the water

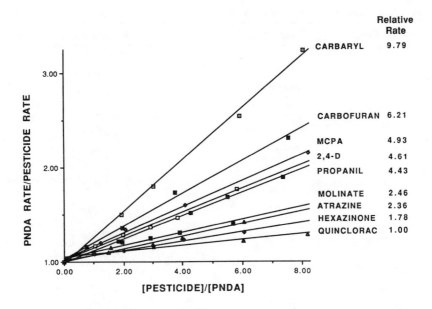

Figure 1. A plot showing the relative reactivities of the pesticides investigated.

column by volatilization and sorption to sediments and degraded by hydrolysis, direct photolysis, and indirect photolysis (mostly reactions with hydroxyl). Our research has focused on the relationship between hydroxyl reactivity and pesticide persistence under rice field conditions and a laboratory system for measuring the reactivity of pesticides with hydroxyl. The reactivity is then compared with the degradation rate under laboratory irradiation conditions in sterile field water and with dissipation under actual rice field cultural practices.

Hydroxyl reactivity was measured using a steady-state competition method that is popular in radiolysis studies.[1] Competition for hydroxyl was compared between a standard competitor (p-nitroso-N,N-dimethylaniline)[34] and a pesticide of interest. Hydroxyl was generated by the photolysis of hydrogen peroxide ($10^{-3}\,M$) via a Light Sources FS 40 fluorescent lamp that has a UV maximum at 310 nm. The pseudo-first-order loss of p-nitroso-N,N-dimethylaniline (PNDA, $10^{-5}\,M$) was monitored by UV diode array. Competition experiments were typically of 150 min duration. The reactivity of a particular pesticide was deduced from the degree to which it competed with PNDA for hydroxyl.

A plot of the relative reactivities is shown in Figure 1. Carbaryl was the most reactive toward hydroxyl, followed by another carbamate insecticide, carbofuran. The herbicides MCPA, 2,4-D, and propanil were of intermediate reactivity. The herbicides atrazine, quinclorac, molinate, and hexazinone were of relatively low reactivity towards hydroxyl. Structures are provided in Table 1.

To assess the relationship between persistence and reactivity, carbaryl, carbofuran, MCPA, and hexazinone were irradiated in field water under various

Table 1. Pesticides Mentioned in the Text

Common Name	Trade Name	Structure
Atrazine	Gesaprim	
Carabaryl	Sevin	
Carbofuran	Furadan	
2,4-D	Fernimine	
Hexazinone	Velpar	
Molinate	Ordram	
Propanil	Stam, Rogue	
Quinclorac	Facet	
Thiobencarb	Bolero	

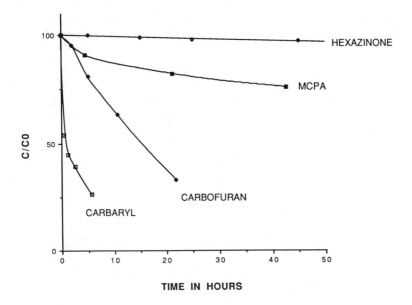

Figure 2. Degradation of model pesticides in the laboratory under simulated sunlight irradiation in filter-sterilized field water.

conditions. Laboratory experiments consisted of irradiating the pesticides in filter-sterilized field water under simulated sunlight conditions (F40-BL lamps filtered by borosilicate glass). Field experiments were conducted with commercial formulations of the pesticides applied by backpack sprayer to experimental rice plots. The pesticides were removed on solid-phase extraction cartridges and analyzed by HPLC with UV (variable wavelength) detection. Laboratory experiments were conducted to measure the Henry's Constant for each compound.

Dissipation of the pesticides from filtered field water in the laboratory under simulated sunlight irradiation is shown in Figure 2. Carbaryl showed rapid loss from field water under all conditions, while hexazinone proved recalcitrant to degradation. The other two probes showed intermediate loss, with carbofuran being more rapidly degraded than MCPA. Measured Henry's Constants for these four compounds were low ($<10^{-6}$ atm $\cdot$ m^3/mol), so volatilization would not be expected to contribute to their loss, and published adsorption coefficients were low (<30). In relative terms, the dissipation of the four pesticides followed the reactivity toward hydroxyl. The contribution to the dissipation by direct photolysis and hydrolysis have not been subtracted and are expected to influence the degradation of carbaryl and carbofuran in particular. Field measurements showed that the four pesticides tested so far also dissipated at rates consistent with the hydroxyl reaction rates.

Despite having comparable field dissipation data for only four of the pesticides, we hypothesize that photochemically generated hydroxyl radicals have an important influence on pesticide persistence in agricultural field water. If this relationship continues to be observed, prediction of the relative persistence of new pesticides would become possible, although the influence of volatilization and adsorption would have to be accounted for. Very recently, Haag and Yao[35] reported hydroxyl reaction rates for 44 common water contaminants based on the photo-Fenton reaction; harmonization of these two types of rate measurements might allow prediction of the persistence of a wide variety of chemical pollutants in natural waters.

REFERENCES

1. Buxton, G. V. and Greenstock, C. L., Critical review of rate constants for reactions of hydrated electrons, hydrogen atoms and hydroxyl radicals ($\cdot$OH/$\cdot$O$^-$) in aqueous solution, *J. Phys. Chem. Ref. Data*, 17, 513, 1988.
2. Barnes, A. R. and Sugden, J. K., The hydroxyl radical in aqueous media, *Pharm. Acta Helv.*, 61, 218, 1986.
3. Zepp, R. G., Hoigné, J., and Bader, H., Nitrate-induced photooxidation of trace organic chemicals in water, *Environ. Sci. Technol.*, 21, 443, 1987.
4. Alif, A. and Boule, P., Photochemistry and environment. Part XIV. Phototransformation of nitrophenols induced by excitation of nitrite and nitrate ions, *J. Photochem. Photobiol.*, 59, 357, 1991.
5. Zafiriou, O. C. and Bonneau, R., Wavelength-dependent quantum yield of OH radical formation from photolysis of nitrite ion in water, *Photochem. Photobiol.*, 45, 723, 1987.
6. Russi, H., Kotzias, D., and Korte, F., Photoinduzierte hydroxylierungsreaktionen organischer chemikalien in natürlichen gewässern, *Chemosphere*, 11, 1041, 1982.
7. Haag, W. R. and Hoigné, J., Photo-sensitized oxidation in natural water via OH radicals, *Chemosphere*, 14, 1659, 1985.
8. Zepp, R. G., Schlotzhauer, P. F., and Sink, R. M., Photosensitized transformations involving electronic energy transfer in natural waters: role of humic substances, *Environ. Sci. Technol.*, 19, 74, 1985.
9. Takahashi, N. M. I., Mikami, N., Matsuda, T., and Miyamoto, J., Identification of reactive oxygen species generated by irradiation of aqueous humic acid solution, *J. Pestic. Sci.*, 13, 429, 1988.
10. Khan, S. U. and Gamble, D. S., Ultraviolet irradiation of an aqueous solution of prometryn in the presence of humic materials, *J. Agric. Food Chem.*, 31, 1099, 1983.
11. Kotzias, D., Huster, K., and Wieser, A., Formation of oxygen species and their reactions with organic chemicals in aqueous solution, *Chemosphere*, 16, 505, 1987.
12. Kotzias, D., Herrmann, M., Zsolnay, A., Beyerle-Pfnur, R., Parlar, H., and Korte, F., Photochemical aging of humic substances, *Chemosphere*, 16, 1463, 1987.
13. Draper, W. M. and Crosby, D. G., The photochemical generation of hydrogen peroxide in natural waters, *Arch. Environ. Contam. Toxicol.*, 12, 121, 1983.
14. Cooper, W. J. and Zika, R. G., Photochemical formation of hydrogen peroxide in surface and ground waters exposed to sunlight, *Science*, 220, 711, 1983.
15. Cooper, W. J., Zika, R. G., Petasne, R. G., and Plane, J. M. C., Photochemical formation of hydrogen peroxide in natural waters exposed to sunlight, *Environ. Sci. Technol.*, 22, 1156, 1988.
16. Dorfman, L. M. and Adams, G. E., *Reactivity of the Hydroxyl Radical in Aqueous Solutions*, NSRDS-NBS-46, National Bureau of Standards, U. S. Government Printing Office, Washington, D.C., 1973.

17. Draper, W. M. and Crosby, D. G., Hydrogen peroxide and hydroxyl radical: intermediates in indirect photolysis reactions in water, *J. Agric. Food Chem.*, 29, 699, 1981.
18. Moffett, J. W. and Zika, R. G., Reaction kinetics of hydrogen peroxide with copper and iron in seawater, *Environ. Sci. Technol.*, 21, 804, 1987.
19. Micinski, E., Moffett, J. M., and Zafiriou, O. C., Fe(III)/Fe(II) photoredox cycling in the Orinoco River: evidence from the photochemical decomposition of $H_2{}^{18}O_2$, *EOS*, 71, 171, 1990.
20. Draper, W. M. and Crosby, D. G., Solar photooxidation of pesticides in dilute hydrogen peroxide, *J. Agric. Food Chem.*, 32, 231, 1984.
21. Draper, W. M. and Crosby, D. G., Photochemistry and volatility of drepamon in water, *J. Agric. Food Chem.*, 32, 728, 1984.
22. Mansour, M., Photolysis of aromatic compounds in water in the presence of hydrogen peroxide, *Bull. Environ. Contam. Toxicol.*, 34, 89, 1985.
23. Mill, T., Hendry, D. G., and Richardson, H., Free-radical oxidants in natural waters, *Science*, 207, 886, 1980.
24. Moza, P. N., Fytianos, K., Samanidou, V., and Korte, F., Photodecomposition of chlorophenols in aqueous medium in presence of hydrogen peroxide, *Bull. Environ. Contam. Toxicol.*, 41, 678, 1988.
25. Ho, P., Photooxidation of 2,4-dinitrotoluene in aqueous solution in the presence of hydrogen peroxide, *Environ. Sci. Technol.*, 20, 260, 1986.
26. Katagi, T., Molecular orbital approaches to the photolysis of organophosphorus insecticide fenitrothion, *J. Agric. Food Chem.*, 37, 1124, 1989.
27. Richmond, R., Halliwell, B., Chauhan, J., and Darbre, A., Superoxide-dependent formation of hydroxyl radicals: detection of hydroxyl radicals by the hydroxylation of aromatic compounds, *Anal. Biochem.*, 118, 328, 1981.
28. Mopper, K. and Zhou, X., Hydroxyl radical photoproduction in the sea and its potential impact on marine processes, *Science*, 250, 661, 1990.
29. Zhou, Z. and Mopper, K., Determination of photochemically produced hydroxyl radicals in seawater and freshwater, *Mar. Chem.*, 30, 71, 1990.
30. Ross, R. and Crosby, D. G., Photooxidant activity in natural waters, *Environ. Toxicol. Chem.*, 4, 773, 1985.
31. Tanji, K., Singer, M., Henderson, D., Whittig, L., and Biggar, J., Nonpoint sediment production in the colusa basin drainage area, First-year annual progress report, EPA Grant No. R805462, EPA, November, 1978.
32. Soderquist, C., Bowers, J., and Crosby, D., Dissipation of molinate in a rice field, *J. Agric. Food Chem.*, 25, 940, 1977.
33. Kawaguchi, H., Steady-state concentrations of hydroxyl radical in titanium dioxide aqueous suspensions, *Chemosphere*, 22, 1003, 1991.
34. Kralgic, I. and Trumbore, C. N., p-Nitrosodimethylaniline as an OH radical scavenger in radiation chemistry, *J. Am. Chem. Soc.*, 87, 2547, 1965.
35. Haag, W. R. and Yao, C. C. D., Rate constants for reaction of hydroxyl with several drinking water contaminants, *Environ. Sci. Technol.*, 26, 1005, 1992.

CHAPTER 11

Effect of Some Natural Photosensitizers on Photolysis of Some Pesticides

Rong Tsao and Morifusa Eto

INTRODUCTION

The fate of pesticides in the natural environment is influenced by various biotic or abiotic factors, of which the ultraviolet (UV) component in the sunlight is one of the most powerful forces. The photodegradation caused by UV light may be strongly affected by the naturally occurring photosensitizers. Thus, studies on the photodegradation of pesticides will not have significance only from the point of view of environmental science, but also are very important in researching and developing new, safer, and effective pesticides. In the present work, the authors investigated the photolytic fate of some typical pesticides, including the bisthiolcarbamate insecticide cartap hydrochloride, the organophosphorus insecticide pyridafenthion, the non-ester pyrethroidal insecticide etofenprox, the anilide herbicide naproanilide, and the benzanilide fungicide flutolanil (Table 1), and the effects of some natural and synthetic photosensitizers on pesticide photolysis (Table 2), using UV lamps (λ = 254 or >300 nm) and a photochemical reactor (λ > 300 nm) in the laboratory. We found that UV light caused a variety of photoreactions on the pesticides, and some photosensitizers did accelerate the photolysis, but with different modes of action.

EXPERIMENTAL

Chemicals

Cartap hydrochloride and nereistoxin oxalate were gifts from Takeda Chemical Industries, Ltd. Pyridafenthion, etofenprox, and naproanilide were provided by

Table 1. List of Pesticides in This Study

Pesticide	Structure	Chemical Name
Cartap hydrochloride	H₃C–N(–CH₂–SCONH₂) trimethylene, H₃C, ·HCl	S,S'-[2-(dimethylamino) trimethylene] bis(thio-carbamate) hydrochloride
Pyridafenthion	EtO–P(=S)(–OEt)–O– pyridazinone (N–N, Ph, =O)	O-(1,6-dihydro-6-oxo-1-phenyl-3-pyridazinyl) O,O-diethyl phosphorothioate
Etofenprox	C₂H₅O–C₆H₄–C(CH₃)₂–CH₂OCH₂–C₆H₄–O–C₆H₅	2-(4-Ethoxyphenyl)-2-methylpropyl 3-phenoxy-benzyl ether
Naproanilide	C₆H₅–NH–C(=O)–CH(CH₃)–O–naphthyl	2-(2-Naphthoxy)-propion-anilide
Flutolanil	CF₃–C₆H₄–C(=O)–NH–C₆H₄–O–iPr	3'-Isopropoxy-2-(trifluoro-methyl) benzanilide

Mitsui Toatsu Chemicals, Inc. Flutolanil was supplied from Nihon Nohyaku Co., Ltd. All other chemicals and solvents were commercially available or synthesized by the authors.

Apparatus and Experiments

The thin layer chromatography (TLC) was done on E. Merck silica gel 60 F_{254} plates. Gas-liquid chromatography (GLC) was carried out on a Shimadzu GC-7A instrument with a glass column (2.6 m × 3.2 mm) packed with 5% SE-30 silicone on Uniport KS (80~100 mesh) by using a flame ionization detector (FID). Mass spectroscopy (MS) was carried out on a Shimadzu GC-MS 9020DF containing a chemical-bonding type of silica capillary column (CBP1, 25 m × 0.33 mm). Infrared spectra (IR) were recorded on a Shimadzu IR-420 instrument. The ¹H-NMR spectra were recorded on a JEOL JNM·FX100 instrument. Samples for NMR were dissolved in chloroform-d, and chemical shifts (δ) are reported downfield from tetramethylsilane.

A rotary photoreactor (Riko Chemical Industries, Ltd., RH400-10W) equipped with a high-pressure mercury lamp (through a Pyrex glass filter, $\lambda > 300$ mn) was used for photoreaction in solution. Photoreactions on the glass and silica gel surface were carried out under irradiation with a 10-W germicidal lamp (Toshiba GL10, $\lambda = 254$ nm) or with a black light fluorescent lamp (Toshiba FL20S·BLB, $\lambda > 300$ nm).

Table 2.   Photosensitization Effect of Different Sensitizers on Pesticides

Sensitizer	Cartap[a]	Pyridafenthion[b]	Etofenprox[c]	Naproanilide[d]	Flutolanil[e]
			Pesticide		
Nonsensitizer	55.1	70.2	1.94	23.1	90.1
Acetone	—	—	—	0.9	82.2
Acetophenone	—	65.7	1.7	0.4	—
Benzophenone	51.2	67.8	1.6	0.4	76.3
Flavone	—	—	—	—	83.1
Xanthone	—	—	—	—	—
Chlorophyllin	ND	50.3	1.1	25.0	98.1
Methylene Blue	—	—	1.3	—	—
Methyl Green	—	—	1.2	—	—
Riboflavin	—	52.2	1.2	37.1	—
Rose Bengal	ND	53.1	1.0	21.3	88.2
Tryptophan	—	—	1.1	—	—
Anthracene	62.0	66.2	2.4	24.1	92.4
Fluorene	58.1	—	—	—	—
Phenanthrene	50.9	—	—	—	—
cis-DME[f]	59.7	ND	1.7	—	—
Flower Extract[g]	—	35.7	1.5	—	—

Note: —: not tested; ND: not detectable.
[a] Percentage of the applied pesticide after 35-hr irradiation on silica gel surface.
[b] Percentage of the applied pesticide after 40-hr irradiation on silica gel surface.
[c] Half-life time (hours) on silica gel surface.
[d] Half-life time (hours) in aqueous solution.
[e] Percentage of the applied pesticide after 8-hr irradiation in aqueous solution.
[f] cis-Dehydromatricaria ester.
[g] Extract of the goldenrod flower.

Concentrations of cartap hydrochloride, pyridafenthion, etofenprox, naproanilide, and flutolanil were 0.1, 1.0, 0.5, 5.0, and 5.0 mg/mL, respectively, in the aqueous solutions and 1.27, 6.40, 140, 570, and 283 $\mu m/cm^2$, respectively, on the thin films.

RESULTS AND DISCUSSION

Photodegradation of Pesticides

Many photoreactions occurred on every pesticide, as shown in Figure 1. Some of the photodegradation reactions were very interesting and have rarely been reported to occur with other pesticides.

The insecticide cartap hydrochloride, S,S'-[2-(dimethylamino)trimethylene] bis(thiocarbamate) hydrochloride, was transformed by UV irradiation to give nereistoxin, 4-N,N-dimethylamino-1,2-dithiolane. The photodegradation occurred both in aqueous solution and on a solid surface such as glass or silica gel as a thin film. This indicates that photodegradation of the insecticide can play an activation role as important as the in vivo metabolism that gives the same activated metabolic product, nereistoxin.[1]

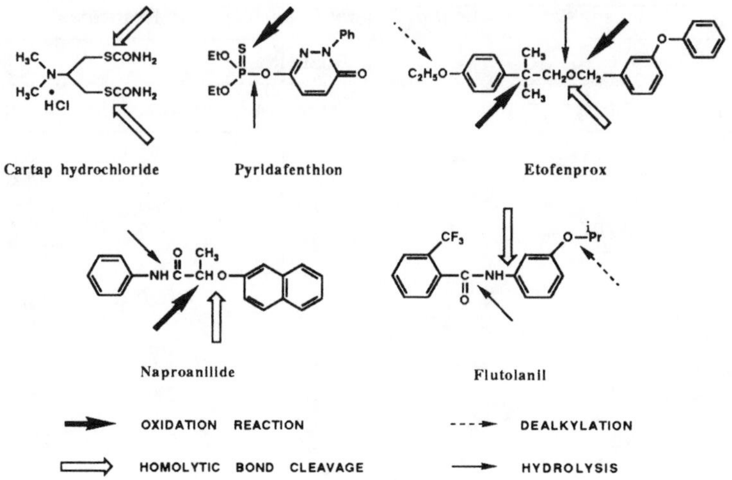

Cartap hydrochloride   Pyridafenthion   Etofenprox

Naproanilide   Flutolanil

→   OXIDATION REACTION   - - -►   DEALKYLATION

⟹   HOMOLYTIC BOND CLEAVAGE   ──►   HYDROLYSIS

Figure 1.   The major types of photodegradation reactions of the pesticides used in the present study.

Pyridafenthion   [*O*-(1,6-dihydro-6-oxo-1-phenyl-3-pyridazinyl)*O,O*-diethyl phosphorothioate] was photodegraded more rapidly on a glass surface than in the aqueous solution. However, two main reactions occurred, giving pyridafenthion-oxon and phenylmaleic hydrazide [6-hydroxy-2-phenyl-3(2*H*)-pyridazinone] in both cases.[2]

The non-ester pyrethroidal insecticide, etofenprox [MTI-500, 2-(4-ethoxy-phenyl)-2-methylpropyl 3-phenoxybenzyl ether], is a new insecticide with high photostability and high insecticidal activity but low fish toxicity. Etofenprox was more photostable ($t_{1/2}$ = 1.94 hour) than an ester pyrethroidal insecticide, allethrin ($t_{1/2}$ = 0.78 hour), under UV irradiation on the glass surface.[3] It was oxidized at the benzylic carbon atom of the ether linkage to form the major ester photoproduct **1** (Figures 1 and 2), which had no insecticidal toxicity against the housefly. Some other photoreactions such as further oxidation (to give photoprod-ucts **1** and **2**), hydrolysis (to give photoproducts **3** and **4**), decarboxylation (to give photoproducts **5** and **6**), and dealkylation (to give photoproduct **7**) also occurred. These photodegradation reactions were observed both in aqueous solution and on a glass surface (Figure 2).[3]

The UV irradiation of the herbicide, naproanilide [2-(2-naphthoxy)propion-anilide, **8**], in aqueous solution resulted in formation of the main photoproducts 2-hydroxypropionanilide **9**, acrylanilide, and pyruvanilide produced from the cleavage of the ether linkage (Figure 1). Naproanilide and its main photoproduct **9** were also photolytically rearranged to their aniline isomers **10**, **11**, **12**, and **13** both in aqueous solution and on solid surface (photo-Fries rearrangement reac-tion) (Figure 3). This interesting photoreaction has rarely been recognized so far in studies of pesticide degradation. Photodegradation of naproanilide on a glass

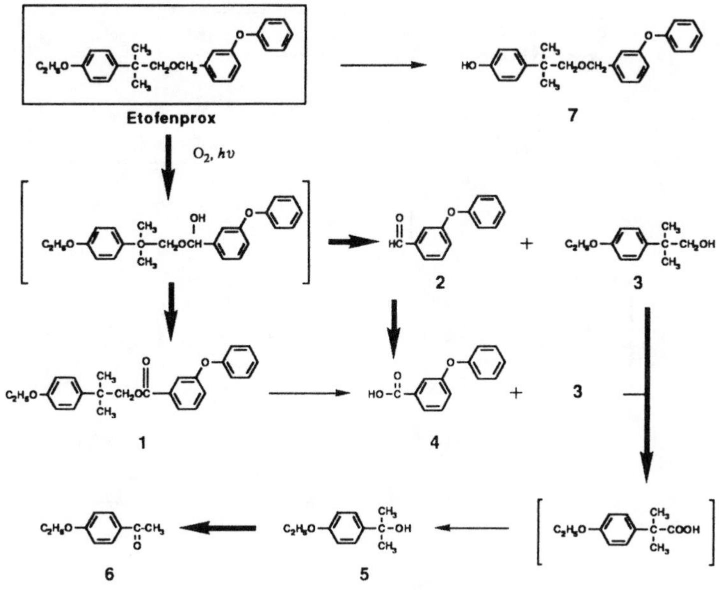

Figure 2. Photodegradation pathways of the insecticide etofenprox.

surface was so much slower than in the aqueous solution that we did not isolate and identify its photodegradation products.[4]

In the ethanolic aqueous solution, the UV irradiation on the fungicide flutolanil, 3'-isopropoxy-2-(trifluoromethyl)benzanilide, resulted in formation of one major product, 2-(trifluoromethyl)benzamide, and some minor products. Flutolanil was not photodecomposed under deaerated conditions in aqueous solution. On the solid surface, the photo-Fries rearrangement and the desisopropylation reaction were recognized in addition to the above photoreactions.[5]

As described above, the UV light may cause various types of reactions, bringing about structural changes or transformation in pesticide molecules that might greatly contribute to their degradation in the environment. In these photochemical changes, the oxidative transformation of cartap hydrochloride into nereistoxin and pyridafenthion to its oxon are clearly toxicological activation processes, whereas the transformation of etofenprox to an ester and photolyses of naproanilide and flutolanil are thought to be detoxification processes. The photo-Fries rearrangement of the anilide or benzanilide compounds may have instructive significance in photodegradation studies of those kinds of pesticides.

On the other hand, chemical oxidants are often used as models of mixed-function oxidase (MFO) in pesticide metabolism studies. The authors used N-bromosuccinamide (NBS), m-chloroperbenzoic acid (MCPBA), and Jones reagent (CrO_3) to oxidize cartap hydrochloride, pyridafenthion, and etofenprox, respectively, because these pesticides mostly incurred oxidations under UV

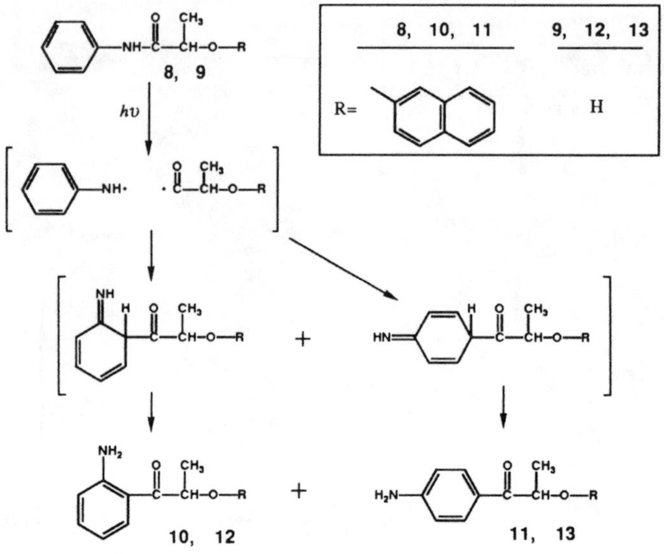

Figure 3. Photo-Fries rearrangement reaction of the herbicide naproanilide and its related compounds.

irradiation. We found that the results were very consistent with those of the photooxidations. We believe that to use these chemical oxidants may help researchers to understand the mechanisms of the pesticide photodegradation reactions.

Photosensitization by Some Natural and Synthetic Photosensitizers

Photodegradation of pesticides will be strongly affected by a variety of photosensitizers, and many substances in the natural environment act as photosensitizers.[6-11] Photodegradation of pesticides applied to control insects, plant diseases, and weeds may be accelerated by the natural photosensitizers in most cases. Different sensitizers have different acceleration effects. The effect of a photosensitizer is dependent upon the wavelength of the irradiation, the reaction media, and the UV absorption of the substrate, as well as the UV absorption of the photosensitizer itself.[12]

In the present study, rose bengal and chlorophyllin showed a great accelerating effect on cartap photodecomposition.[1] Photolysis of pyridafenthion was accelerated by riboflavin, rose bengal, and cis-dehydromatricaria ester.[2] Photosensitizers chlorophyllin, methyl green, tryptophan, rose bengal, and riboflavin enhanced the photodecomposition of etofenprox (Table 2).[3] However, only the photosensitizers with a carbonyl group in the molecule strongly accelerated the photolysis of naproanilide, whereas some dyes showed almost no effect.[4] Photosensitizers that accelerated the photodecomposition of flutolanil showed similar effects on naproanilide photodecomposition (Table 2).[5]

Table 3. Photochemical Data for Sensitizers and Relationship to Pesticides

Photosensitizer	E_s	E_t	Φ_{isc}	$\Phi_{{}^1O_2}$	Pesticide
Acetone	~85	78	1.00	—[a]	
Acetophenone	~79	74	1.00	0.59	Naproanilide
Benzophenone	~75	69	1.00	0.53	Flutolanil
Flavone	—	—	—	—	
Xanthone	77.6	74	1.00	—	
Chlorophyllin	—	29.6	0.88[b]	0.68[c]	
Methylene Blue	—	33.5	0.52	0.52	Cartap hydrochloride
Methyl Green	—	—	—	—	Pyridafenthion
Riboflavin	57.8	50	—	—	Etofenprox
Rose Bengal	—	44.6	0.8	0.83	
Tryptophan	—	—	—	—	
Anthracene	76	47	0.7	—	
Fluorene	94.9	68	0.32	0.11	
Phenanthrene	82.8	61.9	0.82	—	
cis-DME[d]	—	—	—	—	Pyridafenthion Ethofenprox Flutolanil

Note: E_s: Origin of lowest excited singlet state (kcal/mol); E_t: origin of lowest excited triplet state (kcal/mol); Φ_{isc}: quantum yield of intersystem crossing or triplet formation; $\Phi_{{}^1O_2}$: quantum yield of singlet oxygen formation.
Source: All data were gathered from *Handbook of Photochemistry and Photosensitizers*, Tokumaru, K. and Okawara, M., Eds., KOUDANSHA Scientific, Tokyo, 1987 (in Japanese).
[a] Not determined.
[b] Chlorophyll b.
[c] Chlorophyll a.
[d] *cis*-Dehydromatricaria ester.

These results (Tables 2 and 3) show that photolyses of the anilide pesticide naproanilide and flutolanil were mainly accelerated by the carbonyl sensitizers that have higher E_s, E_t, and Φ_{isc} but lower $\Phi_{{}^1O_2}$, whereas that of the other pesticides was mainly enhanced by dye sensitizers having lower E_s, E_t, and Φ_{isc} but higher $\Phi_{{}^1O_2}$.

Photolyses of the anilide pesticides, naproanilide and flutolanil, were more strongly enhanced by the ketone sensitizers than the dye sensitizers, and their photoproducts imply that they may be formed by a radical mechanism. This is probably because aromatic ketones or other carbonyl compounds are efficiently (quantum yields near one) excited to the triplet state, which would abstract hydrogen atoms from the substrate or solvent to produce radical(s).[13,14] These radicals may cause a radical chain reaction and/or a photochemical hydrogen abstraction.

The initial process of photosensitization caused by dye sensitizers has generally been classified into two patterns on the basis of their mode of interrelations among the excited dye molecule (D*), the substrate (R), oxygen ($^3O_2$), and the ground state dye (D). One of the two is the process caused by the energy (E) transference mechanism (D-E) and the other by the electron transference mechanism (D-O, D-R, D-D).[6,13–17] Thus, these sensitizers may affect various kinds of photoreactions. With most dye sensitizers, the lifetimes of the excited singlet state are very short, but the quantum yields of the singlet-triplet interconversion are relatively high (Table 3). Consequently, the main reactive intermediate is the excited triplet state, which possesses a relatively long lifetime, allowing oxidation, bond cleavage, and other complicated photochemical reactions. The photodecomposition enhancement of the insecticide etofenprox by the dye sensitizers may be attributed to this process.

Singlet oxygen ($^1O_2$) plays an important role in the photooxidation reactions.[6,17,18] It is formed by the energy transference from the excited sensitizer to ground state oxygen ($^3O_2$). Potential (energy level) of the singlet oxygen is 22.5–37.5 kcal/mol higher than that of the ground state oxygen, indicating that it is a powerful oxidant. Singlet oxygen is often produced in the presence of dye sensitizers,[6,17,19–21] suggesting it may be the origin of the acceleration effect of the dye sensitizers on the photolysis of cartap and pyridafenthion.

cis-Dehydromatricaria ester (cis-DME) is an allelopathic compound, which exists in the root of goldenrod (Solidago altissima L.). We isolated it first for studying its photoactivated toxicity to plants, but it was later found that this compound shows very strong photosensitization of pesticide photodecomposition. It is needless to say that the large amount of dyes in the plants and other photosensitizers in soils and waters plays an important role in accelerating pesticide photodecomposition. Such compounds as cis-DME, mainly exerting photo-induced activity by producing singlet oxygen, may also play an important part in the pesticide photodecomposition acceleration.[22,23]

REFERENCES

1. Tsao, R. and Eto, M., Chemical and photochemical transformation of the insecticide Cartap hydrochloride into nereistoxin, J. Pestic. Sci., 14, 47, 1989.
2. Tsao, R., Hirashima, A., and Eto, M., Photolysis of the insecticide pyridafenthion and the effect of some photosensitizers, J. Pestic. Sci., 14, 315, 1989.
3. Tsao, R. and Eto, M., Photolytic and chemical oxidation reaction of the insecticide ethofenprox, J. Pestic. Sci., 15, 405, 1990.
4. Tsao, R. and Eto, M., Photoreactions of the herbicide naproanilide and the effect of some photosensitizers, J. Environ. Sci. Health, B25, 569, 1990.
5. Tsao, R. and Eto, M., Photolysis of the fungicide flutolanil and the effect of some photosensitizers, Agric. Biol. Chem., 55(3), 763, 1991.
6. Kutsuki, H., Edoki, A., and Gold, M. H., Riboflavin-photosensitized oxidative degradation of a variety of lignin model compounds, Photochem. Photobiol., 37, 1, 1986.
7. Ivie, G. W. and Casida, J. E., Sensitized photodecomposition and photosensitizer activity of pesticide chemicals exposed to sunlight on silica gel chromatoplates, J. Agric. Food Chem., 19, 405, 1971.
8. Ivie, G. W. and Casida, J. E., Photosensitizers for the accelerated degradation of chlorinated cyclodienes and other insecticide chemicals exposed to sunlight on bean leaves, J. Agric. Food Chem., 19, 410, 1971.
9. Dixon, S. R. and Wells, C. H. J., Chlorophyll sensitized photodegradation of 2-dimethylamino-5, 6-dimethylpyrimidin-4-ol, Pestic. Sci., 21, 155, 1987.

10. Dodge, A. D. and Knox, P., Photosensitizers from plants, *Pestic. Sci.*, 17, 579, 1986.

11. Jensen-Korte, U., Anderson, C., and Spiteller, M., Photodegradation of pesticides in the presence of humic substances, *Sci. Total Environ.*, 62, 335, 1987.

12. Davidson, R. S., Mechanisms of photo-oxidation reactions, *Pestic. Sci.*, 10, 158, 1979.

13. Tokumaru, K., *Photosensitizers*, Tokumaru, K. and Okawara, M., Eds., KODANSHA Scientific, Tokyo, 1987 (in Japanese).

14. Suppan, P., *Principles of Photochemistry*, The Chemistry Society, Burlington House, London, 1972, 49.

15. Nishijima, Y. and Yamamoto, M., *Photosensitizers*, Tokumaru, K. and Okawara, M., Eds., KODANSHA Scientific, Tokyo, 1987, 25–55, 64–99 (in Japanese).

16. Maruyama, K. and Otsuki, T., *Organic Radicals and Photo-Reactions*, Maruyama, K., Ed., Maruzen Co., Tokyo, 1983 (in Japanese).

17. Larson, R. A., Insect defenses against phototoxic plant chemicals, *J. Chem. Ecol.*, 12, 859, 1986.

18. Gohre, K. and Miller, G. C., Photooxidation of thioether pesticides on soil surfaces, *J. Agric. Food Chem.*, 34, 709, 1986.

19. Knox, J. P. and Dodge, A. D., Photodynamic damage to plant leaf tissue by rose bengal, *Plant Sci. Lett.*, 37, 3, 1984.

20. Callaham, M. F., Broome, J. R., Poe, W. E., and Heitz, J. R., Time dependence of light-independent biochemical changes in the boll weevil, *Anthonomous grandis*, caused by dietary rose bengal, *Environ. Entomol.*, 6, 669, 1977.

21. Bezman, S. A., Burtis, P. A., Izod, T. P. J., and Thayer, M. A., Photodynamic inactivation of *E. coli* and *S. cerevisiae* by phenylheptatriyne from bidens pilosa, *Photochem. Photobiol.*, 28, 325, 1978.

22. Wat, C. K., Macrae, W. D., Yamamoto, E., Towers, G. H. N., and Lam, J., Phototoxic effects of naturally occurring polyacetylenes and α-terthienyl on human erythrocyl, *Photochem. Photobiol.*, 32, 167, 1980.

23. Campbell, G., Lambert, J. D. H., Arnason, T., and Towers, G. H. N., Allelopathic properties of α-terthienyl and phenylheptatriyne, naturally occurring compounds from species of asteraceae, *J. Chem. Ecol.*, 6, 961, 1982.

CHAPTER 12

Sorption and Photochemical Transformation of Polycyclic Aromatic Compounds on Coal Stack Ash Particles

E. L. Wehry and Gleb Mamantov

INTRODUCTION

In the atmosphere, "semivolatile organic compounds,"[1] including polycyclic aromatic hydrocarbons (PAHs) and their derivatives, may occur both in gas and particulate phases. PAHs formed in the vapor phase via combustion processes undergo cooling as they enter the atmosphere; as they are cooled, they may deposit onto particulate matter. While some compounds (e.g., low-molecular weight PAHs such as anthracene, phenanthrene, and the benzofluoranthenes) exist in the atmosphere at appreciable equilibrium concentrations in the vapor phase, less volatile compounds (e.g., PAHs containing five or more rings) are encountered primarily in the particulate phase.[2-4] The highest concentrations of organic compounds often are associated with particles in the respirable size range.[5] The atmospheric residence times of these small particles are relatively long (perhaps 4 days or more); hence, organic compounds adsorbed on them may be transported over large distances.[6,7]

Subsequent to release into the atmosphere, PAHs and their derivatives may undergo transformation; both photochemical and nonphotochemical processes may be important. These processes can have important ramifications, especially when the products are more hazardous than the parent compounds (e.g., conversion of PAHs to strongly mutagenic nitro derivatives).[8] The chemical fate of PAHs in the vapor phase has received extensive and careful study, especially by Atkinson and co-workers,[9-11] and appears to be well understood. However, the

0-87371-871-2/94/$0.00+$.50
© 1994 by CRC Press, Inc.

course of chemical transformation of PAHs in the particulate phase is much less well characterized.[12]

The chemical fate of a PAH in the particulate phase may depend on the physical properties and/or chemical composition of the surface on which it is sorbed. For example, photochemical oxidation of PAHs sorbed on wood soot,[13] alumina,[14,15] silica gel,[14,15] glass,[16] and airborne particles sampled via high-volume filtration[17,18] is reported to proceed rapidly, whereas PAHs adsorbed on carbon black[14] and many (but not all) coal stack ash particles[19-21] are quite resistant to photodecomposition.

Our studies have concentrated on the photochemical behavior of PAHs deposited from the vapor phase[22] on coal stack ashes and compositional subfractions obtained therefrom. Most coal ashes contain particles of several different chemical compositions,[23-25] including glassy aluminosilicate particles that may contain numerous impurities including iron, a mullite-quartz "crystalline" phase, a magnetic phase (a "magnetic spinel," $Fe_xAl_yO_4$, or a mixture of Fe_2O_3 and Fe_3O_4), and a carbonaceous phase. Many of the carbonaceous particles[26-28] are elemental carbon (not organic or inorganic carbon) that are morphologically indistinguishable from coked coal.

Physically, coal ash is usually an inhomogeneous mixture of particle shapes, sizes, and colors; 11 major coal ash particle morphologies have been identified.[29] Most coal ash particles are spherical, especially in the smaller particle-diameter size fractions. Glassy spherical particles generally are transparent to visible light, while particles of graphitic carbon are opaque. The carbonaceous particles tend to have much larger specific surface areas than do the magnetic or nonmagnetic mineral particles.[26] Most carbonaceous coal ash particles possess complex internal pore structures within which it may be possible for sorbed organic molecules to be "sequestered."[23,26]

Most coal ashes are sufficiently heterogeneous to be separable into "subfractions" based on differences in particle size, density, and chemical composition.[30] This chapter summarizes our studies of the photochemical reactivity of PAHs deposited from the vapor phase onto "intact" and "fractionated" coal ashes of diverse origin and properties.

EXPERIMENTAL SECTION

Coal ashes are used as received. It has been our experience that the levels of extractable organic compounds present in coal stack ashes obtained from steam plants for electricity generation generally are undetectably small; thus, no "pre-extraction" of ash samples is carried out.

When it is desired to fractionate a coal ash, the following general procedure is used. An ash is first separated into size fractions by placing ca. 200 g of the bulk ash into the top sieve of a stack of four sieves of sizes 150, 125, 75, and 45 μm and shaking vigorously, using a mechanical shaker, for at least 100 min. From each size fraction, the magnetic particles are then removed via use of magnets of various sizes (often, "strongly magnetic" and "weakly magnetic" subfractions can be obtained).[30] Next, the remaining, "nonmagnetic," particles are separated by density into "mineral" and "carbonaceous" fractions using the "float-or-sink" procedure with a suitable pure organic solvent or mixture (the carbonaceous particles tend to exhibit lower densities than the less porous nonmagnetic mineral particles). The most suitable solvent medium varies from one ash to another; often, a mixture of dibromoethane and hexane is used. By use of various solvent mixtures of different density, "light" and "heavy" subfractions of the carbonaceous and mineral fractions can sometimes be obtained.[30] Once the density separation has been completed, great care is taken to remove all traces of solvent from the separated particles. These fractionation procedures are described in further detail by Miller et al.[30]

In preparation of a PAH-containing ash sample, dry nitrogen is first passed over a column of the ash at a temperature of 250°C for 24 hr. Then, vapor of the PAH of interest, generated using a diffusion cell and suitably diluted with nitrogen, is passed through an expanded bed of the ash, as described by Miguel et al.[22] The quantity of PAH that actually deposits onto a particular ash sample is determined by monitoring the PAH concentration in the vapor stream before and after it passes through the fly ash bed using a gas-chromatographic flame ionization detector.[31]

For a coal ash sample on which a PAH has been deposited, a portion is retained in the dark as a "control"; the remainder of the sample is illuminated, using a rotary quartz cell,[32,33] with a Cermax LX3000UV xenon lamp (ILC Technology) whose output is filtered through 10 cm of water (to remove near-infrared radiation and thus to minimize heating of samples). The lamp output is not otherwise filtered and thus contains radiation from 250 nm through the visible region; the power output over the 280 to 380-nm region (within which most PAHs absorb strongly) is virtually flat. The characteristics of this lamp are described fully elsewhere.[34,35] The illuminated area of a sample is ca. 1–3 cm^2, and the radiant intensity is ca. 0.014 W cm^{-2}.

Illuminated and control samples are then subjected to ultrasonic or supercritical fluid extraction to remove PAHs and phototransformation products (if any); the extracts are then analyzed by ultraviolet absorption spectrometry, gas chromatography, and/or mass spectrometry. The relative standard deviations of the quantities of PAH detected and quantified in this manner typically range from 5 to 12%. Details are set forth elsewhere.[21,36]

Surface area measurements are performed by the BET method, using hydrogen in helium, with a Quantachrome model QS16 "Quantasorb" apparatus.

RESULTS AND DISCUSSION

The phototransformation of five PAHs (pyrene, benzo[a]pyrene, anthracene, benz[a]anthracene, and phenanthrene) deposited from the vapor phase onto eight stack ashes has been examined.[19] Six of these ashes were derived from

Table 1. Phototransformation of PAHS Sorbed on Coal Stack Ashes

	TX	AR	WK	NM	ET	IL	EA	KA
Anthracene	++	+	+	+	+	0	0	0
Phenanthrene	+	0	0	+	0	0	0	0
Benz[a]anthracene	++	+	0	0	0	0	0	0
Pyrene	+	+	0	0	0	0	0	0
Benzo[a]pyrene	++	+	+	0	0	0	0	0

Note: 0: little or no reaction ($\leq$10% of PAH consumed); +: moderate reaction (10–50% of PAH consumed); ++: extensive reaction (>50% of PAH consumed).

combustion of eastern Appalachian (denoted EA), east Tennessee (ET), western Kentucky (WK), and Illinois (IL) bituminous coals; New Mexico (NM) subbituminous coal; and Texas (TX) lignite. A seventh ash was produced via the combustion of Kaneb (KA) bituminous coal of unknown origin; the eighth ash (denoted AR) was a commercially available analytical standard produced from an unknown coal.

Phototransformation of all PAHs proceeded less rapidly when sorbed on any of these ashes than when deposited on silica gel, alumina, or controlled-porosity glass. The various PAHs exhibited widely differing photoreactivities when sorbed on the various coal ashes, as noted in Table 1. In Table 1, the following "reactivity designations" are used (assuming 24-hr illumination of each sample):

$$0: \text{little or no reaction} \quad (\leq 10\% \text{ of PAH consumed})$$
$$+: \text{moderate reaction} \quad (10\text{–}50\% \text{ of PAH consumed})$$
$$++: \text{extensive reaction} \quad (> 50\% \text{ of PAH consumed})$$

As noted in Table 1, the extremes of behavior are TX ash, upon which all PAHs exhibited at least moderate photodecomposition in a 24-hr period, and IL, EA, and KA ashes, upon each of which no sorbed PAH exhibited detectable phototransformation upon 24-hr illumination. When the bulk elemental compositions of these ashes were determined, it was noted that the ashes on which PAHs exhibited discernible photoreactivity (TX, AR) tended to be relatively low in carbon and/or iron content, whereas those ashes (IL, EA, KA) that appeared to totally suppress the photodecomposition of sorbed PAHs were relatively high in bulk carbon and/or iron content.[19] This observation suggested that separation of ashes into mineral nonmagnetic, mineral magnetic, and carbonaceous fractions might help to elucidate the efficiency of various types of ash particles on the photochemistry of sorbed PAHs.

Accordingly, such a separation was carried out for two ashes (KA and TX) that represent extremes of behavior indicated in Table 1. Bulk KA ash has a relatively high carbon content (5.5% C for the unfractionated ash); in contrast, TX ash is relatively low in carbon (0.64%). We have examined the photochemical decomposition of pyrene adsorbed on the four major subfractions of both ashes.[19] The major conclusions of this study are listed below.

1. Pyrene adsorbed on graphitic carbon, or particles high in carbon content, is extremely resistant to photochemical transformation. Moreover, gas-solid chromatographic studies of pyrene vapor deposition onto fly ash fractions indicate that pyrene has a much greater affinity for carbonaceous particles than for the other constituent phases of coal ash. Hence, for coal ashes relatively high in carbon, pyrene will tend preferentially to sorb on the carbonaceous particles and thus be very resistant to photodecomposition. The presence of even a small quantity of elemental carbon in stack ash is likely to lead to stabilization of adsorbed PAH, when compared with data for adsorbents that do not contain carbon. However, while the largest size fractions of coal ash tend to contain the highest concentrations of carbon, it is the smaller (respirable) ash particles that are most likely to escape stack collection devices and be released into the atmosphere.

2. For ashes that are very low in carbon but contain both magnetic and nonmagnetic mineral particles, pyrene preferentially adsorbs on the nonmagnetic particles. Pyrene adsorbed on nonmagnetic mineral particles is relatively susceptible to photodegradation. Hence, pyrene adsorbed on a coal ash that is low both in carbon and magnetic mineral phases should exhibit relatively efficient photodegradation.

3. The magnetic subfraction of coal ash is the weakest adsorbent for pyrene. However, pyrene adsorbed on magnetic particles is very resistant to phototransformation. Hence, for ashes that are low in carbon but high in iron content, the magnetic subfraction may play a minor role in stabilizing adsorbed pyrene toward photodegradation.

4. Particles that exhibit very low affinity for pyrene but are dark in color (such as the magnetic fractions of some ashes) may act to suppress the photodegradation of pyrene adsorbed on nonmagnetic mineral particles via the "inner-filter effect" (shielding pyrene molecules adsorbed on the nonmagnetic phases from the incident light).

These conclusions are further amplified by a study of the fractionation of an additional coal stack ash that was unusually high in carbon content (22.5% by weight in the bulk ash).[30] In this highly carbonaceous ash, the demarcation between "magnetic" and "nonmagnetic" mineral particles was not well defined; a variety of subfractions having different magnetic properties was isolated. Moreover, it was virtually impossible to isolate any subfraction that did not contain an appreciable quantity of graphitic carbon.

A small sampling of the data[30] obtained for the photoreactivity of pyrene sorbed on various fractions of this ash is given in Table 2.

Clearly, this ash consists of phases that vary dramatically in surface area and chemical composition; the susceptibility of an adsorbed PAH (pyrene) to photodegradation is strongly dependent on the type of particle on which the PAH is adsorbed. The photochemical reactivity of pyrene adsorbed on subfractions of this ash clearly demonstrates the importance of the carbon content of the ash. For all subfractions having bulk carbon percentages $\geq 4\%$ and specific surface areas ≥ 2.85 m^2/g, minimal phototransformation ($\leq 6\%$) of adsorbed pyrene is detected

Table 2.   Properties of Subfractions from a High-Carbon Coal Ash Sample

Fraction No.	Particle Size Range (μm)	Specific Surface Area (m²/g)	% Cª	% Feᵇ	Pyrene Photo- decomposition (%)ᶜ
SM1	<45	2.85	4.4	37.8	3
LC1	<45	14.0	30.4	4.0	3
NM3	75–124	0.68	0.5	5.3	44
HC3	75–124	7.2	66.4	1.6	3
SM3	75–124	0.50	0.4	37.6	42
LC2	45–74	20.0	65.9	1.73	

[a]   Weight percentage of carbon in fraction.
[b]   Weight percentage of iron in fraction.
[c]   Percentage of adsorbed pyrene destroyed by 24-hr illumination with xenon lamp.

upon 24-hr illumination. However, for subfractions containing lower weight percentages of carbon (ranging from 0.47 to 0.69%C) and having smaller specific surface areas (0.50–0.68 m²/g), the photochemical behavior of adsorbed pyrene is dramatically different, with extensive photodecomposition (35–45%) of adsorbed pyrene being observed for 24-hr illumination. For subfractions of this ash, the only parameters that correlate with the photoreactivity of pyrene are carbon content and specific surface area. The iron content of the fractions (which has tended to correlate with low photoreactivity of adsorbed PAHs in previous studies) appears here to be irrelevant.

For particulate materials that may not contain appreciable quantities of elemental carbon (e.g., some incinerator ashes), the chemistry of adsorbed PAHs, and other semivolatile organic compounds, might be substantially different from that observed on coal ash. Interestingly, a recent report by Koester and Hites, dealing with the photolysis of polychlorinated dibenzodioxins and dibenzofurans sorbed on several incinerator ashes, shows that minimal phototransformation of these compounds takes place.[33] These results, together with those dealing with PAHs sorbed on coal ashes, imply that sorption of semivolatile organic compounds onto fly ash particles has the general effect of stabilizing them — at least with respect to photochemical decay. In very strong contrast, however, Kamens and coworkers have observed very rapid photodecomposition of PAHs sorbed on diesel and wood soot particles.[13] Clearly, the susceptibility of particulate semivolatile organic compounds to phototransformation is strongly influenced by the nature of the particulate substrate.

Numerous aspects of the chemistry of particle-associated PAHs (and other semivolatile organic compounds) remain to be examined. For example, little is known about the effects of sorption on particles of the reactivity of PAHs with reactive atmospheric species (such as nitric acid, oxides of nitrogen, and the hydroxyl radical). Also, few reliable quantitative measures of the binding strengths of PAHs with particles of various chemical compositions have yet been reported.

ACKNOWLEDGMENTS

This research is supported in part by a grant from the U.S. Department of Energy (Advanced Coal Research at U.S. colleges and universities, administered by the Pittsburgh Energy Technology Center).

REFERENCES

1. Bidleman, T. F., Billings, W. N., and Foreman, W. T., Vapor-particle partitioning of semivolatile organic compounds, *Environ. Sci. Technol.*, 20, 1038, 1986.
2. Ligocki, M. P., Leuenberger, C., and Pankow, J. F., Trace organic compounds in rain. III. Particle scavenging of neutral organic compounds, *Atmos. Environ.*, 19, 1619, 1985.
3. Pistikopoulos, P., Masclet, P., and Mouvier, G., A receptor model adapted to reactive species: polycyclic aromatic hydrocarbons, *Atmos. Environ.*, 24A, 1189, 1990.
4. Westerholm, R. N., Almen, J., Li, H., Rannug, J. U., Egeback, K.-E., and Gragg, K., Chemical and biological characterization of particulate-, semivolatile-, and gas-phase-associated compounds in diluted heavy-duty diesel exhausts, *Environ. Sci. Technol.*, 25, 332, 1991.
5. Miguel, A. H. and Friedlander, S. K., Distribution of benzo[a]pyrene and coronene with respect to particle size in Pasadena aerosols in the submicron range, *Atmos. Environ.*, 12, 2407, 1978.
6. Masclet, P., Pistikopoulos, P., Beyne, S., and Mouvier, G., Long range transport and gas/particle distribution of polycyclic aromatic hydrocarbons at a remote site in the Mediterranean Sea, *Atmos. Environ.*, 22, 639, 1988.
7. McVeety, B. D. and Hites, R. A., Atmospheric deposition of polycyclic aromatic hydrocarbons to water surfaces, *Atmos. Environ.*, 22, 511, 1988.
8. Pitts, J. N., Jr., Sweetman, J. A., Zielinska, B., Atkinson, R., Winer, A. M., and Harger, W. P., Formation of nitroarenes from the reaction of polycyclic aromatic hydrocarbons with dinitrogen pentoxide, *Environ. Sci. Technol.*, 19, 1115, 1985.
9. Arey, J., Zielinska, B., Atkinson, R., Winer, A. M., Ramdahl, T., and Pitts, J. N., Jr., The formation of nitro-PAH from the gas-phase reactions of fluoranthene and pyrene with the OH radical in the presence of NO_x, *Atmos. Environ.*, 20, 2339, 1986.
10. Atkinson, R., Arey, J., Zielinska, B., and Aschmann, S. M., Kinetics and nitroarene product yields from the gas-phase reactions of naphthalene, fluoranthene, and pyrene with N_2O_5 and OH radicals, *Int. J. Chem. Kinet.*, 22, 999, 1990.
11. Helmig, D., Arey, J., Harger, W. P., Atkinson, R., and Lopez-Cancio, J., Formation of mutagenic nitrodibenzopyranones and their occurrence in ambient air, *Environ. Sci. Technol.*, 26, 622, 1992.
12. Seinfeld, J. H., Urban air pollution: state of the science, *Science*, 243, 745, 1989.
13. Kamens, R. M., Guo, J., Guo, Z., and McDow, S. R., Polynuclear aromatic hydrocarbon degradation by heterogeneous reactions with N_2O_5 on atmospheric particles, *Atmos. Environ.*, 24A, 1161, 1990.
14. Behymer, T. D. and Hites, R. A., Photolysis of polycyclic aromatic hydrocarbons adsorbed on simulated atmospheric particulates, *Environ. Sci. Technol.*, 19, 1004, 1985.
15. Saucy, D. A., Cabaniss, G. E., and Linton, R. W., Surface reactivities of polynuclear aromatic adsorbates on alumina and silica particles using infrared photoacoustic spectroscopy, *Anal. Chem.*, 57, 876, 1985.
16. Benson, J. M., Brooks, A. L., Cheng, Y. S., Henderson, T. R., and White, J. E., Environmental transformation of 1-nitropyrene on glass surfaces, *Atmos. Environ.*, 19, 1169, 1985.

17. Fox, M. A. and Olive, S. Photooxidation of anthracene on atmospheric particulate matter, *Science*, 205, 582, 1979.

18. Gibson, T. L., Korsog, P. E., and Wolff, G. T., Evidence for the transformation of polycyclic organic matter in the atmosphere, *Atmos. Environ.*, 20, 1575, 1986.

19. Yokley, R. A., Garrison, A. A., Wehry, E. L., and Mamantov, G., Photochemical transformation of pyrene and benzo[a]pyrene vapor-deposited on eight coal stack ashes, *Environ. Sci. Technol.*, 20, 86, 1986.

20. Behymer, T. D. and Hites, R. A. Photolysis of polycyclic aromatic hydrocarbons adsorbed on fly ash, *Environ. Sci. Technol.*, 22, 1311, 1988.

21. Dunstan, T. D. J., Mauldin, R. F., Jinxian, Z., Hipps, A. D., Wehry, E. L. and Mamantov, G., Adsorption and photodegradation of pyrene on magnetic, carbonaceous, and mineral subfractions of coal stack ash, *Environ. Sci. Technol.*, 23, 303, 1989.

22. Miguel, A. H., Korfmacher, W. A., Wehry, E. L., Mamantov, G., and Natusch, D. F. S., Apparatus for vapor-phase adsorption of polycyclic organic matter onto particulate surfaces, *Environ. Sci. Technol.*, 13, 1229, 1979.

23. El-Mogazi, D., Lisk, D. J., and Weinstein, L. H., A review of the physical, chemical, and biological properties of fly ash and effects on agricultural ecosystems, *Sci. Total Environ.*, 74, 1, 1988.

24. Hulett, L. D. and Weinberger, A. J., Some etching studies of the microstructure and composition of large aluminosilicate particles in fly ash from coal-burning power plants, *Environ. Sci. Technol.*, 14, 965, 1980.

25. Kim, D. S., Hopke, P. K., Casuccio, G. S., Lee, R. J., Miller, S. E., Sverdrup, G. M., and Garber, R. W., Comparison of particles taken from the ESP and plume of a coal-fired power plant with background aerosol particles, *Atmos. Environ.*, 23, 81, 1989.

26. Griest, W. H. and Harris, L. A., Microscopic identification of carbonaceous particles in stack ash from pulverized-coal combustion, *Fuel*, 64, 821, 1985.

27. Griest, W. H. and Tomkins, B. A., Influence of carbonaceous particles on the interaction of coal combustion stack ash with organic matter, *Environ. Sci. Technol.*, 20, 291, 1986.

28. Soltys, P. A., Mauney, T., Natusch, D. F. S., and Schure, M. R., Time-resolved solvent extraction of coal fly ash: retention of benzo[a]pyrene by carbonaceous components and solvent effects, *Environ. Sci. Technol.*, 20, 175, 1986.

29. Fisher, G. L., Prentice, B. A., Silberman, D., Ondov, J. M., Biermann, A. H., Ragaini, R. C., and McFarland, A. R., Physical and morphological studies of size-classified coal fly ash, *Environ. Sci. Technol.*, 12, 447, 1978.

30. Miller, V. R., Wehry, E. L., and Mamantov, G., Photochemical transformation of pyrene vapor-deposited on eleven subfractions of a high-carbon coal stack ash, *Environ. Toxicol. Chem.*, 9, 975, 1990.

31. Engelbach, R. J., Garrison, A. A., Wehry, E. L., and Mamantov, G., Measurement of vapor deposition and extraction recovery of polycyclic aromatic hydrocarbons adsorbed on particulate solids, *Anal. Chem.*, 59, 2541, 1987.

32. Korfmacher, W. A., Wehry, E. L., Mamantov, G., and Natusch, D. F. S., Resistance to photochemical decomposition of polycyclic aromatic hydrocarbons vapor-adsorbed on coal fly ash, *Environ. Sci. Technol.*, 14, 1094, 1980.

33. Koester, C. J. and Hites, R. A., Photodegradation of polychlorinated dioxins and dibenzofurans adsorbed to fly ash, *Environ. Sci. Technol.*, 26, 502, 1992.

34. Perchalski, R. J., Winefordner, J. D., and Wilder, B. J., Evaluation of Eimac lamp as excitation source for molecular fluorescence, *Anal. Chem.*, 47, 1993, 1975.

35. Cochran, R. L. and Hieftje, G. M., Spectral and noise characteristics of a 300-watt Eimac arc lamp, *Anal. Chem.*, 49, 2040, 1977.

36. Mauldin, R. F., Vienneau, J. M., Wehry, E. L., and Mamantov, G., Supercritical fluid extraction of vapor-deposited pyrene from carbonaceous coal stack ash, *Talanta*, 37, 1031, 1990.

CHAPTER 13

Photodehalogenation of 1,2,3,4-Tetrachlorodibenzo-*p*-Dioxin

Peter K. Freeman and Susan A. Hatlevig

INTRODUCTION

Polyhalogenated aromatic compounds are environmental contaminants, with some of the most potent toxins being the polychlorinated dibenzo-*p*-dioxins. A knowledge of the mechanism of the photodehalogenation process could lead to efficient methods for decontamination and waste disposal. There have been several studies in our laboratory devoted to understanding the mechanism of photodehalogenation of polyhaloarenes.[1-9] Our attention now has turned to studying the mechanism of photodechlorination of 1,2,3,4-tetrachlorodibenzo-*p*-dioxin (1,2,3,4-TCDD) (5).

Investigations have been conducted on the photochemical decay of dibenzo-*p*-dioxins, especially after the release of a large amount of 2,3,7,8-TCDD over a wide area near Seveso, Italy, in 1976. Choudhry and Hutzinger published a review of these studies in 1982,[10] and some trends became apparent in the photodehalogenation reaction mechanisms. Irradiation of highly chlorinated dibenzo-*p*-dioxins leads to dechlorination at the lateral rather than at the *peri* positions with the exception of the photolysis of octachlorodibenzo-*p*-dioxin (OCDD) on a wood substrate.[11] The lower chlorinated dioxins are decomposed more quickly in light, e.g., than say for OCDD, and photodecomposition requires an effective hydrogen donor.

Polychlorobenzenes exhibit enhanced and efficient photodechlorination in the presence of electron donors.[1,3] Hung and Ingram showed an enhanced degradation of OCDD in hexane with the addition of triethylamine (TEA).[12] Photodechlorination was also shown to proceed substantially faster in the presence of NaBH$_4$.[13] Experiments in our laboratory exploring the photodehalogenation of pentachlorobenzene (1) in CH$_3$CN at 254 nm[3,4] or of three tetrachloronaphthalenes (TCN) in CH$_3$CN at 300 nm[7] show that quantum yields are elevated with increasing TEA or NaBH$_4$.[8]

The presence of an electron donor has a dramatic effect on the regiochemistry of the photodehalogenation products. At low conversion (15–20%) of 1, 1,2,3,5-tetrachlorobenzene (2) is formed as the major product along with 1,2,4,5- and 1,2,3,4-tetrachlorobenzene (3 and 4, respectively) as minor products (Scheme 1). In the presence of TEA, irradiation at 254 nm leads to a reversal in the regiochemistry of dechlorination of 1, so that 3 is now formed as the major product with 2 and 4 as minor components. Generation of a radical anion of 1 with lithium p,p'-di-*tert*-butylbiphenylide (LDBB) gives essentially identical product ratios to those in the latter photodechlorination.[5] The regiochemistry of the photolysis of three TCNs is also sharply dependent upon the presence of TEA.[7]

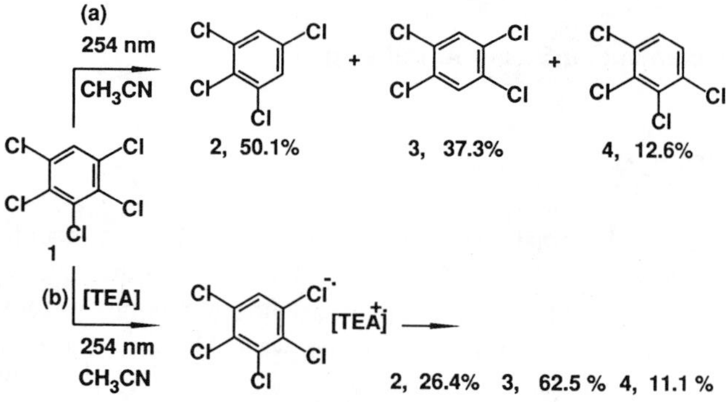

Scheme 1.

PHOTODEHALOGENATION OF 1,2,3,4-TCDD

Irradiation of 5 was carried out in CH$_3$CN at 254 nm both with and without TEA (Scheme 2). The presence of TEA substantially increases the rate, reducing the time for complete loss of 5 from ca. 8 hr to ca. 10 min (Figures 1 and 2). At low conversions (10%), the regiochemistry of the photoproducts in the absence of TEA reveals preferential dechlorination at the *peri* positions rather than the lateral positions in ca. a 4:1 ratio. As with 1, photoreactions with TEA produce a reversal in regiochemistry with a ratio of ca. 0.64 7 to 6. Preliminary experiments with NaBH$_4$ also give a similar ratio of 1,2,3- to 1,2,4-triCDD (ca. 0.75). This points to a radical anion-like intermediate (Scheme 3).

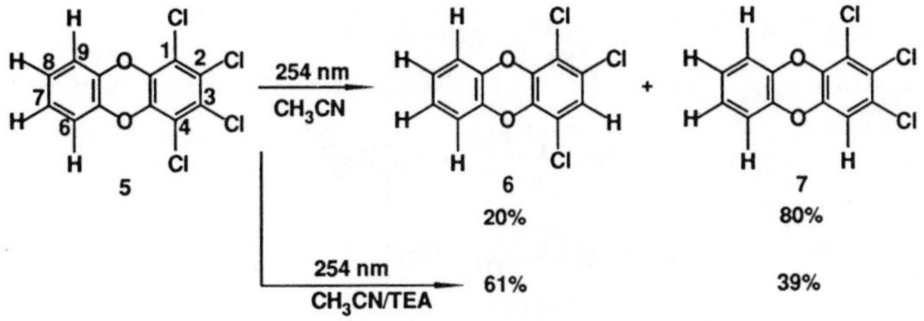

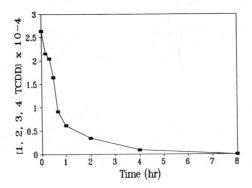

Scheme 2.

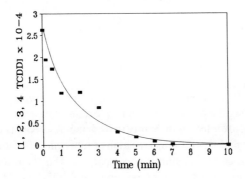

Figure 1. Plot of [1,2,3,4-TCDD] vs time (hr). Irradiation was at 254 nm in CH_3CN. Initial concentration of 1,2,3,4-TCDD $\simeq 3 \times 10^{-4}$ *M*.

Figure 2. Plot of [1,2,3,4-TCCD] vs time (min). Irradiation was at 254 nm in CH_3CN with 0.01 *M* TEA. Initial concentration of 1,2,3,4-TCDD $\simeq 3 \times 10^{-4}$ *M*.

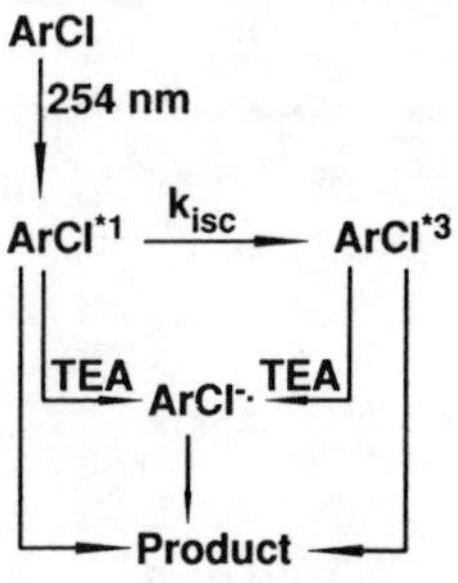

ArCl
$\big|$254 nm
$\big|$
ArCl^{*1} $\xrightarrow{\text{k}_{isc}}$ ArCl^{*3}

$\lfloor$ TEA $\rfloor$ ArCl^{-} $\lfloor$ TEA $\rfloor$

$\big|$
Product

Scheme 3.

Further experiments have been performed in order to explore the applicability of the mechanism in Scheme 3 to the photodehalogenation of **5**. Generation of a radical anion by allowing **5** to react with LDBB in THF gives the two trichlorinated isomers of the dehalogenated dioxin with **6** being the major product (Scheme 4). The product ratio of 1,2,3- to 1,2,4-triCDD shows a slight increase with increasing temperature. The relative ratio at 25°C can be calculated with the aid of an Arrhenius plot. The ln of the ratio of the rates of formation of **6** and **7**, ln (k_7/k_6), was plotted against the inverse of absolute temperature of reaction. The ratio of rates is obtained as the ratio of the percentages of **6** and **7** at the respective temperatures (Figure 3). Extrapolation to 25°C gives a ratio of **7** to **6** $\simeq 0.72$, which agrees very well with the photochemical product ratio.

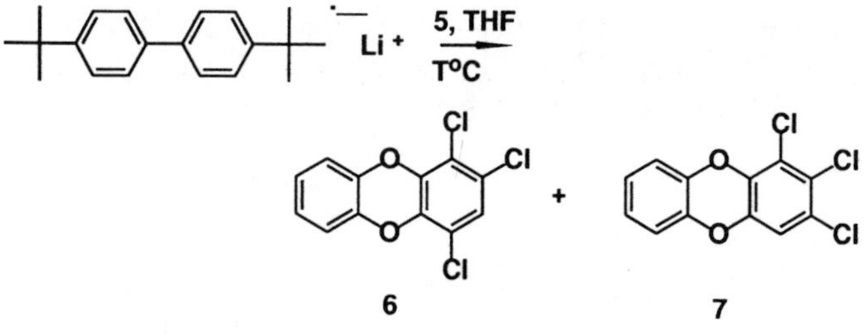

$$\text{(biphenyl diradical anion)} \quad Li^+ \quad \xrightarrow[\text{T°C}]{\text{5, THF}}$$

6   **7**

Scheme 4.

The photodehalogenation of **5** is quenched by piperylene with the Stern-Volmer plot showing the participation of both excited singlet (ca. 10%) and triplet states (Figure 4). A plot of the inverse of the relative quantum yield of photodechlorination of **1** vs the concentration of fumaronitrile demonstrates analogous behavior with ca. 2–6% singlet participation.[4] The fluorescence lifetime of **1** remains constant over a concentration range of 0.005–0.04 M, and the singlet quantum yield does not increase with increasing substrate concentration, thus ruling out the intervention of a singlet excimer. The quantum yield of disappearance of **1** does increase with increasing concentrations of substrate and supports triplet excimer formation.

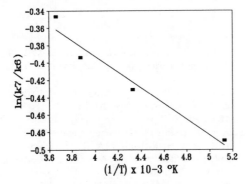

Figure 3.   Plot of the ln of the ratio of the rates of formation of **6** and **7** vs the inverse of the absolute temperature of reaction.

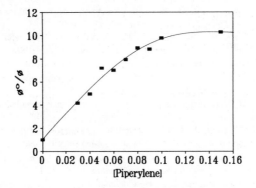

Figure 4.   Plot of the inverse of the relative quantum yield of photodechlorination of 1,2,3,4-TCDD vs the concentration of piperylene.

CONCLUSIONS

The insolubility of 1,2,3,4-TCDD at higher concentrations in CH_3CN is a limiting factor in providing evidence for triplet excimer formation. Even with this limitation, the experimental results from the photodehalogenation reactions of 1,2,3,4-TCDD with and without TEA, with triplet quencher, and with LDBB clearly point to a mechanism analogous to that of the photodechlorination of pentachlorobenzene, and they demonstrate the sensitivity of the regiochemistry to the mechanism.

REFERENCES

1. Freeman, P. K., Srinivasa, R., Campbell, J.-A., and Deinzer, M. L., The photochemistry of polyhaloarenes. 5. Fragmentation pathways in polychlorobenzene radical anions, *J. Am. Chem. Soc.*, 108, 5531, 1986.

2. Freeman, P. K. and Srinivasa, R., Photochemistry of polyhaloarenes. 6. Fragmentation of polyfluoroarene radical anions, *J. Org. Chem.*, 52, 252, 1987.

3. Freeman, P. K. and Ramnath, N., Photochemistry of polyhaloarenes. 7. Photodechlorination of pentachlorobenzene in the presence of sodium borohydride, *J. Org. Chem.*, 53, 148, 1988.

4. Freeman, P. K. and Ramnath, N., Richardson, A. D., Photochemistry of polyhaloarenes. 8. Photodechlorination of pentachlorobenzene, *J. Org. Chem.*, 56, 3643, 1991.

5. Freeman, P. K. and Ramnath, N., Photochemistry of polyhaloarenes. 9. Characterization of the radical anion intermediate in the photodehalogenation of polyhalobenzenes, *J. Org. Chem.*, 56, 3646, 1991.

6. Freeman, P. K. and Jang, J.-S., Ramnath, N., Photochemistry of polyhaloarenes. 10. Photochemistry of 4-Bromobiphenyl, *J. Org. Chem.*, 56, 6072, 1991.

7. Freeman, P. K., Clapp, G. E., and Stevenson, B. K., Photochemistry of three tetrachloronaphthalenes, *Tetrahedron Lett.*, 32, 5705, 1991.

8. Clapp, G. E. Reactive Intermediates: I. Mechanisms of Photodehalogenation of Three Tetrachloro-Naphthalenes II. Structure and Electronic Effects in Some Selected Carbenes, Ph.D. thesis, Oregon State University, 1991.

9. Freeman, P. K. and Lee, Y.-S., Photochemistry of polyhaloarenes. 12. Photochemistry of pentachlorobenzene in micellar media, *J. Org. Chem.*, 57, 2846, 1992.

10. Choudhry, G. G., Hutzinger, G., Photochemical formation and degradation of polychlorinated dibenzofurans and dibenzo-*p*-dioxins, *Residue Rev.*, 84, 113, 1982.

11. Lamparski, L. L., Stehl, R. H., and Johnson, R. L., Photolysis of pentachlorophenol-treated wood. Chlorinated dibenzo-*p*-dioxin formation, *Environ. Sci. Technol.*, 14, 196, 1980.

12. Hung, L.-S. and Ingram, L. L., Effect of solvents on the photodegradation rates of octachlorodibenzo-*p*-dioxin, *Bull. Environ. Contam. Toxicol.*, 44, 380, 1990.

13. Epling, G. A., Qiu, Q., and Kumar, A., Hydride-enhanced photoreaction of chlorodibenzodioxins, *Chemosphere*, 18, 329, 1989.

CHAPTER 14

Photoinduced Reactions on Clay and Model Systems

Xinsheng Liu, Kai-Kong Iu, Yun Mao, and J. Kerry Thomas

INTRODUCTION

Organic molecules adsorb on solid surfaces to varying degrees, depending on the degree of activation of the surface. The prime objective of the present work is to assess the conditions that lead to the photodegradation of molecules adsorbed on clays. To this end, the synthetic clay laponite is chosen as a model system for all clays. Alumina, silica-alumina, and zeolites are also studied as model systems that gradually approach the complexity of a clay system.

EXPERIMENTAL

The experimental techniques have been described previously.[2-4,11,12] They consist of fast-pulse laser excitation with observations of short lived species by reflectance spectroscopy. Electron paramagnetic resonance (EPR), infrared (IR), and standard ultraviolet (UV)-visible spectroscopy are also used for analysis, together with high-pressure liquid chromatography (HPLC).

Laponite Clay

Laponite is a synthetic counterpart of natural hectorite clay with a layered structure.[1] The layer is composed of two SiO_4 tetrahedral sheets and one MgO_6 octahedral sheet arranged in a TOT sandwich (T = SiO_4 tetrahedral sheet and

0-87371-871-2/94/$0.00+$.50

187

O = MgO_6 octahedral sheet). The Mg^{+2} ions in the octahedra are partially substituted by Li^+ ions, which provide the source of the negative charge of the sheet. The negative charge produced by the substitution is compensated by Na^+ cations located in the interlayer space. The interlayer space can also accept water molecules and, as a consequence, is expanded to an extent, depending on the actual conditions of hydration. The surface of a dried laponite particle is built up of an outer surface of the stack that involves siloxane bonds and a lateral surface of the layers that include broken and terminated Si-O and Mg-O and/or Li-O bonds, as well as an interlayer surface partially covered by the Na^+ cations. The hydrated form of laponite exhibits surfaces covered by water.

Our studies[2-4] on surface properties of laponite using pyrene as a photophysical probe molecule have revealed that the adsorption capacity, polarity, and electron-accepting ability of the laponite surface are dependent on the preactivation temperature and on actual conditions that laponite undergoes. In the case of thermal activation, the surface polarity of laponite increases in the temperature range of 100–350°C, while above 440°C the polarity decreases. The adsorption capability of the surface for pyrene molecules increases dramatically at preactivation temperatures higher than 100°C. The electron transfer between the adsorbed pyrene molecule and the surface active sites (which accept electrons from the adsorbed pyrene molecules) only takes place in the temperature range of 100–400°C. Pyrene cation radicals start to form on the lateral surface of laponite at 115°C. The effect increases linearly with the temperature up to 350°C and then decreases dramatically as the temperature increases beyond 440°C. Figure 1A gives a spectrum (dotted line) showing the formation of pyrene cation radical (absorption around 450 nm).

Irradiation of the samples with UV light (370–300 nm) dramatically increases the yield of the pyrene cation radicals, 24 times larger than the sample without irradiation, as shown in Figure 1A (the solid line spectrum). The formation of pyrene cation radicals is shown to be a single-photon ionization process. This was done by varying the intensity of UV light and by irradiation under the presence of O_2. The mean lifetime of the pyrene cation radical created on the surface of Na^+-laponite ($t_{1/2}$) is 74 hr. There is a distribution of electron-accepting sites on the surface that ionize the adsorbed pyrene molecules (Figure 1B). These sites require different photon energies for the adsorbed pyrene molecules to promote the electron transfer. As shown in Figure 1B, UV light with energy as high as 3.4 eV (340 nm) is sufficient for all surface active sites to accept electrons from the adsorbed pyrene molecules. Below this energy, some sites cannot accept electrons from the molecules. Experiments of blocking the lateral surface of the clay with polymetaphosphate molecules and adsorption of water and NH_3 on the surfaces reveal that most of the active sites (82–86%) are located on the lateral surface, while others (14–18%) are located elsewhere in the regions that are unaffected by these molecules.[3]

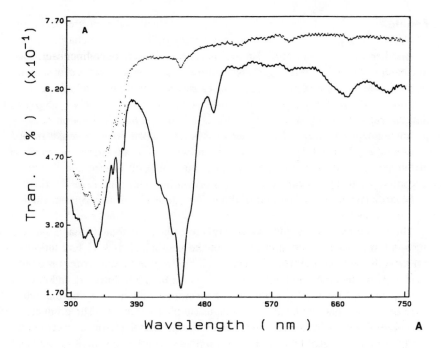

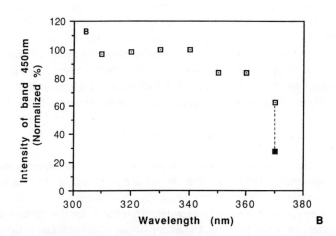

Figure 1. **(A)** Steady-state diffuse reflectance spectra of pyrene/laponite. The laponite was activated at 325°C for 12hr. Dotted line — before irradiation; solid line — after irradiation with UV light of wavelength 300 nm. **(B)** Change in intensity of pyrene cation radicals in their diffuse reflectance spectra as a function of UV-irradiation wavelength. Data were obtained on SLM photofluorimeter (250 W), bandpass, 10 nm, and a heat filter KG-3 was used. The symbol ■ is for the thermal case.

Zeolites

Another model system in our studies is zeolite. Zeolites have long been used in catalysis, petroleum refining, water treatment, and ion and molecular sieving. In contrast to the amorphous nature of most adsorbents and solid catalysts, zeolites exhibit unique uniform ionic frameworks with pores and tunnels (3–8 Å).[5-9] Several reports on the remarkable effects of constrained zeolites on various photochemical and photophysical processes have been documented.[10] In our early studies,[11] we employed both time-resolved and steady-state fluorescence of pyrene to illustrate the restricted and polar nature of the zeolites X and Y. Recently, we have expanded the studies on zeolite with time-resolved diffuse reflectance spectroscopy to monitor the photoinduced ionization of arene inside zeolites X and Y.[12]

These studies suggest that pyrene molecules are distributed in at least two different sites (active and nonactive) on the zeolites X and Y. Excitation of pyrene adsorbed in nonactive sites produces singlet excited states of pyrene with characteristic spectrum and photoionized species by absorbing a second photon and triplet excited states through inter-system crossing; however, pyrene adsorbed in active sites is photoionized through a single-photon process. The products of the photoionization are the cation radical of pyrene and the ionized electron that is captured by the zeolite framework cation (e.g., Na_4^{3+} for sodium-exchanged zeolite Y). The absorption spectrum of the captured electron in the zeolite framework cation depends on the alkali-exchanged cation. Figure 2A shows the time-resolved diffuse reflectance spectra of pyrene in different alkali cation-exchanged zeolite X. The products are triplet pyrene, $^3Py^*$ (415 nm); pyrene cations, $Py^{+\bullet}$ (450 nm); and alkali cation-trapped electron (broad band at the red side of the spectrum). Water coadsorbed onto the sample blocks the oxygen quenching of both $^3Py^*$ and trapped electrons, while the absorption spectrum of the trapped electron shifts to about 620 nm regardless of the alkali-exchanged cation in the zeolite. The spectra of hydrated samples are shown in Figure 2B.

Alumina and Silica-Alumina

Pyrene cation radicals are formed on γ-alumina and silica-alumina surfaces.[13] Spectroscopic study shows that the absorption spectrum of the pyrene cation radicals produced on surfaces fades on contact with water vapor and that pyrene absorption peaks simultaneously increase. This indicates that pyrene is partly restored after reaction with water. Under anhydrous conditions and in the presence of air, however, the pyrene cation radicals are quite stable.

Figure 3 shows the HPLC spectra of the photoinduced chemical products after extraction with aqueous methanol (10% volume water). Samples made on silica-alumina preactivated at 750°C (Ta = 750°C) exhibit HPLC spectra with several peaks, while only one product peak is obtained from the silica-alumina without preactivation (Ta = 20°C) under the same conditions. The retention time of the

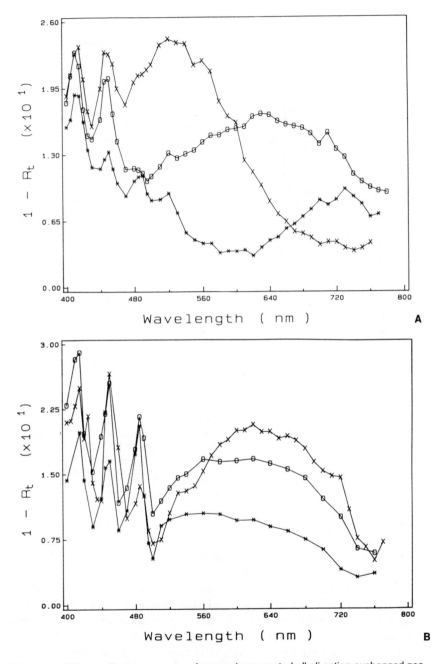

Figure 2. Diffuse reflectance spectra of pyrene impregnated alkali cation-exchanged zeolite X taken 1 μs σ after the laser excitation. **(A)** Samples are dehydrated and under vacuum. **(B)** Samples are hydrated and exposed to air. X symbol with solid line is Na-exchanged zeolite; O symbol with solid line is Li-exchanged zeolite; and * with solid line is K-exchanged zeolite.

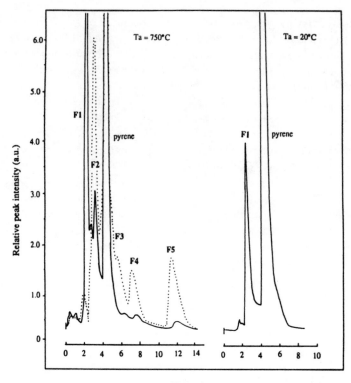

Figure 3. Chromatographic spectra of HPLC analysis for the chemical products on silica-alumina (Ta = 20 and 750°C). The samples were irradiated under approximately 1.5×10^{-3} Einsteins/min/cm^2 at room temperature for 20 min with Pyrex glass filter. After extraction of the photoirradiated samples with methanol solution, the reactant mixture was centrifuged, and the supernatant was subjected to HPLC. Column: C$_{18}$, eluent: methanol, detecting at 300 nm, aufs 0.1. The dotted line shows the products obtained by extraction under aerated conditions.

product peak on γ-alumina (Ta = 20°C) is basically identical to that of the first peak on γ-alumina (Ta = 750°C). Different spectra are obtained from silica-alumina (Ta = 750°C) samples under aerated and deaerated extraction conditions. This result exhibits that oxygen takes part in the extraction reaction.

After HPLC separation of the product mixture was obtained from silica-alumina (Ta = 750°C), the collected fractions were studied in detail. Photospectroscopic studies show a pH dependence of the absorption and fluorescence spectra of the fraction one (F1 in Figure 3). At higher pH values (pH > 10), the absorption spectrum shifts to longer wavelengths with a concomitant increase in the absorption coefficient, while the emission wavelengths are also shifted to longer wavelength.

The photochemical products separated by HPLC are tentatively ascribed to hydroxy pyrene. F1 in Figure 3 is ascribed to dihydroxy pyrene, and F2 is ascribed to monohydroxy pyrene. This assignment is supported by the following observations. The retention time of the hydroxypyrenes on the HPLC column is considerably shorter than pyrene due to the reverse-phase column where more polar compounds show shorter retention time. The hydroxyl group on the pyrene ring generates an additional polarity which causes shortening of its retention time. The assignment of F2 in Figure 3 to monohydroxypyrene was confirmed by a standard compound, 1-hydroxypyrene, obtained from Molecular Probes Company.

The pH dependence of the absorption and emission of F1, which shows the characteristics of a phenol-like compound,[13-15] gives further support for the assignment. A pK_a value of the excited state, $pK_a(S_1^*)$, of F1 was experimentally estimated to be 6.2, which is close to the pK_a value of phenol, 5.7, reported by Bartok et al.[14a] Mass spectrometric measurements give the mass numbers of F1 and F2, which agree with those of di- and monohydroxy pyrene molecules.

The reaction mechanism is proposed below. Nucleophilic attack by water on the pyrene cation radical first creates an intermediate adduct.[16] The intermediate undergoes a rapid equilibrium with water and a deprotonation to form Py(OH)· radical.

$$Py^{+\cdot} + H_2O \rightleftharpoons Py(OH_2)^{+\cdot} \qquad (1)$$

$$Py(OH_2)^{+\cdot} + H_2O \rightleftharpoons Py(OH)^{\cdot} + H_3O^+ \qquad (2)$$

Although the Py(OH)· radical, proposed as an intermediate, was not directly observed from present experiments, such radicals are well known in radiation chemistry. For example, OH· addition to different molecules such as benzene, methylated benzenes, biphenyl,[17-18] and pyrene have been observed spectroscopically.[19] In the case of radiation chemistry, the cation radical of aromatic hydrocarbon was generated through the reverse reactions shown above, i.e., protonation of the hydroxyl radical adduct, followed by elimination of water.[18]

The subsequent electron transfer from Py(OH)· to pyrene cation radical is the rate-determining step, which leads to formation of hydroxypyrene or the Py(OH)$^+$ cation, the latter reacts further with water, leading to dihydroxypyrene.

$$Py(OH)^{\cdot} + Py^{+\cdot} \rightarrow Py(OH) + Py^0 + H^+$$

$$\rightarrow Py(OH)^+ + Py^0 \qquad (3)$$

$$Py(OH)^+ + 2H_2O \rightarrow Py(OH)_2 + H_3O^+ \qquad (4)$$

The reaction of perylene cation radicals with water in a homogeneous solution leads to formation of 3,10-perylenequinone and perylene.[20] The reaction of 9,10-diphenylanthracene cation radicals with water in acetonitrile gives equal amounts of trans-9,10-dihydro-9,10-dihydroxy-9,10-diphenylanthracene and 9,10-dipheylanthracene.[21] A similar reaction mechanism could be expected therefore for pyrene cation radicals on the surfaces of γ-alumina and silica-alumina. Studies with biphenyl and its chlorinated derivatives have been carried out; these compounds exhibit similar photochemical processes, leading to hydroxy compounds. All the solid surfaces show similar photoreactivity with minor but distinct affects unique to each system.

ACKNOWLEDGMENTS

The authors thank the Environmental Protection Agency for financial support of this work.

REFERENCES

1. Grim, R. E., *Clay Mineralogy*, 2nd ed., McGraw-Hill, New York, 1968.
2. Liu, X. and Thomas, J. K., Study of surface properties of clay laponite using pyrene as a photophysical probe molecule, *Langmuir*, 7, 2808, 1991.
3. Iu, K.-K., Liu, X., and Thomas, J. K., Photoinduced ionization of arenes on laponite clay: a single-photon process, *Chem. Phys. Lett.*, 186, 198, 1991.
4. Liu, X., Iu, K.-K., and Thomas, J. K., Studies of surface properties of clay laponite using pyrene as a photophysical probe molecule. II. Photoinduced electron transfer, *Langmuir*, 8, 539, 1992.
5. Breck, D. W., *Zeolite Molecular Sieves*, Wiley, New York, 1974.
6. Katzer, J. R., Ed., *Molecular Sieves II,* ACS Symposium Ser. No., 40, American Chemical Society , Washington, D.C., 1977.
7. Rabo, J. A., Ed., *Zeolite Chemistry and Catalysis*, ACS Monographs, 1976, 171.
8. Barrer, R. M., *Hydrothermal Chemistry of Zeolites*, Academic Press, New York, 1982.
9. Flank, W. H. and Whyte, T. E., Jr., Ed., *Perspectives in Molecular Sieve Science*, ACS Symposium Ser. No. 368, American Chemical Society, Washington, D.C., 1988.
10a. Turro, N. J., Photochemistry of organic molecules in microscopic reactors, *Pure Appl. Chem.*, 59, 1219, 1986.
10b. Kevan, L., Catalytically important metal ion intermediate on zeolites and silica surfaces, *Rev. Chem. Intermed.*, 8, 53, 1987, and references cited therein.
10c. Ramamurthy, V., Photoprocesses of organic molecules included in zeolites, in *Photochemistry in Organized and Constrained Media*, Ramamurthy, V., Ed., VCH Publishers, New York, 1991, 429.
11a. Liu, X., Iu, K.-K. and Thomas, J. K., Photophysical properties of pyrene in zeolites, *J. Phys. Chem.*, 93, 4120, 1989.
11b. Iu, K.-K. and Thomas, J. K., Photophysical properties of pyrene in zeolites, 2: effects of coadsorbed water, *Langmuir*, 95, 471, 1990.
12a. Iu, K.-K. and Thomas, J. K., Single-photon ionization of pyrene and anthracene giving trapped electrons in alkali-cation exchanged zeolites X and Y, *J. Phys. Chem.*, 95, 506, 1991.
12b. Iu, K.-K. and Thomas, J. K., Photophysical properties of pyrene in zeolites, 3: a time-resolved diffuse reflectance study, *Colloids Surf.*, 63, 39, 1992.
13a. Pankasem, S. and Thomas, J. K., Reflectance spectroscopic studies of the cation radical and the triplet of pyrene on alumina, *J. Phys. Chem.*, 95, 6990, 1991,

13b. Pankasem, S. and Thomas, J. K., Pyrene, pyrene derivatives and 1,1'-binaphthyl as luminescent probes for photophysical studies of alumina surfaces, *J. Phys. Chem.*, 95, 7385, 1991.

13c. Pankasem, S. and Thomas, J. K., Photophysical and photochemical studies of N,N,N',N'-tetramethylbenzidine on γ-alumina, *Langmuir*, 8, 501, 1992.

13d. Pankasem, S., Iu, K.-K., and Thomas, J. K., Visible and near-IR fluorescence of aromatic radical Cation in micellar solution, BF_3-trifluoroacetic acid, γ-Al_2O_3 and SiO_2-Al_2O_3, *J. Photochem. Photobiol. A*, 62, 53, 1991.

13e. Mao, Y. and Thomas, J. K., Photochemical reaction of pyrene on γ-alumina and silica-alumina surfaces, *Langmuir*, 8, 2501, 1992.

14a. Bartok, W., Lucchesi, P. J., and Snider, N. S., Protolytic dissociation of electronically excited organic acids, *J. Am. Chem. Soc.*, 84, 1842, 1962.

14b. Foerster, T., Influence of pH on the fluorescence of naphthalene derivatives, *Z. Elekrochem.*, 54, 531, 1950.

15a. Weller, A., Fluorescence shifts of naphthols, *Z. Elektrochem.*, 56, 662, 1952.

15b. Weller, A., General base catalysis for the electrolytic dissociation of excited naththol, *Z. Elektrochem.*, 58, 849, 1954.

16. Bard, A. J., Ledwith, A., and Shine, H. J., Formation, properties and reactions of cation radicals in solution, in *Advances in Physical Organic Chemistry*, Vol. 13, Gold, V. and Bethell, D., Eds., Academic Press, London, 1976, 13, p 135.

17. Dorfman, L. M., Taub, I. A., and Bühler, R. E., Pulse radiolysis studies, I: transient spectra and reaction-rate constants in irradiated aqueous solutions of benzene, *J. Chem. Phys.*, 36, 3051, 1962.

18a. Sehested, K. and Hard, E. J., Formation and decay of the biphenyl cation radical in aqueous acidic solution, *J. Phys. Chem.*, 79, 1639, 1975.

18b. Sehested, K., Holcman, J., and Hard, E. J., Conversion of hydroxycyclohexadienyl radicals of methylated benzenes to cation radicals in acid media, *J. Phys. Chem.*, 81, 1363, 1977.

19. Barber, D. J. and Thomas, J. K., Reactions of radicals with lecithin bilayers, *Radiat. Res.*, 74, 51, 1978.

20. Ristagno, C. V. and Shine, H. J., Ion radicals. XXIII: some reactions of the perylene cation radicals, *J. Org. Chem.*, 36, 4050, 1971.

21. Shida, R. E., Electrolytic oxidation of 9,10-diphenylanthracene and properties of its free radical cation and anion, *J. Phys. Chem.*, 72, 2322, 1968.

CHAPTER 15

Anthracene Photochemistry in Aqueous and Heterogeneous Media

Michael E. Sigman and S. P. Zingg

INTRODUCTION

Anthropogenic and naturally occurring polynuclear aromatic hydrocarbons, PAHs, are found throughout the environment, and some PAHs have been shown to exhibit light-dependent cytotoxicity.[1,2] Large amounts of various PAHs have been determined in water bodies and soils throughout the world.[3,4] Previous studies of PAHs in aqueous systems have focused on determining the quantum yield for the disappearance of these materials, Φ_r,[5] while investigations of PAHs adsorbed onto surfaces or in slurries have primarily been concerned with the photophysical behavior of these chemicals.[6] Due to analytical problems associated with the low solubility of PAHs in pure water (10^{-4}–10^{-9} M), only limited information is available concerning the products resulting from their photolysis.[5] Similarly, the influence of solid-liquid interfaces on the product distribution resulting from the photochemical decomposition of PAHs in slurries has received little attention.[6c] Anthracene (**1**) has been chosen for initial studies of the aqueous and interfacial photochemistries of these common pollutants. This chapter is an overview of our preliminary investigations of anthracene photochemistry in these media as reported elsewhere.[7] The choice of SiO_2-cyclohexane as a heterogeneous model system for initial study was dictated by the near match in refractive indices for these two media and by previous photophysical and spectroscopic studies of **1** in this system.[6a] The initial stable photoproducts from the photolysis of **1** are

The submitted manuscript has been authored by a contractor of the U.S. Government under contract No. DE-AC05-84OR21400. Accordingly, the U.S. Government retains a nonexclusive, royalty-free license to publish or reproduce the published form of this contribution, or allow others to do so, for U.S. Government purposes.

reported, and it is shown that the composition of the product set varies dramatically with oxygen concentration in both aqueous solutions and at SiO_2-cyclohexane interfaces.

PHOTOCHEMISTRY OF ANTHRACENE IN WATER

Experimental (Aqueous)

Solutions were prepared by stirring anthracene crystals in water for a period of 2 days. The solution was filtered through a fine fritted glass funnel to remove undissolved anthracene. Samples were checked for Beer's law behavior to insure that the solutions were homogeneous. The anthracene, and subsequent photo-products, were conveniently removed from solution by passing the aqueous sample through a reverse-phase (C18) solid-phase extraction filter (J&W Accubond, J&W Scientific, Folsom, CA). The filter was subsequently washed with 1 ml of acetonitrile, and the resulting solution was analyzed by HPLC on a reverse-phase (C18) column. Photolysis was carried out in a Rayonet RPR-208 photoreactor equipped with 350-nm bulbs. Samples were photolyzed in uranium glass tubes to prevent secondary photolysis of initial photoproducts containing peroxide func-tional groups. Product yields were determined by HPLC from calibrations with authentic materials.

Photolysis Under Aerated and Oxygenated Conditions

Anthracene is a moderately soluble PAH in pure water, exhibiting a saturation solubility of approximately 3×10^{-7} M at ambient temperatures. The wavelength-averaged quantum yield for the photodepletion of **1** in aqueous solutions has previously been measured by Zepp and Schlotzhauer as 3.0×10^{-3},[5a] and a detailed study of the products arising from the photodecomposition of **1** in pure water has now been reported.[7a] Photolysis of **1** in aerated aqueous solution was shown to lead to two stable photoproducts, endoperoxide (**2**), and 9,10-anthraquinone (**3**). At 10% photolysis, **2** and **3** are formed in a mole ratio of 3:1 (Scheme 1). Conflicting opinions have been expressed concerning the formation of **3** from the photolysis of **2**.[8] Independent photolysis of aqueous solutions of **2** did not lead to the formation of **3** under these conditions. Photolysis of **1** to 70% conversion gave approximately a 7% yield of two minor products, 9-hydroxyanthrone (**4**), and 9,10-dihydro-9,10-dihydroxyanthracene (**5**) (Scheme 1). These minor products are formed in a 1:1 molar ratio. At 70% photolysis of **1**, both **2** and **3** have undergone secondary photochemistry. The secondary photolysis of **2** and **3** was shown not to be the source of **4** and **5**. Under one atmosphere of O_2, the photolysis of **1** to 10% conversion yields **2** and **3** in a 1:1 molar ratio. At 70% conversion under one atmosphere of $^{18}O_2$, all products were determined to have incorporated two ^{18}O atoms.[7a]

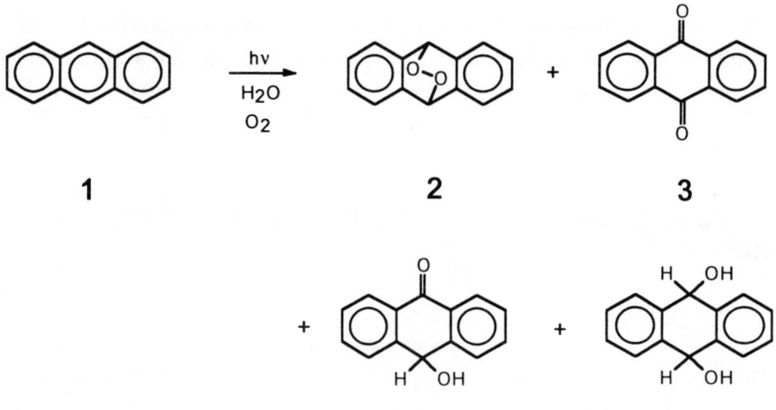

Scheme 1.

Photolysis Under Oxygen-Deficient Conditions

Photolysis of anthracene in aqueous solutions that had been bubbled with argon containing less than 0.5 ppm O_2 gave a totally different product set. Photolysis gave 10,10'-dihydroxy-9,9',10,10'-tetrahydro-9,9'-bianthryl (**6**), as the major product, in 87% yield at 50% photolysis of **1** (Scheme 2). This novel photoproduct, **6**, was identified by low-resolution mass spectrometry, oxidation with Jones reagent to give bianthronyl and 9,10-anthraquinone,[9] by ultraviolet-visible (UV-Vis) spectroscopy, and independent synthesis. The oxygen incorporated into **6** was determined to be derived from water and not from residual oxygen. Photoproduct **6** has not previously been observed as a photoproduct from **1**. Literature precedents,

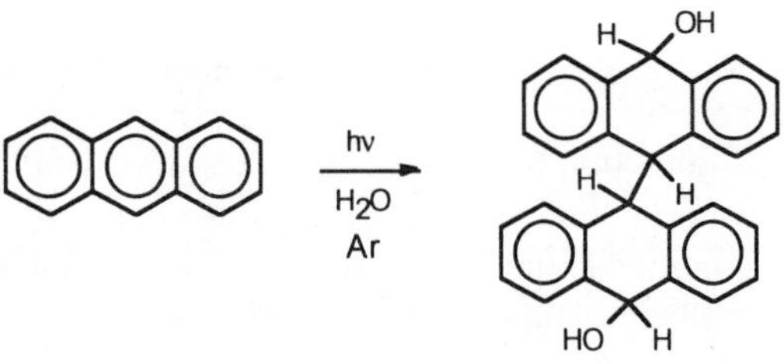

Scheme 2.

involving chemical and electrochemical oxidations of aromatics,[10,11] have led us to suggest that photooxidation of **1** in oxygen-deficient aqueous solution involves attack by water on the cation radical of **1**.[7a] The cation radical of **1** has been postulated to be involved in the aqueous photochemistry of **1**.[5]

PHOTOCHEMISTRY OF ANTHRACENE IN SiO_2/CYCLOHEXANE SLURRIES

Experimental (Heterogeneous)

The silica gel used in these experiments (Aldrich, TLC high purity without binder) was characterized by the following properties: $Fe^{3+} \leq 0.001\%$, $Cl^- \leq 0.003\%$, BET surface area approximately 500 m^2/g; average pore diameter 60 Å, and average particle size 5–25 μm . Activation of the SiO_2 was involved evacuating an ampoule containing the sample at room temperature for 24 hr and subsequently heating the evacuated ampoule to 200°C for 4 hr. The ampoules were sealed under vacuum, and subsequent handling of the SiO_2 was performed in an argon-filled glove box with O_2 and H_2O maintained at 1 ppm maximum. Oxygen-deficient solutions of **1** in cyclohexane were prepared by purging with argon containing less than 0.5 ppm O_2.[7b]

Photochemical experiments were performed in uranium glass tubes using a Rayonet RPR-208 photoreactor equipped with a merry-go-round. A horizontal configuration of the reactor allowed for mixing and even irradiation of the slurries. Slurries comprised of a SiO_2 to cyclohexane ratio of 5.9×10^{-2} g/mL were used for all adsorption isotherm and photochemical studies. An initial ratio of **1** to SiO_2 of 3.4×10^{-7} mol/g was used in all photochemical studies. Products from the photolyses were analyzed by reverse-phase (C18) HPLC. Products were extracted from the SiO_2 with multiple washes of acetone. The acetone was removed under vacuum at room temperature, and the residue was diluted volumetrically with acetonitrile for HPLC analysis.[7b]

Partitioning of Anthracene Between SiO_2 and Cyclohexane

Adsorption of **1** onto SiO_2 from cyclohexane was measured for solutions of **1** where the initial concentration varied over five orders of magnitude (1×10^{-7} M to 1×10^{-2} M). A modified Freundlich adsorption isotherm was required to model the data.[12,13] The isotherm data can be interpreted as revealing a distribution of sites with variable interactions between **1** and the surface.[13] The limiting surface coverage, that coverage reached when the equilibrium concentration of **1** in solution equals the saturation solubility limit, was predicted from the adsorption data to be 1.7×10^{-4} mol/g. This value is less than the predicted monolayer value of 8.2×10^{-4} mol/g for **1** on the surface, and is well above the 3.4×10^7 mol/g (0.04% of a monolayer) employed in the photochemistry discussed in the following sections.[7b]

Photolysis in Oxygen-Deficient Slurries

Photodepletion of **1** was found to follow first-order kinetics in SiO_2-cyclohexane slurries. A knowledge of the irradiance in the slurry is required for calculating the wavelength-averaged quantum yields for the photodepletion.[14] This value is not precisely known due to diffuse scattering of the incident light;[15] therefore, we have limited our discussion to the relative observed rates and photoproducts under varying conditions.[7b]

Anthracene-9,10-photodimer (**7**) was the only product resulting from photolysis of a 2.0×10^{-5} M solution of **1** in argon-purged cyclohexane. In oxygen-deficient SiO_2-cyclohexane slurries, the major product was also photodimer **7**. This product accounted for approximately 68% of the consumed starting material. Trace amounts of anthracene-9,10-endoperoxide (**2**) 9,10-anthraquinone (**3**) 9,10-dihydro-9,10-dihydroxyanthracene (**5**) bianthronyl (**8**) 9-hydroxyanthrone (**4**) and anthrone (**9**) were formed from trace amounts of oxygen which are not removed by argon purging (Scheme 3). The observed rate of photolysis of **1** in oxygen-deficient cyclohexane was enhanced by a factor of 6.5 upon addition of SiO_2. An increase in photolysis rate due to diffuse scattering of the incident light[15] and an increased efficiency of the dimerization reaction under the influence of the surface are possible explanations for the observed behavior.[7b]

In the slurries, approximately 70% of **1**, originally in solution, adsorbed onto the SiO_2. The adsorption process reduces the concentration of solvated **1** by a factor of approximately 3. Dimerization occurring only in the solution phase should show an approximate factor of 3 decrease in the rate as a result of the concentration lowering effect of equilibration with the surface.[16] The experimentally determined rate would represent approximately a factor of 20 enhancement over the anticipated rate for dimerization occurring in solution. Diffuse scattering in heterogeneous systems can lead to enhanced photolysis rates, as compared to those rates observed in homogeneous solution.[15] The theoretical upper limit for the rate enhancement in a diffusely scattering medium is a factor of 2. Enhancement of the rate by a factor of 20 is considerably greater than this theoretical limit. This rate enhancement would also be greater than previously observed effects attributed to diffuse scattering.[15] Therefore, we have concluded that the efficiency of dimerization is enhanced by the surface.[7b] The mechanism of enhancement is not known; however, the possibility of aggregation on the surface appears unlikely. This conclusion is based on the absence of any dimer formation under oxygenated conditions (*vide infra*).[7b]

Photolysis in Aerated and Oxygenated Slurries

Photolysis of a 1×10^{-5} M aerated solution of **1** in cyclohexane resulted in the formation of **2** and **7** in a 1:1 molar ratio and in high yield. A trace amount of quinone **3** was also produced in the reaction. Photolysis of **1** in oxygenated SiO_2-cyclohexane slurries gives a variable product distribution. This variability in product distribution was attributed to high yields of oxygenated products that are

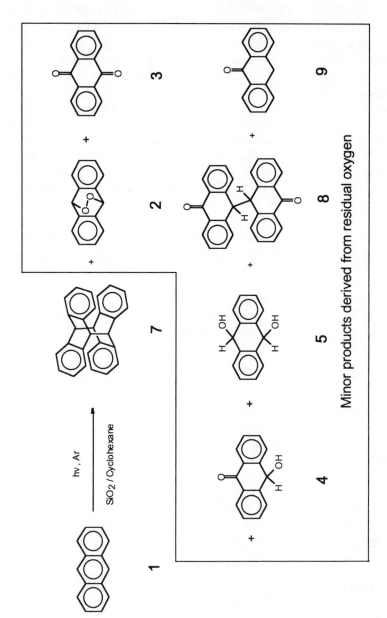

Scheme 3.

thermally unstable in the presence of SiO_2 at room temperature. A typical photolysis to 25% conversion of **1**, approximately 3 min photolysis time, and subsequent analysis within 30 min of the completion of photolysis gave products **2**, **3**, **4**, and **5** (Scheme 4). The endoperoxide, **2**, was found to be unstable in the SiO_2-cyclohexane slurries at ambient temperatures and decomposed to give **3**, **4**, and **5**.[7b]

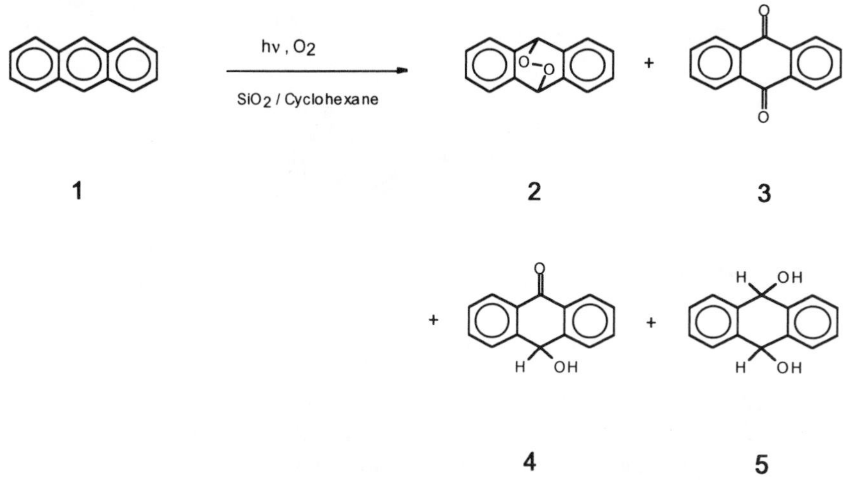

1   **2**   **3**

4   **5**

Scheme 4.

The thermal decomposition of **2** appeared to proceed through two parallel paths. One of the reactions was very rapid (half-life less than 5 min), producing **3** (15%) and **5** (10%); however, only 25% of the starting material was consumed by this path at a loading of 1.68×10^{-8} mol/g. The second reaction proceeded with a half-life of approximately 20 min, leading to the formation of **4** (Scheme 5). At a lower initial surface loading of **2** (3.33×10^{-9} mol/g), all of the starting material was consumed rapidly (in less than 5 min) to produce the same products in very different yields: **3** (14%), **4** (36%), and **5** (49%). A set of highly reactive sites, where preferential adsorption and faster decomposition of **2** occurs, appear to be present on the SiO_2 surface along with a set of less reactive sites. This interpretation is in agreement with the adsorption isotherm data and is consistent with the conclusions of other researchers.[17] The data indicate that the product distributions obtained from the photolysis of **1** in oxygenated SiO_2-cyclohexane slurries arose from the formation of **2** as the only initial photoproduct. Subsequent thermal decomposition of **2** gave **3**, **4**, and **5**.[7b]

Addition of SiO_2 to an oxygenated cyclohexane solution of **1** led to a factor of approximately 56 enhancement in the photolysis rate, from 4.1×10^{-5} s^{-1} to 2.3×10^{-3} sec^{-1}. The rate enhancement for the photolysis of **1** cannot be attributed to an increase in the lifetime of $^1O_2$ in the slurry. In fact, the lifetime of $^1O_2$ in an SiO_2-cyclohexane slurry (13.5 μsec) is shortened relative to the lifetime in cyclohexane

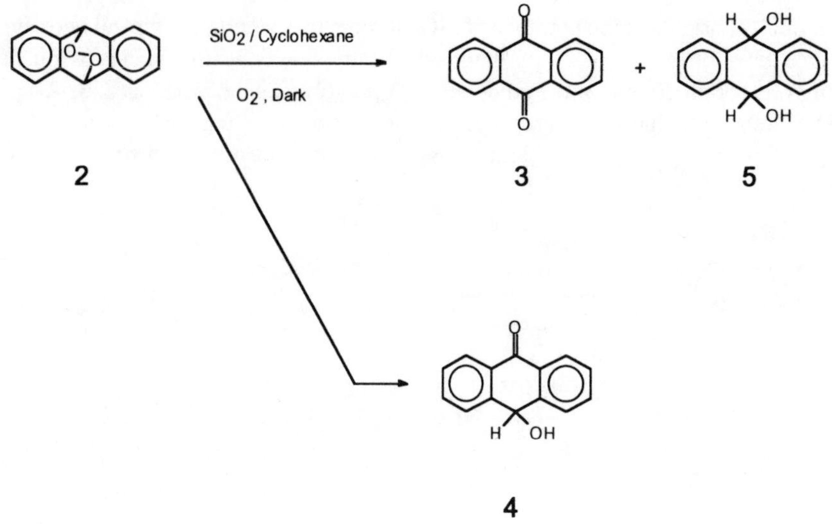

Scheme 5.

(17–23 μsec).[18] The quantum yield for the photodepletion of **1** is known to be enhanced in more polar solvents.[5a,19] The rate enhancement observed in the presence of SiO_2 has been attributed to a medium effect associated with the extreme polarity of the SiO_2-cyclohexane interface.[7b]

Effects of Oxygen and Water on Photolysis Rates

In cyclohexane, the presence of oxygen was seen to decrease the rate of photolysis, relative to the rate of photolysis in an argon-purged solution. This rate decrease is attributable to the quenching of the S_1 state of **1** by O_2 in solution.[19] However, the presence of oxygen (1.14×10^{-2} M) led to a factor of 5.73 ± 0.17 increase in the photolysis rate in SiO_2-cyclohexane slurries. The presence of water in SiO_2-cyclohexane slurries (2.22×10^{-6} M), resulted in only a small enhancement in the rate of photolysis, a factor of 1.59 ± 0.02. There were no significant changes in products for either oxygenated or oxygen-deficient slurries upon addition of water.[7b]

CONCLUSIONS

The photolysis of anthracene in aqueous solutions leads to photooxidation products under both aerated and oxygen-deficient conditions. The photolysis under oxygen-deficient conditions provides support for the involvement of a Type I (electron-transfer) photooxidation mechanism for some PAHs in aqueous solutions. Photolysis of anthracene in SiO_2-cyclohexane slurries demonstrates the profound influence that a solid-liquid interface can exert on photochemical events. The results from this work also indicate the necessity for incorporating these effects into accurate models for predicting environmental residence times.

ACKNOWLEDGMENTS

This research was supported by the Division of Chemical Sciences, Office of Basic Energy Sciences, U.S. Department of Energy under contract DE-AC05-84OR21400 with Martin Marietta Energy Systems, Inc.

REFERENCES

1. Bjørseth, A. and Becher, G., *PAH in Work Atmospheres: Occurrence and Determination*, CRC Press, Boca Raton, FL, 1986, 35.
2. Tuveson, R. W., Wang, G. R., Wang, T. P., and Kagan, Light-dependent cytotoxic reactions of anthracene, *Photochem. Photobiol.*, 52, 993, 1990.
3. Broman, D., Näf, C., Rolff, C, and Zebühr, Y., Occurrence and dynamics of polychlorinated dibenzo-p-dioxins and dibenzofurans and polycyclic aromatic hydrocarbons in the mixed surface layer of remote coastal and offshore waters of the Baltic, *Environ. Sci. Technol.*, 25, 1850, 1991.
4. Jones, K. C., Stratford, J. A., Waterhouse, K. S., Furlong, E. T., Giger, W., Hites, R. A., Schaffner, C., and Johnston, A. E., Increases in the polynuclear aromatic hydrocarbon content of an agricultural soil over the last century, *Environ. Sci. Technol.*, 23, 95, 1989.
5a. Zepp, R. G. and Schlotzhauer, P. F., Photoreactivity of selected aromatic hydrocarbons in water, in *Polynuclear Aromatic Hydrocarbons*, Jones, P. W. and Leber, P., Eds., Ann Arbor Science Publishers, Ann Arbor, MI, 1979, 141–158.
5b. Mill, T., Mabey, W. R., Lan, B. Y., and Baraze, A., Photolysis of polycyclic aromatic hydrocarbons in water, *Chemoshpere*, 10, 1281, 1981.
6a. Bauer, R. K., deMayo, P., Natarajan, L. V., and Ware, W. R., Surface photochemistry: the effect of surface modification on the photophysics of naphthalene and pyrene adsorbed on silica-gel, *Can. J. Chem.*, 62, 1279, 1984.
6b. Ford, W. E. and Kamat, P. V., Photochemistry on surfaces. 3. Spectral and photophysical properties of monomeric and dimeric anthracenesulfonates adsorbed to colloidal alumina-coated silica particles, *J. Phys. Chem.*, 93, 6423, 1989.
6c. Leermakers, P. A., Tomas, H. T., Weis, D. L., and James, F. C., Spectra and photochemistry of molecules adsorbed on silica gel, IV. *J. Am. Chem. Soc.*, 88, 5075, 1966.
6d. Bauer, R. K., Borenstein, R., de Mayo, P., Okada, K., Rafalska M., Ware, W. R., and Wu, K. C., Surface photochemistry: translational motion of organic molecules adsorbed on silica gel and its consequences, *J. Am. Chem. Soc.* 104, 4635, 1982.
7a. Sigman, M. E., Zingg, S. P., Pagni, R. M., and Burns, J. H., Photochemistry of anthracene in water, *Tetrahedron Lett.*, 32, 5737, 1991.
7b. Zingg, S. P. and Sigman, M. E., Influence of a SiO_2/cyclohexane interface on the photochemistry of anthracene, *Photochem. Photobiol.*, 57, 453, 1993.
8a. Schmidt, R., Schaffner, K., Trost, W., and Brauer, H.-D., Wavelength-dependent and dual photochemistry of the endoperoxides of anthracene and 9,10-dimethylanthracene, *J. Phys. Chem.*, 88, 956, 1984.
8b. Rigaudy, J., Defoin and A., Baranne-Lafont, J., *syn*-Anthracene 4a,10:9,9a-dioxide, *Angew. Chem. Int. Ed. Engl.*, 18, 413, 1979.
8c. Sugiyama, N., Iwata, M., Yoshioka, M., Yamada, K., and Aoyama, H., Photooxidation of anthracene, *Bull. Chem. Soc. Jpn.*, 42, 1377, 1969.

9. Fieser, L. F. and Fieser, M., *Reagents for Organic Synthesis*, Vol 1, John Wiley & Sons, New York, 1967, 142.

10a. Walling, C. and Johnson, R. A., Fenton's reagent. V. Hydroxylation and side-chain cleavage of aromatics, *J. Am. Chem. Soc.*, 97, 363, 1975.

10b. Walling, C., Camaioni, D. M., and Kim, S. S., Aromatic hydroxylation by peroxydisulfate, *J. Am. Chem. Soc.,* 100, 4814, 1978.

10c. Eberhardt, M. K., Martinez, G. A., Rivera, J. I., and Fuentes-Aponte, A., Thermal decomposition of peroxydisulfate in aqueous solutions of benzene-nitrobenzene-benzonitrile mixtures. Formation of OH radicals from benzene radical cations and water at room temperature, *J. Am. Chem. Soc.*, 104, 7069, 1982.

11. Majeski, E. J. and Stuart, J. D., Ohnesorge, W. E., Controlled potential oxidation of anthracene in acetonitrile, II, *J. Am. Chem. Soc.*, 90, 633, 1968.

12. Adamson, A. W., *Physical Chemistry of Surfaces*, 3rd ed., John Wiley & Sons, New York, 1990, chap. 11.

13. Urano, K., Koichi, Y., and Nakazawa, Y., Equilibria for adsorption of organic compounds on activated carbons in aqueous solutions. I. Modified Freundlich isotherm equation and adsorption potentials of organic compounds, *J. Colloid Interface Sci.*, 81, 477, 1981.

14a. Zepp, R. G., Quantum yields for reaction of pollutants in dilute aqueous solutions, *Environ. Sci. Technol.*, 12, 327, 1978.

14b. Zepp, R. G. and Cline, D. M., Rates of direct photolysis in aquatic environment, *Environ. Sci. Technol.,* 11, 359, 1977.

14c. Draper, W. M., Determination of wavelength-averaged, near UV quantum yields for environmental chemicals, *Chemosphere*, 14, 1195, 1985.

15. Miller, G. C. and Zepp, R. G., Effects of suspended sediments on photolysis rates of dissolved pollutants, *Water Res.*, 13, 453, 1979.

16. Wei, K. S. and Livingston, R., Reversible photodimerization of anthracene and tetracene, *Photochem. Photobiol.*, 6, 220, 1967.

17. Leffler, J. E. and Zupancic, J. J., Decomposition of azocumene on silica surfaces, *Photochemistry on Solid Surfaces*, Anpo, M. and Matsuura, T., Eds., Elsevier, New York, 1989, 138.

18. Iu, K.-K. and Thomas, J. K., Quenching of singlet molecular oxygen ($^1\Delta_g O_2$) in silica gel/cyclohexane heterogeneous systems. A direct time-resolved study, *J. Am. Chem. Soc.*, 112, 3319, 1990.

19. Bowen, E. J. and Tanner, D. W., The photochemistry of anthracenes: Part 3. Interrelations between fluorescence quenching, dimerization and photo-oxidation, *Trans. Faraday Soc.*, 51, 475, 1955.

CHAPTER 16

HYDROGEN PEROXIDE FORMATION AND DECAY IN LAKE WATERS

David R. S. Lean, William J. Cooper and Frances R. Pick

INTRODUCTION

In addition to the direct effect of ultraviolet (UV) light on organisms and on the environment in which they live, there is the indirect effect of a variety of short-lived free radicals and excited-state species that are photochemically produced by UV light. While UV light does not activate oxygen directly, there are a complex series of reactions that take place in aquatic systems when humic substances are excited by UV light. Some of these transients are reduced oxygen species. Molecular oxygen requires four electrons for its complete reduction to water. One, two, and three electron-reduced intermediates are possible. These are superoxide, hydrogen peroxide, and hydroxyl radical, respectively, with the later being the most toxic and reactive.[1]

While hydrogen peroxide (H_2O_2) is the least reactive of the reduced-oxygen species, it persists longer and may react to form the hydroxyl radical. As such, understanding factors affecting the production and distribution of H_2O_2 may shed light on the entire sequence of reactions initiated by UV light.

Hydrogen peroxide has been measured in many marine and freshwater systems,[2] but prediction of the spacial and temporal patterns has not been achieved. Diel patterns of H_2O_2 concentrations with depth are complicated because three independent processes affect the concentration. One is the production rate, the second results from effects of vertical mixing of the water column, and the third is the decay rate. In the present investigation, complex depth distributions of H_2O_2 concentrations were illustrated; calculations were made to express concentrations on an areal basis;

0-87371-871-2/94/$0.00+$.50
© 1994 by CRC Press, Inc.

207

and when adjusted for decay and the influence of rain, the cumulative daily H_2O_2 production was shown to be a simple function of total radiation.

EXPERIMENTAL

Experiments were conducted in June and July 1988 over continuous 3-day periods. A total of 26 depth profiles were obtained over the three 75-hr experiments conducted at the deepest part of the Eastern and Western Basins of Lake Erie, as well as at Sharpes Bay in Jacks Lake, a temperate lake on the edge of the Canadian Shield, 40 km north of Peterborough, Ontario.

These locations provide a variety of stratification conditions (Figure 1A–1C). Temperature profiles were measured at the same time as water was sampled for H_2O_2, but only values collected at 1500 hr are shown for the three sampling days. The Eastern Basin has a well-defined epilimnion at this time of year to about 11 m, but a temperature gradient of about 1–2° occurs across the epilimnion (0–11 m). During calm periods (e.g., 20 July), an ephemeral "false" thermocline exists at about 4–6 m. Below 11 m, the temperature declines to near 4°C at 30 m then remains constant to the maximum depth at 65 m. Since H_2O_2 changes only occur in surface waters, temperature profiles are not given for greater depths. In contrast, the Western Basin of Lake Erie (Figure 1B) with a maximum depth of 10 m would be considered to be well mixed. (A thermal gradient of 1°C over 1 m is considered to define the bottom of an epilimnetic region.) However, by expanding the temperature scale, some fine detail exists.

Sharpes Bay (Figure 1C) stratified sharply at 5–6 m, and over the 3 days some small changes were observed. With windier conditions on June 16, the profile was steeper than on the more quiescent conditions of June 15 and 17.

Representative profiles of H_2O_2 concentration showed some striking contrasts and provided more detailed information than the temperature values. Of the 36 profiles, only those for the 9:00 a.m., 3:00 p.m., and 9:00 p.m. are provided (Figure 2). Values were obtained across the upper mixed zone every 2–4 hr using the scopoletin-horseradish peroxidase fluorescence decay method.[2-4] It was rainy and relatively calm on July 19 in the Eastern Basin of Lake Erie, but it became more sunny and windy by July 21. This physical influence did not change the temperature profile (Figure 1A) substantially, but the H_2O_2 concentrations (10–100 nM) were quite different (Figure 2A). These data illustrate the difficulty in making useful interpretations from just a few samples taken occasionally in the "upper mixed zone." On July 19, a gradient existed from 0 to 15 m, but by July 21 a truly well-mixed water column was observed. In Sharpes Bay (Figure 2B), concentration changes were limited to the top 7 m, compared to 15 m in Eastern Lake Erie. During calm conditions mid-day on June 14, a high surface value of 602 nM was observed. Later with increased wind, the concentrations were similar at all depths to 6 m. The relatively well-mixed Western Basin water column showed some H_2O_2 stratification at certain times, but this broke down during windier conditions (July 25).

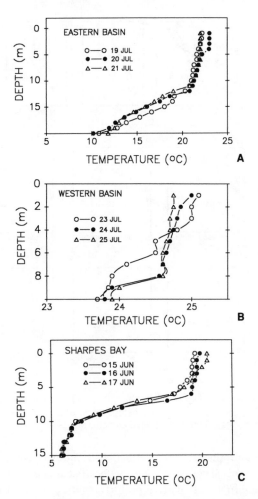

Figure 1. Temperature profiles for **(A)** Eastern Basin of Lake Erie, **(B)** Western Basin of Lake Erie, and **(C)** Sharpes Bay of Jacks Lake over the 3-day experimental period.

Water contained in quartz tubes was held at fixed depths in large translucent tanks supplied with flowing lake water and exposed to direct sunlight. In these experiments, H_2O_2 production rates could be measured in the absence of vertical mixing. It was shown that production was limited to ca. 2 m in Lake Erie and ca. 1 m in Sharpes Bay (data not shown). When these experiments were repeated in other lakes, it was shown that the extinction coefficient for H_2O_2 production was a simple function of the dissolved organic carbon concentration and not substantially influenced by the chlorophyll concentration. The H_2O_2 production rates were obtained on both filtered and unfiltered samples with the observed difference similar to decay rates.

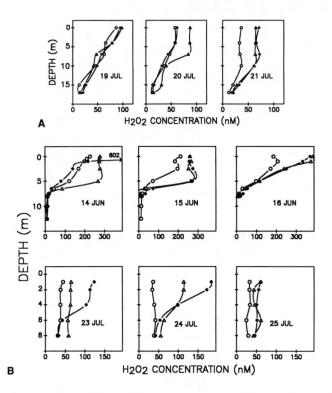

Figure 2. Depth distribution for H_2O_2 concentrations in the **(A)** Eastern Basin of Lake Erie, **(B)** Sharpes Bay (top), and Western Basin of Lake Erie (bottom) at 9:00 a.m. (open circles), 3:00 p.m. (closed circles), and 9:00 p.m. (open triangles).

The third factor influencing the vertical distribution of H_2O_2 is the decay rate. Decay rate constants have been found to be first order[5] with respect to H_2O_2 concentration and first order with respect to abundance of heterotrophic bacteria. Rate constants were obtained here using two methods. The slope of natural logarithm concentration of integrated or areal H_2O_2 concentration measured *in situ* during nighttime plotted as a function of time. These values were similar to those obtained in bottles kept in the dark (Figure 3). Decay rate constants ranged from 0.03 to 0.13 hr^{-1} and were a function of heterotrophic bacteria abundance (obtained using the DAPPI method).[6] This does not mean that bacteria are the sole organisms that contain appropriate enzymes for H_2O_2 breakdown. While they represent only 10–15% of the living biomass in lakes, they have a surface area greater than all other organisms combined. It also means that losses by chemical processes were not important in our experiments.

Further experiments showed that the decay rate constant was similar when H_2O_2 concentrations were increased from ambient levels to 1000 nM (Figure 4). This means that the same rate constant can be applied to all concentrations across the upper mixed zone and that the rate of loss was not at the maximal rate,

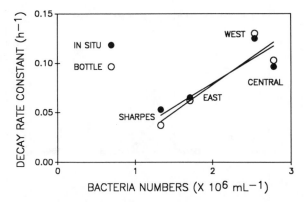

Figure 3. Hydrogen peroxide decay rate constants measured *in situ* during the nighttime (closed circles) and in bottles kept in the laboratory (open circles) plotted as a function of heterotrophic bacteria abundance.

otherwise any increase in H_2O_2 would result in a corresponding decrease in the rate constant.

When both filtered and unfiltered water were incubated in the quartz tubes as above to show the depth of H_2O_2 production, filtered values were always greater than the samples which were not filtered. These experiments suggest that (1) decay rates were similar in the light and in the dark and (2) that there was no substantial production of H_2O_2 by any particle, living or nonliving.

To simplify the complex vertical H_2O_2 distribution patterns (Figure 2), values were integrated with depth to give areal productivity and plotted on the same graph as total incident radiation (Figure 5). It rained on July 19 and 20, and as shown later this has a profound influence on the H_2O_2 concentration. Nevertheless, the clear cloudless day (July 21) illustrates how the observed pattern of H_2O_2 concentration lags the changes in radiation. This results from decay rates being a function of H_2O_2 concentration. The observed vertical inventories are the net

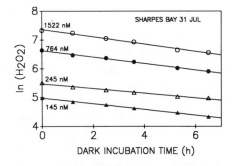

Figure 4. Hydrogen peroxide concentrations added to samples of lake water and measured as a function of time plotted on a natural logarithm scale. The slope provides the first-order decay rate constant and is relatively constant over the range from 145 to 1522 n*M*.

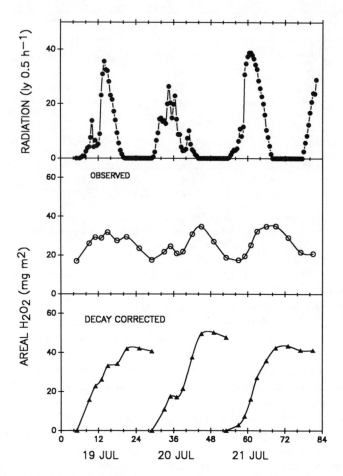

Figure 5.   Total global radiation (top) along with observed vertical inventory of areal H_2O_2 concentrations (middle) and decay corrected or cumulative daily production of H_2O_2 (bottom).

effect of production and decay without the complication due to vertical mixing. Similar results were obtained for the other stations, but are not presented here.

Decay rate constants were used to compute decay corrected areal H_2O_2 production rates, i.e., cumulative daily production (CDP) (Figure 5). To do this, the quantity of H_2O_2 which was lost through decay was summed and added to the observed H_2O_2 production during each sampling interval. The slight decline during the night is likely due to small errors in computing the decay correction. When CDP was plotted as a function of incident light (Figure 6), the relationship was quite good for sunny days (closed circles), but values were too high on rainy days (open circles). When the H_2O_2 concentrations in the rain (30,000–55,000 nM) were multiplied by the depth of rainfall, a quantity was obtained sufficient to

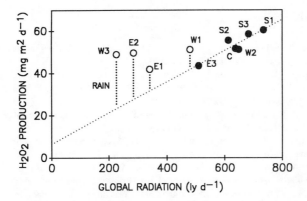

Figure 6. Hydrogen peroxide production plotted as a function of total global radiation for days without rain (closed circles). Values for rainy days (open circles) corrected for the contribution of rain (dotted lines) approximated that predicted from the regression drawn for days without rain. Letters are E (Eastern Basin of Lake Erie), C (Central Basin of Lake Erie), W (Western Basin of Lake Erie), and S (Sharpes Bay) on experimental days 1, 2, and 3.

account for that in excess of that predicted from sunny days. Clearly, further investigations must include a careful budget for rain.

It is well known that H_2O_2 production is related to UV light, but here the areal decay and rain corrected H_2O_2 production was a simple function of total global radiation. At this latitude and time of year, the UV fraction was relatively constant. Clearly, future work should be conducted using precise measurements of light spectral quality and intensity. This relationship also explains the differences in diel patterns between lakes and oceans.[2,5] Changes in diel H_2O_2 concentrations in lakes are from near 0 to 500 nM or more, roughly five times that found in the sea. However, areal production is likely similar in both ecosystems when incident light is similar. Since mixing depths are 100 m or more in oceans, diel concentrations are expected to be small compared to that in lakes where production is distributed across mixing depths of 15 m or less.

While it is well known that dissolved organic carbon is related to production of H_2O_2 in quartz tubes, the areal production of H_2O_2 is the same in lakes of different dissolved organic carbon concentration. In colored lakes, the production rate is higher at the surface, but the extinction coefficient is much steeper making the integrated production similar.

Since production occurs during the day and decay occurs during both daytime and nighttime periods, a dynamic steady state exists for temperate latitudes. At polar latitudes, where daytime production extends over months and bacterial decay is limited by colder waters, profound changes in biogeochemical processes may occur with increased UV-B light resulting from stratospheric ozone depletion. While the basic chemistry has been reasonably well worked out, ecosystem response has been neglected.

REFERENCES

1. Mopper, K. and Zhou, X., Hydroxyl radical photoproduction in the sea and its potential impact on marine processes, *Science*, 50, 661, 1990.
2. Cooper, W. J. and Lean, D. R. S., Hydrogen peroxide dynamics in marine and fresh water systems, *Encyclopedia of Earth System Science*, Vol. 2, Nierenberg, W. A., Ed., Academic Press, San Diego, CA, 1992, 527
3. Kieber, R. J. and Helz, G. R., Two-method verification of hydrogen peroxide determination in natural waters, *Anal. Chem.*, 5, 2312, 1986.
4. Cooper, W. J., Lean, D. R. S., and Carey, J. H., Spatial and temporal patterns of hydrogen peroxide in lake waters, *Can. J. Fish. Aquat. Sci.*, 46, 1227, 1989.
5. Cooper, W. J., Shao, C., Lean, D. R. S., Gordon, A. S., and Scully, F. E., Factors affecting the distribution of H_2O_2 in surface waters, *Environmental Chemistry of Lakes and Reservoirs*, Baker, L. A., Ed., American Chemical Society, Washington, D.C., 1993, in press.
6. Pick, F. R. and Caron, D. A., *Can. J. Fish. Aquat. Sci.*, 44, 2164, 1987.

CHAPTER 17

Singlet Oxygen Formation in Lake Waters from Mid-Latitudes

Chihwen Shao, William J. Cooper, and David R. S. Lean

INTRODUCTION

Although light is efficiently transmitted through distilled water, dissolved organic carbon absorbs light energy in the ultraviolet (UV) region below 400 nm. Particulate materials, such as algae, bacteria, and detritus absorb light in the visible region, but not significantly in the UV region. The absorption of UV light by dissolved organic carbon results in the formation of several reactive oxygen species. The reactive species that is of interest in this study is singlet oxygen, $^1O_2$, the $^1\Delta_g$ excited state of bimolecular oxygen. The major sink for $^1O_2$ is physical quenching by water (quench rate constant = 2.5×10^5 sec^{-1}),[1] which limits the steady-state concentration, $[^1O_2]_{ss}$. Therefore, the steady-state concentration of singlet oxygen is a reflection of its formation rate.[2]

Singlet oxygen production is believed to be related to the concentration of humic substances in natural waters.[2-5] To date, only dissolved organic carbon (DOC, a descriptor for humic substances) has been suggested for predicting the singlet oxygen formation rate. Since fluorescence and absorption spectroscopy can be used to quantify humic substances, it was of interest to determine their relations to $[^1O_2]_{ss}$. Other measurements[2-5] of singlet oxygen were made at various locations using different methods such as addition of photosensitizers. Therefore, no general relationships exist for natural waters. We provide results from a broad range of lakes, which is a start toward extending the existing data to a more generalized relationship.

0-87371-871-2/94/$0.00+$.50

The efficiency of formation of singlet oxygen, the quantum yield, has also been studied.[5-7] However, only wavelengths from 313 to 546 nm have been investigated. Lower and higher wavelength light may be of importance for two reasons. First, lower wavelength light (e.g., <300 nm) may in the future penetrate the atmosphere as the result of a depletion of stratospheric ozone. Therefore, it is important to know the contribution of lower-wavelength light to the formation of singlet oxygen. Second, as sunlight penetrates into a water body, the low-wavelength light is filtered out and only higher wavelength light is present at depth. Therefore, the quantum yield at wavelengths greater than 550 nm may help in predicting photoreactions in deeper waters.

The objectives of this study were (1) to measure the formation rate of $^1O_2$ in several Canadian surface waters and to relate it to humic substances, as characterized by measurements of DOC, UV absorption, and/or fluorescence; (2) to determine $^1O_2$ quantum yield at different wavelengths from 280 nm to 700 nm; and (3) to examine the $^1O_2$ formation with depth.

METHODS AND MATERIALS

Water samples were obtained from several lakes in Ontario, Canada. Samples were collected by submerging an inverted polyethylene bottle about 10 cm below the surface. Immediately after the water sample was obtained it was passed through a 0.2-μm filter and stored in 2-L polyethylene bottles at 4°C. The UV absorption spectra were obtained for each sample using a diode array spectrophotometer (Hewlett Packard, Model 8452A) from 200 nm to 800 nm. Natural water fluorescence was measured in quinine sulfate units (QSU) at $\lambda_{ex} = 350$ nm, $\lambda_{em} = 450$ nm using a spectrofluorometer (Gilford, Model Fluoro IV).

For the measurement of singlet oxygen, furfuryl alcohol (FFA) was used as a trapping agent.[4-5] From the loss of FFA and the second-order reaction rate constant between $^1O_2$ and FFA, $[^1O_2]_{ss}$ can be calculated assuming water is the only physical quencher ($k_d = 2.5 \times 10^5$ sec^{-1}) to $^1O_2$. If other physical quenchers exist (k'_d), the calculation of $[^1O_2]_{ss}$ will be affected by

$$\left[{}^1O_2 \right]'_{ss} = \left(k_d/k'_d \right) \left[{}^1O_2 \right]_{ss}$$

FFA was determined by high pressure liquid chromatography (HPLC). The mobile phase was 40% methanol and 60% water. FFA was separated on a 5-μm, 150 × 4.6-mm, reverse-phase carbon-18 (C18) column at a flow rate of 1.0 mL min^{-1} and detected using a UV detector at 219 nm.

Samples were sealed in 130-ml quartz tubes and attached to a rod that was floated on the surface of water. For depth formation, quartz tubes were attached to a rod standing in the lake. Four depths were chosen for the experiment.

Singlet oxygen quantum yield experiments were obtained using a solar simulator, comprising of a 1000-W xenon lamp with a short-arc power supply, a high-intensity grating monochromator, and a voltage radiometer attached to a computer

Table 1. Lists of Water Characterizations and $[^1O_2]_{ss}$ in Sunlit Surface

Sites	pH	DIC (mg/L)	Chl a (µg/L)	DOC (mg/L)	$[^1O_2]_{ss} \times 10^{-14}$ *M*
Anstruther	7.16	3.2	5.1	5.4	2.42
Bog	5.74	0.1ª	3.4	15.0	8.92
Brooks	7.95	12.1	2.8	7.2	4.27
Clanricarde	6.35	0.1ª	1.5	5.8	2.28
Hamilton Harbor	8.25	23.0	20.2	4.1	3.37
Lake Erie 23	8.20	21.2	2.7	4.2	0.46
Lake Erie 357	8.12	19.6	7.1	3.4	1.45
Lake Ontario	8.20	20.3	4.8	3.4	0.53
Rice	8.05	24.6	34.0	7.9	4.39
Sharpe	8.04	14.5	1.4	5.7	2.30
Wolf	6.90	2.5	5.7	5.9	2.18

Note: All samples were from Ontario, Canada.
ª Means less than.

for data acquisition. Samples were transferred to the quartz cuvette irradiated for 20 hr at the selected wavelength. The concentration of FFA was measured before and after irradiation.

RESULTS AND SUMMARY

Formation of singlet oxygen was observed in all 11 samples. During September in Ontario, Canada, the $[^1O_2]_{ss}$ varied from 0.46 to 8.9×10^{-14} *M* (Table 1). It appeared from the correlation of DOC to $[^1O_2]_{ss}$ that they were linearly related (Figure 1), consistent with the data of others.[2-6] However, we were interested in other, possibly better predictive relationships. Thus, in our study, $[^1O_2]_{ss}$ was not only correlated to DOC, but also to fluorescence as QSU at $\lambda_{ex} = 350$ nm and $\lambda_{em} = 450$ nm (Figure 2); UV absorptivities at 280 nm (Figure 3); and 300 nm, 350 nm, 400 nm, and the integrated area under the absorption curve of the natural waters from 300 to 700 nm (Figure 4). The linear regression correlation coeffi-

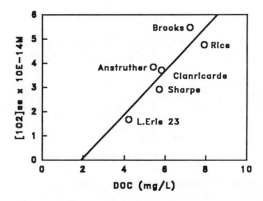

Figure 1. Linear relationship between $[^1O_2]_{ss}$ and natural water DOC ($r^2 = 0.893$).

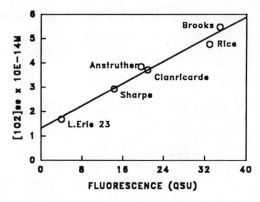

Figure 2. Linear relationship between $[^1O_2]_{ss}$ and natural water fluorescence ($r^2 = 0.988$).

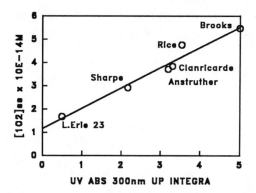

Figure 3. Linear relationship between $[^1O_2]_{ss}$ and natural water UV absorbance at 300 nm above integration ($r^2 = 0.962$).

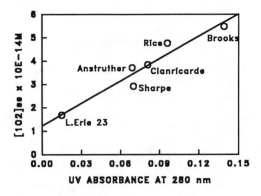

Figure 4. Linear relationship between $[^1O_2]_{ss}$ and natural water UV absorbance at 280 nm ($r^2 = 0.964$).

cients were 0.893, 0.988, 0.964, 0.978, 0.981, 0.978, and 0.962, respectively. It appears from these data and data from other experiments that fluorescence is the best descriptor for predicting the formation of singlet oxygen.

Humic substances (HS), in the ground state, absorb solar energy and may be excited to the singlet state, ^{1}HS. The singlet state may return to the ground state via radiationless (thermal) or radiation (fluorescence) decay. It is also possible that the singlet state may undergo intersystem crossing (ISC) to the triplet state, ^{3}HS. The triplet state humic substance may then react with ground state oxygen through energy transfer to form singlet oxygen, as follows:

$$^0HS \xrightarrow{\text{hv}} {}^1HS \tag{1}$$

$$^1HS \xrightarrow{\text{flourescence}} {}^0HS \tag{2}$$

$$^1HS \xrightarrow{\text{ISC}} {}^3HS \tag{3}$$

$$^3HS + {}^3O_2 \rightarrow {}^1O_2 \tag{4}$$

A possible explanation is that both fluorescence and ^{3}HS result from a ^{1}HS. Therefore, it is reasonable that fluorescence would be a good predictor for the formation of singlet oxygen.

Wavelength-dependent quantum yield data were obtained using a Sharpes Bay water sample (Ontario, Canada) (Figure 5). The results indicate that the quantum yield is highest at 280 nm and decreases with increasing wavelength. This is in agreement with the results of Frimmel et al.,[8] Haag et al.,[7] and Zepp et al.[9] and therefore support their conclusions that humic substance photosensitization is mainly induced by UV and blue radiation. The quantum yield data at wavelengths

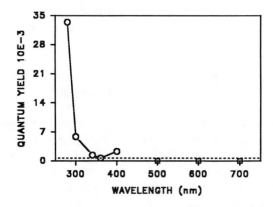

Figure 5. Singlet oxygen quantum yield as the function of wavelength (Sharpe Bay water sample, Ontario, Canada) □ = below detection limit.

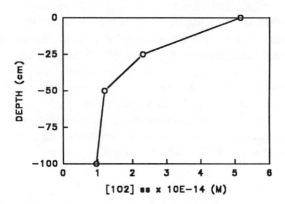

Figure 6. Singlet oxygen formation with depth (Sharpe Bay water was incubated in Sharpe Bay, Ontario, Canada).

greater than 400 nm were below our detection limit. Our results of the significant quantum yield of $^1O_2$ at less than 300 nm indicate that losses in stratospheric ozone will lead to increased formation of $^1O_2$ in the surface water of lakes.

Additional studies have been done to understand the specificity of FFA for $^1O_2$, i.e., is the loss of FFA due to singlet oxygen. A water sample was irradiated using 280 nm light in the presence and absence of sodium azide. In the absence of azide, a loss of 60% of the FFA was observed. In the presence of azide (both 10^{-2} and $10^{-3}\ M$), a loss of 24% of the FFA was observed.[10] Approximately 66% of the loss of FFA was due to singlet oxygen. Therefore, quantum yield at 280 nm (Figure 5) was overestimated by about 33%. Peroxy radicals may contribute to the loss of FFA.[11]

As sunlight penetrates into water, the shorter wavelengths are absorbed and the high-wavelength light penetrates to depth. The wavelengths available at different depths are a function of wavelength and the absorbance characteristics of natural water. Humic substances are principally responsible for the absorption of the lower-wavelength portion of the solar spectrum. To experimentally determine the depth distribution of $^1O_2$, samples were irradiated in a natural water. Due both to the loss of total UV light and the selective loss of shorter wavelengths, photochemical production of singlet oxygen decreases with depth. Figure 6 shows that $[^1O_2]_{ss}$ was highest at the surface and decreased with depth. At a depth of 1 m, the formation of $^1O_2$ was 20% that of the surface. Because of the short lifetime of $^1O_2$, the effects would be localized to the near-surface region of natural waters.

In summary, we have shown that fluorescence ($\lambda_{ex} = 350$ nm, $\lambda_{em} = 450$ nm) is the best predictor of $^1O_2$ formation in mid-latitude lakes. The quantum yield of $^1O_2$ was shown to be highest at 280 nm, indicating that if a loss in stratospheric ozone dose occur then higher surface water concentration of $^1O_2$ will result. Lastly, because the majority of the $^1O_2$ results from absorption of the lower portion of the solar spectrum, its formation is restricted to the upper water column.

ACKNOWLEDGMENTS

This work was supported in part by the Ohio Sea Grant College Program. Project R/PS-4, NA90AA-D-SG496 of the National Sea Grant College Program, NOAA, U.S. Department of Commerce.

REFERENCES

1. Hoigné, J., Faust, B. C., Haag, W. R., Scully, F. E., and Zepp, R. G., Aquatic humic substances as sources and sinks of photochemically produced transient reactants, in *Aquatic Humic Substances: Influence on Fate and Treatment of Pollutants*, Suffet, I. H. and MacCarthy, P., Eds., *Advances in Chemistry Ser. No.* 219, American Chemical Society, Washington, D.C., 1989, 363.
2. Zepp, R. G., Wolfe, N. L., Baughman, G. L., and Hollis, R. C., Singlet oxygen in natural waters, *Nature (London)*, 267, 421, 1977.
3. Wolff, C. J. M., Halmans, M. T. H., and van der Heijde, H. B., The formation of singlet oxygen in surface waters, *Chemosphere*, 10, 59, 1981.
4. Haag, W. R. and Hoigne, J., Singlet oxygen in surface waters. 3. Photochemical formation and steady-state concentrations in various types of waters, *Environ. Sci. Technol.*, 20, 341, 1986.
5. Haag, W. R., Hoigné, J., Gassmann, E., and Braun, A. M., Singlet oxygen in surface waters — Part I: furfuryl alcohol as a trapping agent, *Chemosphere*, 13(5/6), 631, 1984.
6. Kawaguchi, H., Photochemical formation of singlet oxygen in lake waters, *J. Chem. Soc. Jpn.*, 5, 520, 1991, (in Japanese).
7. Haag, W. R., Hoigné, J., Gassmann, E. and Braun, A. M., Singlet oxygen in surface waters — Part II: quantum yields of its production by some natural humic materials as a function of wavelength, *Chemosphere*, 13(5/6), 641, 1984.
8. Frimmel, F. H., Bauer, H., and Putzien, J., Laser photolysis of dissolved aquatic humic material and the sensitized production of singlet oxygen, *Environ. Sci. Technol.*, 21(6), 541, 1987.
9. Zepp, R. G., Baughman, G. L., and Schlotzhauer, P. F., Comparison of photochemical behavior of various humic substances in water: II. Photosensitized oxygenations, *Chemosphere*, 10, 119, 1981.
10. Shao, C., Cooper, W. J., and Lean, D. R. S., Sunlight induced photochemical formation of singlet oxygen in natural water system, *Can. J. Fish. Aquat. Sci.*, 1992, submitted.
11. Faust, B. C. and Allen, J. M., Aqueous-phase photochemical sources of peroxyl radicls and singlet molecular oxygen in clouds and fog, *J. Geophys. Res.*, 97(D12), 12913, 1992.

CHAPTER 18

Aqueous-Phase Photochemical Formation of Peroxides in Authentic Cloud Waters

Cort Anastasio, John M. Allen, and Bruce C. Faust

INTRODUCTION

Peroxides (H_2O_2 and organic peroxides) in atmospheric waters play several significant roles. Perhaps most importantly, H_2O_2 plays a crucial role in the oxidation of SO_2 to H_2SO_4[1] and, thereby, impacts both acid deposition and the global cooling caused by sulfate aerosols.[2,3] There is also evidence that both peroxides and sulfuric acid, formed from the aforementioned oxidation of S(IV) ($SO_2 \cdot H_2O$ and HSO_3^-) by H_2O_2, are components of "acid rain"-mediated damage to forests.[4] Additionally, aqueous-phase peroxides produce hydroxyl radical, a key oxidant in the troposphere, through photolysis and thermal reactions with Fe(II).[5]

Because of these impacts, it is important to understand the sources of peroxides to atmospheric waters. Models of atmospheric water drop chemistry have traditionally focused on gas-to-drop partitioning of peroxides and peroxyl radicals as the sole sources of aqueous-phase peroxides. Since possible aqueous-phase sources of peroxides to atmospheric waters had not been addressed, we sought to determine if there were direct aqueous-phase photochemical sources of peroxides to cloud waters.[6] To date, our research has focused on the following questions: (1) "Are peroxides photochemically produced within cloud drops?", (2) "What are the rates of this aqueous-phase photochemical peroxide production under typical sunlight illumination conditions?", (3) "What are the quantum yields for aqueous-phase peroxide production?", and (4) "How does the initial rate of aqueous-phase peroxide photoproduction depend on the actinic flux?".

EXPERIMENTAL

The results reported here are from 2 years of study. Cloud water samples were collected by collaborators during spring through fall of 1990 and 1991 from high-elevation sites in New York, North Carolina, Virginia, and Washington and by airplane from clouds over Ontario, Canada. As a check for contamination, rinse water controls (distilled water that was sprayed onto the strings of the collector) were collected immediately prior to cloud collection and then handled in the same manner as actual samples. After collection, samples were shipped and stored in the dark at 2–10°C. Upon receipt, samples were filtered (0.2 or 0.5 µm Teflon™*) and their ultraviolet-visible (UV-Vis) absorption spectra and fluorescence values (excitation 313 nm, emission 420 nm) were measured using distilled water as a reference. Sample fluorescence measurements were normalized by dividing the sample fluorescence with the fluorescence of a standard aqueous solution of 0.64 µM quinine sulfate (0.105 M HClO$_4$) measured on the same day. Samples were usually studied within 3 weeks of collection; additional storage time of up to 11 days caused only small changes in the measured peroxide photoproduction rate (range: –18% to +25%, mean: 0.8%).

Samples were studied with four types of illumination: sunlight, simulated sunlight (from the filtered output of a xenon lamp[7]), and monochromatic light of wavelength 313nm or 334nm. Samples were illuminated with simulated sunlight or monochromatic light in Teflon-stoppered quartz cuvettes that were maintained at 20°C and stirred continuously. Sunlight illuminations were carried out in glass-stoppered quartz tubes maintained at 25°C. As a check for artifactual peroxide production, several rinse water controls were also photolyzed using the same procedure. Dark (thermal) controls of the samples were carried out under conditions identical to that of photolysis except for illumination.

Peroxides were measured (as total peroxides) with the fluorescence technique of Kok et al.[8] Initial rates of aqueous-phase peroxide photoproduction were determined using linear regression or least squares fitting of an exponential function, depending on the apparent function of the experimental data. Sunlight and simulated sunlight peroxide photoproduction rates were normalized to clear-sky, mid-day, equinox illumination in Durham, NC, as follows:

$$\text{Rate}_{\text{EQUINOX}} = \text{Rate}_{\text{EXPT}} * \left(j_{\text{EQUINOX}} / j_{\text{EXPT}} \right)$$

where $\text{Rate}_{\text{EQUINOX}}$ is the equinox-sunlight normalized peroxide photoproduction rate (µM/hr), $\text{Rate}_{\text{EXPT}}$ is the measured experimental rate (µM/hr), j_{EQUINOX} is the first-order actinometer rate constant measured in sunlight on the autumnal equinox (sec^{-1}), and j_{EXPT} is the actinometer rate constant measured during the experiment (sec^{-1}). Air-saturated dilute (≤10 µM) aqueous solutions of 2-nitrobenzaldehyde or valerophenone, photolyzed under the same conditions as samples, were used as actinometers. Quantum yield calculations were based on the total absorbance of the cloud water sample at the appropriate wavelength.

---

*Teflon™ is a registered trademark of E.I. du Pont de Nemours and Company, Inc., Wilmington, DE.

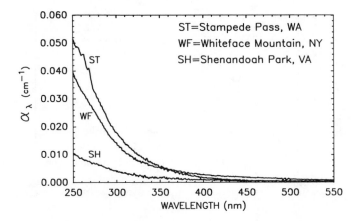

Figure 1. Three typical UV-Vis spectra of authentic cloud waters (filtered, 0.5 μm). α_λ = base 10 absorbance per cm (distilled water reference).

RESULTS AND DISCUSSION

All of the samples absorbed UV-Vis light in the wavelength region of tropospheric solar radiation ($\lambda \geq 290$ nm). Some typical cloud water spectra are shown in Figure 1. Aqueous-phase photochemical peroxide production occurred in samples from all five sites and in nearly all (>90%) of the samples irradiated. This aqueous-phase peroxide photoproduction is a result of the absorption of light by chromophores in the samples and, presumably, the formation of peroxyl radical intermediates (HOO˙ and ROO˙), which are considered to be the dominant precursors of peroxides in atmospheric waters.[9,10] Dark controls showed no peroxide production for any of the samples tested. Photolysis of rinse water controls yielded an average of less than 5% of the initial rate of peroxide photoproduction seen with the associated sample; in half of the tested rinse waters there was no peroxide photo-production. Figure 2 shows one typical set of sample and control experiments consisting of cloud water sample photolysis, cloud water dark control, and rinse water photolysis.

A typical plot of aqueous-phase photochemical peroxide production with simulated sunlight illumination is shown in Figure 3. The mean (±standard deviation) equinox-normalized rate of aqueous-phase peroxide photoproduction for the samples studied was 0.87 (±0.89) μM/hr (35 samples, 5 sites). Photoproduction rates in these samples ranged from 0 to 3.0 μM/hr. It should be noted that the standard deviation of the mean rate is large because of the large differences in sample photochemical activities between different sites and different cloud events, not because of a lack of procedural precision; relative percent differences in rates for replicated experiments were almost always less than 25%.

It is possible to get some measure of the importance of aqueous-phase photoproduction as a source of peroxides to cloud drops by comparing our measured

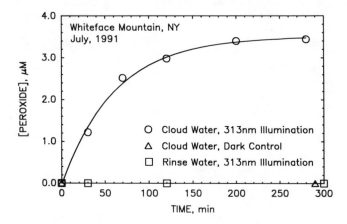

Figure 2. Aqueous-phase photoproduction of peroxides in a cloud water illuminated with 313-nm light (O). There was no peroxide production in the cloud water kept in the dark (△), or in the illuminated (313 nm light) rinse water control (□).

rates of aqueous-phase photoformation with field measurements of peroxides in cloud waters. Reported cloud water peroxide concentrations are ≤5 μM for 100%, 58–72%, and 3–7% of the samples measured in the winter,[11,12] spring and fall,[13,14] and summer,[13,14] respectively. By comparison, our equinox-normalized rates of aqueous-phase peroxide photoproduction ranged up to 3.0 μM/hr. In a more direct comparison, MacDonald et al.[14] reported that 62% of cloud water samples collected by aircraft over Ontario in the spring of 1990 had peroxide concentrations of less than 5 μM. Our samples from Ontario, a subset of the samples reported on by MacDonald et al., had a mean equinox-normalized aqueous-phase peroxide photoproduction rate of 0.75 ± 0.69 μM/hr (17 samples, range: 0.07–2.1 μM/hr).

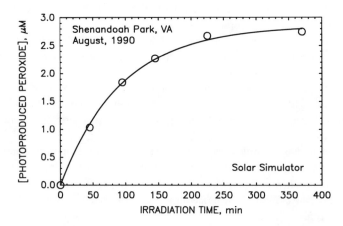

Figure 3. Aqueous-phase photoproduction of peroxides in a cloud water illuminated with simulated sunlight. (The data shown has not been equinox normalized: $j_{EQUINOX}/j_{EXPT} = 0.78$.)

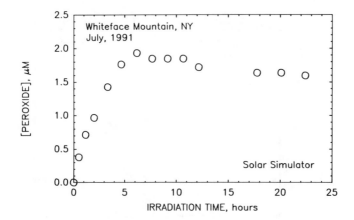

Figure 4. Long-term illumination of a cloud water with simulated sunlight. The sample reaches a pseudo-steady state (after approximately 5 hr of illumination), which then only slowly decays with continued illumination. (The data shown has not been normalized: $j_{EQUINOX}/j_{EXPT} = 3.5$.)

These comparisons argue that aqueous-phase photoproduction is a significant, and in some cases dominant, source of peroxides to cloud drops.

Aqueous-phase peroxide photoproduction was also observed with mono-chromatic illumination (see Figure 2). The mean ($\pm$ standard deviation) quantum yield for the samples studied at 313 nm was $(2.5 \pm 2.7) \times 10^{-3}$ with a range of $(0–10) \times 10^{-3}$ (18 samples from 4 sites). At 334 nm, the mean quantum yield was $(1.3 \pm 1.3) \times 10^{-3}$ and the range was $(0–3.4) \times 10^{-3}$ (7 samples from 3 sites). For the five samples in which both 313- and 334-nm quantum yields were deter-mined, the 334-nm result was lower, generally by a factor of 2.

Longer-term irradiation (approximately 24 hr) suggests that the aqueous-phase production of peroxides is sustained (with a slow net peroxide loss) over many hours as shown in Figure 4. Continued peroxide photoproduction in a sample that was previously irradiated to a peroxide pseudo-steady state, and then treated with S(IV) in the dark to destroy peroxides, is additional evidence that peroxides are still being formed at the pseudo-steady state (Figure 5).

Additional experiments indicate that the initial rate of aqueous-phase peroxide photoproduction is linearly dependent on the actinic flux (Figure 6). By contrast, the gas-phase production of H_2O_2 (from $HO_2^\cdot + HO_2^\cdot \rightarrow H_2O_2 + O_2$) is expected to exhibit a squared dependence on actinic flux in all but the most pristine regions of the troposphere. This suggests a seasonal variation in the relative importance of aqueous-phase photochemical sources compared to gas-phase sources of per-oxides to cloud drops. Aqueous-phase photoproduction will be relatively more important, with respect to gas-to-drop transfer, in the spring, fall, and winter compared to the summer.

Initial aqueous-phase photochemical production rates are correlated with cloud water fluorescence, as shown in Figure 7. This suggests that fluorescent compounds

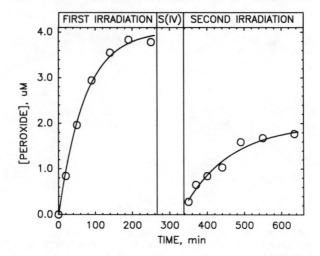

Figure 5. Aqueous-phase peroxide photoproduction is still occurring at the peroxide pseudo-steady state. This cloud water was first illuminated with 313-nm light until it reached a pseudo-stationary state, and then the photoproduced peroxides were destroyed by addition of Na_2SO_3 (S(IV)) in the dark. Additional illumination with 313-nm light demonstrated that the cloud water was still capable of photochemically producing peroxides, although at a slower rate (approximately 25% as fast as the initial photoproduction rate).

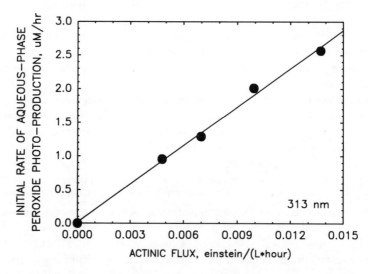

Figure 6. The aqueous-phase photoproduction of peroxides is linearly dependent on actinic flux, as shown by the illumination of this authentic cloud water sample from Stampede Pass, Washington, with 313-nm light.

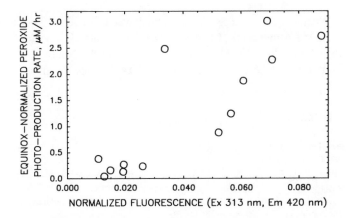

Figure 7. Correlation between the equinox-normalized initial rate of aqueous-phase perox-ide photoproduction and the quinine-sulfate-normalized cloud water sample fluo-rescence (excitation 313 nm, emission 420 nm).

contribute significantly to peroxide photoproduction. Other correlations are cur-rently being sought as potential predictors of aqueous-phase peroxide photopro-duction activity and as evidence of possible aqueous-phase peroxide photoproduc-tion precursors.

CONCLUSIONS

Based on the previously discussed field observations and on the aqueous-phase photoproduction rates that we have reported here, the aqueous-phase photochemi-cal formation of peroxides is a significant, and in some cases probably dominant, source of peroxides to cloud drops. This previously unrecognized *in situ* photo-chemical production of peroxides, combined with its linear dependence on actinic flux, has significant ramifications for models of atmospheric chemistry, especially for predictions of acidic deposition and sulfate aerosol-induced global cooling.

ACKNOWLEDGMENTS

We would like to thank the following people for their generous assistance in this study: J. Kadlecek, J. Lu, R. MacDonald, V. Mohnen, B. Murphy, S. Roychowdhury, P.J. Spink, and S. Virgilio (Whiteface Mountain, NY); J. Galloway, B.M. McIntyre, J. Sigmon, and P. Thompson (Shenandoah Park, VA); A. Basabe, W.L. Chang, R. Cryer, and T. Larson (Stampede Pass, WA); V. Aneja and P. Blankinship (Mount Mitchell, NC); and C. Banic, G.A. Isaac, W.R. Leaitch, and A.M. MacDonald (Ontario). We would also like to thank G.E. Likens and K.C. Weathers for the use of their cloud water collector. This work was supported by the Atmospheric Chemistry Program of the U.S. National Science Foundation, the Andrew W. Mellon Foundation, and the National Institutes of Health through the Duke University Integrated Toxicology Program.

REFERENCES

1. Gunz, D. W. and Hoffmann, M. R., Atmospheric chemistry of peroxides: a review, *Atmos. Environ.*, 24A, 1601, 1990.
2. Charlson, R. J., Schwartz, S. E., Hales, J. M., Cess, R. D., Coakley, J. A., Jr., Hansen, J. E., and Hofmann, D. J., Climate forcing by anthropogenic aerosols, *Science*, 255, 423, 1992.
3. Lelieveld, J. and Heintzenberg, J., Sulfate cooling effect on climate through in-cloud oxidation of anthropogenic SO_2, *Science*, 258, 117, 1992.
4. Masuch, G., Kettrup, A., Mallant, R. K. A. M., and Slanina, J., Effects of H_2O_2-containing acidic fog on young trees, *Int. J. Environ. Anal. Chem.*, 27, 183, 1986.
5. Zepp, R. G., Faust, B. C., and Hoigné, J., Hydroxyl radical formation in aqueous reactions (pH 3–8) of Iron(II) with hydrogen peroxide: the photo-Fenton reaction, *Environ. Sci. Technol.*, 26, 313, 1992.
6. Faust, B. C., Anastasio, C., Allen, J. M., and Arakaki, T., Aqueous-phase photochemical formation of peroxides in authentic cloud and fog waters, *Science*, 260, 73, 1993.
7. Faust, B. C., Generation and use of simulated sunlight in photochemical studies of liquid solutions, *Rev. Sci. Instrum.*, 64, 577, 1993.
8. Kok, G. L., Thompson, K., Lazrus, A. L., and McLaren, S. E., Derivatization technique for the determination of peroxides in precipitation, *Anal. Chem.*, 58, 1192, 1986.
9. Faust, B. C. and Allen, J. M., Aqueous-phase photochemical sources of peroxyl radicals and singlet molecular oxygen in clouds and fog, *J. Geophys. Res.*, 97, 12913, 1992.
10. Allen, J. M., and Faust, B. C., Aqueous-phase photochemical formation of peroxyl radicals and singlet molecular oxygen in cloud water samples from across the United States, in *Aquatic and Surface Photochemistry*, Helz, G., Zepp, R., and Crosby, D., Eds., Lewis Publishers, Boca Raton, FL, 1993, chap. 19.
11. Kadlecek, J., McLaren, S., Mohnen, V., Mossl, B., Kadlecek, A., and Camarota, N., *Winter Cloudwater Chemistry Studies at Whiteface Mountain, 1982–1984*, ASRC/SUNY Publication No. 1008, Atmospheric Sciences Research Center, State University of New York, Albany, 1985.
12. Daum, P. H., Kelly, T. J., Strapp, J. W., Leaitch, W. R., Joe, P., Schemenauer, R. S., Isaac, G. A., Anlauf, K. G., and Wiebe, H. A., Chemistry and physics of a winter stratus cloud layer: a case study, *J. Geophys. Res.*, 92, 8426, 1987.
13. Olszyna, K. J., Meagher, J. F., and Bailey, E. M., Gas-phase, cloud and rain-water measurements of hydrogen peroxide at a high-elevation site, *Atmos. Environ.*, 22, 1699, 1988.
14. MacDonald, A. M., Anlauf, K. G., Wiebe, H. A., Leaitch, W. R., Banic, C. M., Watt, M. F., Puckett, K. J., and Bregman, B., Measurements of aqueous and gaseous hydrogen peroxide in central Ontario, in *Proc. Seventh Joint Conf. Appl. Air Poll. Meteorology with AWMA*, New Orleans, 1991.

CHAPTER 19

Aqueous-Phase Photochemical Formation of Peroxyl Radicals and Singlet Molecular Oxygen in Cloud Water Samples from Across the United States

John M. Allen and Bruce C. Faust

INTRODUCTION

Aqueous-phase photochemical oxidation reactions in clouds are known to play an important role in the overall chemistry of the troposphere.[1] Concerns about acid deposition have fueled much interest in these processes. The atmospheric oxidation of SO_2 occurs largely within cloud droplets. Unfortunately, the present understanding of the sources, rates of formation, and concentrations of oxidants in clouds is limited. Much effort has been directed toward an understanding of the transfer of oxidants from the gas phase to the liquid phase, but comparatively little is known about the aqueous-phase photochemical production of oxidants in clouds. Several investigations have suggested that aqueous-phase photochemical reactions are a significant source of hydroxyl radical in clouds.[2,3] Analogous sources of other oxidants have not been identified. For this reason, a study[4,6] was undertaken to measure the aqueous-phase photochemical formation rates and steady-state concentrations of several oxidant species in cloud waters. We report here on the formation and measured steady-state concentrations of peroxyl radicals (HO_2^- and RO_2^-) and singlet molecular oxygen, $O_2(^1\Delta_g)$, in illuminated authentic cloud water samples. We have performed experiments that demonstrate that aqueous-phase photochemical production of these oxidants can lead to concentrations in the aqueous phase that equal or exceed concentrations predicted by atmospheric models that are based primarily upon gas-to-drop partitioning as the source for aqueous-phase oxidants.

0-87371-871-2/94/$0.00+$.50
© 1994 by CRC Press, Inc.

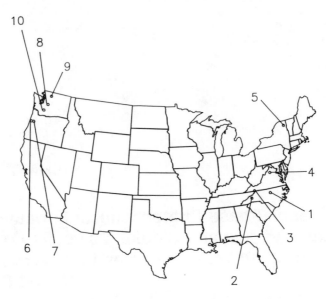

Figure 1.   Locations of cloud and fog water collection sites in the United States: (1) Duke University, Durham, NC (location of irradiations); (2) Mount Mitchell, NC; (3) Whitetop Mountain, VA; (4) Shenandoah Park, VA; (5) Whiteface Mountain, NY; (6) Mary's Peak, OR; (7) Corvallis, OR; (8) Stampede Pass, WA; (9) Cascade Pass, WA; and (10) Burley Mountain, WA.

EXPERIMENTAL

Cloud water samples were obtained from collaborators throughout a 3-year period (1988–1991) from North Carolina, Virginia, New York, Washington, and Oregon (Figure 1). Ultraviolet-visible (UV-Vis) absorption spectra of cloud water samples were measured using distilled water as a reference.

The steady-state concentrations of peroxyl radicals and $O_2(^1\Delta_g)$ were determined using a kinetic probe technique[4,8] in which a small amount of a chemical probe compound was added to the cloud water. The probe compounds chosen were 2,4,6-trimethylphenol (TMP) and furfuryl alcohol (FFA) for measurement of peroxyl radicals and $O_2(^1\Delta_g)$, respectively.[4,8] The probe compounds FFA and TMP were chosen because of their high reactivities toward $O_2(^1\Delta_g)$ and peroxyl radicals, respectively, and because they have been used in studies on natural waters.[4-8] The cloud water sample with the added chemical probe was then illuminated in a closed quartz vessel using sunlight, simulated sunlight, or a single wavelength of light.

The oxidation kinetics of the probe were monitored by high-pressure liquid chromatography. Corrections were made for direct photolysis of the probe and for any dark (thermal) reaction. The kinetic probe model is based upon the assumption that, under constant illumination, a steady-state concentration of the oxidant

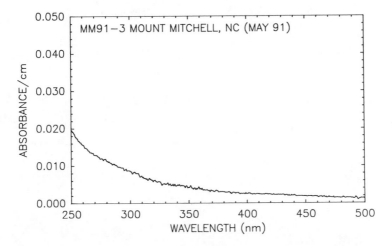

Figure 2.   Ultraviolet-visible (UV-Vis) absorption spectrum for a cloud water sample collected at Mount Mitchell, NC (base 10 absorbance). Distilled water was used in
the reference and sample cuvettes to zero the spectrophotometer. Distilled water
was also used in the reference cuvette during sample measurement in order to
subtract any possible contribution to the absorption spectrum attributable to
water.

will be reached after a very short induction time. The concentration of the probe
was kept low to minimize any perturbations to the overall chemistry of the cloud
water. At low concentrations, the probe has little effect on the steady-state
concentration of the photooxidant because the photooxidant continues to be
produced and consumed almost exclusively via natural pathways (i.e., by pathways that do not involve the added probe compound). The experimentally determined rate expression for the loss of the chemical probe (P) is

$$\frac{d[P]}{dt} = -k[P]$$

where $k = k_2[OX]_{ss}$, k_2 is the bimolecular rate constant for the reaction of the
oxidant with the chemical probe, and $[OX]_{ss}$ is the steady-state concentration of
the oxidant. The steady-state concentration of the oxidant was determined by
measuring k for a chemical probe with a known value for k_2.

RESULTS AND DISCUSSION

Cloud waters were found to absorb radiation at $\lambda > 290$ nm.[4-6] This phenomenon was observed to a greater or lesser degree for every cloud water sample
whose UV-Vis absorption spectrum was measured. Figures 2 and 3 illustrate this
absorption for two cloud water samples. Although the chemical composition of

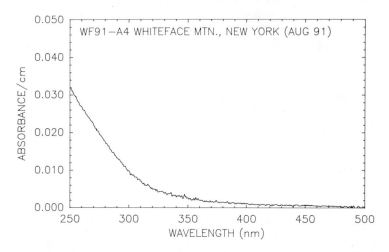

Figure 3. Ultraviolet-visible (UV-Vis) absorption spectrum for a cloud water sample collected at Whiteface Mountain, NY (base 10 absorbance). Distilled water was
used in the reference and sample cuvettes to zero the spectrophotometer.
Distilled water was also used in the reference cuvette during sample measurement in order to subtract any possible contribution to the absorption spectrum
attributable to water.

cloud water has not been well characterized, many chemical species including
aldehydes, ketones, organic fatty acids, polycyclic aromatic hydrocarbons (PAH),
and transition-metal complexes are known to be present in cloud waters. The
presence of these species, which contain chromophoric functional groups, in
cloud water could contribute to the observed UV absorption. This absorption of
UV radiation by the cloud water samples gives rise to the photochemical
reactions that are responsible for the formation of (1) $O_2(^1\Delta_g)$ via a transfer of
energy from excited-state triplets to ground state molecular oxygen, (2) peroxyl
radicals via fragmentation, and (3) other oxidant species via a variety of pathways. Measurements of the UV-Vis spectra for blanks in which pure deionized/
distilled water was used to rinse all surfaces of cloud water collectors and
storage containers did not show the characteristic absorption typical of the cloud
water samples.

   Neither FFA nor TMP reacted appreciably in dark controls or when exposed
to illumination in distilled water. The measured rate constant for the loss of the
probe was observed to be independent of initial probe concentration when the
initial probe concentration was low (approximately 1×10^{-6} *M*). The apparent
first-order plots obtained from the experimental data (e.g., Figures 4 and 5) were
all linear. These two observations support the validity of the kinetic probe model
and the steady-state assumption. The fate of peroxyl radicals was controlled by
natural pathways and not by their reaction with the added probe.

   The photooxidation of FFA was more rapid in air-saturated 3:1 mixtures of
deuterium oxide (D_2O) to cloud water than in 3:1 mixtures of H_2O to cloud water,

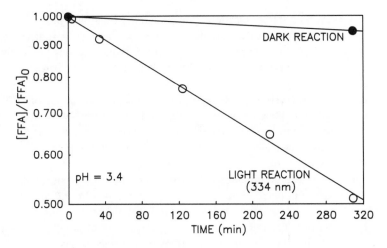

Figure 4. Photooxidation (334 nm) of FFA in a cloud water sample from Whiteface Mountain, NY.

where the pH was adjusted to that of pure cloud water using sulfuric acid. The lifetime of $O_2(^1\Delta_g)$ in the experiments reported here was controlled by solvent (H_2O or H_2O/D_2O) quenching and not by reaction with FFA or other constituents of the cloud water samples. The increased rate of FFA oxidation in the presence of D_2O is attributed to the increased lifetime and therefore the increased steady-state concentration of $O_2(^1\Delta_g)$ in D_2O. The lifetime of $O_2(^1\Delta_g)$ in pure D_2O is a factor of 13 longer than in pure H_2O.[9] The results from these experiments indicate that the fraction of FFA oxidation attributable to $O_2(^1\Delta_g)$ was $36 \pm 16\%$ for cloud

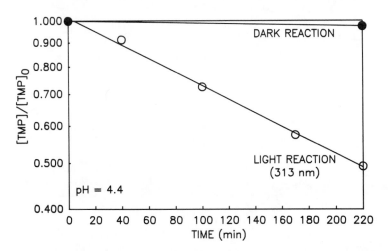

Figure 5. Photooxidation (313 nm) of TMP in a cloud water sample from Whiteface Mountain, NY.

waters from several different locations. The remaining fraction is probably due to reaction with peroxyl radicals. Further evidence for the aqueous-phase photochemical formation of peroxyl radicals is provided by the aqueous-phase photochemical formation of peroxides[10,11] which are known to involve peroxyl radical intermediates.

Hydroxyl radical ($\cdot$OH) was not responsible for the photooxidation of FFA nor TMP. The cloud water photooxidations of FFA and TMP were 3–9 and 4–12 times faster, respectively, than photooxidations of phenol despite the similar diffusion controlled reactions of phenol, FFA, and alkylphenols with $\cdot$OH.[5,6]

Experiments were conducted to assess the effect of particles upon the rates of loss for the probe compounds. The losses for both FFA and TMP were measured in an illuminated cloud water sample without filtering and then the experiment was repeated using the same cloud water after filtering with a 0.5 μm Teflon™* filter. The rate of loss for the probes was not greatly affected by filtration ($<25\%$ in all cases). Thus, reactions on particle surfaces for particles greater than 0.5 μm do not appear to play a significant role in the results reported here.

A comparison was made of the steady-state concentrations of $O_2(^1\Delta_g)$ and peroxyl radicals in several individual sunlit cloud water samples. A linear correlation between the observed steady-state concentrations of $O_2(^1\Delta_g)$ and peroxyl radicals in these samples was observed (Figure 6). A possible explanation for this phenomenon is that the same chromophores that absorb UV light and are responsible for the production of $O_2(^1\Delta_g)$ are also responsible for the production of peroxyl radicals. Upon absorption, some fraction of the excited species may transfer their energy to ground state molecular oxygen forming $O_2(^1\Delta_g)$ or this

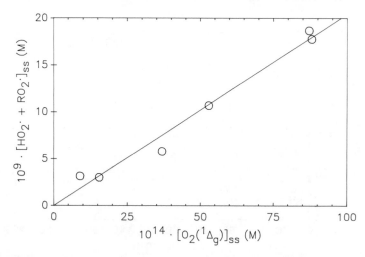

Figure 6. Correlation between photostationary-state concentrations of $O_2(^1\Delta_g)$ and peroxyl radicals in several individual cloud water samples (in sunlight).

---

*Teflon™ is a registered trademark of E.I. du Pont de Nemours and Company, Inc., Wilmington, DE.

acquired energy may lead to molecular fragmentation ultimately giving rise to peroxyl radicals.[12,13]

Due to a lack of information, atmospheric water drop models do not incorporate reactions involving aqueous-phase photochemical sources of HO_2^- and RO_2^-. The aqueous-phase production and reactions of $O_2(^1\Delta_g)$ are not addressed by atmospheric models at all. An upper bound of 5×10^{-15} M for $O_2(^1\Delta_g)$ concentrations in cloud drops can be calculated from equilibrium partitioning of gas-phase $O_2(^1\Delta_g)$ using the Henry's law constant for O_2 and assuming a gas-phase concentration of 1×10^8 molecules per cubic centimeter.[6] We have measured $O_2(^1\Delta_g)$ concentrations ranging from 3×10^{-14} to 1.5×10^{-12} M in sunlit cloud waters.

Model predictions of organic peroxyl radical (RO_2^-) concentrations in cloud drops range from 2×10^{-12} to 2×10^{-10} M and from 2×10^{-12} to 2×10^{-8} M for hydroperoxyl radicals (HO_2^-). We have measured total, unspeciated steady-state peroxyl radical concentrations in cloud waters ranging from 7×10^{-10} to 3×10^{-8} M. These results have been calculated from experiments in closed vessels that did not allow any transfer of species from the gas phase. Therefore, it must be emphasized that the measured aqueous-phase concentrations reported here for both $O_2(^1\Delta_g)$ and peroxyl radicals reflect aqueous-phase photochemical production only.

Fog, cloud, and aqueous aerosol drops contain substantial concentrations of organic matter[11,14] and therefore, it is possible that an organic phase is present within the drop and/or at the drop surface.[15] The presence and concentration of organic compounds in fog waters has been shown to be correlated with measurements of droplet surface tension.[14] The presence of organic films on atmospheric water droplets, especially fog and aquated aerosol drops, could magnify the importance of aqueous-phase oxidant production relative to gas-to-drop transfer of oxidants because the outer organic film could impede the mass transfer of many oxidant species both into and out of the drop. In addition, any $O_2(^1\Delta_g)$, either formed in or transported to an organic phase, would have a greater lifetime and, consequently, a higher steady-state concentration because the lifetime of $O_2(^1\Delta_g)$ in most organic solvents is greater than in H_2O.[16] Moreover, the concentrations of organic chromophores in an organic phase would likely be greater than that in water, which would also contribute to increased concentrations of $O_2(^1\Delta_g)$ in the organic phase. The rates of photooxidation of some organic compounds by $O_2(^1\Delta_g)$ could be more rapid in an organic phase than in water due to the increased concentrations of $O_2(^1\Delta_g)$ and organic compounds in the organic phase. The reactivity of $O_2(^1\Delta_g)$ with an organic compound in water vs that in an organic phase would also need to be considered in an overall assessment of this possible effect. An additional important factor in assessing the overall effect of a separate organic phase on the $O_2(^1\Delta_g)$-mediated oxidation of organic compounds in fog drops is the volume fraction of the organic phase in the drop. Although this quantity is not known, the mole fraction of organic carbon is 10^{-3} for a fog drop with an organic carbon concentration of 667 mg C/L.

CONCLUSIONS

Two important conclusions have been reached as a result of the work presented here. (1) Aqueous-phase photochemical reactions can be an important, and in many cases a dominant, source of $O_2(^1\Delta_g)$ and peroxyl radicals in cloud drops. (2) Atmospheric models that do not include reactions involving $O_2(^1\Delta_g)$ and fail to incorporate aqueous-phase photochemical sources of peroxyl radicals could seriously underestimate the concentrations of these oxidants in clouds and fogs.

ACKNOWLEDGMENTS

We would like to thank the following people for their generous assistance in this study: J. Kadlecek, J. Lu, R. MacDonald, V. Mohnen, B. Murphy, S. Roychowdhury, P. J. Spink, and S. Virgilio (Whiteface Mountain, NY); J. Galloway, B. M. McIntyre, J. Sigmon, and P. Thompson (Shenandoah Park, VA); A. Basabe, M. Böhm, W.-L. Chang, R. Cryer, and T. Larson (Washington); V. Aneja and P. Blankenship (Mount Mitchell, NC); M. Böhm and P. Muir (Oregon); and R. Imhoff, J. F. Meagher, and L. Reisinger (Whitetop Mountain, VA). We would also like to thank G. E. Likens and K. C. Weathers for the use of their cloud water collector. This work was supported by the Atmospheric Chemistry Program of the U.S. National Science Foundation and the Andrew W. Mellon Foundation.

REFERENCES

1. Brimblecomb, P., *Air Composition and Chemistry*, Cambridge University Press, London, 1986, 85.
2. Graedel, T. E., Mandich, M. L., and Weschler, C. J., Kinetic model studies of atmospheric droplet chemistry homogenous transition metal chemistry in raindrops, *J. Geophys. Res.*, 91, 5205, 1986.
3. Faust, B. C. and Hoigné, J., Photolysis of Fe(III)-hydroxy complexes as sources of OH radicals in clouds, fog, and rain, *Atmos. Environ.*, 24A, 79, 1990.
4. Faust, B. C. and Allen, J. M., Aqueous-phase photochemical sources of oxidants in clouds, *Effects of Solar Ultraviolet Radiation on Biogeochemical Dynamics in Aquatic Environments*, Blough, N. V. and Zepp, R. G., Eds., Woods Hole Oceanographic Institution Technical Report WHOI-90-09, Woods Hole, MA, 1990, 48.
5. Allen, J. M., Ph.D. dissertation, Duke University, 1992.
6. Faust, B. C. and Allen, J. M., Aqueous phase photochemical sources of peroxyl radicals and singlet molecular oxygen in clouds and fog, *J. Geophys. Res.*, 97(D12), 12913, 1992.
7. Faust, B. C. and Hoigné, J., Sensitized photooxidation of phenols by fulvic acid and in natural waters, *Environ. Sci. Technol.*, 21, 957, 1987.
8. Haag, W. R., Hoigné, J., Gassman, E., and Braun, A. M., Singlet oxygen in surface waters-part one: furfuryl alcohol as a trapping agent, *Chemosphere*, 13, 631, 1984.
9. Rogers, M. A. J. and Snowden, P. T., Lifetime of $O_2(^1\Delta_g)$ in liquid water as determined by time-resolved infrared luminescence measurements, *J. Am. Chem. Soc.*, 104, 5541, 1982.

10. Anastasio, C., Allen, J. M., and Faust, B. C., Aqueous-phase photochemical production of peroxides in authentic cloud waters, in Aquatic and Surface Photochemistry, Helz, G., Zepp, R., and Crosby, D., Eds., Lewis Publishers, Boca Raton, FL, 1992.

11. Faust, B. C., Anastasio, C., Allen, J. M., and Arakaki, T. Aqueous-phase photochemical formation of peroxides in authentic cloud and fog waters, *Science,* 260, 73, 1993.

12. Gorman, A. A. and Rodgers, M. A. J., The quenching of aromatic ketone triplets by oxygen: competing singlet oxygen and biradical formation?, *J. Am. Chem. Soc.*, 108, 5074, 1986.

13. Kristiansen, M., Scurlock, R. D., Iu, K.-K., and Ogilby, P. R., Charge transfer state and singlet oxygen $O_2(^1\Delta_g)$ production in photoexcited organic molecule-molecular oxygen complexes, *J. Am. Chem. Soc.*, 95, 5190, 1991.

14. Capel, P. D., Gunde, R., Zurcher, F., and Giger, W., Carbon speciation and surface tension of fog, *Environ. Sci. Technol.*, 24, 722, 1990.

15. Gill, P. S., Graedel, T. E., and Weschler, C. J., Organic films on aerosol particles, fog droplets, cloud droplets, raindrops and snowflakes, *Rev. Geophys. Space Phys.*, 21, 903, 1983.

16. Turro, N. J., *Modern Molecular Photochemistry*, Benjamin/Cummings, Menlo Park, CA, 1978, 588.

CHAPTER 20

Examining the Role of Singlet Oxygen in Photosensitized Cytotoxicity

Thomas A. Dahl

BACKGROUND

Photosensitization is the conversion of light energy to chemical reactivity, often with the potential for causing biological damage. Molecular oxygen is a common quencher of organic photosensitizer excited states. Quenching can occur via charge transfer reactions, producing oxygen-centered and carbon-centered radicals along with nonradical reactants such as H_2O_2, and via energy transfer reactions, chiefly with sensitizers that exhibit significant triplet quantum yields and lifetimes. The transfer of energy between organic triplet states and O_2 produces one of the two low-lying singlet excited states for molecular oxygen, $^1\Delta_g$ and $^1\Sigma_g^+$, at near-diffusion-controlled rates. In the condensed phase, this process ultimately produces the $^1\Delta_g$ state. Thus, the term "singlet oxygen," commonly abbreviated as $^1O_2$, generally refers specifically to the $^1\Delta_g$ state. Because $^1O_2$ is one of the most common reactive intermediates formed by photosensitization with appropriate sensitizers under aerobic conditions, toxic effects of photosensitization are frequently assumed to be mediated by $^1O_2$. The reactivity of $^1O_2$ implies serious deleterious effects for living systems, since one or more $^1O_2$-reactive groups can be found in proteins, lipids, and nucleic acids. A high potential for biological damage coupled with ease of formation by a variety of photosensitizers in aerobic environments have led to the general assumption that $^1O_2$ produced by a photosensitizer is the toxic inter-mediate in cellular damage. Consequently, the *in vitro* production of $^1O_2$ is often used as a measure of the potential for sensitizers to cause biological damage.

A number of observations point to this description as an oversimplification of the biologically relevant mechanisms of photosensitization. The low concentration and limited lifetime of extracellular $^1O_2$, the protective effects afforded by quenchers

0-87371-871-2/94/$0.00+$.50

Table 1.   Reactivity of Cellular Components with Bulk-Phase $^1O_2$

Component	Max. Reaction Rate Constant (M^{-1} sec^{-1})[a]	% Cell Mass[b]	Apparent First-Order Rate Constant (sec^{-1})[c]
Protein	6×10^7	15	10^{-5}—10^{-7}
Lipid	1×10^5	2	10^{-9}—10^{-11}
DNA	4×10^4	1	10^{-11}—10^{-13}

[a] Maximum reaction rate constants are considered to be the rate constant for the most reactive moiety of the cellular component, as any differences in composition or accessibility of reactive components would necessarily drop the true value for an individual biomolecule below the maximum.

[b] For a "typical" bacterial cell. From *Molecular Biology of the Cell*, Alberts, B., Ed., Garland, New York, 1983, 92.

[c] Apparent first-order rate constants for the reaction of cellular components with extracellular $^1O_2$ represent the products of the maximum rate constant, the fractional cell mass of the component, and the extracellular $^1O_2$ concentration.

of non-$^1O_2$ reactants, and the different kinetics for lethal cytotoxicity observed during photosensitization vs pure $^1O_2$ exposure argue against $^1O_2$ as the sole cytotoxic species. Additionally, the complexity of photosensitized damage provided by rose bengal, often regarded as a model for $^1O_2$ effects, and the profound influence that the immediate environment can have on pathways for photosensitization imply alternate, and possibly multiple, cytotoxic mechanisms.

EXTRACELLULAR $^1O_2$

With a condensed-phase lifetime on the order of microseconds, the diffusion of $^1O_2$ is highly restricted from its source to potential targets, with a mean pathlength in aqueous environments of approximately 100 nm.[1] In considering the potential for $^1O_2$ involvement in toxic effects, therefore, a distinction must be drawn between $^1O_2$ generated near the site of potential reaction (i.e., intracellular) and that produced at a distance from the biological target (e.g., extracellular). Photosensitization by dissolved organic matter can be used as a model for extracellular generation in surface waters. Maximum (mid-day) surface concentrations of $^1O_2$ generated photochemically have been calculated for both marine and freshwater (10^{-12}–10^{-14} M) by dividing the maximum rates of formation by the rate of removal.[2] Table 1 provides a compilation of the reactivities of cellular components weighted by fractional cell mass as an overall approximation of reactivities with maximum extracellular $^1O_2$ concentrations.

This model presents maximum values and assumes that reactive components have unimpeded access to $^1O_2$ generated in the water. These values are almost certainly overestimates in that calcareous or siliceous coverings may be nonreactive barriers to $^1O_2$ penetration, as might cell walls composed of or overlaid with carbohydrates and carbohydrate derivatives, which are essentially unreactive toward $^1O_2$. Without consideration for these barriers, the maximum reactivity of the cell would lie with proteins, which fall in the range 0.03–3% per hour of sustained exposure at these calculated concentrations and would represent the maximum potential rate of inactivation of unicellular organisms assuming the simplest, "single-hit" kinetic model for cytotoxicity (discussed in more detail below). Exposure to these concentrations is not sustained, however, since the calculations consider only mid-day surface concentrations, which are neither representative throughout the day nor of decreased light penetration below the surface. In contrast, other secondary reactants produced photochemically in surface waters, including H_2O_2, have sufficient lifetimes and rates of production to accumulate in micromolar concentrations[3,4] and provide potential mechanisms for extracellular phototoxicity. Despite potential for biological damage, low external $^1O_2$ concentrations coupled with low reactivities and nonreactive barriers make the probability of damage from $^1O_2$ produced in the bulk phase far less than would be expected from $^1O_2$ produced intracellularly. Intracellular $^1O_2$, then, represents the more critical source in assessing $^1O_2$-mediated damage, although not necessarily in determining potential photosensitized cytotoxicity.

QUENCHING STUDIES

"Specific" quenchers are often used in biological investigations as a first approximation of the mechanism of photosensitized toxicity. In some cases, partial quenching by $^1O_2$-specific agents is taken as evidence of $^1O_2$ involvement, with the less-than-expected protection explained as limitation in the accessibility of $^1O_2$ to the quencher. Based on the same information, an equally valid interpretation would be that $^1O_2$ is only partially responsible for the results. The latter interpretation is supported by protection results with some non-$^1O_2$ quenchers, including superoxide dismutase (SOD), catalase, ascorbate, dimethylsulfoxide (DMSO), and mannitol.[5,6] In one study,[6] an SOD mimic and DMSO each showed greater protective effect than the $^1O_2$ quencher, β-carotene. This implies not only a mixture of reactants involved in photosensitization, but also that $^1O_2$ may not even be the predominant toxic species. This result is especially important in light of results with exposure to "pure" $^1O_2$, i.e., $^1O_2$ in the absence of alternate reactive species, where alteration of intracellular SOD and catalase levels had no effect on bacterial survival.[7]

COMPARATIVE KINETICS

Virtually all cell types examined, from prokaryotic to mammalian, have demonstrated a constant probability of cell death upon exposure to pure $^1O_2$,[8,9] rather than death from an accumulation of injuries, consistent with single-hit kinetics in target theory.* In contrast, photosensitzed cell death typically follows "multi-hit,"

---

*Although these kinetics denote that a single $^1O_2$ reaction with a cell is responsible for killing, calibration of the number of collisions between $^1O_2$ and a cell indicates that this lethal reaction is highly unlikely ($\sim 10^{-13}$). This means that a large number of $^1O_2$-cell collisions will take place — some of which may damage the cell nonlethally — on average, before the *critical* lethal reaction takes place. The exception, gram-negative bacteria, is described in more detail later.

possibly biphasic, kinetics in which the probability of cell death increases with continued light exposure up to a maximum, pseudo-first-order rate. Before examining these points in more detail, it would be helpful to describe some of the methodology used in comparing the effects of $^1O_2$ exposure specifically with the more complex set of processes in photosensitization.

METHODOLOGY

In contrast to photosensitizer solution studies, providing a physical separation between the photosensitizer and the target(s) of reaction precludes any direct interaction between either ground or excited states of the sensitizer and the target substrate(s) (Figure 1). Maintaining the sensitizer in dry form typically prohibits charge-transfer reactions with diffusible species such as O_2, although with sensitizers that self-quench efficiently it also may be necessary to reduce the sensitizer concentration to ensure rigorous inhibition of charge-transfer processes. Oxygen quenching of triplet excited-state photosensitizer through energy transfer is still permitted, however. Thus, $^1O_2$ is the only reactive species produced that is capable of diffusing across the intervening gas-phase separation to react with a target sample placed a short distance (usually less than 1 mm) away. These characteristics are combined in what has been called the separated-surface-sensitizer system[10] for the generation of "pure" $^1O_2$ ($^1O_2$ as the sole reactive species). In principle, this arrangement is extremely similar to Kautsky's original experiments proposing the intermediacy of $^1O_2$ in photosensitized oxidations.[11]

The separated system works well for kinetic studies in many organic solvents in which the extended lifetime and high diffusivity allow extensive penetration of the short-lived intermediate into solution from an external source. Biological samples, however, are restricted to aqueous media: environments less well suited to significant $^1O_2$ diffusion, even into shallow cell suspensions. This creates a reaction zone that is only a minute fraction of the total sample even in a cell suspension of 1 mm depth (discussed in the Appendix). Deposition of cells on membrane filters and removal of the bulk medium from the samples allows the diffusable $^1O_2$ to impinge directly on the target cell surfaces.[7,8,12,13] This arrangement was originally designed for walled cells such as bacteria, which are less sensitive to handling in this manner, and has been refined more recently to permit the inclusion of mammalian cells as targets. These modifications, which are designed primarily to prevent sample desiccation during illumination in the absence of extracellular medium,* affect neither the magnitude nor purity of $^1O_2$ generation in this apparatus,[9,14] although the gas-phase lifetime of $^1O_2$ is certainly reduced in the more humid environment.[15] In addition to improved exposure, other advantages provided by this arrangement are the following:

---

*Beyond specific application to singlet oxygen work, this arrangement is sufficiently versatile to provide for cultured cell exposure to any gaseous or gas-borne toxicant or potential toxic agent for direct *in vitro* assessment in gas-phase toxicology studies.

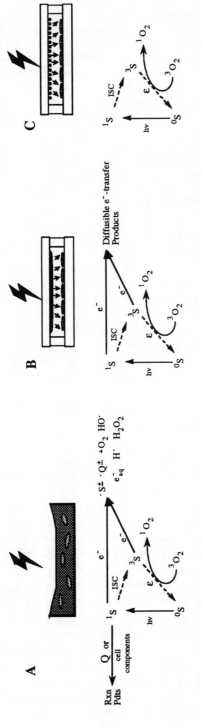

Figure 1. Diagrams of photosensitization experimental techniques with schematics of corresponding photosensitization pathways. **(A)** Illumination of targets in the presence of photosensitizer; the sensitizer (S) may react directly with cellular components, may react indirectly through initial interaction with quenchers (Q), or produce other secondary reactants through electron- or energy-transfer pathways. **(B)** Illumination of solubilized photosensitizer physically separated from targets; direct reaction of sensitizer with cellular targets or potential quenchers is prevented while still allowing the production of diffusible electron- and energy-transfer products (i.e., reactive oxygen species). **(C)** Illumination of dry photosensitizer physically separated from cellular targets; singlet oxygen produced through energy-transfer is the only reactant available for reaction with cellular targets.

1. With $^1O_2$ impinging directly on cells, there is no possibility for $^1O_2$ conversion or reaction to secondary cytotoxic species in extracellular media prior to inter-action with the cells. This may not be a problem initially with simple buffers as suspension media, although it would be a consideration with more complex media, particularly those containing serum, common in cell culture work. Addi-tionally, $^1O_2$ causes cells to "leak."[14,16] As $^1O_2$-initiated damage releases intrac-ellular components, including potential reactants and reducing agents, into the suspension medium, the certainty that $^1O_2$ itself mediates further cellular damage and death diminishes.

2. Direct $^1O_2$ exposure of samples offers additional advantages for quantitation of results. Deposition of cells on filters in a monolayer or less ensures that all cells receive identical exposure; there is no gradient of $^1O_2$ concentration as the metastable reactant decays while diffusing through an extracellular medium (discussed more fully in the Appendix).

In addition to the dry separated-surface-sensitizer system, we have recently worked with separated systems employing solubilized sensitizer in place of the previous dry dye-bead preparations.[17] This system still maintains the physical separation of sensitizer and substrate, yet provides solvent (generally water or methanol) to mediate electron transfer to O_2 in the gas phase, thereby providing mixtures of reduced O_2 species ($\cdot O_2^-$ and H_2O_2) and $^1O_2$ (Figure 1). Coupling solution studies with the two separated-sensitizer techniques allow sample expo-sure to (1A) all possible photosensitization pathways, (1B) mixtures of volatile reactive oxygen species only, and (1C) $^1O_2$ exclusively. By comparing results of the various techniques, we are now able to dissect complex photosensitization pathways into component parts.

RESULTS

Figure 2A depicts representative curves describing cell toxicity photosensi-tized by hematoporphyrin derivative (HpD) in two different cell lines, each exposed at two different rates.[18,19] Studies employing HpD were chosen for comparison to $^1O_2$ specifically because the model proposed for HpD photody-namic action involves the localization of sensitizer in cell membranes with sub-sequent generation of $^1O_2$ in or near the membranes, followed by $^1O_2$ reaction preferentially with membrane components leading directly to cytotoxic damage. Similarly, $^1O_2$ generated extracellulary would reach the cell membrane as the first reactive target. Thus, the proposed mechanism predicts that from this point onward the reaction and consequent cell damage would be identical in the two systems. Singlet oxygen cell killing is shown in Figure 2B.[20]

In the simpler, pure $^1O_2$ case, the overall rate of lethality (change in survival = $-dS/dt$) will be given by the product of the toxic reactant exposure rate ($E = {}^1O_2$ flux) and the proportion of $^1O_2$ molecules leading to cell death (Φ_L).

$$-dS/dt = \Phi_L E \qquad (1)$$

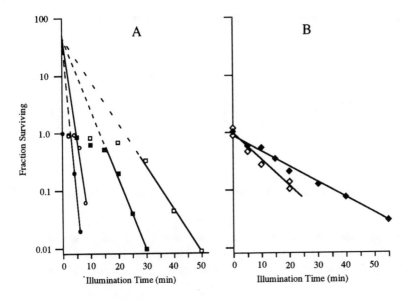

Figure 2. Photosensitized lethal toxicity of cultured mammalian cells. (A) Illumination in the presence of photosensitizer. Circles represent data with CHO cells:[19] ● is HpD 100 μg/mL and ○ is HpD 50 μg/mL. Squares show data with TA-3 cells:[18] ■ is HpD 0.6 μM and □ is HpD 0.6 μM + quencher. (B) Exposure to singlet oxygen only.[20] ◆ is FaDu cells, 0.8 pmol/cell min; and ◇ is RBL cells, 1.1 pmol/cell min.

The proportionality constant Φ_L corresponds to the slope of the survival curve and can be further considered as the product of the proportion of the reactant reacting with cell components, Φ_r, and the proportion of reactions that result in cell death, Φ_T. A logarithmic plot of surviving fraction vs duration of exposure (e.g., illumination time) yields a curve with a y-intercept value of 1.0 and a slope proportional to $\Phi_r\Phi_T = \Phi_L$ (Figure 2B). Varying the $^1O_2$ flux has shown that increases or decreases in E are directly tied to corresponding decreases or increases, respectively, in the duration of exposure required to reach a given survival level, i.e., $\Phi_L E$ is constant (Table 2). This is also reflected in the constant probability of cell death for each $^1O_2$ molecule.*

In the curves describing photosensitized killing, the intial phase depicts accumulation of sublethal damage, while the maximum rate describes the kinetics of the final event leading to cell death. A description of the more complex curves must consider the rates of formation of different reactants, which are also influenced by location within the cells (i.e., immediate environment dependence). The analogous description of E, above, includes the rate of formation of each reactant based on the light fluence (F), the quantum yield for formation of the reactant (Φ), and the number of photons absorbed per time (A).

---

*When cell size is considered, i.e., E is expressed as number of molecules *per cell* per time, the probability of cell death is remarkably constant across various cell types. Cultured mammalian cells require $3–15 \times 10^{12}$ $^1O_2$ molecules per cell killed, and various species of gram-positive bacteria fall in the range $10^{10}–10^{11}$ $^1O_2$ molecules per cell killed.

Table 2.   Apparent First-Order Inactivation of Cultured Mammalian Cells with Pure $^1O_2$.

Cell Type	Assay	Singlet Oxygen Flux		$^1\Delta_gO_2$ Molecules per Cell Killed
		(μmol/cm^2 min)	(pmol/cell min)	
Primary hepatocytes	Leakage	0.39	1.3	3.3×10^{12}
Squamous carcinoma	Colony	0.26	0.2	8.1×10^{12}
	Leakage	0.98	0.8	8.4×10^{12}
	Colony	0.98	0.8	7.9×10^{12}

$$\text{Rate of exposure} = \text{Rate of formation} = \Phi FA \qquad (2)$$

The initial phase of the photosentization curves would be described by the sum of the individual contribution of each reactant to cell death, each with its own proportionality constant reflecting both reactivity and location within the cell (which will influence both the probabilities of reaction and contribution of reaction to death)

$$-dS/dt = \Phi_1 FA\Phi_{r_1}\Phi_{T_1} + \Phi_2 FA\Phi_{r_2}\Phi_{T_2} + \ldots + \Phi_n FA\Phi_{r_n}\Phi_{T_n} \qquad (3)$$

where the numerical subscripts denote different secondary reactants produced by primary photoprocesses, or the same reactant produced in different environments (which may change any or all of the rates of production, reactivities, and consequences of reaction). The curves in Figure 2A show that photosensitized damage accumulates to an apparently first-order maximum rate, reducing the equation for this part of the curves to

$$-dS/dt = \Phi_x FA\Phi_{r_x}\Phi_{T_x} \qquad (4)$$

where x represents the reactant involved in the final stage of inactivation. As with exposure to $^1O_2$, $\Phi FA\Phi_r\Phi_T$ will be constant for this portion of the curve, and extrapolation back to the y-intercept yields an "apparent" initial survival greater than 1. Although this represents something of an artificial construct, in that initial survival cannot literally be greater than 1, this parameter contains information in that the value reflects the number of events leading up to cell death. That is, the larger the intercept the greater the accumulation of sublethal damage contributing to cell death. For a given sensitizer in related cell types, where the mechanism may be reasonably expected to be the same, intercepts should be identical. This is demonstrated in Figure 2A for HpD-sensitized cell toxicity in two different cell lines, each exposed at two different rates (all intercepts $\simeq 40$). Initial "hits" may involve a repartitioning or relocation of toxicants, or the production of additional reactants by reaction with cell components. Either or both of these processes would contribute to the curve shapes shown in Figure 2A, and, in fact, both are believed to be involved in killing gram-negative bacteria by both $^1O_2$ and rose bengal photosensitization, for example, in which initial reactions with the cell's outer membrane provides greater penetration of subsequent reactants, in addition to providing additional sources of toxic reactants.[8,21,22] Markedly different intercepts denote different numbers of events contributing to cell death, and thereby imply different mechanisms of killing. For comparison, pure $^1O_2$ curves yield y-intercepts equal to 1 for a single lethal event (Figure 2B).

Equation 3 cannot be solved without detailed information regarding the identities, reactivities, and immediate cellular environments of all the reactants formed. While not generally known in detail, this information is often approximated by quencher studies, and deuterated solvent studies intended to selectively enhance $^1O_2$ lifetime. These techniques are limited by the imperfect specificity of quenching agents and deuterated solvents[23] and uncertainty in the cellular locations of sensitizers and secondary reactants. However, some information may be gained by focusing on the second, apparently first-order portion of these curves. Analyses of the efficiency of photosensitized damage relative to photons absorbed can provide evidence for or against $^1O_2$ involvement. This is analogous to the methods employed by Allen and co-authors[24] for the distinction of at least two different, wavelength-dependent mechanisms of photosensitization with rose bengal. Care must be taken to note that evidence for $^1O_2$ involvement in this phase of toxicity does not completely describe events, since this represents only the *final* event in lethal toxicity. Similarly, evidence for non-$^1O_2$ processes in this phase do not rule out $^1O_2$ participation in earlier events. In either case, however, the complexity of the curve and the unknown nature of the fundamental events involved dictate challenging the basic assumption of $^1O_2$ as the obligate toxic intermediate for different photosensitizers, and for a given photosensitizer in different cellular environments.

ROSE BENGAL

Rose bengal is often considered the quintessential $^1O_2$ producer in photosensitzing systems, with a quantum yield for $^1O_2$ production of 0.75. While the production of other reactants has been described for illuminated rose bengal,[25] including charge-transfer to O_2, rose bengal photosensitization is frequently cited as a standard of comparison in the demonstration of $^1O_2$-mediated effects. One caveat to use of this approach in complex biological systems is the effect the immediate environment of the sensitizer has on the relative production of excited triplets and fluorescence from excited singlets, and consequent effects on $^1O_2$ yields and non-$^1O_2$ photoprocesses.[26] Even the simple act of solvation of rose bengal complicates identification of reactants by permitting charge-transfer processes, as mentioned above. The impact of this on living cells can be illustrated in the following example. The mechanisms of bacterial killing by rose bengal photosensitization or pure $^1O_2$ exposures cannot be distinguished based on kinetics alone.[8,21] Using a separated-surface-sensitizer apparatus for the production of

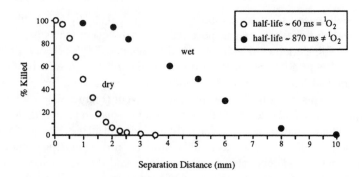

Figure 3. Distance-dependence/lifetime analyses of toxic species produced in the gas phase by illuminated rose bengal physically separated from bacterial targets.

volatile toxicants from illuminated, immobilized photosensitizer, a distance-dependence analysis provides a lifetime measurement for the reactive inter-mediate(s) formed at a distance from the target.[10,12] The fraction of reactant reaching depth x from the surface of a solution or cell suspension depends on quenching (k_q) and diffusion (D) constants according to the expression

$$N_x/N_0 = e^{[-k_q x^2/2D]} \tag{5}$$

where N_0 = the amount at the surface of the sample, and N_x = the amount reaching depth x.

A comparison of measurements made with dry photosensitizer, in which only $^1O_2$ is formed, with solubilized rose bengal, in which non-$^1O_2$ products are possible (Figure 3), revealed a gas-phase half-life of approximately 900 ms for one lethal toxicant with solubilized rose bengal,[17] or approximately ten times greater than the gas-phase lifetime of $^1O_2$ in air at 1 atm. Therefore, with the solubilized sensitizer, $^1O_2$ could not possibly be solely responsible for cytotoxicity. Since $^1O_2$ diffuses almost 3 mm from the sensitizer plate, lethal effects at the shortest separation distances with solubilized sensitizer will reflect this mixture. Equation 5 would thus become

$$\ln \sum \left(N_{x_i}/N_{0_i}\right) = -x^2 \sum \left(k_{q_i}/2D_i\right) \tag{6}$$

in this range of x (0–3 mm).

A more accurate lifetime determination may be made for the non-$^1O_2$ reactant by including in the calculation only those values outside this range (>3 mm in distance), i.e., in which $^1O_2$ cannot be involved. (The reasons for believing there is only one are discussed under "Singlet Oxygen vs Other O_2 Species".) This calculation provides a lifetime of 939 ms. With dry photosensitizer, the reactive intermediate half-life for lethal toxicity to the bacteria was determined to be about 60 ms, consistent with $^1O_2$ involvement. At separation distances ≥3 mm, where $^1O_2$ effects have dropped to near zero, killing must be due to photo-induced electron transfer, which could give rise to $\cdot O_2^-$, and probably diffusable H_2O_2.[17] At distances ≤ 2 mm, toxicity may be due to a combination of $^1O_2$ and the product of electron transfer; at distances greater than 3 mm, there is no possibility that $^1O_2$ is contributing to lethal photosensitized cytotoxicity. This provides unambiguous evidence that the cellular effects of rose bengal photo-sensitization are not restricted to $^1O_2$ intermediacy. This is further supported by recent work showing that biochemical effects of rose bengal photosensitization display different mechanisms depending on the wavelength of illumination, while the $^1O_2$ quantum yield is wavelength-independent.[24]

EFFECTS OF SENSITIZER ENVIRONMENT

As previously described for rose bengal, mechanisms of photosensitized bio-chemical and cellular damage may be profoundly influenced by the environment of the sensitizer. This is especially important when considering that the reducing nature of a cell's interior would tend to promote charge-transfer processes rather than energy-transfer processes. One striking example of an environment-dependent mechanism is demonstrated by the fungal photosensitizer cercosporin, which produces abundant $^1O_2$ extracellularly. The presence of cellular reducing agents, however, virtually shuts down $^1O_2$ production, while simultaneously enhancing reduction of O_2.[27] This compound possesses the potential for multiple predomi-nant pathways of phototoxicity, with the "switch" in pathways a function only of sensitizer environment. Another example of a sensitizer displaying environment-specific reaction pathways is the plant product, curcumin, which has the highly respectable $^1O_2$ quantum yield of 0.5 in aprotic solvents. This yield drops an order of magnitude or more in protic environments, while simultaneously reducing O_2 sequentially to $\cdot O_2^-$ and H_2O_2.[28,29] The mechanism(s) of *in vivo* photosensitization would be difficult to predict based on these *in vitro* measurements, since cell studies have shown that a small fraction of photosensitizer binds to cell mem-branes, while most of it remains in the aqueous medium.[29,30] Although this sets the stage for either or both pathways to be involved in cytotoxic effects, reactant lifetime measurements ($t_{1/2} = 27$ sec) indicate clearly that $^1O_2$ is not involved.[30] In this example, the *in vitro* measurements of $^1O_2$ production are not instructive in elucidating the mechanism of photosensitized cytotoxicity.

EXAMPLIS GRATIS

The biological significance of the considerations discussed above can be especially clearly seen in the context of $^1O_2$-induced genotoxicity, an area that has

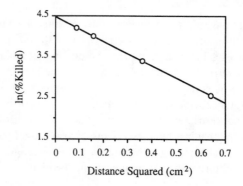

Figure 4.   Plot of the solubilized sensitizer data from Figure 3 in the range 3 mm < x < 10 mm, according to Eq. 5.

developed into significant controversy over the past decade.[13] Photosensitization is genotoxic. Singlet oxygen has been shown to react with DNA *in vitro*. Yet *in vivo* studies with pure $^1O_2$ have not provided evidence for either mutagenicity or other genotoxic effects, while still demonstrating cell inactivation. Coupled with kinetic analysis, comparisons of photosensitization and separated systems using either dry or solubilized sensitizer have allowed the resolution of a complicated set of events into individual components or groupings based on mechanistic considerations. We were able to distinguish singlet oxygen-mediated events from reactions of other reactive oxygen species, and even to resolve singlet oxygen reactions into direct and indirect components. The following discussion summarizes an extensive body of information that has been recently submitted for publication by Camoirano et al.[31]

Singlet Oxygen vs Other O₂ Species

Mixtures of volatile reactive oxygen species produced by illumination of the solubilized sensitizer proved genotoxic by preferential killing of various DNA repair-deficient *Escherichia coli* strains relative to the repair-proficient wildtype. This genotoxic effect was efficiently blocked by non-$^1O_2$ quenchers. In contrast, with pure $^1O_2$ generated by dry photosensitizer, the initial rate of killing of all repair-deficient strains was indistiguishable from the wildtype. These results together indicate that $^1O_2$ itself was not genotoxic, although at least one diffusable reactant generated during photosensitization caused genotoxicity. This is further supported by the distance-dependence and gas-phase lifetime measurements for the respective toxic species (presented under the previous section Rose Bengal). Furthermore, values for killing at separation distances x > 3 mm are proportional to the square of the distance, with a correlation coefficient $r^2 = 1.00$ (Figure 4). This is exactly as predicted by Equation 5 for diffusion of the short-lived intermediate, strongly suggesting that (1) only one toxic species is involved in bacterial killing at these separation distances and (2) killing is proportional to the amount of toxicant over an order of magnitude reduction in survival, which is not true of $^1O_2$ lethal toxicity for gram-negative bacteria.[8,12] Photosensitization produces at least one diffusable non-$^1O_2$ reactive species, therefore, which both contributes to cell death and can cause genotoxic damage.

Direct vs Indirect Reactions

In addition to distinguishing $^1O_2$ from non-$^1O_2$-mediated processes, this approach has allowed the distinction of direct from indirect $^1O_2$-mediated toxicity. Singlet oxygen itself failed to exert direct genotoxic effects, although secondary reactants produced by its reaction with cell components enhanced lethality in some repair-deficient bacteria. In gram-negative bacterial survival curves, virtually all (>90%) of the killing in the late stages of $^1O_2$ exposure is due to formation of secondary reactants. This is also the range where repair-deficient strains evince greater sensitivity than the wildtype (Figure 5), indicative of genotoxic damage. In contrast, all strains were equally sensitive to the pure $^1O_2$ phase of inactivation (Figure 5). These strains should not differ from the wildtype in the rates of *formation* of any of the reactants, since in surface area, penetrability, and composition they are identical; $\Phi_i FA$ (from Eq. 2) will be the same for each strain. The differences in sensitivity between the repair-deficient strains and the wildtype, therefore, are due to genotoxic effects of secondary reactants (differences in Φ_{L_i} depending on DNA repair capability). It is also important to recognize that the secondary reactants formed intracellularly are not simply reduction products of $^1O_2$ (e.g., $\cdot O_2^-$, H_2O_2), as the order of sensitivity is reversed among the repair-deficient strains with dry sensitizer relative to the solubilized sensitizer experiment, in which molecular oxygen was reduced in the gas-phase. Secondary reactants could include endoperoxides, hydroperoxides, dioxetanes, and thioperoxides, which can be formed by $^1O_2$ reaction with various biomolecules (reviewed in Reference 32). While $^1O_2$ has been reported to react directly with naked DNA in solution, producing mutagenic damage (e.g., Reference 33), the current description is mechanistically distinct in that genotoxicity is an indirect result of $^1O_2$ interaction with the bacteria rather than a direct reaction between $^1O_2$ and DNA.

CONCLUSIONS

Intracellular photosensitization pathways are complex and often mixed. Based on the variety of potential toxicants produced concomitantly with $^1O_2$, or alternatively in different environments, and the variety of potential cellular targets involved in toxic effects, the presumption of $^1O_2$ involvement in O_2-dependent photosensitization is insupportable, even with sensitizers that are known to produce abundant $^1O_2$ under *in vitro* and some *in vivo* conditions. Mechanistic discussions of photosensitization require rigorous testing of this assumption with the individual sensitizers in the systems under investigation and consideration of

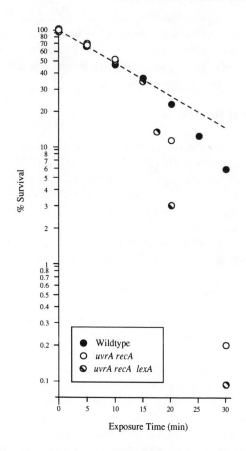

Figure 5. Comparison of singlet oxygen-initated killing of wildtype and repair-deficient *E. coli* strains. All strains behave identically in the initial, apparently first-order phase where singlet oxygen alone is responsible for killing (dashed line). In the late portion, where the data deviate from proportionality, killing is believed to be mediated by secondary reactants produced through singlet oxygen reaction with cell components.[8,17]

alternate — and possibly multiple — reaction pathways. Comparisons among the results of sample exposure to complex photosensitization pathways, mixtures of volatile reactive oxygen species only, or $^1O_2$ exclusively, have allowed the beginnings of dissection of complex photosensitization pathways into component parts.

ACKNOWLEDGMENTS

Financial support was provided in part by a grant from the Earl P. Charlton Fund. Neera K. Dahl, Robert W. Redmond, and Richard M. Kream provided helpful discussion and insight.

APPENDIX

Penetration of gas-phase $^1O_2$ into an aqueous suspension or solution is shown in a series of scaled drawings in Figure 6. Both plots show the curve of Eq. 5 for a situation in which all the $^1O_2$ quenching in the system is assumed to be due exclusively to H_2O ($k_q = 2.5 \times 10^5$ sec^{-1};[1] $D = 2 \times 10^{-5}$ cm^2 sec^{-1})[34]. The curve in Figure 6A is plotted on a nanometer scale, relevant to subcellular distances. This curve suggests that a significant fraction of $^1O_2$ reaching a cell from external sources penetrates to intracellular targets, which is also supported by the work of Nye and co-authors.[35] The plot in Figure 6B is drawn on the order of 1 mm, the scale of the shallowest cell suspensions adapted to gas-phase $^1O_2$ generation.[16,36] From this plot is is clear that $^1O_2$ penetration into the suspension medium is highly restricted, relative to its depth, so that the bulk of the sample remains unexposed.

Taking the mean $^1O_2$ pathlength ($\bar{x}_\Delta = 100$ nm) to represent the mean reaction zone, the relative exposure (R) can be approximated by the penetration of $^1O_2$ into a sample of total depth x_t :

$$R = \bar{x}_\Delta/x_t = 10^{-4} \tag{7}$$

This means that 99.99% of cells in this shallow suspension are not exposed to $^1O_2$. Considering the *total* reaction zone, extending down nearly 300 nm (Figure 6A), rather than the mean reaction zone, the estimate of fraction sample exposed increases to 0.03%.*

Substitution of water with D_2O would enhance the $^1O_2$ lifetime and increase the depth reaction zone. However, complete replacement with D_2O, again assuming all quenching in the system is due to solvent ($k_q = 1.9 \times 10^4$ sec^{-1}),[1] would increase the mean reaction zone to 380 nm and the total reaction zone to about 1000 nm (0.1% sample exposure).

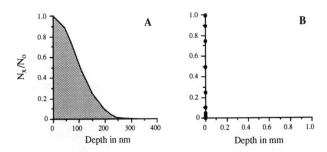

Figure 6. Theoretical plot of $^1O_2$ penetration into aqueous media, assuming all quenching is due to H_2O and that the diffusivity of O_2 into the medium is the same as in pure H_2O. Both plots show the same points expressed on a nanometer **(A)** or millimeter **(B)** x-axis scale.

---

*Stirring the sample, assuming all cells are thereby equally exposed, converts these estimates to the fraction of total exposure each cell receives.

As an alternate estimation of relative exposure, the "dose" of $^1O_2$ the total sample receives will be given by summation of $^1O_2$ in the system by analogy with other cytotoxicants.[37] Unlike a true dose determination, however, the $[^1O_2]$ is steady-state and therefore constant in the sample. Rather than integrated concentration over time, this dose becomes the $^1O_2$ input (E = $^1O_2$ flux times the surface area of the sample, S_r) integrated over the sample depth that contains it. The calculation of penetration, above, indicates that this depth is essentially infinite compared to the $^1O_2$ pathlength, so

$$\text{"dose"} = E \cdot S_r \int_0^\infty e^{-k_q x^2/2D} \, dx \tag{8}$$

Relative exposure, then, compares this quantity to the flux reaching the total sample volume, V_t.

$$R = \frac{E \cdot S_r \int_0^\infty e^{-k_q x^2/2D} \, dx}{E \cdot V_t} \tag{9}$$

Since flux to the sample is defined as that reaching the surface[12] and the exposed surface area of the reaction zone is the same as that of the entire sample, $V_t = S_r x_t$, reducing this equation to

$$R = \frac{\int_0^\infty e^{-k_q x^2/2D} \, dx}{x_t} \tag{10}$$

The integral is solvable in the form

$$\int_0^\infty e^{-a^2 x^2} \, dx = \frac{\sqrt{\pi}}{2a} \tag{11}$$

by setting $a = (k_q/2D)^{1/2}$.

Again assuming water in the system is the sole influence on $^1O_2$ diffusion and quenching, $\sqrt{\pi}/2a = 112$ nm, and $R = 112$ nm/10^6 nm $\simeq 10^{-4}$. This value is in good agreement with the value obtained by the mean pathlength method (Equation 7).

These calculations consider only the *fraction* of samples receiving *any level* of exposure within the reaction zone. The magnitude of exposure, however, varies among the cells, depending on their location within the reaction zone. From the area under the curve in Figure 6, which represents the total amount of $^1O_2$ in the sample medium, it can be seen that the extent of exposure in the interval 200–300 nm depth is approximately $1/12$, and $1/2$ in the interval 100–200 nm, that of the top 100 nm of the sample. These calculations do not consider that diffusion for cells in suspension tends to be unidirectional, *away* from the reaction zone (i.e., settling). This indicates that the duration of illumination of such a system is not an adequate determination of the duration of exposure to $^1O_2$, as required for quantitative studies. Quantitative results require a system in which all cells in the sample can be exposed to the toxic agent, all at the same level of exposure, and the level quantified. While these conditions could be met with a homogenous sensitizer solution-cell suspension system, failure to maintain a physical separation between the dry sensitizer and target abolishes the specificity of the generation scheme for $^1O_2$ production. These conditions are met, however, by the exposure of cells deposited in a single-depth planar array to direct contact with the toxicant in the heterogenous $^1O_2$-generation system.

REFERENCES

1. Lindig, B. A. and Rodgers, M. A. J., Laser photolysis studies of singlet molecular oxygen in aqueous micellar dispersions, *J. Phys. Chem.*, 83, 1683, 1979.
2. Haag, W. R. and Mill, T., Survey of sunlight-produced transient reactants in surface waters, in *Effects of Solar Ultraviolet Radiation on Biogeochemical Dynamics in Aquatic Environments*, Blough, N. V. and Zepp, R. G., Eds., Woods Hole Oceanographic Institute Technical Report, WHOI-90-09, Woods Hole, MA, 1990, 82.
3. Zika, R. G., Hydrogen peroxide as a relative indicator of the photochemical reactivity of natural waters, in *Effects of Solar Ultraviolet Radiation on Biogeochemical Dynamics in Aquatic Environments*, Blough, N. V. and Zepp, R. G., Eds., Woods Hole Oceanographic Institute Technical Report, WHOI-90-09, Woods Hole, MA, 1990, 61.
4. Cooper, W. J., Saltzman, E. S., and Zika, R. G., The contribution of rainwater to variability in surface ocean hydrogen peroxide, *J. Geophys. Res.*, 92, 2970, 1987.
5. Gulliya, K. S., Matthews, J. L., Fay, J. W., and Dowben, R. M., Effect of free radical quenchers on dye-mediated laser light induced photosensitization of leukemic cells, in *New Directions in Photodynamic Therapy*, Neckers, D. C., Ed., SPIE, 847, Bellingham, WA, 1987, 163.
6. Athar, M., Elmets, C. A., Zaim, M. T., Lloyd, J. R., Bickers, D. R., and Mukhtar, H., Involvement of superoxide anions in cutaneous porphyrin photosensitization: an *in vivo* study, in *New Directions in Photodynamic Therapy*, Neckers, D. C., Ed., SPIE, 847, Bellingham, WA, 1987, 193.
7. Midden, W. R., Dahl, T. A., and Hartman, P. E., Cytotoxicity but no mutagenicity in bacteria with externally generated singlet oxygen, in *New Directions in Photodynamic Therapy*, Neckers, D. C., Ed., SPIE, 847, Bellingham, WA, 1987, 122.
8. Dahl, T. A., Midden, W. R., and Hartman, P. E., Comparison of killing of gram-negative and gram-positive bacteria by pure singlet oxygen, *J. Bacteriol.*, 171, 2188, 1989.
9. Dahl, T. A., Direct exposure of mammalian cells to pure gas-phase singlet oxygen ($^1\Delta_g O_2$), *Photochem. Photobiol.*, 53S, 119S, 1991.
10. Midden, W. R. and Wang, S. Y., Singlet oxygen generation for solution kinetics: clean and simple, *J. Am. Chem. Soc.*, 105, 4129, 1983.
11. Kautsky, H. and deBruijn, H., Die Aufklärung der Photoluminescenzfilgung fluorescierender Systeme durch Sauerstoff: Die Bildung activer diffusionsfähiger Sauerstoffmoleküle durch Sensibilisierung, *Naturwissenschaften*, 19, 1043, 1931.
12. Dahl, T. A., Midden, W. R., and Hartman, P. E., Pure singlet oxygen cytotoxicity for bacteria, *Photochem. Photobiol.*, 46, 345, 1987.
13. Dahl, T. A., Midden, W. R., and Hartman, P. E., Pure exogenous singlet oxygen: nonmutagenicity in bacteria, *Mutat. Res.* 201, 127, 1988.
14. Dahl, T. A., Midden, W. R., Klaunig, J. E., and Ruch, R., Pure singlet oxygen toxicity in mammalian cells, in *Effects of Solar Ultraviolet Radiation on Biogeochemical Dynamics in Aquatic Environments*, Blough, N. V. and Zepp, R. G., Eds., Woods Hole Oceanographic Institute Technical Report, WHOI-90-09, Woods Hole, MA, 1990, 162.

15. Midden, W. R. and Dahl, T. A., Biological inactivation by singlet oxygen: distinguishing $^1\Delta_g O_2$ and $^1\Sigma_g^+ O_2$, *Biochim. Biphys. Acta*, 1117, 216, 1992.

16. Wang, T. P., Kagan, J., Lee, S., and Keiderling, T., The hemolysis of erythrocytes by singlet oxygen generated in the gas phase, *Photochem. Photobiol.*, 52, 753, 1990.

17. Camoirano, A., De Flora, S., and Dahl, T. A., unpublished data, 1992.

18. Weishaupt, K.R., Gomer, C. J., and Dougherty, T. J., Identification of singlet oxygen as the cytotoxic agent in photo-inactivation of a murine tumor, *Cancer Res.*, 36, 2326, 1976.

19. Musselman, B. and Chang, C. K., Characterization of oligomeric hematoporphyrin tumoricidal reagents for photodynamic therapy, in *New Directions in Photodynamic Therapy*, Neckers, D. C., Ed., SPIE, 847, Bellingham, WA, 1987, 96.

20. Dahl, T. A., unpublished data, 1992.

21. Dahl, T. A., Midden, W. R., and Neckers, D. C., Comparison of photodynamic action by rose bengal in gram-positive and gram-negative bacteria, *Photochem. Photobiol.*, 48, 607, 1988.

22. Dahl, T. A., Valdes-Aguilera, O., Midden, W. R., and Neckers, D. C., Partition of rose bengal anion from aqueous medium into a lipophilic environment in the cell envelope of *Salmonella typhimurium*: implications for cell-type targetting in photodynamic therapy, *J. Photochem. Photobiol. B: Biol.*, 4, 171, 1989.

23. Valezeno, D. P., Photomodification of biological membranes with emphasis on singlet oxygen mechanisms, *Photochem. Photobiol.*, 46, 147, 1987.

24. Allen, M. T., Lynch, M., Lagos, A., Redmond, R. W., and Kochevar, I. E., A wavelength dependent mechanism for rose bengal-sensitized photoinhibition of red cell acetylcholinesterase, *Biochim. Biphys. Acta* 1075, 42, 1991.

25. Srinivasan, V. S., Podolski, D., Westrick, N. J., and Neckers, D. C., Photochemical generation of O_2^- by rose bengal and Ru(bpy)$_3^{2+}$, *J. Am. Chem. Soc.*, 100, 6513, 1978.

26. Bilski, P., Dabestani, R., and Chignell, C. F., Influence of cationic surfactant on the photoprocesses of eosine and rose bengal in aqueous solution, *J. Phys. Chem.*, 95, 5784, 1991.

27. Hartman, P. E., Dixon, W. J., Dahl, T. A., and Daub, M. E., Multiple modes of photodynamic action by cercosporin, *Photochem. Photobiol.*, 47, 699, 1988.

28. Chignell, C. F., Reszka, K., Bilski, P., Motten, A., Sik, R., and Dahl, T. A., Singlet oxygen generation and radical reactions photosensitized by curcumin, *Photochem. Photobiol.*, 51S, 6S, 1990.

29. Dahl, T. A., McGowan, W., Shand, M. A., and Srinivasan, V. S., Photokilling of bacteria by the natural dye curcumin, *Arch. Microbiol.*, 151, 183, 1989.

30. Bilski, P., Chignell, C. F., and Dahl, T. A., unpublished data, 1991.

31. Camoirano, A., De Flora, S., and Dahl, T. A., Genotoxicity of volatile and secondary reactive oxygen species generated by photosensitization, *Environ. Mol. Mutagen.*, 21, 219, 1993.

32. Dahl, T. A., Pharmacological implications of photosensitization and singlet oxygen toxicity, *Int. J. Immunopathol. Pharm.*, 5, 57, 1992.

33. Ribeiro, D. T., Madzak, C., Sarasin, A., DiMascio, P., Sies, H., and Menck, C. F. M., Singlet oxygen induced DNA damage and mutagenicity in a single-stranded SV40-based shuttle vector, *Photochem. Photobiol.*, 55, 39, 1992.

34. *CRC Handbook of Chemistry and Physics*, 71st ed., CRC Press, Boca Raton, FL, 1990, 6–151.

35. Nye, A. C., Rosen, G. M., Gabrielson, E. W., Keana, J. F. W., and Prabhu, V. S., Diffusion of singlet oxygen into human bronchial epithelial cells, *Biochim. Biophys. Acta*, 928, 1, 1987.

36. Decuyper-Debergh, D., Piette, J., Laurent, C., and van de Vorst, A., Cytotoxic and genotoxic effects of extracellular generated singlet oxygen in human lymphocytes in vitro, *Mutat. Res.* 225, 11, 1989.

37. Walum, E., Stenberg, K., and Jenssen, D., Toxicokinetics in cell culture: principles and practice, in *Understanding Cell Toxicology*, Ellis Horwood, Ltd., London, 1990, chap. 4, 64.

PART II

Photochemistry in Water Treatment

CHAPTER 21

Photocatalytic Treatment of Waters

D. Bahnemann, J. Cunningham, M. A. Fox, E. Pelizzetti,* P. Pichat, and
N. Serpone

INTRODUCTION

The increasingly clear need for new and effective methods for cleaning polluted air and water streams has recently resulted in a renewed interest in developing environmentally benign methods for detoxification, possibly by complete
mineralization of a wide range of organic compounds. The most currently used
methods for cleaning liquid or gaseous streams involve stoichiometric chemical
treatment with oxidizing reagents, usually either chlorine or ozone to oxidatively
degrade the organic contaminant. Although the chemical treatments employed in
these methods are well known, they are limited in effectiveness in attaining only
incomplete purification and by the need for large quantities of the oxidizing
reagent consumed in the operation. The possibility that improved detoxification
might result, possibly with lower cost, by using other methods for this oxidative
cleanup is therefore particularly appealing. Several of these new methods, called
in aggregate "advanced oxidation processes,"[1] employ a high-energy source to
induce chemical redox reactivity in a range of organic pollutants. For example,
both high energy-electron beam irradiation[2] and ultraviolet (UV) light
photoexcitation have been suggested as possible energy sources to initiate chemical reactions that attack and ultimately destroy undesirable components of air or
water mixtures.

Ultraviolet light can be used in several ways: (1) direct photolysis of contaminants, (2) in conjunction with treatment of a solution with ozone or hydrogen

---

*To whom correspondence should be addressed.

peroxide, or (3) by heterogeneous photocatalysis. In direct photolysis, the contaminant to be destroyed must absorb the incident light and suffer degradation in its photochemically excited state. This can occur only when the contaminant efficiently absorbs the light and is highly active photochemically, a situation which is difficult to achieve under typical practical conditions, i.e., when the contaminant absorbs light only weakly and is present only at very low concentrations. Light absorption by the pollutant is not absolutely required in the UV-ozone or UV-H_2O_2 processes, where oxidative degradation occurs at least partially through the chemical generation of highly reactive hydroxyl radicals. When the solution being subjected to UV photolysis also contains ozone or hydrogen peroxide, photochemical cleavage forms ·OH, which is known to attack most organic compounds. But here too, absorption by some sensitizer must initiate the reaction, and limited absorption by the solute or the additive restricts the efficiency of the photoinduced component of the degradation. Furthermore, these mixtures often still require large quantities of the added oxidant. In contrast, in the last method (heterogeneous photocatalysis), dispersed solid particles absorb larger fractions of the UV spectrum efficiently and generate the chemical oxidant *in situ* from dissolved oxygen or water. These advantages make heterogeneous photocatalysis a particularly attractive method for environmental detoxification. For this reason, the remainder of this chapter deals only with this method.

In heterogeneous photocatalysis,[3-7] a suspension of a particulate metal oxide (or other insoluble inorganic semiconductor powder) is irradiated with natural or artificial UV light. This excitation promotes an electron from a bonding or no bonding level in the solid to a highly delocalized level, creating a localized oxidizing site (a "hole") and a mobile reducing site (an "electron"). This photogenerated electron-hole-pair can then be captured by reagents present on the surface of the solid photocatalyst.[8] For example, the hole can be filled by electron transfer either from an adsorbed pollutant molecule or from an adsorbed water molecule. In the latter case, the same hydroxyl radical is formed as was generated in the homogenous UV-ozone or UV-peroxide photolyses. Alternatively, the photogenerated electron can be captured by oxygen adsorbed onto the surface of the particle, completing the first step in the formation of superoxide radical anion, hydroperoxide radical, hydrogen peroxide, or a hydroxyl radical. Subsequent chemical reactions by any of these species can induce oxidative degradation. Most frequently, TiO_2 or ZnO have been used as the photocatalyst, since they both efficiently absorb long-wavelength UV light and are chemically stable to the reaction conditions. Other semiconductor particles, e.g., CdS or GaP, absorb larger fractions of the solar spectrum and can form chemically activated surface-bound intermediates, but unfortunately these photocatalysts are degraded during the repeated catalytic cycles involved in heterogeneous photocatalysis. Mechanistic work on heterogeneous photocatalysis conducted on non-aqueous suspensions of TiO_2 has clearly shown that photoexcitation of the solid photocatalyst results in interfacial electron transfer, producing surface-bound intermediates which undergo a sequence of secondary chemical reactions.[9,10] Furthermore, most organic compounds can be oxidized directly by direct band gap photoexcitation of these particles.[11-16] Sufficient information concerning relative reactivity in such oxygenations and oxidative cleavages is now available to confidently predict the course of as-yet-unexplored photocatalytic reactions, although not their quantum efficiencies.

In water, similar processes lead to nearly complete oxidative degradation of a number of organic compounds.[17-21] As in the reactions occurring in nonaqueous solutions, the chemical principles governing these reactions correspond exactly with those involved with photocurrent generation in photoelectrochemical cells.[22] That is, in both aqueous and nonaqueous solutions, the primary photoprocess involves electron-hole pair formation followed by interfacial electron transfer to form surface-bound oxidants. Further details on photocatalytic oxidation mechanisms occurring in aqueous solutions are more difficult to establish, however, both because secondary photoreactions in water occur faster than in nonaqueous suspensions by virtue of the higher polarity of the aqueous-solid interface and because of the difficulty in isolating small quantities of oxidizable organic intermediates from an aqueous phase.

In both aqueous and nonaqueous suspensions, however, the key issue governing the efficiency of photocatalytic oxidative degradation is minimizing electron-hole recombination by maximizing the rate of interfacial electron transfer to capture the photogenerated electron and/or hole. In turn, these rates are significantly influenced by the densities of Lewis acidic and basic sites on the surface of the photocatalyst,[23] the state of hydration of active –OH groups on the surface,[24] and the hydrophobic-hydrophilic character of the surface.[25,26] These variables, in turn, are influenced by reaction conditions and the method used in the preparation of the photocatalyst. Such issues are discussed in more detail in later sections of this chapter.

ACTIVE SPECIES

The principles of semiconductor photoelectrochemistry require formation of an electron-hole pair in the primary photoprocess of heterogeneous photocatalysis, but they make no clear distinction between the rates of the next step in the sequence which leads to oxidative degradation. It is known that electrons are trapped in colloidal TiO_2 within about 30 psec of the formation of the electron-hole pair and that the hole is trapped in a period shorter than 250 nsec.[27] That is, inhibition of electron-hole recombination can be accomplished by either electron or hole trapping. For example, within a series of differentially substituted aromatic hydrocarbons, the relative oxidative reactivity varied with the substrate oxidation potential, implying the involvement of the organic reagent in a kinetically decisive step.[28] Presumably, this step involves either the formation of a singly oxidized organic ion on the surface of the irradiated photocatalyst (hole trapping) or the subsequent chemical reactivity of this surface-bound intermediate. On the other hand, such reactions were shown to fail in the absence of oxygen, which presumably implicates electron trapping as a limiting step. Although electron acceptors besides oxygen can be used successfully as electron traps,[29] the chemical efficiency of the photoinduced oxidation is usually lower than in the presence of air. The involvement of surface-adsorbed oxygen similarly seems to be decisive in reactions conducted in aqueous suspension.[30] In the absence of air, it is possible to charge a TiO_2 particle negatively so that it migrates under the influence of an electrical field[31,32] and shows the characteristic absorption spectrum of a trapped electron at a Ti^{3+} site.[27] It may be that energy-dissipative recombination can be suppressed by either electron or hole trapping under varying reaction conditions and with varying substrates.

Irrespective of which step is responsible for suppression of electron-hole recombination, the species ultimately resulting from hole trapping (formulated as hydroxyl radical or as an oxidized surface –OH group) is critical for the net oxidation chemistry observed in environmental detoxification. The ultimate route to complete mineralization of an organic substrate (to CO_2) must begin with reaction with this oxidant, leading to intermediate formation of C–O bonds. These can be formed either from the attack of a hydroxyl radical on the organic skeleton or by hydration of a chemically oxidized organic radical ion intermediate. It is known, for example, that oxygenated products are often encountered in nonaqueous TiO_2 suspensions, implying molecular oxygen as the source for new C–O bonds at high catalytic turnover.[10-12] Furthermore, hydroxylated aromatics and quinones are often detected in the photocatalytic degradation of organic substrates,[33-35] with hydrogen peroxide[36] and formate also being detected. An understanding of how such products of intermediate oxidation level are formed is very important in environmental applications because it emphasizes the fact that monitoring the disappearance of a contaminant is insufficient to demonstrate full detoxification. Sometimes, these partially oxidized products have comparable or even greater toxicity than the initial pollutant.

Another controversial question relates to the site of the secondary redox reactivity. Both direct observation with magnetic resonance techniques[37] and indirect kinetic evidence[33] point to the presence of a surface-bound hydroxyl radical on irradiated aqueous TiO_2 suspensions. Many times, the same intermediates formed from pulse radiolytic generation of hydroxyl radicals are also involved in heterogeneous photocatalytic transformations.[39,40] Nonetheless, time-resolved diffuse reflectance spectroscopic measurements indicate the transient formation of singly oxidized (or reduced) intermediates derived from a range of organic and inorganic reagents on irradiated aqueous suspensions of TiO_2 powder.[41] That this implies the formation of a surface-bound single electron oxidant was reinforced by kinetic studies by Lawless and co-workers.[42] The rates of decay of transient intermediates formed in homogeneous solution seem to differ from those of the same intermediates formed on irradiated TiO_2 suspensions.[43] Thus, the chemical character of the surface seems to influence the rate of dark secondary reaction of the transients formed there. In other cases, for example, in the photocatalytic oxidation of a herbicide in an aqueous suspension of TiO_2, at least some of the attack by the photogenerated ·OH radical on carbetamide seems to occur within the contacting homogeneous solution rather than at the photocatalyst surface.[44] An insightful study describing the difficulty in distinguishing these two paths by kinetic methods has been described,[38] and conflicting evidence concerning the site for photoinduced redox activity continues to appear.[45-49] Further details concerning this question appear later in this chapter.

KINETICS

Apparent Kinetics

The presence of TiO_2 particulates presents a formidable challenge in describing the kinetics in heterogeneous photocatalysis. The difficulty originates with the heterogeneous surface that interacts with solvent and substrates to differing degrees. Some of these interactions have yet to be assessed quantitatively in photo-oxidative heterogeneous catalysis. Complicating the picture is the effect photons have on the surface properties, for example, on adsorption-desorption equilibria or on the nature of the catalytic sites, among others. Clearly then, the observed kinetics in a heterogeneous photocatalysis must be considered, for the moment, as *apparent kinetics*. Our approach to describe these is summarized below.

The photocatalytic process for the oxidation of various aromatic substrates, on the basis of the above constraints, has been modeled in its most simplistic form by the phenomenological consecutive and parallel reactions in Scheme 1,

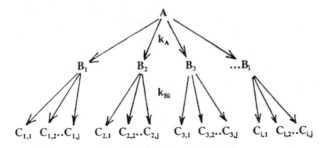

Scheme 1. Kinetic scheme for the initial steps of the photocatalyzed transformation of a compound A.

where k_A represents Σk_{Ai} (i = 1, 2, ..., i), where k_{Ai} represents the rate constant for the reaction from A to B_i (i is the number of possible sites available on the target compound A with consequent formation of i intermediates, denoted as B in Scheme 1) for the attack by the primary oxidizing species; and k_B denotes the macroscopic rate constant for consecutive and parallel reactions for the disappearance of the various stable intermediate products B_i. Where the initial organic substrate is a haloaromatic, we use k_X to represent the observed rate constant of formation of the halide, irrespective of its form (X^- or HX) and its source. The implicit assumption has also been made that k_{Ai} is nearly identical irrespective of the position of attack by the $\cdot$OH radical.[50] The disappearance of the initial substrate, A, together with the formation and disappearance of B can be described by[51]

$$\left[A_{(t)}\right] = \left[A_{(o)}\right] \exp\left\{-k_A t\right\} \tag{1}$$

$$\left[B_{i(t)}\right] = \frac{k_A\left[A_o\right]}{k_B - k_A} \left\{\exp\left(-k_A t\right) - \exp\left(-k_B t\right)\right\} \tag{2}$$

The formation of the final products, CO_2 and anions when heteroatoms are present (see later), follows, in nearly all cases, simple exponential growth kinetics: [Product] = $c\{1 - \exp(-k_P t)\}$.[3,6]

A few additional points need to be noted. Although some of the organic compounds examined may be adsorbed only slightly to the particle surface in the dark, the extent of photoadsorption may be a difficult parameter to evaluate depending on adsorption-desorption rates and on whatever other process(es) that takes place subsequent to adsorption. In describing a rate expression for the photooxidation, the implicit assumption is often made that there is a constant fraction, however small, of the organic substrate on the oxidative active surface sites of the catalyst, Ti^{IV}-OH.[38] Following $\cdot$OH attack on the aromatic ring of the substrate S_{ads}, one or more intermediates (I_{ads} and/or I_{sol}) form, which subsequently or nearly simultaneously undergo dehalogenation and fragmentation to aliphatic species which ultimately also degrade to produce stoichiometric quantities of CO_2 (Equation 3).

$$
\begin{array}{ccccc}
A & \rightarrow & B & \rightarrow & C
\end{array}
$$

$$
(site) + S \underset{k_d}{\overset{k_a}{\Leftrightarrow}} S_{ads} \overset{k_s}{\rightarrow} \{I_{ads} + I_{sol}\} \overset{k_I}{\underset{\rightarrow}{}} \ldots \rightarrow \ldots \rightarrow CO_2 \tag{3}
$$

$$
\begin{array}{cc}
\quad \downarrow & \quad \downarrow \\
X^- & X^-
\end{array}
$$

where k_a and k_d are the rate constants for the adsorption and desorption processes, respectively; k_S denotes a sum of rate constants for the formation of various intermediate species, and k_I represents a sum of rate constants in the fragmentation of these intermediates.

By employing the formalism often used in enzyme kinetics, it may be shown that the rate of formation of product is given by:[52,53]

$$\text{rate} = \frac{k_S K_S N_s[S]}{1 + \alpha K_S[S]} \tag{4a}$$

where $\alpha = (k_S + k_I)/k_I$, K_S $\{= k_a/(k_d + k_S)\}$ is the photoadsorption coefficient for the various substrates, and N_s is the number of oxidative active sites. Equation 4 is reminiscent of a Langmuirian rate equation of the general form

$$rate = kKC/(1 + KC) \tag{4b}$$

where k is the specific rate constant, K the adsorption coefficient of the reactant, and C its concentration.

Further modification of Eq. 4 is needed to take account of the formation of the reactive $\cdot$OH species. The quantum yield of $\cdot$OH formation is $\phi_{OH} = \phi I_a^n k_f \tau$ (at low light fluxes $n = 1$), where ϕ is the quantum yield of e^-/h^+ generation, I_a the rate of light absorption, k_f the rate constant of the reaction of H^+ with $Ti-OH_2$, and τ is the lifetime of the photogenerated h^+. If the rate of recombination of trapped electrons and holes k_{rec} is greater than k_f,[27] then $\phi_{OH} \cong \phi I_a \beta (A_p) k_f/k_{rec}$, where β A_p is the fraction of the irradiated surface, and A_p is the particle surface area. As well, the rate of formation of products depends on the lifetime of the bound $\cdot$OH radicals, τ_{OH}. This leads to

$$Rate = \frac{\left(\dfrac{\phi I_a^n \beta A_p k_f \tau_{OH}}{k_{rec}}\right) k_s K_s N_s[S]}{\{1 + \alpha K_S[S] + K_W[H_2O] + K_I[I] + K_{ions}[ions]\}} \frac{K_{O_2}[O_2]}{(1 + K_{O_2}[O_2])} \tag{5}$$

The additional terms in the denominator indicate the influence of the solvent water, the intermediates, and the various anions and cations present in the system. Through adsorption, any one or all may act as inhibitors by blocking some of the active surface sites on the photocatalyst.[54] The adsorption isotherm expression for oxygen is included in Eq. 5 to consider the effect of molecular oxygen on the rate of product formation.[55]

Equation 5 has the same analytical form as the equations reported by Okamoto and co-workers[56,57] and more recently by Turchi and Ollis.[38] These latter authors noted that with minor variations the expression for the rate of photooxidation of organic substrates on irradiated TiO_2 presents the same saturation-type kinetic behavior as portrayed by the Langmuir-Hinshelwood[51,156] rate law.[51] This is true whether or not (1) reaction occurs while both reacting species are adsorbed, (2) a "free" radical species reacts with the adsorbed substrate, (3) a surface-bound radical reacts with a free substrate in solution, or (4) the reaction takes place between the free reactants in solution. Assigning an operational mechanism for reactions taking place in heterogeneous media to a Langmuir-type process, to an Eley-Rideal pathway,[156] or to an equivalent-type process therefore is not possible on the basis of observed kinetics alone,[58,59] unless specific values for the relevant rate constants are known independently. For example, an equation analogous to the Langmuirian expression is obtained in homogeneous media for the reduction of cis-Ru(NH$_3$)$_4$Cl$_2{}^+$ by Cr^{2+}(aq).[60] Clearly, although the analytical expression obtained for the rate of photooxidation may be analogous to the Langmuir-Hinshelwood relationship, *nothing can be concluded* about the operational mechanism in a heterogeneous photocatalysis experiment. Other experiments must be undertaken to unravel the complex events that take place. Unraveling the events that occur on an illuminated semiconductor presents a formidable mechanistic challenge.

ROLE OF ADSORPTION

Monolayer-Limited, Langmuir-Type Adsorption Model in Relation to Photocatalyzed Degradation (PCD)

In experimental studies of the variation of the initial rates $(R_{PCD})_o$ of TiO$_2$-photocatalyzed degradations vs the initial concentration of the pollutant, C_o (see Eq. 4b), a set of experimental conditions can often be found such that the observed data for the dependence of the PCD rate on concentration are well linearized by double-reciprocal type plots [of $(R_{PCD})_o - 1$ vs $(C_o) - 1$). The slope and intercept of such plots yield values for two of the parameters related to PCD rates in those conditions. For the reasons reported above and also discussed later, these parameters have often been denoted by k_{LH} (compare with k of Equation 4b) and K' (compare with K of the same equation)[61-63] although their physical-mechanistic significance remains unclear. However, parallels were drawn in early PCD papers between such double-reciprocal linearizations of data for PCD rate over TiO$_2$ photocatalysts and analagous linearizations of data for heterogeneously catalyzed conversions of gaseous reactants. For the latter conversions, such linearization is commonly taken to indicate that *their rate-determining process* involves reaction events between reactant species adsorbed onto the surface of the catalyst, as in the Langmuir-Hinselwood (LH)- or Michaelis-Menten (MM)-type mechanisms. Such parallels led to similar interpretations being given by early workers to parameters deduced from double-reciprocal linearizations of PCD data, viz., that the intercept equaled the reciprocal of a rate coefficient k_{LH} for reaction events on the surface, whereas the slope equaled $(k_{LH} K') - 1$, in which K' represented a constant for a fast adsorption-desorption equilibrium between surface-adsorbed and fluid-dispersed reactant. Numerical values denoted by K' and k_{LH} were thus often deduced from double-reciprocal linearizations in early papers and sometimes treated as fundamental and independent constants defining, respectively, the preequilibrium adsorption and the rate-determining reaction events within an adsorbed monolayer at the TiO$_2$-aqueous solution interface.

Recent papers by Cunningham and Al-Sayyed[64] and Cunningham and Srijaranai[65] describe several experimental approaches taken to examine the validity of such treatments. Adsorption isotherms were directly determined by experiment to delineate the actual dependence on C_{eq}, the concentration of reactant at equilibrium for the extent of adsorption of three substituted benzoic acids from dilute solution onto TiO_2 (P25) in the absence of UV illumination. The extent of adsorption data were normalized as n_2, the number of micromoles adsorbed per gram of TiO_2. Excellent linearization of data on such curved, Langmuir-like isotherms was achieved by replotting in accordance with the linearized form of an isotherm equation applicable to a solid-liquid interface featuring a competition between solute and solvent species for adsorption into a surface-solution monolayer.[66] The extent of adsorption (in the 10–200 ppm region) investigated thus behaved qualitatively in the manner expected if the measured adsorption corresponded predominantly to adsorption into a surface-solution monolayer. The linearized plots were originally interpreted as yielding values for the actual adsorption constants, K_{ads}, of the benzoic acids, which differed by an order of magnitude from values of the apparent adsorption constants K' deduced from $(R_{PCD}) - 1$ vs $(C_o) - 1$ plots, and they were also assumed as upper limit of 0.03 on the mole fraction of adsorbed solute attainable in the surface-solution monolayer, thereby indicating much greater occupancy of monolayer sites at the TiO_2-aqueous solution microinterface by water molecules than by solute species. Holes photogenerated within TiO_2 and arriving at such microinterfaces therefore would have lower probability for direct encounter and charge transfer with organic adsorbates than by a surface-bound water molecule present in the surface-solution monolayer. This should lead, in turn, to a lower contribution to PCD by charge-transfer to solute, unless compensated for in some way in later steps in the overall mechanism of PCD.

One possibility for such compensation is illustrated in Figure 1 for chlorohydroxybenzoic acid (CHBA) when present at its limiting saturation concentration as a monolayer on TiO_2, viz., the rapid secondary transfer of a hole from a surface H_2O^+, which is first produced by a statistically favored initial encounter with $(H_2O)_s$, onward across the surface monolayer to an adjacent $(CHBA)_s$. As indicated in Figure 2, the adsorption isotherm data for CHBA point to favorable conditions for such intramonolayer charge-transfer events, since the donor and acceptor cannot be further than 0.7 nm away if $(CHBA)_s$ at its adsorption limit is uniformly distributed throughout the monolayer. Very rapid charge transfer within the solution monolayer, e.g., before completion of H_2O^+ dissociation to $\cdot OH$, would strongly influence the outcome of a question raised in the Introduction section, viz., whether the main secondary redox activity at the interface is a transient one-electron oxidized intermediate or formation of a hydroxyl radical adduct.

On this basis, the possibility of hole transfer to solutes in the second or third solution layer cannot be discounted, although most workers envisage hole transfer as involving only solutes present in the surface monolayer. On the basis of their recent studies of the PCD of 3-chloro salicylic acid, Anderson and co-workers have envisaged an important role for hole transfer at the TiO_2-aqueous solution interface.[67,68]

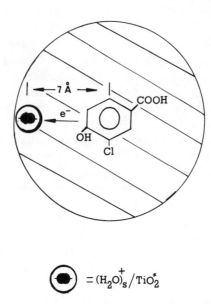

$$\bullet = (H_2O)^+_s / TiO^*_2$$

Figure 1. Relative areas depicted schematically (1) for a molecule of CHBA adsorbed flat
on the TiO_2 surface and (2), by the large circle, for the mean area of TiO_2 surface
available per adsorbed CHBA species (plus surrounding H_2O molecules) when
CHBA attained its saturation concentration in the surface-solution monolayer.
Also indicated is the feasibility for activation of such adsorbed CHBA species by
charge transfer along the surface-solution monolayer between it and any H_2O^+
species formed by initial localization of a TiO_2 hole on H_2O.

Charge-transfer possibilities of the type schematically represented in Figure 1
may also introduce dependencies upon separation distance (and hence upon
concentration), quite different from any surface-diffusion contribution contained
within the pseudo-rate coefficient k_{LH} of the LH-type rate expression that is
frequently assumed to apply to PCD. Figure 2 illustrates discrepancies between
the observed concentration dependence of the initial PCD rates of CHBA (solid
line) and those expected on the basis of various differing (but concentration
independent) values of k_{LH}.[64]

Together with UV photons and TiO_2, the presence of dissolved dioxygen is
essential for ongoing photocatalytic degradation of organic pollutants. Data from
experimental studies of the dependence of PCD rates upon Po_2 in the gas phase
(see Eq. 5) have been shown to be adequately linearized by double-reciprocal
plots of $(R_{PCD})_o - 1$ vs $(Po_2) - 1$.[56,57] This has been interpreted as evidence for
monolayer adsorption of O_2 at the TiO_2-oxygenated aqueous solution interface
and have been used to deduce numerical values for $(k_{LH})_{O_2}$, an apparent rate
constant for LH-type reactions of oxygen in the surface monolayer, and for $(K')_{O_2}$,
an apparent equilibrium constant for adsorption-desorption of O_2 at the interface.
As yet the validity and physical significance of such values in PCD have not been
critically assessed as was done for substituted benzoic acids. However it is
apparent that the $(K')_{O_2}$ value is much more likely to be the equivalent of a Henry's

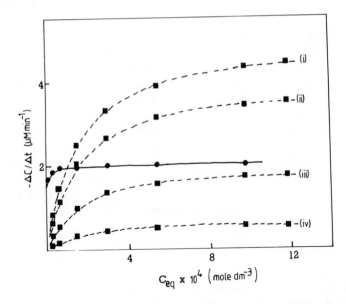

Figure 2.   Illustration of discrepancies between (1) experimentally obtained data for solute concentration dependence of initial rate, $\Delta C/\Delta t^*$, for TiO_2-photocatalyzed degradation of CHBA (as denoted by solid line); and (2) the solute concentration dependence predicted by inserting into the LH-type rate equation, rate = $k_{LH}KC_{eq}/(1 + KC_{eq})$. Values for K and C_{eq} were determined from equilibrium dark-adsorption measurements, plus the following empirically selected values for k_{LH}: (i) 5 M min^{-1}; (ii) 4 M min^{-1}; (iii) 2 M min^{-1}; (iv) 0.5 M min^{-1}.

law-type equilibrium constant than of a Raoult's law or Langmuir-type constant based upon ideal behavior.

Influence of Aqueous Phase pH and Surface Heterogeneity

An isoelectric point at pH ca. 6 is widely reported for titanium dioxide surfaces in contact with aqueous solution.[69] On this basis, the surfaces of TiO_2 particles suspended in aqueous solutions having higher pH should carry significant net negative charge associated with a substantial excess of adsorbed surface hydroxide ions. Upward band bending within regions of the semiconducting solid adjacent to the solid-solution interface thus may be expected at pH higher than 6.3, leading in turn to expectations for improved transfer of photogenerated holes to the increased surface density of OH$^-$ adsorbed thereon. Increased efficiency in resultant oxidations of adsorbed species thus may be expected at pH >> 6, but such expectations have been seldom unequivocally realized by experiment for TiO_2-sensitized photodegradations (see below).

However, evidence for an opposite effect of high pH upon the extent of adsorption of organic acids onto TiO_2 has emerged from several adsorption studies. Thus, Foissy et al. in 1983 showed a strongly inhibiting effect of high pH upon the adsorption of polyacrylic acid onto TiO_2 and interpreted this in terms of unfavorable electrical forces, viz., repulsion between the negatively charged surface and anionic forms of the polyacrylic acid.[70] Subsequent work with this and other polyelectrolytes indicates that the effect of pH on their molecular equilibrium conformation may exert a stronger influence on adsorption than was fully appreciated in the earlier work.

However, molecular conformational complications should not unduly affect results recently reported[71] for the marked pH dependence on the extent of adsorption of the relatively small salicylic acid or chlorohydroxy benzoic acids onto TiO_2 (Degussa P25). In aqueous solutions having 200 ppm initial concentrations of these acids, a sixfold decrease in the equilibrium adsorption onto TiO_2 (Degussa P25) was observed at a pH of 10 relative to that at a pH of 5. Values for the equilibrium adsorption for three intermediate pH values demonstrate that the decrease with increasing pH was progressive but nonlinear.[71] In another recent study of the extent of adsorption of salicylic acid and 3-chlorosalicylic acid onto TiO_2 prepared by a sol-gel technique, Tunesi and Anderson have reported a higher extent of adsorption at pH = 3.7 than at pH = 5.7.[67] In explanation of their observation of a bright yellow color upon the sol-gel-derived ceramic TiO_2, they assumed the formation of a chemisorbed charge-transfer complex between the singly charged salicylate anions and positively charged titanium ion at the oxide surface. They considered that the decrease in adsorption at pH 5.7 relative to that at pH 3.7 could be understood as decreased availability of positively charged titanium centers for charge-transfer complex formation. Indications for contributions to adsorption from nonchemisorbed bond formation were recognized on the basis of appreciable adsorption at pH 6.4, where both the oxide surface and the salicylate ion are negatively charged. Differences in the nature of adsorption at high and low pH were likewise noted by Lamarche et al.[72] in their studies of polyacrylic acid adsorption by calorimetry and microelectrophoresis. Enthalpy for displacement of solvent by this adsorbate was endothermic and constant at a pH of ca. 9.5, in contrast to being exothermic and giving evidence for two modes of adsorption at pH 3.5.

Such studies clearly point to significant effects of pH and of surface heterogeneity on the extent and nature of adsorption of ionizable solutes onto TiO_2. In some recent work, attempts have been made to take advantage of such pH dependence, and especially of the suppression of anion adsorption onto TiO_2 expected at high pH, to greatly alter the surface-adsorbed to fluid-dispersed ratio for such anionic species. In cases where such species are reactive toward ·OH radicals, tests can then be made as to whether such radicals are confined to the surface monolayer or can escape and react with the fluid-dispersed anions.

It has increasingly been recognized in recent years that surfaces of real solids seldom consist solely of the perfect array of identical sites envisaged in the Langmuir model, with each site displaying favorable enthalpy for binding an adsorbate. Nonuniformity can take many forms on real surfaces, but in theoretical considerations of adsorption onto a heterogeneous surface it frequently suffices to differentiate between various possible distributions of adsorption energies, e.g., whether according to some continuous distribution of energy, or discontinuously into a limited number of subsets. Theoretical studies of monolayer-limited liquid adsorption on heterogeneous surfaces characterized by continuous energy distribution have been numerous[73,74] in contrast to a limited number of theoretical studies of multilayer liquid adsorption.[75]

Surfaces of high surface area TiO_2 materials have been shown to retain subsets of surface hydroxyls having varying degrees of acidity and a net surface density of 4–5 hydroxyls per square nanometer.[69] Unequivocal experimental differentiation between the effects of other sources of surface heterogeneity and those of subsets of strongly retained surface hydroxyls has not yet been possible in relation to solute adsorption from dilute aqueous solution because of the strong competition by water. However, suspensions of TiO_2 in acetonitrile solutions of benzyl alcohol show some involvement of surface hydroxide ion as locations for the primary photooxidation processes.[65] Heterogeneity of adsorption sites on anatase TiO_2 has also been demonstrated by infrared (IR) studies of carbon monoxide adsorption.[76] Furthermore, Chandrasekaran and Thomas have reported differing adsorptive properties for crystalline and amorphous TiO_2.[77] Thus, a changeover to continuous distribution of adsorption sites upon porous, sol-gel-derived TiO_2 to a more limited variety of adsorption subsets on crystalline nonporous anatase would not be surprising. For sol-gel-derived TiO_2, Tunesi and Anderson[67] and Sabate et al.[68] have reported a "best fit" of their adsorption isotherm data for salicylic acid from aqueous solution with a Freundlich isotherm, which they considered consistent with continuous distributions of adsorption energies.

Operationally, important questions about the magnitude of the adsorption onto TiO_2 on the efficiency of TiO_2-photocatalyzed photoremediation of contaminated water are whether and to what extent the effects of pH and TiO_2 surface heterogeneity influence rate and efficiency of PCD. Over their ex-sol-gel TiO_2 membranes, Tunesi and Anderson[67] report substantial decreases in ongoing PCD rates for increases of pH in the range from 4 to 9.5 and interpret these as an enhancement at the lower pH of direct electron transfer from salicylic acid to TiO_2 mediated by charge-transfer complexes in the interfacial monolayer.

The absence of any enhancement in initial PCD across the pH range from 3.3 to 10.5, however, has been demonstrated by Cunningham et al.[71] in studies made with TiO_2 particles (Degussa P25) suspended and UV illuminated in aqueous solutions of salicylic or chlorohydroxybenzoic acids at an initial concentration of 80 ppm. From this, it was concluded that PCD in those systems, at the photon flux used, was mainly determined by factor(s) other than the extent of dark-equilibrated adsorption.

MULTILAYER ADSORPTION IN RELATION TO PCD

In relation to solid-liquid interfaces, Everett[78] and Brown et al.[79] have expressed the view that "whatever their detailed structure, surface layers, even in simple systems, must extend in a direction normal to the surface for distances

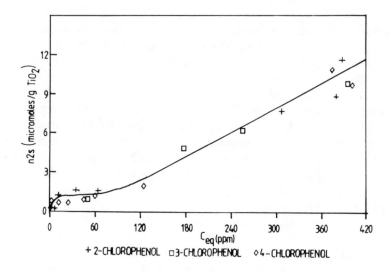

Figure 3. Data for three monochlorophenol isomers showing similar dependencies of equilibrium extent of adsorption (expresssed as n^s2, the number of micromoles adsorbed per gram of TiO_2 [Degussa P25]) upon the equilibrium concentration remaining in the aqueous phase of TiO_2 suspensions after equilibration for 3 hr at 290 K in the absence of illumination.

corresponding to several molecular diameters." The proportion of solute existing in such multilayers — relative to those in a surface monolayer or in the bulk solution — will depend on the relative magnitudes of the respective free energies of $\Delta G_{(multi)}$, $\Delta G_{(mono)}$ and $\Delta G_{(soln)}$. Predominance of monolayer adsorption should be favored if $\Delta G_{(mono)}$ is much more strongly exergonic than $\Delta G_{(multi)}$ or $\Delta G_{(soln)}$, as appeared to be the case for the substituted benzoic acids. On the other hand, multilayer formation can be significant if $\Delta G_{(multi)} \cong \Delta G_{(mono)} > G_{(soln)}$, as implied by the common shape of the adsorption isotherm shown in Figure 3 for the three monochlorophenols onto TiO_2 from aqueous solution.

The overall shape of this isotherm — with the amount adsorbed rising in Region I (from 0 to 20 ppm) toward the start of a rather low plateau, followed by a very slow increase in Region II (from 20 to 100 ppm), and eventually showing a rise in the high concentration Region III (100–400 ppm) M — strongly resembles those reported earlier by Hansen and Craig[80] for adsorption of alcohols and acids onto graphite. These observations were taken as evidence for growth in multilayer adsorption at the higher concentrations. Discontinuities qualitatively similar to those shown in Figure 3 for the 2-chlorophenol (2-CP) adsorption isotherm, however, were not observed when careful measurements were made across Regions I, II, and III for the dependence of initial 2-CP PCD rates upon initial 2-CP concentration. Instead, the initial PCD rates, under a flux of 2×10^{18} near-UV photons per minute incident on 30 ml of dark-preequilibrated suspension in each case, failed to exhibit any significant increase with increasing C_0. Such zero-order behavior in the concentration dependence of initial PCD rates

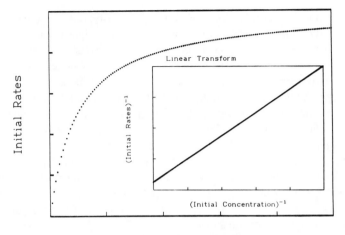

Initial Concentration

Figure 4. Plot of initial rates vs reactant concentration (see Eq. 24 in the text) for arbitrary units; the insert shows the linear transform of Eq. 24 from which the rate constant k and the adsorption coefficient K can be estimated from the intercept and slope, respectively.

represents another significant discrepancy from the concentration dependence of dark adsorption observed for the mono at the indicated photon flux chlorophenols over the same concentration range. This implies that a process other than adsorption-desorption equilibrium operated as the principal efficiency-limiting factor in the initial PCD rates of monochlorophenols at the indicated photon flux.

In view of the demonstrated multilayer nature of the monochlorophenol adsorption isotherm (Figure 4), possibilities for some Eley-Rideal-type contribution to primary photooxidation events merit consideration, namely for rapid inward motion of monochlorophenol from multilayer locations for reactive encounters with h$^+$ or ˙OH photogenerated in the surface monolayer. Since the approximate first-order dependence on C_{eq}, expected from any such contributions, was not observed at high photon flux, further experiments are needed to determine if an altogether different process, such as electron transfer to oxygen, may be efficiency limiting and the origin of the observed zero-order dependencies of initial PCD rates on C_{eq}.

EFFECTS OF VARIOUS FACTORS

Effects of Inorganic Ions

Common inorganic ions present in natural waters are the anions HCO_3^-, Cl^-, SO_4^{2-}, and NO_3^- and the cations Ca^{2+}, Na^+, and Mg^{2+}. Therefore, these ions are also present in industrial waters. Their effects on the TiO_2-photocatalyzed water treatment should accordingly be determined. Obviously, only those ions that are bound to TiO_2 or are very close to its surface can have significant effects if it is assumed that the chemical events occur predominantly at the surface. Consequently, the point of zero charge (PZC) of this semiconductor should be a determining property. As well, the chemical affinities of the ions for TiO_2 PZCs of several TiO_2 samples have been measured; for Degussa P25, it is 6.3 pH.[81]

To our knowledge, no study has been reported concerning the effects of the common cations cited above. Chloride, sulfate, nitrate, or hydrogen carbonate with sodium as a counterion have been found to have no effect on the initial rate of disappearance of 3-chlorophenol (0.155 mM; pK_A ca. 9) at pH 8 (adjusted with NaOH) at concentrations up to 0.1 M, i.e., at concentrations much higher than both the usual values in natural waters and in the pollutant concentration.[82,83] Similarly, the addition of 1 M NaCl to a TiO_2 suspension at pH 13.6 (also adjusted with NaOH) was found to decrease the disappearance rate of phenol by a factor of only 1.1.[84] That is not surprising, since at pH greater than PZC (6.3) the surface is negatively charged so that the concentration in anions is lower near the TiO_2 surface than in the bulk of the solution.[85] In particular, hydrogen carbonate, which is detrimental for the treatment of water by H_2O_2-UV, H_2O_2-O_3, and O_3-UV homogeneous systems[86,87] because it efficiently scavenges ·OH radicals (k = 1.5 × 10^7 M^{-1} s^{-1}), has no effect on the rate of the photocatalytic disappearance of 3-chlorophenol. That practical advantage has also been presented as a hint that ·OH radicals attack the organic pollutant within the layer where the concentration of hydrogen carbonate is substantially decreased with respect to that in the bulk of the solution.[83]

Studies at pH less than PZC include the effects of Cl^-, SO_4^{2-}, and NO_3^- (pH = 4.5–4.8 without acid addition)[82,83] and of $H_2PO_4^-$, ClO_4^-,[88] again with Na^+ as the counterion and with several organics: 3-chlorophenol,[82,83] salicylic acid, aniline, and ethanol.[88] The amount of remaining organic pollutant was measured as a function of illumination time.[82,83] Low et al. found that evolved CO_2 could be determined kinetically by use of a conductivity cell.[88] In both cases, the maximum concentration of the inorganic salt was 0.1 M. No effect of nitrate and perchlorate was observed within experimental error. Chloride, sulfate, and phosphate have detrimental effects. The decrease in efficiency of pollutant oxidation is more pronounced for phosphate[88] and is significant at lower concentrations of both phosphate and sulfate (ca. 1 mM) than with chloride, especially for ethanol[88] or 3-chlorophenol.[82]

Nevertheless, although these effects should be considered in a water-treatment process, they are of secondary importance, since the highest decrease reported was fourfold in phosphate (0.1 M) -salicylic acid couple.[88]

At pH less than PZC, the surface is positively charged, as is expressed by the equation

$$TiOH + H_3O^+ \rightarrow TiOH_2^+ + H_2O \tag{6}$$

The concentration $[A^-]_s$ of monovalent anions at the surface is related to the surface charge σ and to the concentration in the bulk $[A^-]_b$ by[85]

$$[A^-]_s = \sigma^2/2\varepsilon\varepsilon_o \, kT + 2[A^-]_b \tag{7}$$

where the symbols have their usual meanings. At pH 4.5 (HCl) for TiO_2 Degussa P25 in 10 mM NaCl,[81] $\sigma \cong 6 \times 10^{-2}$ C per square meter and therefore $[A^-]_s = 6.3 \times 10^{26}$ anions per cubic meter + $2[A^-]_b$ = ca. 1.05 M + 2 $[A^-]_b$ (at 293 K), which is considerably more than the highest $[A^-]_b$ in the cases reported.[82,83,88] Although this coulombic treatment shows a marked excess of anions close to the surface, it does not predict differences between the various monovalent anions considered. Also, differences in $[A^-]_b$ should have minute effects in the range and at the pH investigated, which is not the case for chloride.

The contrasting effects of the anions at acidic pH indicate either a competition between the inorganic salt and the organic compound for the surface oxidizing species (holes, ˙OH radicals, etc.) and/or a competitive adsorption. In the latter case, the LH kinetic model (see sections entitled Kinetics and Role of Adsorption) indicates that the reciprocal of the initial rate, r_o, of disappearance of the organic substrate increases linearly with the concentration of anion, which is the case for chloride for the pollutants cited above[88] or with various halogenated hydrocarbon with one or two carbon atoms.[89] However, the LH model is not compatible with the fact that the surface concentration of Cl^- ions does not depend much on the bulk concentration if only electrostatic forces are taken into account. Also, a competition between chloride and a neutral organic pollutant for the same sites seems somewhat unlikely given the different chemical natures of these entities, and it might be that the LH model only yields an overall expression for complex phenomena.

By contrast, a LH equivalent relationship was not found for phosphate and sulfate.[88] The differences in the interactions of TiO_2 with these various anions are reflected by the fact that the photocatalytic activity was restored when TiO_2 was rinsed with water in the case of chloride, whereas an alkali or a $NaHCO_3$ wash was needed after the use with phosphate or sulfate.[88]

The TiO_2 surface is apparently chemically modified by these latter anions and exhibits a different photocatalytic activity. X-ray photoelectron spectroscopy (XPS) has been used to examine the species remaining in the surface layers of TiO_2 after exposure to aqueous solutions of various sodium salts at pH 6–7, i.e., close to or slightly higher than the PZC, after subsequent drying in a desiccator.[69] From these analyses, it was inferred that phosphate interacts strongly with TiO_2, presumably by replacing adsorbed OH groups, a conclusion which is consistent with the observed effect of the phosphate on pollutant degradation. Conversely, no Cl (from Cl^-) or S (from SO_4^{2-}) could be detected by XPS, or N (from NO_3^- or Cl (from ClO_4^-), at least for unreduced anatase powder. For sulfate, the discrepancy between this result and the above-mentioned necessity of using a base to restore the TiO_2 surface may be a result of the close-to-PZC pH used to prepare the XPS samples. Indeed, adsorption measurements carried out by the radioactive-tracer method with labeled anions showed the following order of preferential adsorption on TiO_2 at pH less than PZC:

$$H_2PO_4^- > HSO_4^- > Cl^- > ClO_4^-$$

Unlike the other anions, phosphate exhibited still significant extent of adsorption in neutral and basic solutions and was only slightly desorbed by pure water. These results are consistent with the effects of these anions on pollutant removal at pH less than PZC.

Alternatively, Low et al.[88] have explained the effect of these anions (by their competitive reaction with the photoproduced holes) to give radicals capable of oxidizing the pollutant, but at rates lower than with the ˙OH radicals (or the direct interaction of the pollutant with the hole).

Effect of pH

Experimental data on the effect of pH are numerous.[40,53,56,57,82-84,88,90-92] At first sight and for the pollutants whose pK_a is outside the range of 1–13, a very acidic solution pH appears to be detrimental and a very basic solution pH to be favorable, whereas the variations are modest or nonexistent around neutrality. Because even at the extreme pHs the change in the rate of photocatalytic rate degradation is generally less than one order of magnitude, the TiO_2-photocatalyzed water treatment definitively possesses an advantage over biological, H_2O_2-Fe^{2+}, H_2O_2-UV, O_3-UV, and O_3-H_2O_2 processes.

To exclude the pH effect unambiguously (see also the section on Substituted Monoaromatics), the reagents used to change the pH must contain counterions that have no effect on the rate of water treatment, at least at the concentrations employed. Sodium hydroxide has generally been chosen to produce basic pH. Because sodium ions are inert (*vide infra*), the results thus obtained are relevant. By contrast, hydrogen chloride or sulfate have been utilized most frequently[40,53,56,57] to attain acidic pH. Because Cl^- and SO_4^{2-} ions interact with TiO_2 (*vide infra*), however, the effects found under these conditions are inconclusive.

For acidic pH, the results obtained in $HClO_4$ or HNO_3 are significant, since perchlorate and nitrate have no effect by themselves. The rates, R, of mineralization of phenol, ethanol, and 2-propanol have been measured in the pH range 2–5.5 with $HClO_4$ as the acidifying agent.[88] For phenol and 2-propanol, R increases from pH 2 to pH 3.4 and then remains constant, whereas for ethanol it increases from pH 2 to pH 2.8, remains constant up to pH 3.6, and then decreases. In all cases, the variations are relatively small. Therefore, for these compounds with no heteroatom apart from oxygen, acidification below pH $\cong$ 3 seems to be slightly detrimental to the total mineralization. The initial rates R_o of disappearance of benzamide, nitrobenzene, and 3-chlorophenol have been determined down to pH ca. 1.3 using HNO_3.[83] Initial rate studies relate directly to the pollutant studied, unlike the rates of complete mineralization; however, they are not very precise. Within experimental accuracy, no change in R_0 was found on acidification for the two nitrogen-containing monoaromatics, despite the fact that protonation of nitrobenzene (pK_a ca. 4) should hinder adsorption of this compound on the positively charged surface. By contrast, a decrease in R_0 was observed at the most acidic pH when HCl was employed. This can be attributed, at least qualitatively, to the unfavorable role of chloride discussed in the preceding section. Even in aqueous HNO_3, the R_0 of 3-chlorophenol was significantly decreased at pH below 4. The number of Cl^- ions that were analyzed in the solution within the time required to determine R_0 was thought to be too small to explain this decrease. Therefore, the effect of acidic pH seems to depend on the pollutant, unless almost all the chlorophenol destroyed had released chloride which remained at the surface, despite the much larger concentration of nitrate. Further data are clearly needed.

Very high pHs have been found to be favorable even when the anionic form of the pollutant should hamper adsorption on the negatively charged surface. In this pH range, OH^- ions determine the potential at the TiO_2 surface.[66]

$$TiOH + OH^- \Leftrightarrow TiO^- + H_2O \qquad (8)$$

It has been proposed that a higher concentration of $\cdot OH$ radicals is produced by the neutralization of adsorbed OH^- ions by photogenerated holes. Alternatively, if the primary steps of the photocatalyzed oxidation were the formation of the organic radical cation, the holes would be trapped by the $Ti-O^-$ entities and subsequently transferred to the organic compound.

Added Oxidants

It was pointed out earlier that the presence of O_2 is essential for the photomineralization of organic compounds. Adsorbed oxygen traps e^-_{CB}, thereby delaying electron-hole recombination. Oxygen may also be required to oxidize the elements to their highest oxidation number. The possibility of adding other reducible species has been considered in order to (1) scavenge e^- of the conduction band (2) generate additional active species, e.g. $\cdot OH$, $SO_4^{\cdot-}$, I(VI), ..., and (3) increase the rate of oxidation of intermediates.

The most obvious candidate is H_2O_2, which is a possible intermediate in the photocatalytic sequence. Hydrogen peroxide can generate $\cdot OH$ through the reaction

$$H_2O_2 + e^-_{CB} \rightarrow \cdot OH + OH^- \qquad (9)$$

The mechanism of this reaction on irradiated TiO_2 has recently been elucidated.[36] Under conditions in which H_2O_2 is not directly excited,

$$H_2O_2 \xrightarrow{h\nu} 2\ ^\cdot OH \tag{10}$$

and when such ions as Fe^{2+} and Cu^+ (Fenton's generating sequence) are not present,[93] the effect of H_2O_2 on the degradation rates is not so straightforward. In fact, the initial decomposition rates depend on the concentration of H_2O_2, and very often an optimum range of concentrations may be observed. Even more interesting, in some cases, enhancement of the photocatalyzed oxidation is realized, while in others a detrimental effect is observed.

It must be recalled that H_2O_2 may also participate in the following reactions:

$$H_2O_2 + \ ^\cdot OH \rightarrow H_2O + HO_2^\cdot \tag{11}$$

$$HO_2^\cdot + \ ^\cdot OH \rightarrow H_2O + O_2 \tag{12}$$

which consume OH.[49]

In the photocatalyzed oxidation of chlorinated compounds over TiO_2 (chloroethylenes, chloral hydrate), a maximum enhancement is observed in a H_2O_2 concentration range of 10^{-3}– 10^{-2} M. There is little effect at 10^{-4} or 10^{-1} M.[94,95] It is noteworthy that a detrimental effect is observed when H_2O_2 is added to ZnO dispersions.[95] Examples of positive effects are in the degradation of phenol,[96,97] organophosphorous derivatives at 0.1 M H_2O_2,[98] and dioxins (slightly beneficial at 0.1 M).[99] Negative effects have also been observed on chloroacetic acids,[100] and a slightly negative influence was observed for atrazine.[99] There is a small effect of the H_2O_2 presence in the degradation of a series of organochlorine solvents.[101]

Other peroxycompounds, e.g., peroxydisulfate and other strong oxidants like IO_4^-, BrO_3^-, and ClO_3^-, have been examined. Peroxydisulfate and periodate have a beneficial effect on the degradation of organophosphorous derivatives,[98] dioxins,[99] chlorophenols, and atrazine.[99] Bromate enhances the rate of organophosphorous decomposition,[98] whereas chlorate has been found to have little effect, if any, on dioxin and atrazine degradation.[99]

Peroxydisulfate, for example, can act through the initial trapping of the photogenerated electrons according to

$$S_2O_8^= + e^- \rightarrow SO_4^= + SO_4^{\cdot -} \tag{13}$$

The sulfate radical anion is a very strong oxidant ($E^\circ \cong 2.6$ V) and engages in at least three reaction modes with organic compounds: (1) by abstracting a hydrogen atom from saturated carbon, (2) by adding to unsaturated or aromatic carbon and (3) by removing one electron from carboxylate anions and from certain neutral molecules. As a result, when 2-chlorophenol is irradiated in the presence of TiO_2 in an aqueous solution of peroxydisulfate, the degradation rate and the rate of CO_2 evolution are remarkably increased.[99]

The effect of chlorine has also been investigated.[102] A slightly beneficial effect can be observed at the highest Cl_2 concentration on 2-chlorophenol and atrazine degradation. An interesting aspect of water treatment is that chlorine (at the concentration found after chlorination) has virtually no effect on the photocatalytic treatment.

Finally, the effect of such cations as Fe^{3+}, Fe^{2+}, and Ag^+ has been extensively investigated for the photocatalytic degradation of phenol.[103,104] These studies are of interest for mechanistic details.

Effect of Photon Flux and Reaction Temperature

Although the main research aim in photocatalytic detoxification has been to demonstrate the feasibility of the method, detailed mechanistic studies have been carried out in many laboratories during the past 10–15 years. In particular, the effects of reaction temperature and light flux have been studied extensively, as the knowledge of the influence of these parameters is a prerequisite for efficient reactor design. Often, yields of reaction products increase only with the square root of the increasing illumination intensity.[57,105-107] Ollis has recently presented a generalized model to account for the influence of the light intensity on photocatalytic mineralization in a solar parabolic trough reactor.[106] While a linear dependence of the reaction rate on the intensity (i.e., a constant quantum efficiency) is expected at low fluxes, the described square root law is valid in an intermediate regime. Finally, at very high fluxes (>100 suns), transport limitation (a reactor design parameter) causes the rate to remain constant upon even further increases of the illumination intensity. This method, then, does not benefit from expensive solar concentrators.

It should be noted, however, that these laboratory studies, each of which has been carried out with a special test molecule, cannot simply be generalized to account for the behavior of other chemicals. Halmann et al., for example, have observed constant quantum efficiency even at relatively high photon fluxes when they studied the photocatalytic degradation of various chlorinated alkanes.[108] Bahnemann et al. showed that while chloroform exhibited a square root dependence of its degradation rate upon the illumination intensity over a wide pH regime,[21] an investigation of the photocatalytic degradation of dichloroacetate yielded similar results only at pH 5.[109] Quantum efficiency[110] independent of the illumination intensity were obtained with the latter molecule at all other pHs studied. It was suggested from a related flash photolytic study that the nonlinearity of the product yield as a function of the photon fluence could be explained by a model based on depletion of surface-adsorbed reactants.[111] Alternatively, a kinetic model has been developed that predicts a square root dependence of the degradation rate in photocatalytic systems as a result of a competition of the oxidative attack on the pollutant molecule A (reaction in Eq. 14) with the combination of surface-bound hydroxyl radicals, $^{\cdot}OH_s$:[21]

$$A + \cdot OH_S \rightarrow degradation \tag{14}$$

$$\cdot OH_S + \cdot OH_S \rightarrow H_2O_2 \tag{15}$$

Only few studies have focused on the temperature dependence of the photocatalytic detoxification. While most authors obtained data exhibiting an Arrhenius-type behavior with reasonable accuracy (activation energies of 10 kJ/mol for phenol,[57] 11 kJ/mol for salicylic acid,[112] and 16.2 kJ/mol for dichloroacetate[109] have been reported), the reaction rate vs temperature plot was linear when chloroform was the test molecule.[21] Moreover, the degradation rate in the latter case decreased with increasing temperature, especially at high light fluxes, indicating that it might be beneficial (in special cases) to incorporate a cooling system into a photocatalytic reactor.

NEW CLASSES OF POLLUTANTS PHOTOCATALYTICALLY DEGRADED

Photocatalytic Degradation of Carboxylic Acids

Even though at first sight they may appear to be rather harmless, increasing attention has lately been focused on the photocatalytic degradation of carboxylic acids.[113-121] Halogenated derivatives of acetic acid, e.g., dichloroacetate (DCA), are carcinogenic and have been detected as intermediates during the photocatalyzed mineralization of chlorinated ethanes.[113,118] Following an earlier report by Ollis et al.,[114] Hilgendorff has shown that DCA is readily degraded on illuminated titanium dioxide following the overall stoichiometry:[119]

$$CHCl_2COO^- + O_2 \rightarrow 2CO_2 + H^+ + 2Cl^- \tag{16}$$

The unusual light intensity dependence of this reaction has already been described above.

Acetic acid (HAc) itself has been the subject of several photoelectrochemical investigations. Kraeutler and Bard reported that the so-called photo-Kolbe decarboxylation constitutes the main reaction path when HAc is oxidized on TiO_2 (rutile) electrodes:[120]

$$CH_3COO^- + h^+ \rightarrow CH_3CO_2\cdot \rightarrow \cdot CH_3 + CO_2 \tag{17}$$

followed by

$$\cdot CH_3 + \cdot CH_3 \rightarrow 2C_2H_6 \tag{18}$$

However, it has been shown that the reaction paths can be quite different when acetic acid is photooxidized on TiO_2 particles.[120,121,172] Here, the initial reaction step seems to involve formation of surface-bound hydroxyl radicals ($\cdot$OH$_s$) through the trapping of holes (h$^+$) generated in the bulk of the semiconductor particle upon bandgap illumination.

Subsequently, $\cdot$OH$_s$ preferentially abstracts hydrogen from the α-carbon atom of the acetate molecule:

$$\cdot OH_S + CH_3COO^- \rightarrow \cdot CH_2COO^- + H_2O \qquad (19)$$

Glycolate (HOCH$_2$COO$^-$) and glyoxylate (OCHCOO$^-$) have been identified as reaction products when molecular oxygen was present in the reaction mixture.[121] Indeed, photoelectrochemical studies using TiO$_2$ electrodes have confirmed that OH$_s$ is formed through the anodic reaction (Eq. 20) and not primarily as a product of the cathodic reduction of O$_2$ (Eqs. 21 and 22) which in particle suspensions parallels the oxidation process (cf. section on Added Oxidants).[110]

$$OH_S^- + h^+ \rightarrow \cdot OH_S \qquad (20)$$

$$O_2 + 2\,e^- + 2\,H^+ \rightarrow H_2O_2 \qquad (21)$$

$$H_2O_2 + e^- \rightarrow \cdot OH_S + OH^- \qquad (22)$$

Substituted Monoaromatics

Monoaromatics are the pollutants whose degradation by photocatalysis has been the most investigated. Consequently, they appear as model compounds in every section of this chapter.

Lists of these pollutants have already been published[19,61] with the corresponding references. Therefore, only more recent studies are indicated here. Monoaromatics recently investigated or reinvestigated include benzene;[92] dimethoxybenzenes;[122] chloro- and bromobenzene;[92,123] nitrobenzene;[83] chlorophenols;[40,82,83,92,124-127] fluorophenols;[128] nitrophenols;[129,130] cresols;[53] benzamide;[82,83,131,132] aniline;[88] phenylalanine;[130] salicylic, 4-aminobenzoic, and 3-chloro-4-hydroxybenzoic acids;[64] benzyl alcohol; [65] and phthalate esters.[133,134]

For several of these compounds, the beneficial effect of the semiconductor on their disappearance rate has been easily demonstrated by comparing the results obtained with and without the semiconductor in a wavelength range where the aromatic absorbs. Depending on the molecular structure of the pollutant, the aromatic intermediates formed by direct photolysis or by photocatalysis are different, as is expected from the very distinct natures of the excitation processes. Even though the photocatalyst plays the role of an internal filter and, accordingly, almost suppresses the direct photolysis, it is recommended to perform the fundamental studies at wavelengths where the aromatic pollutant does not absorb if one is to clearly discriminate between these processes. Nevertheless, in practice, there is no need to filter the excitation radiation in order to treat water.

Hydroxylated aromatic intermediates have been identified in many cases and are very unstable, with a regiochemistry as that expected from the known orientation effects of the substituents. Deactivating substituents, such as $-NO_2$, do not hinder the formation of these intermediates. As shown in the section Correlation between the Photocatalytic Degradability and other Properties of Pollutants, the reactivity of the aromatic pollutant does not depend only on the substituents, but also on the tendency of the pollutant to approach the surface of the photocatalyst or adsorb onto it (see the section Role of Adsorption). Total mineralization is achieved, i.e., monocyclic aromatics are completely destroyed by the photocatalytic treatment. In particular, carbon-fluorine bonds are completely broken (see the section Mineralization for the Role of Heteroatoms).

Miscellaneous

In addition to the previous classes of compounds, several other molecules have been investigated as substrates for photocatalytic degradation. Two exhaustive reviews have described these studies covering contributions up to the late 1980s.[20,61] Other studies have appeared in the last few years, which have dealt with molecules of particular interest in possible water treatment.

It is not the aim of this chapter to exhaustively treat these. They have been collected here for their general interest and have been classified into the following classes:

1. Haloaliphatics: in particular, halomethanes, haloethanes, and halo-ethenes[92,101,105,113,135] — This class is important because many of these compounds have been released into the environment and contaminate surface and groundwaters. Some also originate by chlorination as a water treatment.
2. Water-miscible solvents: ethanol, alkoxyethanol[136-138] and small polar molecules as chloroethylammonium salts.[105] — The compounds of this class are very difficult to detoxify since they are resistant to several treatments and are poorly adsorbed on activated carbon.
3. Pesticides: — This class contaminates surface and groundwaters where agricultural runoff is important. Interestingly, some of the pesticides examined are destroyed even when present at parts per billion levels. Among recently investigated compounds are: s-triazine derivatives;[140,141] carbammides;[44] organophosphorous derivatives;[98] and such molecules as monuron,[142] bentazon,[143] and permethrin.[144]
4. Surfactants: — Surface active agents such as nonionic and ionic surfactants reach domestic and industrial wastewaters in increasing amounts. Because their biodegradability may be one of the more important constraints in their use in detergent formulations, photocatalytic degradation has received increasing attention.[50,145,146]
5. Dyes: — Strongly colored compounds present in industrial effluents can be removed by adsorption or can be destroyed by oxidation. Destructive oxidation by photocatalysis has recently received attention.[147,148]
6. High molecular weight compounds: — These include natural compounds, as well as contaminants present in effluents from the pulp and paper mill industries.[149,150]

CORRELATION BETWEEN PHOTOCATALYTIC DEGRADABILITY AND OTHER PROPERTIES OF THE POLLUTANTS

It can be assumed reasonably that the photocatalytic degradability of an organic pollutant depends on both its ease of oxidation and the probability of the surface-formed oxidizing species to encounter the toxic substance. For aromatic compounds, the former property can be tentatively quantified by the Hammett coefficient σ_H.[151,152] The extent of adsorption or surface coverage by the pollutant should contribute to the photocatalytic degradation rate, at least for adsorbed species that are reactive. In a number of cases, the extent of adsorption is not easy to measure directly because it is too low, and the values derived from the LH kinetic model can be questioned, as has been emphasized in the section Role of Adsorption of this chapter. Therefore, the octanol-water partition coefficient, K_{ow}, was thought to provide an acceptable substitute, although it does not take into account the nature of the adsorbent, but only the hydrophobicity of the substrate, i.e., its tendency to approach the semiconductor surface or to adsorb onto it.

The validity of these hypotheses has been tested for various chlorophenols (CP) with one, two, or three chlorine atoms. The disappearance of each CP as a function of illumination time obeys an apparent first-order kinetics. The corresponding apparent rate constants, k_{app}, have been correlated to σ_H and K_{ow} by a simple linear relationship

$$k_{app} = A\,\sigma_H + B\log K_{OW} + C \qquad (23)$$

in which A, B, and C are adjustable coefficients, σ_H and K_{ow} having been taken from Reference 151. The differences between the calculated and observed values are generally small,[83] except in the case of 2,4,6-trichlorophenol for which the calculated value was much too large. The correlation coefficient is 0.973 if this compound is omitted. 2,4,6-Trichlorophenol was the only CP in which the ortho and para positions activated by the phenolic functionality were all occupied by chlorine atoms. Despite this restriction, this study shows that known characteristics of aromatic compounds, such as σ_H and K_{ow}, may be reasonably good predictors of photocatalytic degradability. Clearly, other data are needed to more firmly establish the validity of this type of calculation or perhaps to improve it. Such calculations could be useful to predict the relative efficiency of photocatalysis in the treatment of mixtures of known pollutants.

DEGRADATION PATHWAYS

Surface vs Solution Reactions

An issue often raised in discussions of photocatalyzed mineralization of organic substrates is whether the initial oxidation of the organic substrate occurs on the surface of the photocatalyst or in solution.

As pointed out in the section Influences of Aqueous-Phase pH and Surface Heterogeneity, in analyzing the kinetic data of photocatalyzed oxidations (and reductions), mediated by photoactivated semiconductor particles, several studies in the literature of the 1980s have fitted the results to the simple rate expression of the form of Eq. 4b.[40,153,154]

Determination of initial rates of photocatalyzed oxidation as a function of increased concentration of organic substrate, for a given photocatalyst loading, commonly yields the type of plots illustrated in Figure 4. The similarities of Eq. 4b, or its more complex form for a multicomponent system (Eq. 5; see also Reference 40), to the Langmuir kinetic model[155] have led to inferences that the mineralization takes place on the surface of the photocatalyst.[126,157] (Note that the values of k and K are apparent empirical constants that describe the rate of degradation under a given set of experimental conditions covering a range of solute concentrations; they have no absolute meaning.) Yet these parameters continue to be reported and detailed interpretations to be given.[157] Similar plots are obtained if one considers a bimolecular reaction between reactants X and Y. The initial rate will increase with increase in the concentration of the Y (or X)-substrate; the kinetics become pseudo-first order. Thus, a Langmuirian-type behavior does not guarantee that the process of interest occurs on the surface.

A rigorous treatment[38] of the kinetics involved in the photocatalyzed oxidations of organic substrates on an irradiated semiconductor under a variety of conditions has recently examined whether it was possible to delineate surface vs solution bulk reactions. As already mentioned in the section on Kinetics, the derived kinetic model considered four cases implicating ·OH radical attack of the organic substrate: (1) that the reaction occurs while both species are adsorbed, (2) that the reaction occurs between the adsorbed substrate and the free radical, (3) that the reaction occurs between surface-bound ·OH radical and the substrate in solution, and (4) that the reaction occurs while both species are in solution. In all cases, the analytical forms of the derived complex rate expressions were identical and were similar to that expected from the Langmuir model. Clearly, kinetic studies alone are silent as to whether photooxidations are surface or solution processes. An empirical model based on enzyme kinetics confirmed this difficulty.[53,128] In fact, insofar as the observed kinetics are the same as those generally observed in saturation-type experiments, the kinetic behavior normally observed in heterogeneous photocatalysis may have nothing to do with the Langmuir model or any other equivalent model.

Some studies have sought chemical evidence and inferences to ascertain whether or not the oxidation is a surface process. The selective inhibiting influence of isopropanol in the oxidation of furfuryl alcohol by ·OH radicals over ZnO dispersions suggested a homogeneous-phase process.[158] The relative importance of the formation of glycolate and glyoxylate via ·OH oxidation of acetate increases with increasing pH.[121,159] Because little adsorption of acetate takes place in alkaline media, the hydroxyl radicals must diffuse away from the surface of the photocatalyst to oxidize acetate in solution. Kinetic arguments, based on adsorption coefficients (K), have recently been presented[157] to the effect that at high phenol surface coverage of the TiO_2 particles, the ·OH radical reacts at the surface, while at low coverage the ·OH radical freely diffuses into the solution where it contributes significantly to the overall photooxidation. Pulse radiolytic and electron paramagnetic resonance (EPR) evidence that disputes the latter conclusion were not considered.[157]

Turchi and Ollis[38] inferred in their analysis that the photooxidative process need not occur at the surface of the catalyst, as the reactive ·OH species can diffuse several hundred angstroms into the solution bulk. Other workers have suggested that the diffusion length of an ·OH radical from the surface of TiO_2 may only be a few atomic distances or less.[160-162]

A recent photoelectrochemical study[49] is strongly supportive of a solution ·OH reactive species in some photocatalyzed oxidations. The two important anodic and cathodic reactions on the working metal electrode are embodied in Eq. 24 and 25. In a slurry photochemical reactor containing an organic substrate,

$$TiO_2^- \rightarrow TiO_2 + e^- \quad \text{(anodic)} \tag{24}$$

$$\cdot OH \rightarrow OH^- + h^+ \quad \text{(cathodic)} \tag{25}$$

the initial cathodic photocurrent rapidly became anodic under steady-state irradiation (Figure 5). Immobilization of the TiO_2 onto a conducting carbon paste in a nonuniform manner, in which presumably no TiO_2 comes in contact with the electrode, gave similar results as the slurry cell reactor. Peterson et al.[49] inferred

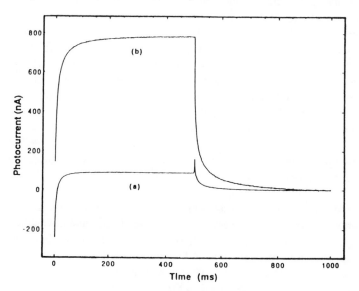

Figure 5. Transient response for TiO_2 P25 (a) without HCOOH and (b) with 0.1 M HCOOH. (Reprinted with permission from Peterson, M. W., et al., *J. Phys. Chem.*, 95, 221, 1991. Copyright © 1991 American Chemical Society.)

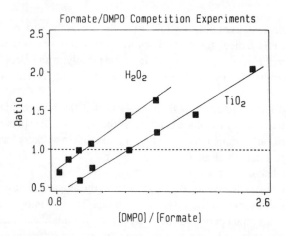

Figure 6. Formate competition analysis. Ratio denotes the ratio of the ESR peak heights of the DMPO-OH to the DMPO-CO$_2$ spin adducts. The relative [DMPO] was taken as 1. (Adapted from the data of References 160–162.

from the results that photogenerated surface-originating ·OH radicals must diffuse into solution to generate some species that induces the observed cathodic photocurrent.

By contrast, another (ESR) study[160-162] concludes that the ·OH radical does most of its work on the surface of the catalyst, and photooxidation is mostly a surface process. Irradiated H$_2$O$_2$ is a well-known source of OH radicals in homogeneous phase. Variations in the ratio [DMPO-·OH] to [DMPO-CO$_2$·] as a function of [Formate] to [DMPO] in a competition experiment between formate and the spin trap used (DMPO) for the ·OH radical, which generates ·CO$_2$- radicals[163], were followed by the ESR technique. A similar competition experiment was carried out in a heterogeneous system containing TiO$_2$, formate, and DMPO. If the ·OH radical produced by illumination of TiO$_2$ reacted in a homogeneous phase, then the variation in [DMPO-OH] to [DMPO-CO$_2$] would have to overlap that from the H$_2$O$_2$ experiment. As noted in Figure 6, the reference system H$_2$O$_2$ and the TiO$_2$ system showed different behavior, taken as evidence that the ·OH radicals react heterogeneously on the surface of the TiO$_2$ particles.[160-162]

Unfortunately, these competition experiments remain inconclusive. Considerations of the several complex events (Scheme 2) occurring on the TiO$_2$ particle and the various other species present on the surface lead us to conclude that the behavior embodied in Figure 6 should not have been unexpected. Certainly, if the two curves overlapped, all the processes indicated in Scheme 2 would probably be taking place in the bulk phase. However, the alternative, as experimentally observed, does not necessarily lead to the conclusion that the processes occur at the surface of the catalyst and that no ·OH radical leaves the surface. It suffices only that one of the events described by the $\phi'_{DMP-·OH}$ to $\phi_{DMPO-·CO_2}$ ratio take place on the surface of the catalyst for the curves of Figure 6 to not overlap completely;

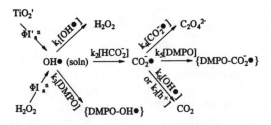

Scheme 2. Schematic representation of the reactions involving ·OH radicals in the presence of formate ions and DMPO.

the ·OH radical may still have left the surface. Thus, identical variation of the ratios would only have been observed if the parameters describing the steps were identical in both homogeneous and heterogeneous phases. The presence of a TiO_2 particle-solution interface will, no doubt, influence the kinetics (k') of the various stages in Scheme 2, as one or more may occur on the particle surface:

where

$$\phi'_{DMPO-\cdot OH}/\phi'_{DMPO-\cdot CO_2-} = k_3/\left(k_2 k_5 \left[HCO_2^-\right] \tau_{\cdot CO_2^-}\right)$$

and

$$\phi'_{DMPO-\cdot OH}/\phi'_{DMPO-\cdot CO_2-} = k'_3/\left(k'_2 k'_5 \left[HCO_2^-\right] \tau_{\cdot CO_2^-}\right)$$

It is clear that one technique alone cannot provide unambiguous conclusions and that other evidence is required to achieve a reasonable understanding of the complex events in photocatalyzed oxidation.

Further indications that the ·OH radical is surface bound, and unlikely to desorb into the solution, emanates from a recent study,[48] which notes that decafluoro-biphenyl (DFBP) is tenaciously adsorbed (>99%) on metal oxide particle surfaces (Al_2O_3 and TiO_2) and does not undergo facile exchange between the two oxide materials (<5%). When adsorbed on the alumina surface in dispersions into which H_2O_2 or a TiO_2 colloidal sol (particle size ca. 0.05 μm) is added, followed by UV irradiation, the DFBP is photodegraded. This indicates that the ·OH · radicals from H_2O_2 and TiO_2 sols (particles adsorbed on alumina) are in contact with the reaction site on the DFBP-Al_2O_3 system to initiate the photo-oxidative events. By contrast, if TiO_2 beads (size ca. 1000 μm) were used in lieu of H_2O_2 or the TiO_2 sol to generate the oxidizing species, the photodegradation is nearly suppressed and is identical to the behavior of the DFBP-Al_2O_3 system alone, irradiated with UV light under otherwise identical conditions. Pentafluorophenol, which readily exchanges between the two metal oxide surfaces, undergoes facile photodegradation under the same conditions. Thus, the photogenerated oxidizing species (·OH radical) does not migrate far from the photogenerated active sites on TiO_2, and the degradation process must occur at the photocatalyst surface or within a few atomic distances from the surface.[164]

Recent time-resolved microwave-conductivity studies[165] on TiO_2 (Degussa P25) showed a definite increase in the lifetime of the mobile charge carrier (electron) in the presence of isopropanol, resulting either from (1) scavenging of surface $OH^{\cdot}$ radicals by i-PrOH or (2) displacement of a deep surface trap by the alcohol. A similar process explained the observed photocurrent doubling effect by alcohols,[163,166,167] a process which could only take place at the TiO_2 surface.

Intermediates

Stable

Oxidative photocatalytic degradation over semiconductor metal oxides has been shown to implicate oxygenated radicals (or the addition of water to transient radicals). The consequence of this is the formation of more hydrophilic compounds with respect to the original one. However, there are some exceptions to this. For example, quinones are formed from phenols or dihydroxybenzenes, or less hydrophilic compounds may originate from a decarboxylation process or from breaking an ethoxylated chain present in nonionic surfactants.

Also, the degradation chain generally involves formation of products with a decreasing number of carbon or heteroatoms (with the exception of oxygen). Some minor pathways, however, may be operating that may lead to condensation products, for example, (hydroxy)chlorobiphenyls from chlorobenzenes[123,168] and chlorophenols.[169] These may pose serious concerns in water treatment plants.

Another feature of the degradation mechanism is that intermediates are sometimes degraded as fast or even faster than the initial compounds. In some cases (e.g., hexafluorobenzene[170] and dodecane[171]), only traces of intermediate products are detectable. Some exceptions to this are worth mentioning; s-triazine herbicides disappear rapidly, but the overall conversion to the final degradation product takes longer times, and several intermediate species have been isolated. In some cases, the final degradation steps involving small organic acids may also require longer times to form the end product.

In order to analyze completely a reaction mixture, it is necessary to have available a series of highly specialized instrumentation and techniques (GC, HPLC, IC, MS) that will permit separation of compounds with very different hydrophilic-hydrophobic characteristics and spanning very different concentration ranges (from sub-parts per billion to parts per million).

Very few detailed and quantitative degradation schemes have so far been described so far. In most cases, only the initial stage(s) has been examined and only the first stable intermediates have been detected. Secondary pathways have been generally neglected or have eluded detection altogether. Below, we illustrate some degradation schemes operating under photocatalytic conditions (different catalysts).

Even different preparations and treatments of the same catalyst may affect the quantitative distribution of the intermediates formed in a given process. This is also underlined by the importance of the different degradation routes that may prevail on different photocatalysts.

Short-Chain Aliphatics

Acetic acid on platinized titanium dioxide[172] — Scheme 3 reports the proposed reaction scheme for the photocatalytic reactions of acetic acid in the presence of TiO_2-Pt under illumination. The most characteristic aspects are hydrogen evolution and the formation of reduction products, as well as the formation of C3 and C4 compounds.

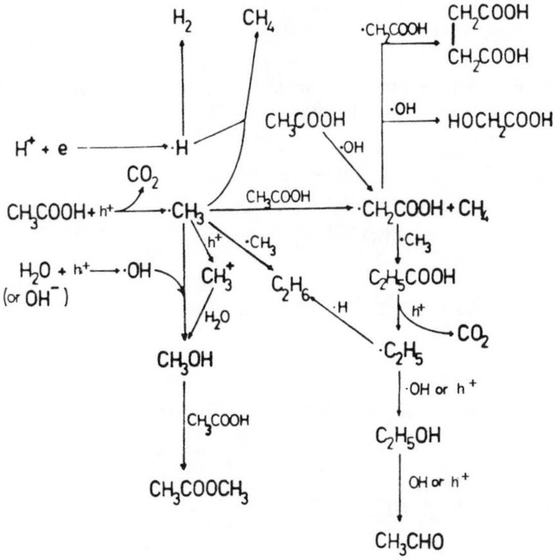

Scheme 3. Reaction scheme proposed for the photocatalytic reactions of acetic acid in aqueous medium with powdered TiO_2/Pt photocatalyst. (Adapted from Sakata, T., et al., *J. Phys. Chem.*, 88, 2344, 1984.)

Trichloroethylene on titanium dioxide — Trichloroethylene is among the most investigated compounds under photocatalytic conditions. Detection of intermediates has been carried out only in few reports, and a tentative scheme collecting the available data is presented in Scheme 4.[62,118,170]

Substituted Benzene Derivatives

Fluorophenols on titanium dioxide — The degradation mechanism of mono- and difluorophenols has been investigated in detail, including the identification of

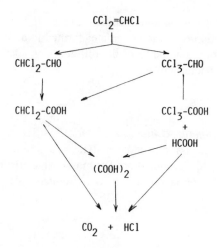

Scheme 4. Intermediates detected in the photocatalytic degradation of trichloroethylene over TiO_2. (Based on References 62, 118, and 170).

the intermediates still retaining the aromatic ring. Scheme 5 summarizes the proposed scheme for the early part of the photocatalytic degradation of ortho and para substituted mono- and difluorophenols.[128]

Dichlorobenzenes on zinc oxide — Scheme 6 presents the identified oxidation (up to the compounds retaining the ring moiety) and condensation products originating from irradiation of 1,3-dichlorobenzene over ZnO.[123] Interestingly, no condensation products were detected for irradiation of chlorophenols under the same conditions.[125]

Heteroaromatic Compounds

s-*Triazine derivatives on titanium dioxide* — As mentioned above, the photocatalytic degradation of this class of compounds presents some unusual characteristics. Many of the intermediates formed are more stable than the initial compounds. This allows a detailed identification and quantitative description of the most important intermediates, as depicted in Scheme 7.[140,141]

Transients

The ˙OH radical is a significant oxidant in the TiO_2-assisted photomineralization of many organic compounds in aqueous environments,[40,128,154,173,174] as is shown by the large number of highly hydroxylated intermediates detected for many of the phenols examined. Thus, phenol yields hydroquinone, catechol, 1,2,4-trihydroxybenzene, pyrogallol, benzoquinone, and hydroxybenzoquinone.[56,57,84] However, at least in principle, these species can also originate from the hydrolysis of the oxidized cation radical produced by direct interfacial electron transfer to an

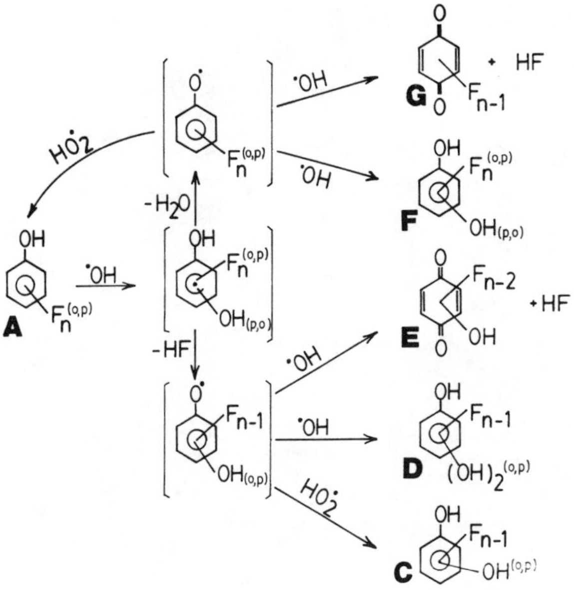

Scheme 5. Proposed reaction scheme for the initial transformations of mono- and difluorophenols (ortho and para substituted) over irradiated TiO_2. (Reprinted with permission from Minero, C., et al., *Langmuir*, 7, 928, 1991. Copyright © 1991 American Chemical Society.)

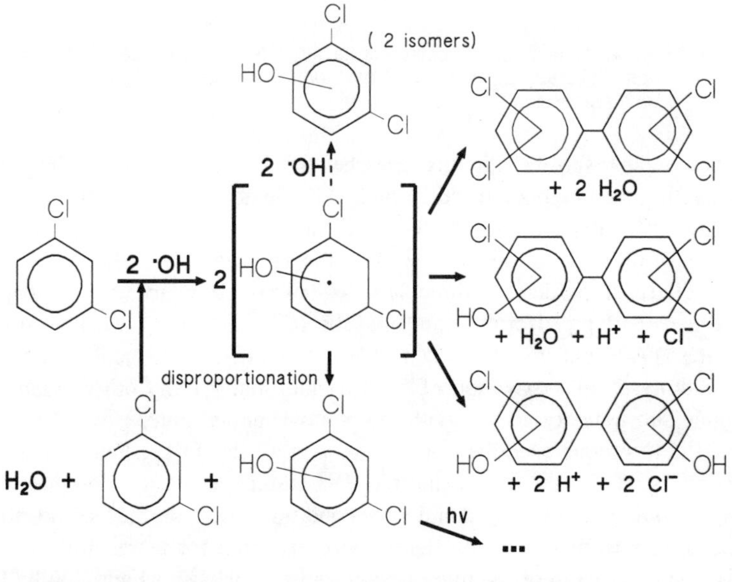

Scheme 6. Proposed reaction scheme for the initial transformations of 1,3-dichlorobenzene over irradiated ZnO. (Adapted from Sehili, T., et al., *J. Photochem. Photobiol. A: Chem.*, 50, 103, 1989. With permission.)

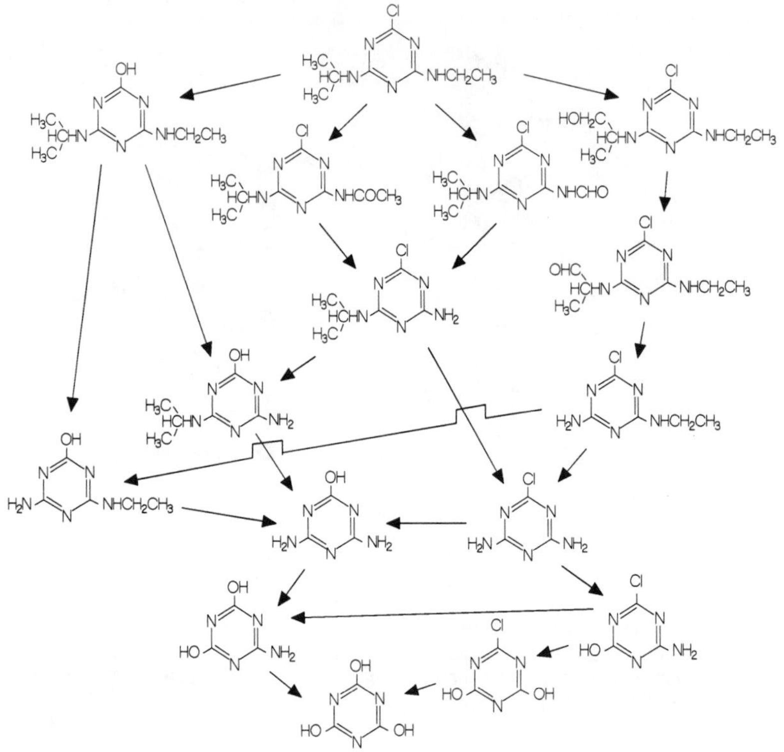

Scheme 7. Proposed degradation pathways for atrazine over irradiated TiO$_2$. (Reprinted from Pelizzetti, E., et al., *J. Photochem. Photobiol. A: Chem.*, 50, 103, 1989. With permission.)

adsorbed organic substrate, in this case phenol.[41] Details concerning the primary processes initiated by photon activation of TiO$_2$ in aqueous media continue to be debated.[42] The major source of ·OH radicals is from the valence band hole oxidation of the OH$^-$ groups (and H$_2$O) at the TiO$_2$ particle surface.[175]

The relative importance of direct hole oxidation of organic substrates depends on the extent of preadsorption and may be insignificant for weakly adsorbed substrates.[38] A recent study[41] has concluded, from diffuse reflectance nanosecond flash photolysis, that oxidation of 2,4,5-trichlorophenol and other organic and inorganic substrates originates with the photogenerated hole. However, the experimental technique used does not preclude at all the ·OH radical as the major oxidizing agent. To the extent that the ·OH adduct of the trichlorophenol, for example, could not be detected under the conditions used, the surface-bound ·OH radical differs from the chemical reactivity expected for a free fully solvated species. This question of the role of holes vs free ·OH radical addition by light-activated TiO$_2$ in aqueous media has been addressed recently.[42]

The role of electrons produced concurrently with the holes has not yet been established in the mineralization of organic substrates. However, it is not inconceivable that electrons react via a parallel process, which reduces easily reducible organic substrates, for example, the haloaliphatics. Reductive dehalogenations of haloaromatics are known to occur rapidly in steady-state radiolysis.[176]

Hydroxyl radicals add to halogenated phenols to form dihydroxycyclohexadienyl radicals (the so-called ˙OH adducts) in competition with H atom abstraction and electron transfer.[177] Table 1 summarizes the observed formation and decay kinetics of transient radicals formed for the pentahalophenol substrates.[178]

$$(26)$$

$$(27)$$

Table 1. Observed Kinetics of Formation and Decay of Transients in Various Reaction Systems

Reaction System	Formation		Decay	
	λ (nm)	k_{obs} (sec⁻¹)	λ (nm)	$2k/\epsilon$ (sec⁻¹)
PBP-O⁻ + N₃⁻ + OH·	470	$(3.5$—$12.7) \times 10^5$	470	$(1.3 \pm 0.3) \times 10^6$
			460	$(1.2 \pm 0.2) \times 10^6$
			350	$(3.7 \pm 0.7) \times 10^6$
PBP-O⁻ + N₃⁻ + ascorbate	390	$(1.0 \pm 0.2) \times 10^5$	470	$(1.1 \pm 0.2) \times 10^5$
PBP-O⁻ + OH·	460	1.6×10^6	460	$(1.3 \pm 0.2) \times 10^6$
			330	$(1.4 \pm 0.3) \times 10^6$
PBP-O· + OH⁻ + ascorbate	380	$(6.9 \pm 2.0) \times 10^4$	460	$(5.5 \pm 1.0) \times 10^4$
PBP-O· + e⁻ₐq	480	4.3×10^5	715	5.1×10^6
	330	8×10^5	460	8.6×10^5
			330	5.4×10^5
PBP-O· + e⁻ₐq + ascorbate			480	5.5×10^4
			380	9×10^4
PCP-O⁻ + N₃⁻ + OH·	440	$(6.1$—$83) \times 10^4$	440	$(7.6 \pm 1.6) \times 10^6$
PCP-O⁻ + N₃⁻ + ascorbate	360	$(7.4 \pm 1.5) \times 10^4$	440	$(7.4 \pm 1.6) \times 10^4$
PCP-O⁻ + OH·			450	$(6.5 \pm 1.0) \times 10^5$
			340	$(3.0 \pm 0.6) \times 10^6$
PCP-O· + OH⁻ + ascorbate	360	$(3.0 \pm 0.9) \times 10^4$	450	$(2.1 \pm 0.6) \times 10^4$
PFP-O⁻ + OH·	430	$(3.0$—$24) \times 10^5$	430	$(7.0 \pm 1.4) \times 10^5$
	300	$(3.1$—$21) \times 10^5$		
PFP-O⁻ + N₃⁻ + OH·	320	1.9×10^4	300	$(2.8 \pm 0.3) \times 10^5$

PXP-O = pentohalophenol; X = B, bromine; X = C, chlorine; X = F, fluorine.

Reprinted with permission from Terzian, R., et al., *Langmuir*, 7, 3081, 1991. Copyright © 1991 American Chemical Society.

Table 2. Comparison of the Kinetic Properties of PXP-O with Values Published for Similar Radicals

Radical	$k_f(M^{-1}\ s^{-1})$	$k_d(M^{-1}\ s^{-1})$
phenoxyl radical	4.3×10^9	—
2,4,5-trichlorophenoxyl radical	4.3×10^9	7.7×10^8
4-fluorophenoxyl radical	4.6×10^9	—
2,4-dibromophenoxyl radical	—	1×10^8
pentachlorophenoxyl radical	3.7×10^9	9.1×10^8
pentabromophenoxyl radical	5.9×10^9	2.8×10^9

Reprinted with permission from Terzian, R., et al., *Langmuir*, 7, 3081, 1991. Copyright © 1991 American Chemical Society.

Tables 2 and 3 compare the kinetic properties of pentahalophenoxyl radicals and of ·OH adducts for pentahalophenols with various other phenolic substrates (A).

Phenoxide ions react with electrons to give the corresponding electron adduct, which loses halogen to produce isomeric hydroxyphenyl radicals (Eq. 29).[176,179,180] Their bimolecular rate of formation, $k_f = 2.6 \times 10^{10}\ M^{-1}s^{-1}$ is one to two orders of magnitude greater than those of 2-bromophenoxide and 4-chlorophenoxide anions (Table 4).[179,180]

Reaction of the OH radical with pentafluorophenol yields exclusively the ·OH adduct, HO-PFP-O·, while pentabromo- and pentachlorophenol react with ·OH· to form mainly the respective phenoxyl radicals, i.e., electron-transfer products (ca. 75% PBP-O· and ca. 77% PCP-O·). The nature of the halogen substituents bears directly on the reaction pathway of pentahalophenols, ranging from simple addition (PFP-O⁻) to mainly electron transfer (PCP-O⁻ and PBP-O⁻). The addition of ·OH radicals to aromatic rings is not the sole reaction mechanism for

Table 3.   Comparison of the Kinetic Properties of HO-PXP-O⁻ with Values Published for Similar Radicals

Radical	$k_f(M^{-1}\ s^{-1})$	$k_d(M^{-1}\ s^{-1})$
	1.2×10^{10}	—
	1.2×10^{9}	5.6×10^{8}
	9.6×10^{9}	—
	1.4×10^{9}	—
	4.6×10^{9}	5.4×10^{8}
	3.8×10^{8}	—

pentahalophenols in homogeneous phase, and the nature and extent of the products formed vary with the halogen. In heterogeneous media (TiO$_2$ dispersions), photooxidations may proceed via analogous pathways, even though analysis of products at intermediate stages of oxidation would suggest OH additions to aromatic ring.[56,57,84,128] Recent work[41,42] has begun to address the details of these pathways. It would seem that direct oxidation by valence-band holes and oxidation via the intermediacy of OH radicals cannot be distinguished in aqueous media.[42] The nature of the halogen influences the rates of formation of the various radical products in pentachlorophenols; for OH adducts, k_{add} varies in the order F > Br > Cl, while for the phenoxyl radicals, k_{et} changes as Br > Cl; the reverse is true for the production of semiquinone radicals (Cl > Br).

Electrons react reductively with haloaromatics (as noted for PBP-O⁻ $k_e = 2.6 \times 10^{10}\ M^{-1}\ s^{-1}$) to yield isomeric tetrahalophenoxyl radicals. The possible direct role of conduction-band electrons as a route to adsorbed radical anion in TiO$_2$-catalyzed photodegradation cannot be dismissed for all substrates.[113]

Table 4.   Comparison of the Rate of Reaction of PBP-O⁻ with e⁻_eq with Published Values for Similar Compounds

Compound	$k(M^{-1}\ s^{-1})$
	2.3×10^9
	6.4×10^8
	2×10^9
	2.6×10^{10}

MINERALIZATION

Because complete mineralization is usually desirable in a water-treatment process, increased attention has been focused in demonstrating the importance of not only the disappearance of the initial reactant, but also the formation of inorganic carbon (CO_2). This can be followed either through the appearance of CO_2 and/or the disappearance of total organic carbon. Thus, a complete mass balance of photocatalytic degradation processes is imperative and is generally reported in studies that envisage application of heterogeneous water treatment.

Some cases of interest that may be of more immediate concern are the low- to medium- molecular weight compounds. These are noted below. Aromatic compounds containing heteroatoms also present some specific problems.

Compounds Containing C, H, O

In these cases the following stoichiometric equation generally holds:

$$C_n H_m O_p + xO_2 \rightarrow nCO_2 + m/2H_2O \tag{28}$$

Examples of this class are aliphatic hydrocarbons (e.g., dodecane[171]) and their derivatives (e.g., dodecanol, dodecanoic acid[171]), as well as other shorter aliphatic oxygenated compounds[136-139], aromatic hydrocarbons and oxygenated aromatics (e.g., phenols[96] and cresols[53]), and even nonionic surfactants (ethoxylated nonylphenol, average molecular weight up to 800[50]).

Organic Compounds Containing Halogen Atoms

Chlorinated aliphatic and aromatic derivatives have received the most attention. For chlorine- and fluorine-containing derivatives, the quantitative formation of the corresponding halide has been assessed according to the stoichiometry:

$$C_n H_m O_p X_q + xO_2 \rightarrow nCO_2 + yH_2O + zHX \tag{29}$$

Examples are the haloaliphatics (chloromethanes, -ethanes, and -ethenes,[62,92,101,113]) and haloaromatics (chlorobenzenes,[181] DDT,[90] chlorophenols,[35,127] and fluorophenols[128]).

A few cases have concerned the degradation of bromine-containing compounds: 1-bromododecane,[171] 1,2-dibromoethane and bromomethanes,[182] and halothane.[183] In these, bromide is recovered quantitatively.

More intriguing is the case of iodine-containing compounds. To our knowledge, no data are available. It is worth noting here that iodide is oxidized to I_2^- under photocatalytic conditions.[41,184,185]

Sulfur-Containing Compounds

Under photocatalytic oxidative conditions, sulfur is recovered as sulfate from many of these compounds, irrespective of its initial oxidation state, according to

$$C_n H_m O_p S_r + xO_2 \rightarrow nCO_2 + yH_2O + zH_2SO_4 \tag{30}$$

Equation 30 has been verified for, as examples, sodium dodecylsulfate ($-SO_4Na$),[171] sodium benzenesulfonate ($-SO_3Na$),[186] bentazon ($-SO_2-$),[143] and prometryn ($-S-$).[140]

Other compounds reported are thioridazine hydrochloride (S included in heterocyclic moiety), penicillamine,[130] and parathion.[98]

Phosphorous-Containing Compounds

Even for these heteroatomic substances, under photocatalytic oxidative conditions, the highest oxidation state of phosphorous, i.e., PO_4^{3-}, has been recovered quantitatively. Some of the molecules that have been examined include phosphorous in different oxidation states, for example, the organoposphorous derivatives (phosphates, phosphinates, and thiophosphates)[98,130] and some in which phosphorous is included in the aromatic moiety, e.g., cyclophosphamide.[130]

Nitrogen-Containing Compounds

Low et al. recently reported an extensive study on the behavior of nitrogen-containing substances transformed under photocatalytic oxidative conditions.[130]

Titanium dioxide-mediated illumination of organic compounds containing nitrogen heteroatoms yields both ammonia and nitrate ions. The relative concentration of these species depends largely on the nature of the nitrogen atom in the compound, illumination time, and initial solute concentration. Aliphatic amines produce a higher ammonium to nitrate concentration ratio than compounds containing ring nitrogen. This showed that reactions involving ring-nitrogen compounds are more complex and that there are significant reaction paths leading to nitrate ions in addition to ammonium ions. The nitrate ion formed at significant rates in the photocatalytic oxidation of ammonia and at much lower rates than from ammonium ions. Nontrivial quantities of ammonium ions were found with the photocatalytic oxidation of compounds containing nitro and nitrate groups. Although the conversion of ammonia to nitrate was facile, the reverse process was not. The conversion of nitrate to ammonium ion was enhanced when the reacting solution contained a hole scavenger.

Incomplete Mineralization

A particular case in which complete transformation into inorganic ions is not realized is represented by the class belonging to s-triazine derivatives. Under oxidative photocatalytic conditions, the trihydroxy derivative, i.e., cyanuric acid, is formed. This product is remarkably stable to further degradation at wavelengths of irradiation above 300 nm.

With the exception of these triazines, all the elements present in low- to medium-molecular weight organics are converted to their inorganic forms under TiO_2-mediated photocatalytic oxidative conditions, e.g., carbon is converted to carbon dioxide, sulfur to sulfate, phosphorous to phosphate, halogens (F,Cl,Br) to halides, and nitrogen to both ammonium and nitrate ions.

PHOTOCATALYST FORM

Different Preparations of Photocatalysts

Wherever different semiconductor materials have been tested under comparable conditions for the degradation of the same compound, TiO_2 has generally exhibited the highest activity.[174,181,187] Only ZnO possesses an activity similar to that of TiO_2, which has also been observed for chlorofluoromethanes[188] and chloroaromatic derivatives.[123,125,189] However, zinc oxide dissolves in acidic and alkaline solutions[190] and therefore cannot be used for most technical applications. The anatase modification of titanium dioxide shows much higher photocatalytic activity than the rutile form,[84,97,191] which has been explained by a faster recombination of charge carriers in rutile and also by a considerably lower amount of reactants adsorbed on the surface of the semiconductor particle.[84] The latter effect can be rationalized by a lower number of surface hydroxyl groups present on rutile.[33] The role of adsorption on efficient PCD has been discussed in the section entitled Role of Adsorption.[191-193] It should be pointed out at this point that only relative values for the efficiency of a particular photocatalytic system are given in the majority of the literature cited above. Absolute determinations of quantum yields are usually not presented, and the available experimental details are normally not sufficient to estimate yields. Therefore, it is almost impossible to compare data from different laboratories. However, a detailed discussion of this argument is available.[110]

Titanium dioxide, the material with the highest photocatalytic detoxification efficiency, is a wide band gap semiconductor ($E \cong 3.2$ eV). Thus, only light below 400 nm is absorbed and capable of forming e^--h^+ pairs.[194] Therefore, only less than 5% of the solar energy reaching the surface of the Earth, in principle, could be utilized when TiO_2 is used as a photocatalyst. Hence, it is evident that, especially for solar applications, other materials have to be found or developed that exhibit similar efficiencies as anatase, but possess spectral properties more closely adapted to the terrestrial solar spectrum. Moreover, even reactors using artificial light sources will be much cheaper in their set-up, and running cost of lamps with a radiant flux in the visible could be employed.

Unfortunately, the choice of convenient alternatives to substitute for titanium dioxide in photocatalytic detoxification systems is limited. The appropriate semiconducting material should

- not contain any toxic constituents
- be stable in aqueous solutions containing highly reactive and/or toxic chemicals
- not be subject to photocorrosion under band gap illumination

Generally, metal oxides fulfill these requirements rather well and have been shown to be extremely stable in an aqueous environment even under continuous illumination.[195] Unfortunately, most metal oxides are large bandgap semiconductors or insulators.[196] Iron(III) oxide is one of the few exceptions, with a band gap energy of ca. 2.2 eV (for hematite, α-Fe_2O_3).[196] Hence, α-Fe_2O_3 absorbs already in the visible below 560 nm. However, hematite particles only exhibit a satisfactory photocatalytic activity in a limited number of cases, e.g., for the oxidation of sulfite.[197] For most other reactions which have been studied, α-Fe_2O_3, which exhibits by far the highest photolytic activity of all Fe(III) oxide modifications, shows very little, if any, photocatalytic action as compared with TiO_2 or ZnO.[198] Often, photochemical transformations in the presence of Fe(III) oxides are well explained by the formation of light-absorbing surface complexes capable of LMCT or MLCT transitions rather than by their solid state properties.[199]

Following the early observations by MacNevin and Ogle who studied the phototropy of titanates in the presence of transition-metal impurities,[200] various attempts have been made to extend the action spectrum of the otherwise "ideal" photocatalyst, TiO_2, into the visible part of the absorption spectrum.[201-219] In particular, the incorporation of transition-metal dopants at relatively high concentrations (0.01, ..., 10%) results in the anticipated visible light response of the material.[201-209] Goodenough has presented a model using localized, metastable, occupied states to explain these observations.[205] Transition-metal doping did not result only in a change of the absorption spectrum of the material, but also in its photochemical activity upon excitation in the visible. Several authors observed the generation of hydrogen and oxygen.[202,209] Wong and Malati detected the reduction of carbon dioxide,[208] and Augugliaro et al.,[210] Conesa et al.,[211] and Marin et al.[212] reported that molecular nitrogen could be reduced to ammonia. While most of the authors agreed that chromium-doped TiO_2 was the most active material for the above transformations, Martin et al. showed that the addition of chromium ions is detrimental to the photodegradation of phenol.[212] Following an earlier observation by Moser et al. that doping with 0.5% Fe(III) augments the charge-carrier lifetime in titanium dioxide from 30 nsec to minutes or even hours,[213] Navio et al. have shown recently that although the presence of less than 0.5% iron ions is beneficial to the photoreduction of methylviologen (MV^{2+}), an increased amount of iron sharply reduces the MV^+ yield.[214] Bickley and co-workers have prepared various iron-titanium photocatalysts and studied their structural and morphological properties extensively.[215-217] They were able to show that well-dispersed biphasic solids are formed when the iron content is around 1 atom%.

Investigations of the effect of transition-metal doping on the activity of the material as a potential photocatalyst for the detoxification of polluted water have only been started recently. Using Nb-doped membranes, Sabate et al. have observed a drop in the rate of degradation of 3-chlorosalicylic acid, as compared with an undoped membrane.[219] On the other hand, Bahnemann and co-workers have reported a considerable increase in the rate of degradation of dichloroacetate when they employed Ti(IV)-Fe(III) mixed-oxide colloids, instead of pure TiO_2 colloids, as photocatalysts.[109] Photodissolution of these mixed-oxide colloids presents, however, still a serious drawback of these novel catalysts.

Another approach to improve the photocatalytic activity of titanium dioxide involves the deposition of minute amounts of noble metal on its surface, e.g., silver[220] or gold.[221] This treatment does not extend the spectral response above the limit determined by the bandgap energy, but efficient visible light sensitization of TiO_2 can be achieved by surface complexation with, e.g., transition-metal cyanides.[222,223]

Effect of Supported Catalysts

The vast majority of the studies reported so far have been carried out in aqueous suspensions of small semiconductor particles. From a practical viewpoint, however an immobilized catalyst would be preferred (if it exhibits a similar photocatalytic activity), since its separation from treated water would be simplified. Several different methods for the immobilization of the photocatalyst have been described in the literature.[1,68,112,154,224-229] While most of the authors coated the surfaces of their plates,[154] coils,[224] tubes,[225] or glass beads[226,227] with TiO_2 particles, Sabate and co-workers prepared TiO_2 membranes.[68,228] Generally, the efficiency of the overall process decreases as the catalyst is immobilized, e.g., suspended particles exhibit a fourfold higher rate for the reduction of Cr(VI) than the respective membrane system.[228] It is not clear whether mass-transfer limitations can account for all these differences.[1,224]

SOLAR-DRIVEN PILOT PLANTS

Various reactor designs have been tested, initially by Matthews, for their applicability in solar-driven pilot plants.[126,230] He compared the efficiency of titanium dioxide attached to glass mesh with that of a thin film coated on the inside of a glass coil using a parabolic trough solar concentrator. In his study with the model compound salicylic acid, Matthews achieved the maximum destruction rate with the glass coil set-up.[230] The latter reactor type has, in fact, been chosen for the first solar detoxification loops constructed in the United States at Sandia National Laboratories in Albuquerque, NM[231,232] and in Europe at the Plataforma Solar de Almeria, Spain.[233] While this glass-tube trough system indeed exhibits considerable activity for the destruction of the model compounds salicylic acid,[231] trichloroethylene,[232] and pentachlorophenol,[233] an alternative falling film-heliostat system tested under comparable conditions permitted even higher degradation rates.[231] A similar falling film reactor has recently been constructed at the University of Campinas, Brazil and showed a very high activity for the degradation of dichloroacetate.[109] However, much remains to be done to understand the events during degradation processes that employ concentrated sunlight.[234] Very little is known about high-temperature and/or high light intensity photochemistry.

COMPARISON WITH OTHER OXIDATION TREATMENTS

It seems premature to compare the economy of photocatalysis with those of the other so-called "advanced oxidation processes" (AOPs) of water treatment. However, pertinent arguments have been advanced.[96,131,235-238]

We have indicated in the section entitled Effects of Various Factors the relative insensitivity of the photocatalytic method, as compared with the other AOPs, to the pH and the common salts at the usual concentrations in the natural waters. The absorption of TiO_2 in the near-UV region and the high absorption coefficient of this solid is another advantage of the photocatalytic method over the H_2O_2-UV and O_3-UV processes which require light of shorter wavelengths and therefore cannot use sunlight. In addition, the absorption coefficient of hydrogen peroxide is very low, and the solubility of ozone in water limits the efficiency of the O_3-based treatments.

Until now, direct comparisons of the efficiencies of photocatalysis with other AOPs have been rare.[96,131,235,238] These comparisons have no absolute meaning, since their conclusions depend not only on the pollutant, but also on factors that differently affect the treatments. For instance, the relationship between the initial pollutant concentration and the initial rate of removal is not the same for the heterogeneous system (Langmuir form) as for a treatment in a homogeneous phase (linear relationship). Changes in the UV-lamp emission have different effects on the dissociation of H_2O_2 or on the excitation of TiO_2. The choices of the titania specimen and the concentrations of H_2O_2 and Fe^{2+} ions obviously influence the results, as well as the pH and the presence of certain ions (see "Effects of Various Factors") not to mention any pretreatment of TiO_2.

However, under reasonable conditions for each method, useful indications can be obtained, since in some cases substantial improvements are observed that cannot be overcome by adjusting the above-mentioned parameters. For example, as compared with the photocatalytic method, the Fenton reagent does not readily destroy benzamide and is not as efficient for the complete dearomatization of 4-hydroxybenzamide.[131] By contrast, if the results obtained by the TiO_2-UV and H_2O_2-UV systems do not differ much for a chosen set of conditions, the assessment of these treatments to eliminate the pollutant considered (such as trichloro-2,2,2 ethanol hydrate,[57] dimethyl-2,2-dichlorovinyl phosphate,[238] phenol,[96] and benzamide[131]) requires finer comparisons.

In conclusion, these comparative studies are of interest to determine at an early stage which method(s) is the most appropriate.

ACKNOWLEDGMENTS

Our work was generously supported by several agencies: EC (DB,JC,EP,PP); CNR and ENIRICERCHE (EP), CIEMAT (EP,DB); NSERC (NS); the U.S. Army Research Office (MAF); BMFT and Stifterverband der deutschen Wissenschaft (DB). MAF, EP, PP, NS are also grateful to NATO for an exchange grant.

REFERENCES

1. Turchi, C. S. and Ollis, D. F., Photocatalytic reactor design: an example of mass-transfer limitations with an immobilized catalyst, *J. Phys. Chem.*, 92, 6852, 1988.
2. Cooper, W. J., Nickelsen, M. G., Waite, T. D., and Kurucz, C. N., High energy electron beam irradiation: an advanced oxadition process for the treatment of aqueous based organic hazardous wastes, *J. Environ. Sci. Health*, A27, 219, 1991.
3. Pelizzetti, E. and Serpone, N., Eds., *Homogeneous and Heterogeneous Photocatalysis*, D. Reidel, Dordrecht, 1986.
4. Schiavello, M., Ed., *Photocatalysis and Environment*, D. Reidel, Dordrecht, 1988.
5. Serpone, N. and Pelizzetti, E., Eds., *Photocatalysis. Fundamentals and Applications*, Wiley, New York, 1989.

6. Pelizzetti, E. and Schiavello, M., Eds., *Photochemical Conversion and Storage of Solar Energy*, Kluwer, Dordrecht, 1991.

7. Schiavello, M., Ed., *Photoelectrochemistry, Photocatalysis and Photoreactors*, D. Reidel, Dordrecht, 1985.

8. Bard, A. J., Spin trapping and electron spin resonance detection of radical intermediates in the photodecomposition of water at TiO_2 particulate system, *J. Phys. Chem.*, 83, 3146, 1979.

9. Fox, M. A. and Chen, C. C., Mechanistic features of a semiconductor photocatalyzed olefin-to-carbonyl oxidative cleavage, *J. Am. Chem. Soc.*, 103, 6757, 1981.

10. Fox, M. A., Organic heterogeneous photocatalysis: chemical conversion sensitized by irradiated semiconductors, *Acc. Chem. Res.*, 16, 314, 1983.

11. Fox, M. A., Photoinduced electron transfer in arranged media and on semiconductor surfaces, *Pure Appl. Chem.*, 60, 1013, 1988.

12. Fox, M. A., Photocatalytic oxidation of organic substrates, in *Photocatalysis and Environment*, Schiavello, M., Ed., D. Reidel, Dordrecht, 1988, 445.

13. Fox, M. A., Selective formation of organic compounds by photoelectrosynthesis, *Top. Curr. Chem.*, 142, 72, 1987.

14. Fox, M. A., Charge injection into semiconductor particles. Importance for photocatalysis, in *Homogeneous and Heterogeneous*, Pelizzetti, E. and Serpone, N., Eds., D. Reidel, Dordrecht, 1986, 361.

15. Fox, M. A., Chen, C. C., Park, K., and Younathan, J. N., Controlled organic redox reactivity on irradiated semiconductor surfaces, *Am. Chem. Soc. Symp. Ser.*, 278, 69, 1985.

16. Fox, M. A., Organic photoelectrochemistry, *Top. Org. Electrochem.*, 4, 177, 1985.

17. Kaeuttler, B. and Bard, A. J., Heterogeneous photocatalytic synthesis of methane from acetic acid. A new Kolbe reaction pathway, *J. Am. Chem. Soc.*, 100, 2339, 1978.

18. Ollis, D. F., Pelizzetti, E., and Serpone, N., Photocatalytic destruction of water contaminants, *Environ. Sci. Technol.*, 25, 1522, 1991.

19. Fox, M. A., Environmental decontamination with sunlight: photocatalysis on irradiated semiconductor suspensions, *Chemtech*, 22, 680, 1992.

20. Pelizzetti, E., Minero, C., and Maurino, V., The role of colloidal particles in the photodegradation of organic compounds of environmental concern in aquatic systems, *Adv. Colloid Interface Sci.*, 32, 271, 1990.

21. Bahnemann, D., Bockelmann, D., and Goslich, R., Mechanistic studies of water detoxification in illuminated TiO_2 suspensions, *Sol. Energy Mater.*, 24, 564, 1991.

22. Bard, A. J., Photoelectrochemistry, *Science*, 207, 139, 1980.

23. Pichat, P., Mozzanega, M. N., and Courbon, H., Investigation of the mechanism of photocatalytic alcohol dehydrogenation over Pt/TiO_2 using poisons and labelled ethanol, *J. Chem. Soc. Faraday Trans. I*, 83, 697, 1987.

24. Kobayakawa, K., Nakaszawa, Y., Ikeda, M., Sato, Y., and Fujishima, A., Influence of the density of surface hydroxyl groups on TiO_2 photocatalytic activities, *Ber. Bunsenges. Phys. Chem.*, 94, 1439, 1990.

25. Heller, A., Degani, Y., Johnson, D. W., Jr., and Gallagher, P. K., Controlled suppression and enhancement of the photoactivity of titanium dioxide, *J. Phys. Chem.*, 91, 5987, 1987.

26. Becker, W. G., Truong, M. M., Ai, C. C., and Hamel, N. N., Interfacial factors that affect the photoefficiency of semiconductor sensitized oxidations in nonaqueous media, *J. Phys. Chem.*, 93, 4882, 1989.

27. Rothenberger, G., Moser, J., Graetzel, M., Serpone, N., and Sharma, D. K., Charge carrier trapping and recombination dynamics in small semiconductor particles, *J. Am. Chem. Soc.*, 107, 8054, 1985.

28. Fox, M. A. and Chen, C. C., Electronic effects in semiconductor-photocatalyzed oxidative cleavage of olefins, *Tetrahedron Lett.*, 24, 547, 1983.

29. Muzyka, J. and Fox, M. A., Oxidative photocatalysis in the absence of oxygen: methyl viologen as an electron trap in the TiO_2-mediated photocatalysis of the Diels-Alder dimerization of 2,4-dimethyl-1,3-pentadiene, *J. Photochem. Photobiol. A: Chem.*, 57, 27, 1991.

30. Gerischer, H. and Heller, A., Photocatalytic oxidation of organic molecules at TiO$_2$ particles by sunlight in aerated water, *J. Electrochem. Soc.*, 139, 113, 1992.

31. Ward, M. D. and Bard, A. J., Photocurrent enhancement via trapping of photogenerated electrons of TiO$_2$ particles, *J. Phys. Chem.*, 86, 3599, 1982.

32. Dunn, W. W., Aikawa, Y., and Bard, A. J., Characterization of particulate titanium dioxide photocatalysts by photoelectrophoretic and electrochemical measurements, *J. Am. Chem. Soc.*, 103, 3456, 1981.

33. Sclafani, A., Palmisano, L., and Schiavello, M., Influence of the preparation methods of TiO$_2$ on the photocatalytic degradation of phenol in aqueous dispersion, *J. Phys. Chem.*, 94, 829, 1990.

34. Al-Sayyed, G., D'Oliveira, J.-C., and Pichat, P., Semiconductor- sensitized photodegradation of 4-chlorophenol in water, *J. Photochem. Photobiol. A: Chem.*, 58, 99, 1991, and references therein.

35. D'Oliveira, J.-C., Al-Sayyed, G., and Pichat, P., Photodegradation of 2- and 3-chlorophenol in TiO$_2$ aqueous suspensions, *Environ. Sci. Technol.*, 24, 990, 1990.

36. Jenny, B. and Pichat, P., Determination of the actual photocatalytic rate of H$_2$O$_2$ decomposition over suspended TiO$_2$. Fitting to the Langmuir-Hinshelwood form, *Langmuir*, 7, 947, 1991.

37. Jaeger, C. D. and Bard, A. J., Spin-trapping and electron spin resonance detection of radical intermectiates in the photodecomposition of water at TiO$_2$ particulate systems, *J. Phys. Chem.*, 83, 3146, 1979, and references cited therein.

38. Turchi, C. S. and Ollis, D. F., Photocatalytic degradation of organic water contaminants: mechanisms involving hydroxyl radical attack, *J. Catal.*, 122, 178, 1990.

39. Draper, R. B., Fox, M. A., Pelizzetti, E., and Serpone, N., Pulse radiolysis of 2,4,5-trichlorophenol: hydroxytrichlorocyclohexadienyl, trichlorophenoxyl and dihydroxy-trichlorocyclohexadienyl radicals, *J. Phys. Chem.*, 93, 1938, 1989.

40. Serpone, N., Al-Ekabi, H., Patterson, B., Pelizzetti, E., Minero, C., Pramauro, E., Fox, M. A., and Draper, R. B., Kinetic studies in heterogeneous photocatalysis. II. The TiO$_2$ mediated degradation of 4-chlorophenol alone and in a three-component mixture of chlorophenol, 2,4-dichlorophenol, and 2,4,5-trichlorophenol, *Langmuir*, 5, 250, 1989.

41. Draper, R. B. and Fox, M. A., Titanium dioxide photosensitized reactions studied by diffuse reflectance flash photolysis in aqueous suspensions of TiO$_2$ powder, *Langmuir*, 6, 1396, 1990.

42. Lawless, D., Serpone, N., and Meisel, D., Role of OH radicals and trapped holes in photocatalysis. A pulse radiolysis study, *J. Phys. Chem.*, 95, 5166, 1991.

43. Draper, R. B. and Fox, M. A.,Titanium dioxide phocatalyzed oxidation of thiocyanate. (SCN)$_2^-$ studied by diffuse reflectance flash photolysis, *J. Phys. Chem.*, 94, 4628, 1990.

44. Pouyet, B., personal communication, 1992.

45. Ceresa, E. M., Burlamacchi, L., and Visca, M., An ESR study on the photoreactivity of TiO$_2$ pigments, *J. Mater. Sci.*, 18, 289, 1983.

46. Gonzales-Elipe, A. R., Munuera, G., and Soria, J., Photo-adsorption and photo-desorption of oxygen on highly hydroxylated TiO$_2$ surfaces, *J. Chem. Soc. Faraday Trans. I*, 75, 748, 1979.

47. Boonstra, A. H. and Mutsaers, C. A. H. A., Adsorption of hydrogen peroxide on the surface of titanium dioxide, *J. Phys. Chem.*, 79, 1940, 1975.

48. Minero, C., Catozzo, F., and Pelizzetti, E., Role of adsorption in photocatalyzed reactions of organic molecules in aqueous TiO$_2$ suspensions, *Langmuir*, 32, 451, 1992.

49. Peterson, M. W., Turner, J. A., and Nozik, A. J., Mechanistic studies of the photocatalytic behavior of TiO$_2$. Particles in a photoelectrochemical slurry cell and the relevance to photodetoxification reactions, *J. Phys. Chem.*, 95, 221, 1991.

50. Pelizzetti, E., Minero, C., Maurino, V., Sclafani, A., Hidaka, H., and Serpone, N., Photocatalytic degradation of nonylphenol ethoxylated surfactants, *Environ. Sci. Technol.*, 23, 1385, 1989.

51. Laidler, K. J., *Chemical Kinetics*, 3rd ed., Harper & Row, New York, 1987.

52. Terzian, R., Serpone, N., Minero, C., Pelizzetti, E., and Hidaka, H., Kinetic studies in heterogeneous photocatalysis. 4. The photomineralization of a hydroquinone and a catechol, *J. Photochem. Photobiol. A: Chem.*, 55, 243 1990.

53. Terzian, R., Serpone, N., Minero, C., and Pelizzetti, E., Photocatalyzed mineralization of cresols in aqueous media with irradiated titania, *J. Catal.*, 128, 352, 1991.

54. Abdullah, M., Low, G. K.-C., and Matthews, R. W., Effects of common inorganic anions on rates of photocatalytic oxidation of organic carbon over illuminated titanium dioxide, *J. Phys. Chem.*, 94, 6820, 1990.

55. Schrauzer, G. N. and Guth, T. D., Photolysis of water and photoreduction of nitrogen on titanium dioxide, *J. Am. Chem. Soc.*, 99, 7189, 1977.

56. Okamoto, K., Yamamoto, Y., Tanaka, H., Tanaka, M., and Itaya, A., Heterogeneous photocatalytic decomposition of phenol over TiO_2 powder, *Bull. Chem. Soc. Jpn.*, 58, 2015, 1985.

57. Okamoto, K., Yamamoto, Y., Tanaka, H., and Itaya, A., Kinetics of heterogeneous photocatalytic decomposition of phenol over anatase TiO_2 powder, *Bull. Chem. Soc. Jpn.*, 58, 2023, 1985.

58. Pichat, P. and Herrmann, J.-M., Adsorption-desorption, related mobility and reactivity in photocatalysis, in *Photocatalysis. Fundamentals and Applications*, Serpone, N. and Pelizzetti, E., Eds., Wiley, New York, 1989, 217.

59. Al-Ekabi, H. and Serpone, N., Mechanistic implications in surface photochemistry, in *Photocatalysis. Fundamentals and Applications*, Serpone, N. and Pelizzetti, E., Eds., Wiley, New York, 1989, 457.

60. Movius, W. G. and Linck, R. G., Studies on the rate of reduction of ruthenium(III) complexes by chromium(II) and vanadium(II), *J. Am. Chem. Soc.*, 92, 2677, 1970.

61. Matthews, R. W., Environment: photochemical photocatalytic processes. degradation of organic compounds, in *Photochemical Conversion and Storage of Solar Energy*, Pelizzetti, E. and Schiavello, M., Eds., Kluwer, Dordrecht, 1991, 427 and references therein.

62. Pruden, A. L. and Ollis, D. F., Photoassisted heterogeneous photocatalysis: the degradation of trichloroethylene in water, *J. Catal.*, 82, 404, 1983.

63. Matthews, R. W., Carbon dioxide formation from organic solutes in aqueous suspension of ultraviolet-irradiated TiO_2. Effect of solute concentration, *Aust. J. Chem.*, 40, 667, 1987.

64. Cunningham, J. and Al-Sayyed, G., Factors influencing efficiencies of TiO_2-sensitized photodegradations. Part 1. Substituted benzoic acids: discrepancies with dark-adsorption parameters, *J. Chem. Soc. Faraday Trans.*, 86, 3935, 1990.

65. Cunningham, J. and Srijaranai, S., Sensitized photo-oxidations of dissolved alcohols in homogeneous and heterogeneous systems. Part 2. TiO_2-sensitized photodehydrogenations of benzyl alcohol, *J. Photochem. Photobiol. A: Chem.*, 58, 361, 1991.

66. Heimenz, P. C., *Principles of Colloid and Surface Chemistry*, Marcel Dekker, New York, 1986.

67. Tunesi, S. and Anderson, M., Influence of chemisorption on the photodecomposition of salicylic acid and related compounds using suspended TiO_2 ceramic membranes, *J. Phys. Chem.*, 95, 3399, 1991.

68. Sabate, J., Anderson, M., Kikkawa, H., Edwards, M., and Hill, G. G., A kinetic study of the photocatalysed degradation of 3-chlorosalicylic acid over TiO_2 membranes supported on glass, *J. Catal.*, 127, 167, 1991.

69. Augustynski, J., Aspects of photo-electrochemical and surface behaviour of titanium (IV) oxide, *Struct. Bonding*, 69, Berlin, 1988, chap. 1.

70. Foissy, A., El-Attar, A., and Lamarche, J. M., Adsorption of polyacrylic acid on titanium dioxide, *J. Colloid. Interface Sci.*, 96, 275, 1983.

71. Cunningham, J., Al-Sayyed, G., Srijaranai, S., and Debauge, Y., Adsorption of model pollutants onto TiO_2 particles in relation to photoremediation of contaminated water, *Aquatic and Surface Photochemistry*, Helz, G. R., Zepp, R. G., and Crosby, D. G., Eds., Lewis Publishers, Boca Raton, FL, 1993, chap. 22.

72. Lamarche, J. M., Foissy, A., and Robert, G., Adsorption d'un polymere de l-acide acrylique sur le dioxyde de titane, in *Adsorption at the Gas/Solid and Liquid/Solid Interface*, Rouguerol, J. and Sing, K. S. W., Eds., Elsevier, Amsterdam, 1982, 117.

73. Derylo-Murczewska, A. and Jaroniec, M., Adsorption of organic solutes from dilute solutions on solids, in *Surf. Colloid Sci.*, 14, 1986.

74. Jaroneic, M., Gas adsorption on heterogeneous microporous solids, in *Physical Adsorption on Heterogeneous Solids*, Jaroneic, M. and Madey, T., Eds., Elsevier, Amsterdam, 1988, chap. 8.

75. Dabrowski, A., Jaroniec, M., and Oscik, J., Multilayer and monolayer adsorption from liquid mixtures of non-electrolytes on solid surfaces, *Monatsh. Chem.*, 112, 175, 1981.

76. Garrone, E., Bolis, V., Fubini, B., and Morterra, C., Thermodynamic and spectroscopic characterization of heterogeneity among adsorption sites: CO on anatase at ambient temperature, *Langmuir*, 5, 892, 1989.

77. Chandrasekaran, K. and Thomas, J. K., Photochemical reactions of amorphous and crystalline titanium dioxide powder suspensions in water, *J. Chem. Soc. Faraday Trans. I*, 80, 1163, 1984.

78. Everett, D. H., Physical adsorption in condensed phases, *Faraday Discuss. Chem. Soc.*, 59, 9, 1975.

79. Brown, C. E., Everett, D. H., Powell, A. V., and Thorne, P. E., Adsorption and structuring phenomena at solid/liquid interface, *Faraday Discuss. Chem. Soc.*, 59, 97, 1975.

80. Hansen, R. S. and Craig, R. P., The adsorption of aliphatic alcohols and acids from aqueous solutions by non-porous carbons, *J. Phys. Chem.*, 58, 212, 1954.

81. Jaffrezic-Renault, N., Pichat, P., Foissy, A., and Mercier, R., Study of the effect of deposited Pt particles on the surface charge of TiO_2 aqueous suspensions by potentiometry, electrophoresis, and labeled ion adsorption, *J. Phys. Chem.*, 90, 2733, 1986.

82. D'Oliveira, J.-C., Maillard, C., Guillard, C., Al-Sayyed, G., and Pichat, P., Decontamination of aromatics-containing water by heterogeneous photocatalysis, in *Proceedings of the International Conference on Innovation, Industrial Progress and Environment*, M.C.I., Paris, 1991, p. 421.

83. D'Oliveira, J.-C., Guillard, C., Maillard, C., and Pichat, P., Photocatalytic destruction of hazardous chlorine- or nitrogen containing aromatics in water, *J. Environ. Sci. Health A: Toxic Hazardous Substance Control*, A28, 941, 1993.

84. Augugliaro, V., Palmisano, L., Sclafani, A., Minero, C., and Pelizzetti, E., Photocatalytic degradation of phenol in aqueous TiO_2 dispersions, *Toxicol. Environ. Chem.*, 16, 89, 1988.

85. Israelachvili, J. N., *Intermolecular and Surface Forces*, 2nd ed., Academic Press, San Diego, 1992, chap. 12.

86. Peyton, G. R. and Glaze, W. H., Destruction of pollutants in water with ozone in combination uith ultraviolet radiation. 3. Photolysis of aqueous ozone, *Environ. Sci. Technol.*, 22, 761, 1988.

87a. Guittonneau, S., de Laat, J., Doré, M., Duguet, J. P. and Suty, H. Modélisation de la cinétique de degradation du parachloronitrobenzène par photolyse du peroxyde d'hydrogène en milieu aqueux et en réacteur dynamique, *Environ. Technol. Lett.*, 11, 57, 1990.

87b. Paillard, H., Brunet, R., and Doré, M. Conditions optimales d'application du système oxydant O_3-H_2O_2, *Water Res.*, 10, 153, 1988.

88. Low, G. K-C., McEvoy, S. R., and Matthews, R. W., Effect of common inorganic anions on rates of photocatalytic oxidation of organic carbon over illuminated TiO_2, *Environ. Sci. Technol.*, 25, 460, 1991.

89. Ollis, D. F., Contaminant degradation in water. Heterogeneous photocatalysis degrades halogenated hydrocarbon contaminants, *Environ. Sci. Technol.*, 19, 480, 1985.

90. Borello, R., Minero, C., Pramauro, E., Pelizzetti, E., Serpone, N., and Hidaka, H., Photocatalytic degradation of DDT mediated in aqueous semiconductor slurries by simulated sunlight, *Environ. Toxicol. Chem.*, 8, 997, 1989.

91. Hidaka, H., Kubota, H., Graetzel, M., Pelizzetti, E., and Serpone, N., Photodegradation of surfactants. II. Degradation of sodium dodecylbenzene sulphonate catalyzed by TiO_2 particles, *J. Photochem.*, 35, 216, 1986.

92. Sabin, F., Turk, T., and Vogler, A., Photo-oxidation of organic compounds in the presence of TiO_2: determination of the efficiency, *J. Photochem. Photobiol. A: Chem.*, 63, 99, 1992.

93. Wei, T. Y., Wang, Y. Y., and Wan, C. C., Photocatalytic oxidation of phenol in the presence of hydrogen peroxide and titanium dioxide powders, *J. Photochem. Photobiol. A: Chem.*, 55, 115, 1990.

94. Tanaka, K., Hisanaga, T. and Harada, K., Photocatalytic degradation of organohalide compounds in semiconductor suspension with added hydrogen peroxide, *New J. Chem.*, 13, 5, 1989.

95. Tanaka, K., Hisanaga, T. and Harada, K., Efficient photocatalytic degradation of chloral hydrate in aqueous semiconductor suspension, *J. Photochem. Photobiol. A: Chem.*, 48, 155, 1989.

96. Augugliaro, V., Davì, E., Palmisano, L., Schiavello, M., and Sclafani, A., Influence of hydrogen peroxide on the kinetics of phenol photodegradation in aqueous titanium dioxide dispersion, *Appl. Catal.*, 65, 101, 1990.

97. Sclafani, A., Palmisano, L., and Davì, E., Photocatalytic degradation of phenol by TiO$_2$ aqueous dispersions: rutile and anatase activity, *New J. Chem.*, 14, 265, 1990.

98. Graetzel, C. K., Jirousek, M., and Graetzel, M., Decomposition of organophosphorous compounds on photoactivated TiO$_2$ surfaces, *J. Mol. Catal.*, 60, 375, 1990.

99. Pelizzetti, E., Carlin, V., Minero, C., and Gratzel, M., Enhancement of the rate of photocatalytic degradation on TiO$_2$ of 2-chlorophenol, 2,7-dichlorodibenzodioxin and atrazine by inorganic oxidizing species, *New J. Chem.*, 15, 351, 1991.

100. Chemseddine, A. and Bohem, H. P., A study of the primary step in the photochemical degradation of acetic acid and chloroacetic acids on a TiO$_2$ photocatalyst, *J. Mol. Catal.*, 60, 295, 1990.

101. Hisanaga, T., Harada, K., and Tanaka, K., Photocatalytic degradation of organochlorine compounds in suspended TiO$_2$, *J. Photochem. Photobiol. A: Chem.*, 54, 113, 1990.

102. Carlin, V., Minero, C., and Pelizzetti, E., Effect of chlorine on photocatalytic degradation of organic contaminants, *Environ. Technol. Lett.*, 11, 919, 1990.

103. Sclafani, A. and Palmisano, L., Phenol photo-oxidation over aqueous dispersions of oxygenated titanium dioxide mediated by the Fe^{3+}/Fe^{2+} redox system, *Gazz. Chim. Ital.*, 120, 599, 1990.

104. Sclafani, A., Palmisano, L., and Davì, E., Photocatalytic degradation of phenol in aqueous polycrystalline TiO$_2$ dispersions: the influence of Fe^{3+}, Fe^{2+} and Ag$^+$ on the reaction rate, *J. Photochem. Photobiol. A: Chem.*, 56, 113, 1991.

105. Kormann, C., Bahnemann, D. W., and Hoffmann, M. R., Photolysis of chloroform and other organic molecules in aqueous TiO$_2$ suspensions, *Environ. Sci. Technol.*, 25, 494, 1991.

106. Ollis, D. F., Solar-assisted photocatalysis for water purification: issues, data, questions, in *Photochemical Conversion and Storage of Solar Energy*, Pelizzetti, E. and Schiavello, M., Eds., Kluwer, Dordrecht, 1991, 593.

107. Blake, D. M., Webb, J., Turchi, C., and Magrine, K., Kinetic and mechanistic overview of TiO$_2$-photocatalyzed oxidation reactions in aqueous solution, *Sol. Energy Mater.*, 24, 584, 1991.

108. Halmann, M., Hunt, A. J., and Spath, D., Photodegradation of dichloromethane, tetrachloroethylene and 1,2-dibromo-3-chloropropane in aqueous suspensions of TiO$_2$ with natural, concentrated and simulated sunlight, *Sol. Energy Mater.*, 26, 1, 1992.

109. Bahnemann, D. W., Bockelmann, D., Goslich, R., and Hilgendorff, M., Photocatalytic detoxification of polluted aquifers: novel catalysts and solar applications, *Aquatic and Surface Photochemistry*, Helz, G. R., Zepp, R. G., and Crosby, D. G., Eds., Lewis Publishers, Boca Raton, FL, 1993, chap. 23.

110. Serpone, N., Pelizzetti, E., and Hidaka, H., Heteogeneous photocatalysis. Issues, questions, answers and successes, in *Photochemical and Photoelectrochemical Conversion and Storage of Solar Energy*, Tian, Z. W. and Cao, Y., Eds., Int. Academic Publ., Beijing, 1993, 33.

111. Grabner, G. and Quint, R. M., Pulsed-laser-induced charge-transfer reactions in aqueous TiO$_2$ colloids. A study of the dependence of transient formation on photon fluence, *Langmuir*, 7, 1091, 1991.

112. Matthews, R. W., Photooxidation of organic impurities in water using thin films of titanium dioxide, *J. Phys. Chem.*, 91, 3328, 1987.

113. Glaze, W. H., Kenneke, J. F., and Ferry, J. L., Chlorinated byproducts from the TiO$_2$-mediated photodegradation of trichloroethylene and tetrachloroethylene in water, *Environ. Sci. Technol.*, 27, 177, 1993.

114. Ollis, D. F., Hsiao, C-.Y., Budiman, L., and Lee, C.-L., Heterogeneous photoassisted catalysis: conversions of perchloroethylene, dichloroethane, chloroacetic acids, and chlorobenzenes, *J. Catal.*, 88, 89, 1984.

115. Matthews, R. W., Kinetics of photocatalytic oxidation of organic solutes over titanium dioxide, *J. Catal.*, 111, 264, 1988.

116. Bideau, M., Claudel, B., Faure, L., and Kazouan, H., The photo-oxidation of acetic acid by oxygen in the presence of titanium dioxide and dissolved copper ions, *J. Photochem. Photobiol. A: Chem.*, 61, 269, 1991.

117. Park, K. H. and Kim, J. H., Photocatalytic decompositions of carboxylic acid by semiconductors, *Bull. Korean Chem. Soc.*, 12, 438, 1991.

118. Mao, Y., Schoneich, C., and Asmus, K.-D., Identification of organic acids and other intermedi-
      ates in oxidative degradation of chlorinated ethanes on TiO$_2$ surfaces en route to mineralization.
      A combined photocatalytic and radiation chemical study, *J. Phys. Chem.*, 95, 10080, 1991.

119. Hilgendorff, M., Mechanistische Untersuchungen zum photokatalytischen Abbau von
      Dichloressigsaure, Diplomarbeit, Universität Hannover, 1991.

120. Kraeutler, B. and Bard, A. J., The photoassisted decarboxylation of acetate on n-type rutile
      electrodes. The photo-Kolbe reaction, *Nouv. J. Chim.*, 3, 31, 1979.

121. Wolff, K., Bockelmann, D., and Bahnemann, D., Mechanistic aspects of chemical transforma-
      tions in photocatalytic systems, *Proceedings of the Symposium on Electronic and Ionic
      Properties of Silver Halides. Common Trends with Photocatalysis*, Levy, B., Ed., The Society
      for Imaging Science and Technology, Springfield, VA, 1991, 259.

122. Amalric, L., Guillard, C., Serpone, N., and Pichat, P., Water treatment: degradation of
      dimethoxybenzenes by the TiO$_2$-UV combination, *J. Environ. Sci. Health, A: Toxic Hazardous
      Subst. Control*, A28, 1393, 1993.

123. Sehili, T., Boule, P., and Lemaire, J., Photocatalyzed transformation of chloroaromatic
      derivatives on zinc oxide. II. Dichlorobenzenes, *J. Photochem. Photobiol. A: Chem.*, 50, 103,
      1989.

124. D'Oliveira, J. C., Minero, C., Pelizzetti, E., and Pichat, P., Photodegradation of dichlorophenols
      and triclorophenols in TiO$_2$ aqueous suspensions: kinetic effect of the positions of the Cl atoms
      and identification of the intermediates, *J. Photochem. Photobiol. A: Chem.*, 72, 261, 1993.

125. Sehili, T., Boule, P., and Lemaire, J., Photocatalyzed transformation of chloroaromatic deriva-
      tives on zinc oxide. II. Chlorophenols, *J. Photochem. Photobiol. A: Chem.*, 50, 117, 1989.

126. Matthews, R. W., Photo-oxidation of organic material in aqueous suspensions of titanium
      dioxide, *Water Res.*, 20, 569, 1986.

127. Barbeni, M., Pramauro, E., Pelizzetti, E., Borgarello, E., Graetzel, M., and Serpone, N.,
      Photodegradation of 4-chlorophenol catalyzed by titanium dioxide particles, *Nouv. J. Chim.*,
      8, 547, 1984.

128. Minero, C., Aliberti, C., Pelizzetti, E., Terzian, R., and Serpone, N., Kinetic studies in
      heterogeneous photocatalysis. 6. AM1 simulated sunlight photodegradation over titania in
      aqueous media: a first case of fluorinated aromatics and identification of intermediates,
      Langmuir, 7, 928, 1991.

129. Augugliaro, V., Palmisano, L., Schiavello, M., Sclafani, A., Marchese, L., Martra, G., and
      Miano, F., Photocatalytic degradation of nitrophenols in aqueous titanium dioxide dispersion,
      Appl. Catal., 69, 323, 1991.

130. Low, G. K.-C., McEvoy, S. R., and Matthews, R. W., Formation of nitrate and ammonium ions
      in titanium dioxide mediated photocatalytic degradation of organic compounds containing
      nitrogen atoms, *Environ. Sci. Technol.*, 25, 460, 1991.

131. Maillard, C., Guillard, C., and Pichat, P., Comparative effects of the TiO$_2$-UV, H$_2$O$_2$-UV,
      H$_2$O$_2$-Fe^{2+} systems on the disappearance rate of benzamide and 4-hydroxybenzamide in water,
      Chemosphere, 24, 1085, 1992.

132. Maillard, C., Guillard, C., Pichat, P., and Fox, M. A., Photodegradation of benzamide in TiO$_2$
      aqueous suspensions, *New J. Chem.*, 1993, in press.

133. Hallmann, M., Photodegradation of di-n-butyl-ortho-phthalate in aqueous solutions, *J.
      Photochem. Photobiol. A: Chem.*, 66, 215, 1992.

134. Hustert, K. and Moza, P. N., Photokatalytischer Abbau von Phthalaten an Titandioxid in
      wassinger Phase, *Chemosphere*, 17, 1751, 1988.

135. Tanguay, J. F., Suib, S. L., and Coughlin, R. W., Dichloromethane photodegradation using
      titanium catalysts, *J. Catal.*, 117, 335, 1989.

136. Iseda, K., Oxygen effect on photocatalytic reaction of ethanol over some titanium dioxide
      photocatalysts, *Bull. Chem. Soc. Jpn.*, 64, 1160, 1991.

137. Minero, C., Maurino, V., Campanella, L., Morgia, C., and Pelizzetti, E., Photodegradation of
      2-ethoxy- and 2-butoxyethanol in the presence of semiconductor particles or organic conduct-
      ing polymer, *Environ. Technol. Lett.*, 10, 301, 1989.

138. Yamagata, S., Baba, R., and Fujishima, A., Photocatalytic decomposition of 2-ethoxyethanol on titanium dioxide, *Bull. Chem. Soc. Jpn.*, 62, 1004, 1989.

139. Brezova, V., Vodny, S., Vesely, M., Ceppan, M., and Lapcik, L., Photocatalytic oxidation of 2-ethoxyethanol in a water suspension of titanium dioxide, *J. Photochem. Photobiol. A: Chem.*, 56, 125, 1991.

140. Pelizzetti, E., Maurino, V., Minero, C., Carlin, V., Pramauro, E., Zerbinati, O., and Tosato, M. L., Photocatalytic degradation of atrazine and other s-triazine herbicides, *Environ. Sci. Technol.*, 24, 1559, 1990.

141. Pelizzetti, E., Minero, C., Carlin, V., Vincenti, M., Pramauro, E., and Dolci, M., Identification of photocatalytic degradation pathways of 2-Cl-s-triazine herbicides and detection of their decomposition intermediates, *Chemosphere*, 24, 891, 1992.

142. Pramauro, E., Vincenti, M., Augugliaro, V., and Palmisano, L., Photocatalytic degradation of monurom on irradiated TiO_2 aqueous suspensions, *Environ. Sci. Technol.*, 27, 1790, 1993.

143. Pelizzetti, E., Maurino, V., Minero, C., Zerbinati, O., and Borgarello, E., Photocatalytic degradation of bentazon by TiO_2 particles, *Chemosphere*, 18, 1437, 1989.

144. Hidaka, H., Nohara, K., Zhao, J., Serpone, N., and Pelizzetti, E., Photo-oxidative degradation of the pesticide permethrin catalysed by irradiated TiO_2 semiconductor slurries in aqueous media, *J. Photochem. Photobiol. A: Chem.*, 64, 247, 1992.

145. Hidaka, H., Zhao, J., Pelizzetti, E., and Serpone, N., Photodegradation of surfactants. VIII. Comparison of photocatalytic processes between anionic sodium dodecylbenzenesulfonate and cationic benzyldodecyldimethylammonium chloride on the TiO_2 surface, *J. Phys. Chem.*, 96, 2226, 1992.

146. Hidaka, H., Zhao, J., Kitamura, K., Nohara, K., Serpone, N., and Pelizzetti, E., Photodegradation of surfactants. IX. The photocatalysed oxidation of polyoxyethylene alkyl ether homologues at TiO_2-water interfaces, *J. Photochem. Photobiol. A: Chem.*, 64, 103, 1992.

147. Matthews, R. W., Photooxidative degradation of coloured organics in water using supported catalysts. TiO_2 on sand, *Water Res.*, 25, 1169, 1991.

148. Hustert, K., Zepp, R. G., and Schultz-Jander, D., Photocatalytic degradation of dilute azo dyes in water, in *Abstracts First International Conference TiO_2 Photocatalytic Purification and Treatment of Water and Air*, Al-Ekabi, H., Ed., London, Ontario, Canada, 1992, 149.

149. Vesely, M., Ceppan, M., Brezova, V., and Lapcik, L., Photocatalytic degradation of hydroxyethylcellulose in aqueous Pt-TiO_2 suspension, *J. Photochem. Photobiol. A:Chem.*, 61, 399, 1991.

150. Duran, N., Dezotti, M. and Rodriguez, J., Biomass photochemistry. XV. Photobleaching and biobleaching of Kraft effluent, *J. Photochem. Photobiol. A. Chem.*, 62, 269, 1991.

151. Beltrame, P., Beltrame, P. L., Carbiniti, P., Guardione, D., and Lanzetta, C., Inhibiting action of chlorophenols on biodegradation of phenol and its correlation with structural properties of inhibitors, *Biotechnol. Bioeng.*, 31, 821, 1988.

152. Davis, A. P. and Huang, C., The photocatalytic oxidation of toxic organic compounds, in *Physicochemical and Biological Detoxification of Hazardous Wastes, Vol. 1*, Wu, Y. C., Ed., Technomic, Lancaster, 1989, 337.

153. Ollis, D. F., Pelizzetti, E., and Serpone, N., Heterogeneous photocatalysis in the environment: application to water purification, in *Photocatalysis. Fundamentals and Applications*, Serpone, N. and Pelizzetti, E., Eds., Wiley, New York, 1989, 603.

154. Al-Ekabi, H. and Serpone, N., Kinetic studies in heterogeneous photocatalysis. 1. Degradation of chlorinated phenols in aerated aqueous solutions over TiO2 supported on a glass matrix, *J. Phys. Chem.*, 92, 5726, 1988.

155. Pichat, P. and Herrmann, J. M., Adsorption-desorption, related mobility and reactivity in photocatalysis, in *Photocatalysis. Fundamentals and Applications*, Serpone, N. and Pelizzetti, E., Eds., Wiley, New York, 1989, 218.

156. Clark, A., *The Theory of Adsorption and Catalysis*, Academic Press, New York, 1970.

157. Matthews, R. W. and McEvoy, S. R., Photocatalytic degradation of phenol in the presence of near-uv illuminated TiO_2, *J. Photochem. Photobiol. A: Chem.*, 64, 231, 1992.

158. Richard, C. and Lemaire, J., Analytical and kinetic study of the phototransformation of furfuryl alcohol in aqueous ZnO suspensions, *J. Photochem. Photobiol. A: Chem.*, 55, 127, 1990.

159. Bahnemann, D. W., *Proceedings of the Symposium on Semiconductor Photoelectrochemistry*, Koval, C., Ed., The Electrochemical Society, Inc., Pennington, NJ, 1991.

160. Bolton, J. R., Mechanism of the photochemical degradation of chlorohydrocarbon pollutants on TiO₂ in aqueous solutions, *Proceedings of the 13th D.O.E. Solar Energy Conference*, U.S. Department of Energy, Argonne National Laboratory, Argonne, IL, 1989, Vol. 13.

161. Bolton, J. R., Advanced Oxidation Symposium, Toronto, Canada, June 4–5, 1990.

162. Sun, L., Schindler, K.-M., Hoy, A. R., and Bolton, J. R., Spin-trap EPR studies of intermediates involved in photodegradation reactions on TiO₂: is the process heterogeneous or homogeneous?, in *Proc. Symp. Environmental Aspects of Surface and Aquatic Photochemistry*, ACS meeting, San Francisco, CA, April 5–10, 1992, 259.

163. Morrison, S. R. and Freund, T., Chemical role of holes and electrons in ZnO photocatalysis, *J. Chem. Phys.*, 47, 1543, 1967.

164. Hug, G. L., *Optical spectra of nonmetallic inorganic transient species in aqueous solution*, *Natl. Stand. Ref. Data Ser.*, NSRDS-NBS 69, 1981.

165. Warman, J. M., de Haas, M. P., Serpone, N., and Pichat, P., The effect of iso-propanol on the surface localization and recombination of conduction band electrons in Degussa P25 TiO₂; a pulse radiolytic time-resolved-microwave-conductivity study, *J. Phys. Chem.*, 95, 8858, 1991.

166. Nosaka, Y., Sasaki, H., Norimatsu, K., and Miyama, H., Effect of surface compound formation on the photoinduced reaction of polycrystalline TiO₂ semiconductor electrodes, *Chem. Phys. Lett.*, 105, 456, 1984.

167. Miyake, M., Yoneyama, H., and Tamura, H., Two step oxidation reactions of alcohols on an illuminated rutile electrode, *Chem. Lett.*, 635, 1976.

168. Pelizzetti, E., Minero, C., Sega, M., and Vincenti, M., 1992, in preparation.

169. Pelizzetti, E., Minero, C., Vincenti, M., and Pichat, P., 1992, in preparation.

170. Minero, C., Pelizzetti, E., Terzian, R., and Serpone, N., Reactions of hexafluorobenzene and pentafluorophenol catalyzed by irradiated TiO₂ in aqueous suspensions, submitted, 1993.

171. Pelizzetti, E., Minero, C., Maurino, V., Hidaka, H., Serpone, N., and Terzian, R., Photocatalytic degradation of dodecane and some dodecyl derivatives, *Ann. Chim. (Rome)*, 80, 81, 1990.

172a. Sakata, T., Kawai, T., and Hashimoto, K., Heterogeneous photocatalytic reactions of organic acids and water. New reaction paths besides the photo-Kolbe reaction, *J. Phys. Chem.*, 88, 2344, 1984.

172b. Yoneyama, H., Takao, Y., Tamura, H., and Bard, A. J., Factors influencing product distribution in photocatalytic decomposition of aqueous acetic acid on platinized TiO₂, *J. Phys. Chem.*, 87, 1417, 1983.

173. Barbeni, M., Morello, M., Pramauro, E., Pelizzetti, E., Vincenti, M., Borgarello, E., and Serpone, N., Sunlight photodegradation of 2,4,5-trichlorophenoxyacetic acid and 2,4,5-trichlorophenol on TiO₂. Identification of intermediates and degradation pathway, *Chemosphere*, 16, 1165, 1987.

174. Barbeni, M., Pramauro, E., Pelizzetti, E., Borgarello, E., and Serpone, N., Photodegradation of pentachlorophenol catalyzed by semiconductor particles, *Chemosphere*, 14, 195, 1985.

175. Matthews, R. W., Hydroxylation reactions induced by near-ultraviolet photolysis of aqueous titanium dioxide suspensions, *J. Chem. Soc. Faraday Trans. I*, 80, 457, 1984.

176. Getoff, N. and Solar, S., Radiolysis and pulse radiolysis of chlorinated phenols in aqueous solutions, *Radiat. Phys. Chem. (Int. J. Radiat. Appl. Instrument. Part C)*, 28, 443, 1986.

177. Alfassi, Z. B. and Schuler, R. H., Reaction of azide radicals with aromatic compounds. Azide as a selective oxidant, *J. Phys. Chem.*, 89, 3359, 1985.

178. Terzian, R., Serpone, N., Draper, R. B., Fox, M. A., and Pelizzetti, E., Pulse radiolytic studies of the reaction of pentahalophenols with OH radicals: formation of pentahalophenoxyl, dihydroxypentahalocyclohexadienyl, and semiquinone radicals, *Langmuir*, 7, 3081, 1991.

179. Schuler, R. H., Neta, P., Zemel, H., and Fessenden, R. W., Conversion of hydroxyphenyl to phenoyl radicals: a radiolytic study of the reduction of bromophenols in aqueous solution, *J. Am. Chem. Soc.*, 98, 3825, 1976.

180. Anbar, M. and Hart, E. J., The reactivity of aromatic compounds toward hydrated electrons, *J. Am. Chem. Soc.*, 86, 5633, 1964.

181. Pelizzetti, E., Barbeni, M., Pramauro, E., Serpone, N., Borgarello, E., Jamieson, M. A., and Hidaka, H., Sunlight photodegradation of haloaromatic pollutants catalyzed by semiconductor particulate materials, *Chim. Ind. (Milan)*, 67, 623, 1985.

182. Nguyen, T. and Ollis, D. F., Complete heterogeneously photocatalyzed transformation of 1,1- and 1,2-dibromoethane to CO_2 and HBr, *J. Phys. Chem.*, 88, 3386, 1984.

183. Bahnemann, D. W., Monig, J., and Chapman, R., Efficient photocatalysis of the irreversible one-electron and two-electron reduction of halothane on platinized colloidal titanium dioxide in aqueous suspension, *J. Phys. Chem.*, 91, 3782, 1987.

184. Harvey, P. R. and Rudham, R., Photocatalytic oxidation of iodide ions by titanium dioxide, *J. Chem. Soc. Faraday Trans. I*, 84, 4181, 1988.

185. Moser, J. and Gratzel, M., Photoelectrochemistry with colloidal semiconductors: laser studies of halide oxidation in colloidal dispersions of TiO_2 and Fe_2O_3, *Helv. Chim. Acta*, 65, 1436, 1982.

186. Hidaka, H., Kubota, H., Graetzel, M., Pelizzetti, E., and Serpone, N., Photodegradation of surfactants. II. Degradation of sodium dodecylbenzenesulfonate catalyzed by titanium dioxide particles, *J. Photochem.*, 35, 219, 1986.

187. Barbeni, M., Pramauro, E., Pelizzetti, E., Borgarello, E., Serpone, N., and Jamieson, M. A., Photochemical degradation of chlorinated dioxins, biphenyls, phenols and benzene on semi-conductor dispersions, *Chemosphere*, 15, 1913, 1986.

188. Filby, W. G., Mintas, M., and Glisten, H., Heterogeneous catalytic degradation of chlorofluo-romethanes on zinc oxide surfaces, *Ber. Bunsenges. Phys. Chem.*, 85, 189, 1981.

189. Sehili, T., Bonhomme, G., and Lemaire, J., Transformation des composes aromatique chlores photocatalysee par de l'oxyde de zinc. I. Comportement du chloro-3 phenol, *Chemosphere*, 17, 2207, 1988.

190. Bahnemann, D. W., Kormann, C., and Hoffmann, M. R., Preparation and characterization of quantum size zinc oxide: a detailed spectroscopic study, *J. Phys. Chem.*, 91, 3789, 1987.

191. Pichat, P. and Hermann, J. M., Adsorption-desorption, related mobilities and reactivity in photocatalysis, in *Photocatalysis. Fundamentals and Applications*, Serpone, N. and Pelizzetti, E., Eds., Wiley, New York, 1989, 217.

192. Matthews, R. W., An adsorption water purifier with in situ photocatalytic regeneration, *J. Catal.*, 113, 549, 1988.

193. Matthews, R. W., Adsorption photocatalytic oxidation: a new method of water purification, *Chem. Ind.*, 28, January 4, 1988.

194. Kormann, C., Bahnemann, D. W., and Hoffmann, M. R., Preparation and characterization of quantum-size titanium dioxide, *J. Phys. Chem.*, 92, 5196, 1988.

195. Memming, R., Photoelectrochemical solar energy conversion, *Top. Curr. Chem.*, 143, 79, 1988.

196. Bahnemann, D. W., Mechanisms of organic transformations on semiconductor particles, in *Photochemical Conversion and Storage of Solar Energy*, Pelizzetti, E. and Schiavello, M., Eds., Kluwer, Dordrecht, 1991, 251.

197. Faust, B. C., Hoffmann, M. R., and Bahnemann, D. W., Photocatalytic oxidation of sulfur dioxide in aqueous suspensions of α-Fe_2O_3, *J. Phys. Chem.*, 93, 6371, 1989.

198. Kormann, C., Bahnemann, D. W., and Hoffmann, M. R., Environmental photochemistry: is iron oxide (hematite) an active photocatalyst? A comparative study: α-Fe_2O_3, ZnO, TiO_2, *J. Photochem. Photobiol. A: Chem.*, 48, 161, 1989.

199. Siffert, C. and Sulzberger, B., Light-induced dissolution of hematite in the presence of oxalate: a case study, *Langmuir*, 7, 1627, 1991.

200. MacNevin, W. M. and Ogle, P. R., The phototropy of the alkaline earth titanates, *J. Am. Chem. Soc.*, 76, 3846, 1954.

201. Ghosh, A. K. and Maruska, H. P., Photoelectrolysis of water in sunlight with sensitized semiconductor electrodes, *J. Electrochem. Soc.*, 124, 1516, 1977.

202. Maruska, H. P. and Ghosh, A. K., Transition-metal dopants for extending the response of titanate photoelectrolysis anodes, *Sol. Energy Mater.*, 1, 237, 1979.

203. Matsumoto, Y., Kurimoto, J., Amagasaki, Y., and Sato, E., Visible light response of polycrystalline TiO_2 electrodes, *J. Electrochem. Soc.*, 127, 2148, 1980.

204. Matsumoto, Y., Kurimoto, J., Shiniizu, T., and Sato, E., Photoelectrochemical properties of polycrystalline TiO_2 doped with 3d transition metals, *J. Electrochem. Soc.*, 128, 1040, 1981.

205. Goodenough, J. B., Photo responses of pure and doped rutile, *Adv. Chem. Ser.*, 186, 113, 1980.

206. Salvador, P., The influence of niobium doping on the efficiency of n-TiO_2 electrode in water photoelectrolysis, *Sol. Energy Mater.*, 2, 413, 1980.

207. Lam, R. U. E., de Haart, L. G. J., Wiersma, A. W., Blasse, G., Tinnemans, A. H. A., and Mackor, A., The sensitization of $SrTiO_3$ photoanodes by doping with various transition metal ions, *Mater. Res. Bull.*, 16, 1593, 1981.

208. Wong, W. K. and Malati, M. A., Doped TiO_2 for solar energy applications, *Sol. Energy*, 36, 163, 1986.

209. Kutty, T. R. N. and Avudaithai, Photocatalytic activity of tin-substituted TiO_2 in visible light, *Chem. Phys. Lett.*, 163, 93, 1989.

210. Augugliaro, V., D'Alba, F., Rizzuti, L., Schiavello, M., and Sclafani, A., Conversion of solar energy to chemical energy by photoassisted processes. II. Influence of the iron content on the activity of doped titanium dioxide catalysts for ammonia photoproduction, *Int. J. Hydrogen Energy*, 7, 851, 1982.

211. Conesa, J. C., Soria, J., Augugliaro, V., and Palmisano, L., Photoreactivity of iron-doped titanium dioxide powders for dinitrogen reduction, in *Structure and Reactivity of Surfaces*, Morterra, C., Zecchina, A., and Costa, G., Eds., Elsevier Science Publishers B. V., Amsterdam, 1989, 307.

212. Martin, C., Martin, I., Rives, V., Palmisano, L., and Schiavello, M., Structural and surface characterization of the polycrystalline system Cr_xO_y-TiO_2 employed for photoreduction of dinitrogen and photodegradation of phenol, *J. Catal.*, 134, 434, 1992.

213. Moser, J., Graetzel, M., and Gallay, R., Inhibition of electron-hole recombination in substitutionally doped colloidal semiconductor crystallites, *Helv. Chim. Acta*, 70, 1596, 1987.

214. Navio, J. A., Marchena, F. J., Roncel, M., and De la Rosa, M. A., A laser flash photolysis study of the reduction of methylviologen by conduction band electrons of TiO_2 and Fe-Ti oxide photocatalysts, *J. Photochem. Photobiol. A: Chem.*, 55, 319, 1991.

215. Bickley, R. I., Gonzalez-Carreno, T., and Palmisano, L., The preparation and the characterization of some ternary titanium oxide photocatalysts, in *Preparation of Catalysts IV*, Delmon, B., Grange, P., Jacobs, P. A., and Poncelet, G., Eds., Elsevier, Amsterdam, 1987, 297.

216. Bickley, R. I., Gonzalez-Carreno, T., and Palmisano, L., A study of the interaction between iron(III)oxide and titanium(IV)oxide at elevated temperatures, *Mater. Chem. Phys.*, 29, 475, 1991.

217. Bickley, R. I., Lees, J. S., Tilley, R. J. D., Palmisano, L., and Schiavello, M., Characterization of iron/titanium oxide photocatalysts,.Part I. Structural and magnetic studies, *J. Chem. Soc. Faraday Trans.*, 88, 377, 1992.

218. Luo, Z. and Gao, Q-H., Decrease in the photoactivity of TiO_2 pigment on doping with transition metals, *J. Photochem. Photobiol. A: Chem.*, 63, 367, 1992.

219. Sabate, J., Anderson, M. A., Kikkawa, H., Xu, Q., Cervera-March, S., and Hill, C. G., Jr., Nature and properties of pure and Nb-doped TiO_2 ceramic membranes affecting the photocatalytic degradation of 3-chlorosalicylic acid as a model of halogenated organic compounds, *J. Catal.*, 134, 365, 1992.

220. Sclafani, A., Mozzanega, M.-N., and Pichat, P., Effect of silver deposits on the photocatalytic activity of titanium dioxide samples for the dehydrogenation or oxidation of 2-propanol, *J. Photochem. Photobiol. A: Chem.*, 59, 181, 1991.

221. Gao, Y.-M., Lee, W., Trehan, R., Kershaw, R., Dwight, K., and Wold, A., Improvement of photocatalytic activity of titanium(IV) oxide by dispersion of Au on TiO_2, *Mater. Res. Bull.*, 26, 1247, 1991.

222. Vrachnou, E., Vlachopoulos, N., and Graetzel, M., Efficient visible light sensitization of TiO_2 by surface complexation with $Fe(CN)_6^{4-}$, *J. Chem. Soc. Chem. Comm.*, 868, 1987.

223. Vrachnou, E., Graetzel, M., and McEvoy, A. J., Efficient visible light photoresponse following surface complexation of titanium dioxide with transition metal cyanides, *J. Electroanal. Chem.*, 258, 193, 1989.

224. Matthews, R. W., Response to the comment: photocatalytic reactor design: an example of mass-transfer limitations with an immobilized catalyst, *J. Phys. Chem.*, 92, 6853, 1988.

225. Low, G. K.-C. and Matthews, R. W., Flow-injection determination of organic contaminants in water using an ultraviolet-mediated titanium dioxide film reactor, *Anal. Chim. Acta*, 231, 13, 1990.

226. Serpone, N., Borgarello, E., Harris, R., Cahill, P., Borgarello, M., and Pelizzetti, E., Photocatalysis over TiO_2 supported on a glass substrate, *Sol. Energy Mater.*, 14, 121, 1986.

227. Rosenberg, I., Brock, J. R., and Heller, A., Collection optics of TiO_2 photocatalyst on hollow glass microbeads floating on oil slicks, *J. Phys. Chem.*, 96, 3423, 1992.

228. Sabate, J., Anderson, M. A., Aguado, M. A., Gimenez, S., Cervera-March, S., and Hill, C.G., Jr., Comparison of TiO_2 powder suspensions and TiO_2 ceramic membranes supported on glass as photocatalytic systems in the reduction of chromium (VI), *J. Mol. Catal.*, 71, 57, 1992.

229. Pelizzetti, E., Minero, C., Tinucci, L., Borgarello, E., and Serpone, N., Photocatalytic activity and selectivity of titania colloids and particles prepared by the sol-gel technique: Photooxidation of phenol and atrazine, *Langmuir*, 1993, in press.

230. Matthews, R. W., Solar-electric water purification using photocatalytic oxidation with TiO_2 as a stationary phase, *Sol. Energy,* 38, 405, 1987.

231. Pacheco, J. E. and Holmes, J. T., Falling-film and glass-tube solar photocatalytic reactors for treating contaminated water, in *Emerging Technology in Hazardous Waste Management, Washington, D.C.,* ACS Symp. Ser. No. 422, 1990, 40.

232. Alpert, D. J., Sprung, J. L., Pacheco, J. E., Praitie, M. R., Reilly, H. E., Nfilne, T. A., and Nimlos, M. R., Sandias National Laboratories'work in solar detoxification of hazardous wastes, *Sol. Energy Mater.*, 24, 594, 1991.

233. Minero, C., Pelizzetti, E., Malato, S., and Blanco, J., Large solar plant photocatalytic water decontamination: degradation of pentachlorophenol, *Chemosphere,* 26, 2103, 1993.

234. Bard, A. J., Heller, A., Lambert-Bates, J., Garmire, E. M., Goldstein, A. L., St.Clair Kirby, J., Ollis, D. F., Sarofim, A. F., Serpone, N., Tenhover, M. A., and Vaida, V., *Potential Applications of Concentrated Solar Photons*, National Academy Press, Washington, D.C., 1991.

235. Ollis, D.F., Process economics for water purification. A comparative assessment, in *Photoelectrochemistry, Photocatalysis and Photoreactors*, Schiavello, M., Ed., O. Reidel, Dordrecht, 1985, 663.

236. Braun, A. M., Progress in the applications of photochemical conversion and storage, in *Photochemical Conversion and Storage of Solar Energy*, Pelizzetti, E. and Schiavello, M., Eds., Kluwer, Dordrecht, 1991, 551.

237. Ollis, D. F., Comparative aspects of advanced oxidation processes, submitted.

238. Harada, K., Hisanaga, T., and Tanaka, K., Photocatalytic degradation of organophosphorous insecticides in aqueous semiconductor suspensions, *Water Res.*, 24, 1415, 1990.

CHAPTER 22

Adsorption of Model Pollutants onto TiO$_2$ Particles in Relation to Photoremediation of Contaminated Water

Joseph Cunningham, Ghassan Al-Sayyed, and Somkiat Srijaranai

INTRODUCTION

There is general agreement that the strong oxidizing capability generated by ultraviolet (UV) illumination at liquid solution-semiconductor particle microinterfaces originates from the photogeneration of electronic holes (h$^+$) in the semiconductor valence band. At microinterfaces involving aqueous solutions and TiO$_2$, these h$^+$ species can produce surface OH radicals. Less clear, however, at the outset of this work were details of the processes via which such photogenerated holes or OH radicals interact with solutes added to the liquid phase, thereby bringing about their photocatalytic degradations. Particularly lacking were convincing interpretations of the roles of solute adsorption onto surfaces of semiconducting particles and of the influence of such adsorption upon rates/efficiency of photocatalyzed degradation (PCD) of the solute. Often the raw data on which literature interpretations have been based take the form of experimentally determined values showing the dependencies of PCD rates, (*R), upon added solute concentration, ([S]). A popular treatment of such values has been their attempted linearization by means of double-reciprocal plots of *R^{-1} vs [S]$^{-1}$. Values of slope and intercept from such plots frequently then have been made the basis, as follows, of Langmuir-Hinshelwood (LH)-type interpretations of the nature and role of solute adsorption in PCD.[1-4]

First, it is assumed that Θ_2, the fractional coverage of adsorption sites of the solid surface by solute (species 2) at equilibrium (also often termed the "extent of

0-87371-871-2/94/$0.00+$.50

adsorption"), is concentrated within a monolayer at the microinterfaces, i.e., Θ_2, and that this increases with equilibrium bulk-solute concentration, C_{eq}, in accordance with a Langmuir-type adsorption isotherm,

$$\Theta_2 = K_{ads} \cdot C_{eq} / \left(1 + K_{ads} C_{eq}\right) \tag{1}$$

where K_{ads} is truly an equilibrium constant for fast adsorption-desorption processes between surface monolayer and bulk solution in the absence of illumination. Second, the assumption is made that the rate-determining process (rdp) in PCD is one involving a reaction between an adsorbed solute species having surface monolayer concentration, Θ_2, and a strongly oxidizing intermediate, ox, photogenerated at the interface with photostationary surface concentration, Θ_{ox}. Thus, according to a central LH assumption the rate of PCD should be given by

$$*Rate = -d[S]/dt = k_{LH} \cdot \Theta_2 \cdot \Theta_{ox} \tag{2a}$$

which upon insertion of Equation 1 on the basis of an implicit assumption that adsorption-desorption equilibria are unaffected by illumination becomes

$$*Rate = -d[S]/dt = k_{LH} \cdot \Theta_{ox} K_{ads} C_{eq} / \left(1 + K_{ads} C_{eq}\right) \tag{2b}$$

Often the further assumptions are made that Θ_{ox} represents the photostationary concentration of surface OH radicals at interfaces involving aqueous solutions and that this is independent of added solute concentration. Incorporation of these assumptions into Equation 2a makes it pseudo-first order in Θ_2 with an effective rate coefficient ($k_{LH} \cdot \Theta_{ox}$) denoted by k'. Incorporation of that into Equation 2b and its inversion leads to Equation 2c.

$$*Rate^{-1} = \left(k'\right)^{-1} + \left(k'K\right)^{-1} \cdot C_{eq}^{-1} \tag{2c}$$

This indicates that intercept and slope values determined from double-reciprocal plots of data for the solute-concentration dependence of PCD can yield values for the pseudo-first-order rate coefficient, k', and for a PCD "kinetics-related" adsorption-desorption equilibrium constant. Hereinafter, the latter will be denoted by K^* to distinguish it from the K_{ads} values obtained directly from adsorption measurements. The extent to which values for K^* and K_{ads} agree will be used as one pointer to whether or not a pseudo-first-order LH model is adequate to account for observed concentration dependence of PCD.

In our studies, an inverse of the above approach was adopted in efforts to clarify the interrelationships between equilibrium adsorption constants, K_{ads}, for pollutants onto TiO_2 and pseudo-first-order rate coefficients, k', for their PCD degradation over TiO_2.[5,6] In this alternative approach, values for K_{ads} are first deduced from experimentally determined adsorption isotherms for each pollutant onto TiO$_2$, thereby providing a secure basis for knowing the values of Θ_2 in equilibrium with various solution-phase concentrations, C_{eq}, in the absence of illumination. Second, values for initial rates of PCD under UV illumination are determined experimentally for a set of TiO$_2$ suspensions, each preequilibrated in the dark with a different initial C_{eq} value. Finally, the concentration dependences thus separately determined for extent of adsorption and for "initial PCD rates" serve as the basis for a critical examination of the correlation, or lack of correlation, between dark-adsorption and initial PCD rates. Adoption of this alternative approach in the present study was prompted by recognition of the following over simplifications underlying the three assumptions outlined in previous paragraphs: (1) insufficient consideration of the roles of solvent molecules at the microinterfaces since, in reality, polar solvent molecules are likely to compete strongly against solute species for adsorption sites,[7] thereby resulting in competitive solvent vs solute adsorption as per Equation 3a, in which superscripts s and b distinguish between location at the surface or in the bulk solution, respectively

$$\text{Solvent}^s + \text{Solute}^b \rightleftharpoons \text{Solute}^s + \text{Solvent}^b \qquad (3a)$$

(2) neglect of possibilities for adsorption within mixed (solvent + solute) multilayers, rather than solely within a mixed-solution monolayer;[8,9a-c] (3) insufficient consideration of possibilities that Θ_{ox} may not be independent of C_{eq} or Θ_2, but rather inversely related to them, e.g., in analogous fashion to the decreasing photostationary concentrations of luminescent excited states, and related OH radicals, surviving in UV illuminated homogeneous systems in the presence of increasing concentrations of added quencher;[10a,10b] (4) neglect of possibilities that photosorption of pollutant and/or of oxygen may cause the photostationary extent of adsorption (and an associated adsorption-desorption equilibrium constant K*) to differ significantly from those in the absence of UV illumination.[11a,11b]

Thus, a constant theme running through much of the present work has been the search for accurate data with which to establish whether more satisfactory interpretations of the concentration dependence of PCD rates and their relationship to adsorption become possible by dropping one or more of oversimplifications (1) to (4) above. The relevance of this theme to practical applications of pollutant photodegradation over TiO$_2$ may be appreciated in relation to the following operationally important questions: (1) "Can TiO$_2$-sensitized PCD of pollutants be expected to retain useful efficiency down to very low pollutant concentration levels, such that the extent of adsorption of pollutants into the mixed (solvent + solute) monolayer must decrease to very low values in the manner required for Langmuir-type adsorption isotherms?" and (2) "What are the implications for the PCD process design if a change from "mixed-monolayer" to "mixed-multilayer"-type adsorption can switch the rdp of the PCD process away from surface reaction events (i.e., Langmuir-Hinshelwood in character) to reaction events within the mixed-solution multilayer (i.e., Eley-Rideal in character) or even to mass-transfer processes between multilayers and the bulk solution?".[12a,12b]

EXPERIMENTAL

Materials

Titanium dioxide powder, having mainly the anatase crystal structure and a surface area of 50 $m^2 g^{-1}$ (Degussa P25), was used "as supplied" for all adsorption or photocatalytic degradation experiments involving contact with aqueous solutions. Prior to corresponding experiments involving contact with acetonitrile, the TiO_2 was dried overnight at 373 K in a vacuum oven. The structurally simple organic chemicals employed to model the likely adsorption and photodegradation processes of more complex environmental pollutants possessing similar functional groups fell mainly into the following subgroups: Group I — substituted benzoic acids including 3-chloro-4-hydroxy (CHBZA), 4-amino (ABZA), and 2-hydroxy (i.e., salicyclic acid SA) benzoic acids; Group II — substituted phenols including chlorophenols (2-CP, 3-CP, and 4-CP) and amino phenol; Group III — aromatic and pseudo-aromatic alcohols, including benzyl alcohol (BzOH), furfuryl alcohol (FyOH), and chlorobenzyl alcohol (CBzOH). All these model pollutants were obtained from Aldrich in highest available purity (98%), as were the chemicals purchased to assist in HPLC identification of initial products from PCD, e.g., benzaldehyde and 4-hydroxy benzaldehyde from benzyl alcohol. Milli-Q water was used throughout for the preparation of aqueous solutions or as a component of the mixed water-methanol or water-acetonitrile mobile phases employed for HPLC analysis. The acetonitrile used for corresponding processes was HPLC grade.

"Dark" Adsorption

Equilibrium extents of adsorption onto powdered TiO_2 were initially evaluated after its equilibration under constant shaking for times up to 2l hr with aqueous solutions of the chosen pollutant. Small temperature variations of ±3 K about a mean temperature of 295 K occurred during overnight equilibrations. However, comparison of those results with extent of equilibrium dark adsorption at shorter times confirmed that full equilibrium could likewise be attained after l hr with magnetic stirring, during which time temperatures remained at 295 ± 1 K. Measurements directed toward quantitative characterization of the adsorption isotherms for the well-adsorbing Group I solutes were made on suspensions prepared by admixing 20 cm^3 aliquots of solution of various initial concentration C_0 at natural pH with a fixed weight (40 mg) of TiO_2 (Degussa P25). This was to facilitate direct comparison with later PCD studies on suspensions having the same "concentration" (2 g dm^{-3}) of TiO$_2$ particles. Extent of equilibrium adsorption was evaluated from ΔC, the decrease in solute concentration detected after magnetic stirring for 1 hr to promote adsorption-desorption equilibrium. Aliquots of the suspensions were syringe-filtered through Millipore filters, and the residual equilibrium concentration, C_{eq}, of pollutant in the clear filtrate was measured by HPLC analysis (Varian HPLC fitted with Polychrome 9065 detector and μ-Bondapak Cl8 column). Partial reversibility of the adsorption process was demonstrated by flushing pure water through the TiO$_2$ powder retained on the Millipore filter and measuring by HPLC the amount of solute thereby washed off. The various initial concentration values, C_o, were checked experimentally by HPLC on solution aliquots that had been taken through the full sequence of handling, filtration, and analytical procedures, with the omission only of added TiO$_2$. For Group II solutes, which adsorbed only to a slight extent from aqueous solutions onto TiO$_2$, more "concentrated" suspensions (10 g TiO$_2$ dm^{-3}) were used in efforts to characterize their adsorption isotherms. In all cases, however, the extent of adsorption is expressed as n_2^s, the number of moles of solute adsorbed per gram of TiO$_2$. This had elsewhere been termed the "reduced surface excess."[9b]

Photodegradation

In order to ensure that the illumination-induced degradations of pollutants were truly TiO$_2$-sensitized PCD processes, i.e., arising from absorption of photons by TiO$_2$, rather than from direct photoabsorption by the organic pollutants as would be the case for photochemical processes, photon wavelengths were restricted to $\lambda \geq 340$ nm by suitably filtering the output of a 125-W medium-pressure mercury lamp (supplied by Hanovia and fitted by a Pyrex annular water-cooling jacket and an OV1 cutoff filter transmitting only wavelengths greater than 340 nm). Prior to illumination, suspensions containing 40 mg of TiO$_2$, initially in 20 cm^3 of solution having an accurately known C_o of pollutant, were magnetically stirred continually at constant speed for the 1 hr needed to achieve adsorption-desorption equilibration at 295 ± 1 K (e.g., see Plot 2 of Figure 5B). After commencing UV illumination, small aliquots of the stirred suspension were withdrawn by syringe from the UV-illuminated suspension at 5–10 min intervals, filtered through Millipore 0.22-μm filter discs, and analyzed by the reverse-phase HPLC procedure using a μ-Bondapak C18 spherisorb column. Due to the basic nature of the amino group in 4-chloroaniline, the technique of paired ion chromatography (PIC) was used to achieve good separation, the PIC reagent used being the sodium salt of 1 pentane sulfonic acid. The spectral capabilities of the photodiode array detector allowed determinations not only of the remaining concentration of unreacted solute molecules, but also of the intensity and spectral distribution of absorbances due to photoproducts or intermediates that appeared in the suspensions as primary products from TiO$_2$-sensitized photodegradation. Periodic sampling and duplicate HPLC analysis in this manner was continued in some cases, not only until the organic solute had been photodegraded below detection level (ca. 1 ppm), but also

Table 1.   Equilibrium Extents of Dark Adsorption[a] from Dilute Aqueous Solutions Compared for Chlorinated Aromatic Acid, Alcohol, and Phenol

C_o	CHBZA		2-CP		CBzOH	
	n_2^s	(%)	n_2^s	(%)	n_2^s	(%)
10 ppm	4.8	(88)	1.2	(17)	0.8	(12)
30 ppm	14.3	(86)	3.0	(13)	1.4	(9)
90 ppm	35.5	(71)	3.5	(5)	1.6	(2)

[a] Measured after l5-hr equilibration at 295 K with TiO_2 added at l0 g dm^{-3}; CHBZA denotes 3-chloro-4-hydroxy benzoic acid, yielding suspensions of pH 4.7; 2-CP denotes 2-chlorophenol, yielding suspensions of pH = 6.5; CBzOH denotes 4-chlorobenzyl alcohol, yielding suspensions of pH ~ 6.5. n_2^s denotes the number of moles of solute adsorbed per gram TiO_2.

until any organic intermediates initially produced therefrom had photodegraded. Particular attention, however, was given to $(\Delta C / \Delta t)_i^*$, the initial rate of decrease in concentration of the pollutant substrate during the first 20 min of UV illumination. During this time, the temperature rise of the suspensions due to the weak flux of UV photons was ≤ 5 K if nonwater-cooled cells were used, but was negligible when water-cooled photoreactors were used.

RESULTS AND INTERPRETATIONS

Part A: Adsorption Studies

Table 1 summarizes representative data for the extent of adsorption, from aqueous solutions having initial concentrations, C_o, of l0, 30, or 90 ppm of model pollutant, as measured after overnight equilibration with TiO_2 particles added at a concentration/loading of l0 g dm^{-3}. Data are listed for 3-chloro-4-hydroxy benzoic acid (CHBZA) as a representative of group I pollutants, and for ortho-chlorophenol (2-CP) and 4-chlorobenzyl alcohol (CBzOH) as representatives of Group II. Data are given not only in terms of n_2^s, the number of moles adsorbed per gram of TiO_2, but also as the percentage of C_o removed from solution by adsorption, the latter data being shown in parenthesis. Comparison of the values of n_2^s for CHBZA with those for either 2CP or CBzOH from solutions having equal initial C_o demonstrates the much greater adsorption of the substituted benzoic acid, e.g., for C_o of 90 ppm the CHBZA values are greater by ca. 10 times than for 2-CP and by 20 times for CBzOH. Such differences appear less marked for aqueous solutions of lower initial C_o, but values in the table for percentage of C_o adsorbed make it clear that a factor contributing to this for $C_o = 30$ ppm was the very high depletion of the well-adsorbing CHBZA from the aqueous phase, whereas such depletion was much less pronounced for the poorly adsorbing 2-CP or CBzOH.

A1: Adsorption Studies for Substituted Benzoic Acids from Aqueous Suspensions onto TiO₂ at a Loading of 2 g dm⁻³

Studies of the adsorption of salicylic acid (SA) or 4-amino benzoic acid (ABZA) or 3-chloro-4-hydroxy benzoic acid (CHBZA) from aqueous solution onto TiO_2 using the indicated experimental procedures yielded readily measurable values for $(C_o - C_{eq}) = \Delta C$, the adsorption-induced decrease in molarity of the bulk-solution phase (e.g., a 52% reduction in [SA] resulted when 40 mg TiO_2 were equilibrated overnight with 20 cm³ of 10^{-4} mol dm⁻³ SA). The curved adsorption isotherm-type plot obtained for these solutes, showing how n_2^s, the extent of adsorption, depended upon equilibrium concentration of the substituted benzoic acid solute remaining in aqueous solution for nonilluminated TiO_2 suspensions at 295 ± 1 K, is illustrated in Figures 1A, 1B, and 1C for SA, CHBZA, and ABZA, respectively, on the plots labeled (a) in each of those figures. Those plots summarize results in terms of n_2^s, the number of moles of solute adsorbed per gram of TiO_2, vs C_{eq}, the equilibrium molar concentration remaining in 20 ml of bulk-solution phase after their dark-equilibration with 40 mg of TiO_2. ($n_2^s = V \cdot \Delta C/W$, where V is the volume of solution aliquot in milliliters, W is the weight of TiO_2 in milligrams, and ΔC is the adsorption-induced decrease in molarity.) Values for n_2^s may be treated in terms of the equilibrium in Equation 3a which describes the partitioning of solute and solvent molecules between the bulk-solution phase and a monomolecular surface-solution phase (s). Hiemenz has represented such competitive adsorption by the short-form notation in Equation 3b.[7]

$$a_1^s + a_2^b \rightleftharpoons a_1^b + a_2^s \tag{3b}$$

which has associated with it an equilibrium constant K′. Here, the symbol **a** represents activity; the subscripts 1 and 2 distinguish between solvent and solute molecules, respectively; and the superscripts s and b distinguish between the monolayer surface-solution and bulk-solution phases, respectively. Assuming the two-dimensional surface-solution phase to be ideal, K′ can be expressed as $[a_1^b X_2 / X_1 a_2^b]$, where X_1 and X_2 represent mole fractions in that phase. It follows from $X_1 + X_2 = 1$, and from the fact that (K'/a_1^b) remains effectively constant for dilute solution and so can be represented by K, that X_2^s, the mole fraction of solute in the monomolecular surface-solution phase, can be expressed by

$$X_2^s = K \cdot a^b \Big/ \left(1 + K \cdot a_2^b\right) \tag{3c}$$

Assumptions of equal accessibility to surface sites by solute and solute molecules, and of equal surface areas occupied when adsorbed (but see below), mean that the right-hand side of Equation 3c also equals the fractional occupancy, Θ_2, of the surface-solution monolayer. Setting X_2 and Θ_2 equal to n_2^s/n^s, where n^s represents the total number of adsorption sites accessible in the monolayer, allows

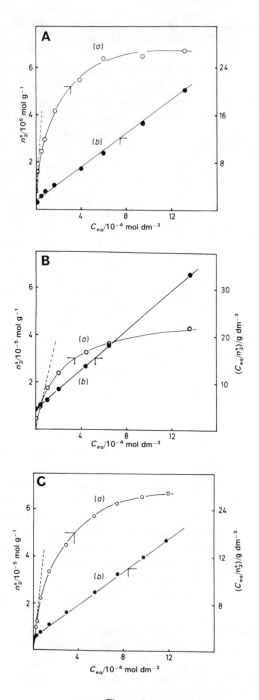

Figure 1.

Equation 3c to be reorganized, as in Equation 3d to express the manner in which the measurable quantity, n_2^s, varies with activity, a_2^b, and concentration, c_2^b, of solute in the bulk-solution phase. Approximation of a_2^b by c_2^b is generally adopted for dilute solutions. Hereinafter, the latter may also be expressed as C_{eq}.

$$n_2^s = \frac{n^s \cdot K \cdot a_2^b}{\left(1 + K \cdot a_2^b\right)} \approx \frac{n^s \cdot K \cdot c_2^b}{\left(1 + K \cdot c_2^b\right)} \approx \frac{n^s \cdot K \cdot C_{eq}}{\left(1 + K \cdot C_{eq}\right)} \qquad (3d)$$

It will emerge below that the physical significance of n^s may differ for high and low values of C_{eq} and that simple approximation of n^s by a geometrically estimated number of TiO$_2$ sites per grams of the adsorbent is not appropriate. When the shapes of the curved n_2^s vs C_{eq} plots in Figures 1A, 1B, and 1C are considered in the context of Equation 3d, two limiting cases with respect to C_{eq} initially appear relevant: (1) A low C_{eq} limit, such that $K \cdot C_{eq} \ll 1$ with the result that, as $C_{eq} \rightarrow 0$, plots should tend towards linearity with an initial slope $\approx n_s \cdot K_o$. Tangents drawn as dashed lines in Figures 1A–1C represent such initial slopes and yield the estimates of $n_s \cdot K_o$ given in column 2 of Table 2. It is this limit which appears physically reasonable for present results obtained at solute concentrations of 10–200 ppm. Since this "dilute-solution monolayer" limit approximates to a two-dimensional analog of the Henry's law limiting case for nonideal dilute homogeneous solutions (see below), a similar view may be appropriate for the K_o values in Column 3 of Table 2 which were calculated from the initial slope values together with a value of $n^s = 4 \times 10^{-M}$ mol g^{-1} taken for the number of surface O-Ti-O sites on TiO$_2$. (ii) A high solute-mole-fraction limit, such that n_2^s approaches n^s for $K \cdot C_{eq} \gg 1$, under the control of an adsorption-desorption equilibrium constant not necessarily the same as for case (1) above. Intuitively, and for reasons elaborated on in later paragraphs, attainment of this limit — implying almost complete occupancy of all sites on TiO$_2$ by solute molecules to the exclusion of solvent molecules — appears physically improbable for the low solute concentrations represented in Figures 1A, 1B and 1C.

The tendency clearly evident for plots labeled (a) in Figure 1 to level off to a plateau at moderate C_{eq} values may rather be interpreted in similar manner to that given by Hansen and Craig, who likewise observed plateaus at moderate concentrations for adsorption isotherms of organic acids (and alcohols) from aqueous solutions onto high surface area carbons.[13] They considered that the plateau value corresponded to attainment of limiting solubility for solute in the solution

---

Figure 1. Isotherms for adsorption onto TiO$_2$ from dilute aqueous solutions for three substituted benzoic acids exhibiting good adsorption at room temperature, viz., salicyclic acid **(A)**; 3-chloro-4-hydroxy benzoic acid **(B)**, and 4-amino benzoic acid **(C)**. Within each figure the open circle data points on Plot (a) correspond to the extent of adsorption, n_2^s, in equilibrium with various residual concentrations, C_{eq}, in solution. Plots labeled b illustrate the successful linearization of that data when plotted in accordance with Equation 3e. (From Cunningham, J. and Al-Sayyed, G., *J. Chem. Soc. Faraday Trans.*, 86, 3935, 1990. With permission.)

Table 2. Adsorption Parameters for Substituted Benzoic Acids from Dilute Aqueous Suspensions of TiO$_2$[a]

Solute	Slope = $n^s \cdot K$[b]	K_{est} dm^3 mol^{-1} [c]	$n^s_{2(max)}$ mol g^{-1} [d]	$\sigma^\circ_{(lim)}$ nm^2 [e]	K_{ads} dm^3 mol^{-1} [f]
SA	0.72	1.8×10^3	7.04×10^{-5}	1.18	1.05×10^4
CHBZA	0.44	1.1×10^3	7.13×10^{-5}	1.16	6.60×10^3
ABZA	0.218	5.5×10^2	4.77×10^{-5}	1.74	4.96×10^3

[a] Suspensions contained TiO$_2$ particles at a loading of 2 g dm^{-3} of solution.
[b] Initial slope of Plot (a) on Figures IA, IB, and IC, see dashed lines.
[c] K_{est} = initial slope ÷ n^s, where n^s = 4 × 10^{-4} mol of sites per gram of TiO$_2$, represents a geometric estimate of the surface density of sites.
[d] Values of $n^s_{2(max)}$ deduced from inverse of slope of Plots (b) in Figures IA, IB, and IC.
[e] Values obtained on the basis of Equation 3f for average area of surface per solute adsorbate.
[f] Values of K_{ads} derived according to Equation 3e.

monolayer. This can be expressed as $n^s_{2(max)}$ and will replace n^s in Equations 3e and 3f. Support for this idea came from the normalization they achieved between adsorption isotherms of different acid solutes, in the low C_{eq} range, whenever plotted vs "reduced concentration," C_{eq}/C_{sat}.

Other observations by Hansen and Craig that appear relevant to the present adsorption studies include (1) evidence for progression of adsorption into a solution-multilayer regime when solute concentration was increased above the moderate concentrations giving rise to the first plateau in adsorption; (2) their recognition that the measured extent of adsorption is a "surface excess" and not necessarily the total quantity of solute in the adsorption region; and (3) their suggestion that, since saturation of the organic solute with water is more readily attained for less soluble organics, adsorption-desorption equilibrium for them can be obtained at a greater distance from the surface and for weaker adsorption forces.

A more searching method for analyzing extent of agreement of the n^s_2 vs C_{eq} data sets with the model involving a mixed solute-solvent monomolecular surface-solution layer is to plot them in accordance with the following linearized forms of Equation 3d:

$$\frac{C_{eq}}{n^s_2} = \frac{1}{K \cdot n^s_{2(max)}} + \frac{C_{eq}}{n^s_{2(max)}} \tag{3e}$$

$$\frac{C_{eq}}{n^s_2} = \frac{N_A \cdot \sigma^\circ}{A_{sp} \cdot K} + \frac{N_A \cdot \sigma^\circ}{A_{sp}} \cdot C_{eq} \tag{3f}$$

In the latter equation, σ° is the average area per solute molecule in the surface-solution monolayer phase, A_{sp} is the specific surface area of the adsorbent, and N_A is Avogadro's Number. It follows from Equations 3e and 3f that plots of C_{eq}/n^s_2

vs C_{eq} should be linear if adsorption of the substituted benzoic acids proceeded via a monomolecular layer in accordance with a Langmuir-type adsorption-desorption model. The plots marked (b) on each of Figures 1A, 1B, and 1C illustrate that plotting adsorption data for the substituted benzoic acids in this way did yield linear plots. Values of $n^s_{2(max)}$ evaluated from the reciprocal of the slope of such plots are listed in column 4 of Table 2. If the above Langmuir-type model for adsorption into a two-dimensional surface solution in accordance with the various forms of Equation 3 is valid, these values have the correct units for, and physical significance of, the limiting number of moles of solute ($n^s_{2(max)}$) that can be adsorbed onto a gram of TiO$_2$ from aqueous solution into a mixed (solvent + solute) monolayer at the TiO$_2$ solution interfaces under the control of an equilibrium constant related to its saturation solubility in that solution monolayer. Combining these with the known value of $A_{sp} = 50$ m^2 g^{-1} for the TiO$_2$ material used throughout the study and using Equation 3f then yielded the values listed in column 5 of Table 2 for $(\sigma^\circ)_{lim}$, the minimum value for "average area per solute molecule" attainable in the mixed (solute + solvent) surface-solution monolayer from aqueous solution. These σ° values are much larger than the geometric cross section of a nonsolvated solute molecule, or the size of a surface TiO$_2$ group; for example, the value calculated for σ° of SA from our adsorption data for the SA-H$_2$O-TiO$_2$ system (1.2 nm^2) is six times greater than the expected geometric cross-section (0.2 nm^2) of a nonsolvated SA molecule, even assuming that these acids are lying flat on the surface and being adsorbed through the oxygen lone pair. Using space-filling models, the actual size of each SA molecule lying flat on the surface has been estimated to be ca. 0.2 nm^2 (0.54×0.35 nm). Assuming that each H$_2$O molecule requires a surface space of 0.02 nm^2 (0.25×0.12 nm), the total cross section of 1.2 nm^2 deduced per adsorbed SA molecule at saturation indicates that each SA molecule could be surrounded by ca. 40 H$_2$O molecules. Furthermore, it is of interest that the $(\sigma^\circ)_{lim}$ values listed for the three substituted benzoic acids all fall in the range 1.45 ± 0.3 nm^2 when determined in this self-consistent manner.

Equations 3e and 3f also recognize, as a self-consistent method for evaluating K_{ads}, the accurate identification of values for slopes and intercepts of the linearized representations shown as plot (b) on each of Figures 1A–1C, followed by calculation of K_{ads} from (slope-intercept) for each of those plots. Values of K_{ads} determined in this way are listed in column 6 of Table 2 and fall in the range $5 \rightarrow 11 \times 10^3$ dm^3 mol^{-1}. Relative to those self-consistent values, it is interesting to note that an order of magnitude decrease in K — down toward values similar to those for K'_0 in column 3 of Table 2 — would result if Equation 3d and a value $n^s = 4 \times 10^{-4}$ mol g^{-1} of TiO$_2$ were used instead of $n^s_{2(max)}$. Thus, the values for adsorption constants, K_{ads}, as deduced from different treatments of Langmuir-type isotherms at low solute concentrations are seen to be strongly dependent on whether n^s or $n^s_{2(max)}$ are adopted, implicitly or explicitly, as the maximum density of interface locations accessible for solute adsorption. In later subsections dealing with experimentally observed concentration dependences of PCD, it will be important to keep in mind such interdependence between the value of K and n^s

or $n_{2(max)}^s$ — especially if the surface density of locations capable of initiating PCD differ from that of sites for adsorption into the surface-solution monolayer.

Everett has pointed out the many difficulties facing attempts to use/interpret K_{ads} and Θ_2 values derived from liquid solution-solid adsorbent cases in analogous fashion to the uses/interpretations made of K_g and Θ_2 values for Langmuir-type gas-solid adsorption.[9a] The following divergences between interpretations of the two situations emerge from the interpretations developed above: (1) that between the Langmuirian view of $(\Theta_g)_{(max)}$ as representing a complete monolayer featuring occupancy by adsorbate of each surface site at the gas-solid interface (but without lateral interactions between adsorbate molecules), and the quite different significance of $n_{2(max)}^s$ as a saturation solubility of solute in the solution monolayer; (2) the strong indication given by the large values of σ° for lateral adsorbate-solvent interactions in cases of competitive solute-solvent adsorption equilibria in the manner of Equation 3a. Even for the well-adsorbed benzoic acids, the latter situation appears to lead to a submonolayer fractional coverage by solute, limited not only by the solvent competing for its share of surface sites, but also by solvation of adsorbed solute by adsorbed solvent in the solution monolayer.

A2: Dark Adsorption for Group II Solutes from Dilute Aqueous TiO$_2$ Suspensions

Attempts to accurately characterize "dark-adsorption" of these materials, using adsorption measurement procedures identical to those successfully used for the substituted benzoic acids, met with much less success. This was due partly to the fact that equilibrium extent of their dark adsorption onto TiO$_2$ from aqueous suspensions in the concentration range 10^{-5}–10^{-3} M was an order of magnitude lower than for the substituted benzoic acids — as already illustrated by the representative values listed for group II materials in Table 1. Figure 2 illustrates, in a manner representative of these systems, the concentration dependence of the low adsorption values measured for benzyl alcohol. A likely additional source of difficulties responsible for limiting reproducibility of n_2^s measurements to $\pm 25\%$ in the Group II systems was the greater volatility of these materials. In marked contrast to the success achieved in Figure 1 in linearizing adsorption data for Group I pollutants on the basis of Equation 3e, little success attended similar efforts to achieve linearization of n_2^s values for extent of adsorption into TiO$_2$ from aqueous benzyl alcohol suspensions of various initial concentrations. The scatter of data points in Figure 2B illustrates the relative failure of such adsorption values to be adequately linearized when plotted as per Equation 3e. Least square analysis of this data indicated a range of 320 ± 180 dm^3 mol^{-1} for K_{ads} of this system. Qualitatively similar failure to be adequately linearized when plotted in the manner of Equation 3(e) resulted with data for the concentration-dependence of dark adsorption of other Group II pollutants from aqueous solutions. More adequate linearization of such data was achieved by plotting $(\ln n_2^s)^{1/2}$ vs $\ln(C_{eq})$ — in analogous fashion to linearization procedures often applied to adsorption onto

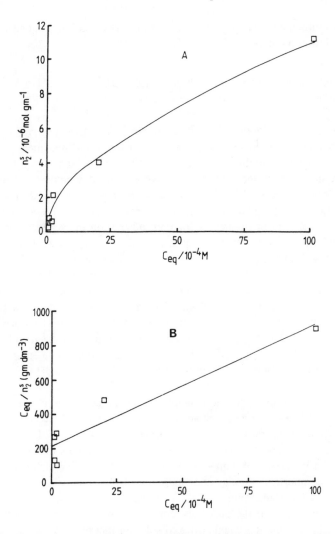

Figure 2. Isotherm for adsorption onto TiO$_2$ from dilute aqueous solutions for a poorly adsorbing solute, viz. benzyl alcohol. **(A)** Data for extent of adsorption, n_2^s, in equilibrium with indicated residual concentrations, C_{eq}, in solution. **(B)** Illustration of unsatisfactory outcome of attempted linearization of that data when plotted in accordance with Equation 3e. (From Cunningham, J. and Srijaranai, S., *J. Photochem. Photobiol. A: Chem.*, 58, 361, 1991. With permission.)

porous solids on the basis of the Dubinin-Radushevich (DR) isotherm equation. In the context of the outline given in Subsection A1 of the ideas of Hansen and Craig,[13] the working hypothesis here adopted to rationalize such failure to be adequately linearized on the basis of a solution-monolayer-type isotherm, but to be better linearized by the DR isotherm, is that the extent of equilibrium dark adsorption in these small-adsorption cases does not proceed in accordance with a mixed-solution monolayer model, but rather involves distribution of adsorbate in mixed-solution multilayers.

A3: Effects of Solvent Polarity Upon Nature and Extent of Dark Adsorption

Within the context of the mixed-solution monolayer model described above in relation to competitive solute-solvent adsorption and formulated under Equations 3a–3f, a change from highly polar water ($\varepsilon \sim 81$) to less polar acetonitrile ($\varepsilon \sim 38$) as the liquid in which the TiO_2 particles were suspended could be expected to result in less competition by solvent molecules for adsorption into the monolayer and therefore to greater dark adsorption of solute into a solution monolayer. Comparison of data in Figure 3A for dark adsorption of benzyl alcohol from acetonitrile solution with data from aqueous solutions (see Figure 2A) show trends consistent with this expectation. Such comparisons showed order of magnitude increases for extent of dark adsorption of benzyl or furfuryl alcohol onto TiO_2 particles suspended in acetonitrile solutions of various initial alcohol concentrations relative to those measured onto TiO_2 particles suspended in equimolar aqueous solutions. Naturally, the larger adsorption values observed from acetonitrile could be measured with greater precision. Figure 3 illustrates the good linearization of such data when plotted as per Equation 3e, thereby indicating adequate modeling of dark adsorption-desorption equilibria in $C_6H_5CH_2OH\text{-}CH_3CN\text{-}TiO_2$ suspensions by a mixed (solvent + solute) monolayer description. This is in marked contrast to the apparent nonapplicability (noted above) of such a model in the case of benzyl alcohol dark adsorption in $C_6H_5CH_2OH\text{-}H_2O\text{-}TiO_2$ suspensions. This solvent-induced contrast in adsorption behavior for the same solute (BzOH) onto the same solid surfaces (TiO_2) lends support to the validity of Equation 3a in respect of competitive adsorption from dilute solutions. It also points to an important interplay of enthalpy for displacement of adsorbed solvent from the surface by adsorbing solute with enthalpies for solvation of solute in the surface monolayer and the bulk solvation. Such interplay would appear to be important in determining whether mixed-solution monolayer or mixed-solution multilayer-type adsorption predominates. Lamarche et al., demonstrated the importance of an exothermic enthalpy of solvent displacement, ΔH_d, for the enhanced adsorption of acrylic acid polymer onto TiO_2 observed at pH 3.5.[14] Their results also demonstrated that this enhancement was most marked for the first 30% of the limiting extent of adsorption, thus providing evidence for nonuniformity in the nature of adsorption sites on TiO_2 (Degussa P25).

Figure 3.   Isotherms for enhanced adsorption of benzyl alcohol onto TiO_2 from acetonitrile solutions. **(A)** Data for extent of adsorption, n_2^s, in equilibrium with various residual concentrations, C_{eq}. **(B)** Adequate linearization of data when plotted as per Equation 3e. (From Cunningham, J. and Srijaranai, S., *J. Photochem. Photobiol. A: Chem.*, 58, 361, 1991. With permission.)

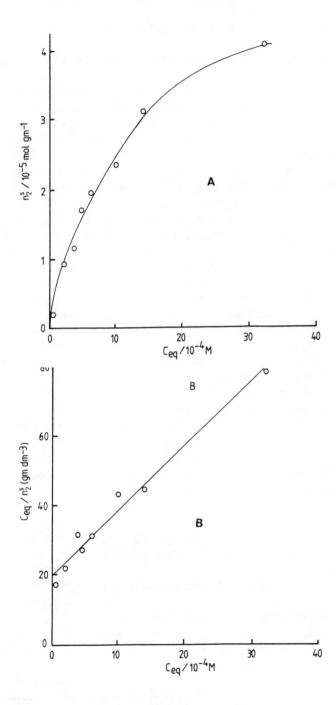

Figure 3.

A4: Tests of pH Dependence of Dark Adsorption

These were carried out in the same manner as described above for suspensions having their natural pH of ca. 5.0, except that requisite small amounts of NaOH or $HClO_4$ solutions were added to the freshly-prepared suspensions to attain higher or lower pH values, respectively. An overnight period with continuous shaking was then allowed for equilibration of dark adsorption, after which ΔC ($=C_o-C_{eq}$) was measured and a check made of the extent to which the preset pH value was significantly changed. Such changes were detected only from preset pH of 10 and 11 which decreased to 7.7 and 9.7. (The HPLC mobile phase used in evaluating C_o and C_{eq} values in these cases was buffered.) Unequivocal evidence for a marked effect of suspension pH upon extent of equilibrium dark adsorption was obtained only for Group 1 materials, e.g., CHZBA and SA, for both of which adsorption was an order of magnitude greater at pH 3 than at pH 10 as illustrated in Figure 4A. A very similar result over TiO_2 (Degussa P25) was reported by Lamarche et al. for adsorption of a polymer of acrylic acid with the limiting extent of adsorption at pH 3.5 being a factor of 7 greater than at pH 9.5 in that system.[14] Neither for 2-CP, nor for BzOH and FyOH, were changes in pH here found to induce variations greater than ca. 25% from their already low adsorption values at natural pH. This difference in pH sensitivity of Group 1 dark adsorption from that of Group II may also be rationalized in terms of the working hypothesis suggested above, viz., the predominance of mixed-solution multilayer-type adsorption for Group II solutes, whereas mixed-solution monolayer-type adsorption predominated for Group I solutes from aqueous solutions onto TiO_2. Adsorption of Group I solutes into a surface monolayer should naturally be strongly inhibited by existence of similar negative charge on the solute species and on the surface of the TiO_2 particles — as would be the case for pH values of 9–11 which are much higher than the pK of the substituted benzoic acids (ca. 3 $\rightarrow$ 4.6) and the point of zero charge (pzc) of TiO_2.[15] On the other hand, at pH 3, nondissociated acid molecules would predominate, while the surface would bear a small (positive) charge, with the result that adsorption would then not be inhibited by like-charge repulsions. Near-surface excesses of Group II solute species dispersed in mixed-solution multilayer regions farther out from the TiO_2 surface would be less sensitive to extent and sign of surface charge, furthermore their pK values (9–13) are high.[15]

Part B: TiO_2-Sensitized Photocatalytic Degradations

B1: "Initial Rates" of Additional Decreases in Model Pollutant Concentrations Caused by Exposure of 20 ml of Dark-Equilibrated Aqueous TiO_2 Suspensions to Fluxes of ca. 3×10^{18} Photons per Minute at $\lambda \approx 340$ to 390 nm

The dark adsorption results presented in Part A made quite evident the need to differentiate such adsorption-induced decreases in solute concentration from

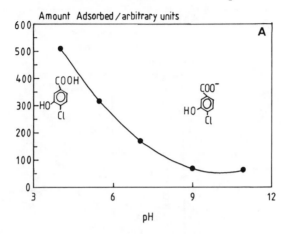

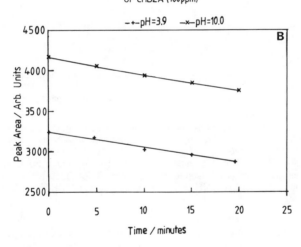

Figure 4. Influence of solution pH upon **(A)** extent of adsorption of chloro-4-hydroxy benzoic acid at initial concentration 200 ppm at 295 K onto equal aliquots of nonilluminated TiO₂. **(B)** Initial rates of TiO₂-sensitized photodegradation of CHBZA in two dark equilibrated CHBZA-H₂O-TiO₂ suspensions prepared from solutions of same C_o, but with pH adjusted to 3.5 or 10.0, respectively (flux of photons having λ 340–390 nm ≈ 3 × 10¹⁸ min⁻¹ onto 20 cm³ of suspension).

whatever additional concentration decreases would appear upon UV illumination. This was achieved by preequilibrating all suspensions for at least 1 hr in the dark within the photoreaction cell immediately prior to commencing UV illumination. Figures 5A and 5B depict the distinction thereby achieved between progress of the suspensions containing, respectively, 2-CP or CHBZA to dark

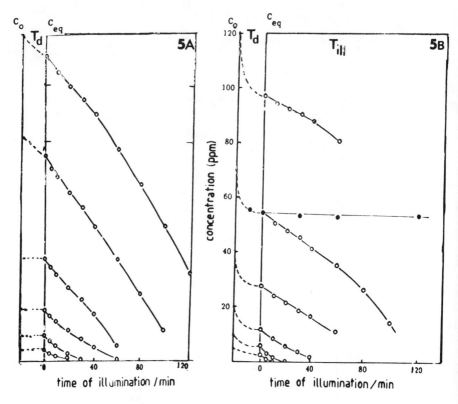

Figure 5. Comparisons of (1) the decreases in bulk solution concentration of model pollutants observed during Td by preillumination adsorption onto TiO$_2$ for I hr (cf. broken curves C$_o$ → C$_{eq}$) and (2) additional illumination-induced decreases below C$_{eq}$, and relative initial rates, (ΔC*/Δt), originating from TiO$_2$-sensitized photodegradation (cf. slopes of solid lines at T$_{i11}$ 0–40 min illumination). **(B)** Shows data for CHBZA/H$_2$O/TiO$_2$. **(A)** Shows data for 2-CP-H$_2$O-TiO$_2$ suspensions.

adsorption-desorption equilibrium (cf. data to the left of zero on the bottom axis indicating preillumination time) and the additional decreases in solute concentration, ΔC*, caused by subsequent UV illumination of the preequilibrated suspensions. (Note also data points shown as filled-in circles on plot (ii) of Figure 5B confirming that, for dark-equilibration times greater than I hr, no additional decrease in [CHBZA] occurred in the absence of UV illumination.) Two contrasting features can be recognized from comparisons of data for well-adsorbed substituted benzoic acid, as shown in Figure 5B, with data for a poorly adsorbed monochlorophenol in Figure 5A. First, from data to the left of zero illumination time, a much greater extent of dark adsorption is evident for the substituted benzoic acid. Second, it may be seen from initial slopes of the progressive decrease in concentration caused by UV illumination, that rates of TiO$_2$-sensitized photodegradation of the poorly adsorbed monochlorophenol are greater than for a suspension prepared from the corresponding initial concentration of the

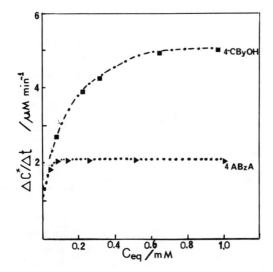

Figure 6. Contrasting profiles observed in similar conditions of preadsorption and illumination between variations of experimentally determined initial rates ($\Delta C^*/\Delta t$) of TiO$_2$-sensitized photodegradations of one well-adsorbed, 4-ABzA, and one poorly adsorbed, CByOH, pollutant with C_{eq}, the equilibrium concentration of pollutant remaining in the bulk-aqueous solution of TiO$_2$ suspensions after preillumination attainment of adsorption-desorption equilibria. (Flux of photons having λ 340–390 nm ≈ 3 × 10^{18} min^{-1} onto 20 cm^3 suspension).

substituted benzoic acid. Figure 6 summarizes, as a function of C_{eq}, the higher "initial rates of photodegradation" (evaluated from initial slopes of plots of the type shown in Figure 5 during the first 40 min of illumination) observed for one group II and one group I pollutant. The trend 4-CByOH > 4ABZA evidenced in Figure 6 for observed initial PCD rates is contrary to the trend in extent of dark adsorption evidenced in Table 1 and to the direct correlation with extent of adsorption which might be expected within LH, pseudo-first-order mechanisms (cf. Equation 2). Such discrepancies emphasized the need to give consideration to alternative mechanistic pathways for PCD, including those in which the locations for PCD events differ from those for dark adsorption.

B2: PCD of Solutes Well-Adsorbed onto TiO$_2$ from Aqueous Solution

In the cases of the well-adsorbed substituted benzoic acids, more searching tests for possible correlations across a range of solute concentrations — whether direct, as expected in pseudo-first-order LH-type mechanisms, or inverse, as expected for self-inhibiting mechanisms — could be made on the basis of the values for $n^s_{2(max)}$ and K_{ads}, which were shown to be applicable across a range of concentrations by dark adsorption experiments. Those values made possible the calculation of fractional occupancy values, Θ_2, of a mixed (solute + solvent)

monolayer by SA, CHZBA or ABZA species when in adsorption-desorption
equilibrium with aqueous solutions of varying C_{eq}. Insertion of such calculated
values for Θ_2 into the simple pseudo-first-order LH rate expression $^*R = k'_{LH}\Theta_2$
— together with various assumed values for k'_{LH} (chosen to cover the range of k'_{LH}
values deduced from $^*R^{-1}$ vs $[S]^{-1}$ plots of PCD rates) — predicts that PCD rates
should increase with C_{eq} in the manner indicated by the curved plots passing
through the open circle points on Figure 7A–7C. However, the experimentally
measured rates, as marked with closed circle points, do not increase with measured C_{eq} in accordance with any one of the LH-predicted curves across the
indicated concentration range. Indeed, the only semblance of convergence between calculated and actual PCD rates, at C_{eq} values ca. $10^{-3}\,M$, occurs with
assumed k'_{LH} values of 1.44, 1.0, and 1.0 for SA, CHBZA, and ABZA, respectively.
However, for each of those acid solutes, the PCD rate measured for concentrations
$4 \times 10^{-4}\,M$ remains very much higher than required if those values for k'_{LH} remained
valid at those lower C_{eq} values. Figure 8 provides an alternative illustration of this
non-Langmuirian persistence, down to low values of Θ_2, of readily measurable
PCD rates for each of the three substituted benzoic acids.

In relation to possible practical applications of PCD,[2,16] the above observations
encouraged the view that useful PCD efficiencies can be attained even in the
absence of appreciable adsorption into the surface monolayer. In relation to
acceptable mechanisms for PCD, they again indicated the inadequacy of an
unmodified LH-type mechanism and pointed to the need to consider, inter al,
alternatives that do not limit reaction possibilities solely to bimolecular events in
which both the substrate pollutant and photogenerated oxidizing species are held
in the surface monolayer. Such considerations will be especially important with
respect to poorly adsorbing pollutants (see below). Prior to that a final question
concerning the concentration dependence observed for the well-adsorbed benzoic
acids first requires attention: viz., "Why did the PCD rates observed at C_{eq} greater
than $4 \times 10^{-4}\,M$ not increase markedly in line with the large increases in Θ_2
demonstrated to occur at such concentrations by the dark-adsorption measurements?". In other words, if an LH-type mechanism contributes to PCD in these
cases, why did it not add on to the mechanism which apparently operates at near-
zero Θ_2 values (cf. Figure 8)? One possible explanation, to this puzzle would be
if, (a) the rate observed at low Θ_2 values stemmed mainly from a number of
oxidizing species photogenerated at the TiO_2-aqueous solution interface but
having the capability to escape to react with solute species in near-surface
regions of the solution, whereas (b) upon increasing Θ_2 toward limiting saturation

---

Figure 7.   Comparisons of dependences upon C_{eq} observed experimentally for rates of
TiO_2-sensitized photodegradations of substituted benzoic acids in aqueous suspensions (cf. solid data points in A, B, and C) with the dependences predicted,
respectively, for SA, CHBZA and AMBZA on the basis of Equation 2, together
with Θ_2^m values from dark-adsorption studies, plus the following range of
assumed values for k'_{LH} 0.65, 1.0, 1.44, and 2.0 and 4.0 (cf. dashed lines labeled
a,b,c,d, and e). (From Cunningham, J. and Al-Sayyed, G., *J. Chem. Soc. Faraday
Trans.*, 86, 3935, 1990. With permission.)

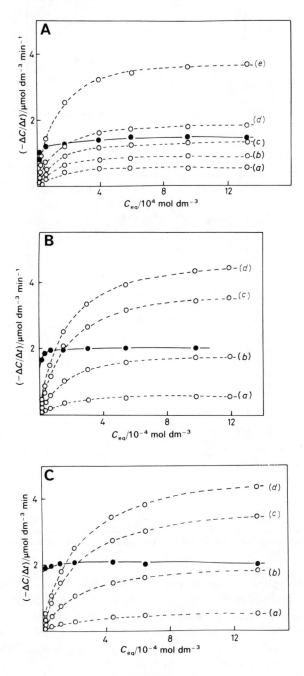

Figure 7.

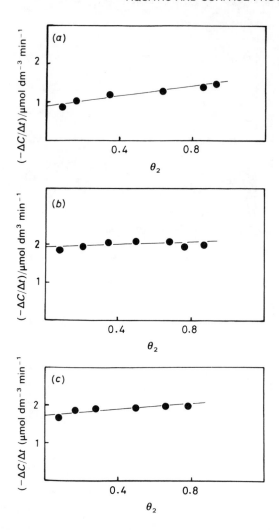

Figure 8. Tendency of initial PCD degradation vs Θ_2^m plots to extrapolate to non-zero PCD rates as $\Theta_2 \to 0$, in the cases of the well-adsorbing benzoic acids SA **(A)**, CHBZA **(B)**, and AMBZ **(C)**. (From Cunningham, J. and Al-Sayyed, G., *J. Chem. Soc. Faraday Trans.*, 86, 3935, 1990. With permission.)

concentration in the monolayer, that type of process did not increase further because the majority of the such photogenerated oxidizing species were intercepted at source by solute species in the surface-solution monolayer, i.e., at higher Θ_2, such LH-type events came to *replace* rather than to *supplement* the process operating at low Θ_2. A second possible explanation, Explanation B, would be if two opposing effects upon PCD accompanied the increases in Θ_2: viz., an increased probability of LH-type events proportional to the increase in Θ_2, but a compensatory decrease in the k'_{LH} term. Since the latter equals $k_{LH}\Theta_{ox}$, some decreases might arise from processes such as site blocking or from promotion of electron-hole recombination events, thereby diminishing Θ_{ox}, the surface concentration of highly oxidizing species photogenerated in the surface monolayer.[17,18] A rather fortuitous balance between the opposing effects would seem to be required, however, to result in the plateau values shown by the PCD rates in Figures 7A–7C at C_{eq} greater than $4 \times 10^{-4} M$. Additional evidence disfavoring Explanation B, but consistent with Explanation A, emerged as follows from pH-effect studies.

For CHBZA and SA aqueous solutions of fixed initial concentration (200 ppm), measurements were made of the extent to which modification of the pH of TiO₂ suspensions made with these solutions could affect initial rates of PCD of the benzoic acids. Recall that varying the pH of such suspensions from 3 to 10 was shown (Figure 4A) to cause an order of magnitude decrease in extent of dark adsorption of the acids. However, despite the large effect of pH upon the extent of adsorption, the virtual identity of slopes of the two PCD plots in Figure 4B illustrates an absence of significant pH-induced difference between the initial PCD rates observed for two CHBZA-H₂O-TiO₂ suspensions prepared from solutions of equal C_o, but which had their pH values respectively preadjusted to 3.3 and 10.0. (The vertical displacement evident between the two plots on Figure 4B reflects an observation already made and explained in subsection A4 and Figure 4A, viz. that "extent of dark-adsorption" onto TiO₂ was an order of magnitude lower at pH approximately 10 than at pH approximately 3.) The contrast illustrated in Figure 4 between the insignificant effect of pH changes from 3 to 10 upon initial PCD rate, but a large decrease in dark adsorption was also reproduced with SA suspensions. The observed similarity in PCD rates would be explicable in the context of Explanation A, since similar rate coefficients have been reported in homogeneous aqueous solutions for OH radical attack upon benzoate anions (the predominant species at pH 10) and upon nonionized benzoic acid molecules present at pH approximately 3.[19,20a,20b] Conversely, the fact that an order of magnitude increase in extent of adsorption occurred from pH 10 to pH 3 without significant effect upon PCD rate argues against Explanation B, since the maintenance of an unchanged PCD rate while surface excess of solute increased would require an improbable balancing act involving exact compensation of increasing Θ_2 by some inhibiting effect(s) of adsorbate species.

B3: PCD of Group II Solutes

A point of interest in respect of poorly adsorbing Group II pollutants was whether the evidence already given in Subsection A2 for non-Langmuirian trends in concentration dependence of their dark adsorptions would translate into unusual trends in the concentration dependence of their PCD rates. However, application of the usual linearization procedure, as $^*Rate^{-1}$ vs $([S])^{-1}$ plots, to data for TiO₂-sensitized PCD of poorly adsorbing benzyl alcohol enjoyed adequate linearization success (regression coefficients 0.993 and 0.998 respectively). Table 3

Table 3. Parameters Deduced from Intercept and Slope of Linear
          "Double-Reciprocal" Plots of (Initial PCD Rates)$^{-1}$ vs
          (Initial Pollutant Concentration)$^{-1}$

System	PCD Process Monitored	Parameter from Intercept[a] (μM^{-1} min^{-1})	Parameter from Intercept Slope[b] (M^{-1})	Parameter from Dark-Adsorption Studies[c] (M^{-1})
$C_6H_5CH_2OH$-H_2O-Ti$_2^*$	$-\Delta[C_6H_5CH_2OH]/\Delta t$	1.73	4080	320 ± 180
$C_6H_5CH_2OH$-H_2O-TiO$_2^*$	$+\Delta[C_6H_5CHO]/\Delta t$	0.92	647	320 ± 180
$C_6H_5CH_2OH$-CH_3CN-TiO$_2^*$	$+\Delta[C_6H_5CHO]/\Delta t$	16.0	2140 ± 450	930 ± 160
$ClC_6H_4CH_2OH$-H_2O-TiO$_2^*$	$-\Delta[ClC_6H_4CH_2OH]/\Delta t$	7.3	4630	

a Often interpreted in the literature, on the basis of Equation 2, as a pseudo-first-order,
   Langmuir-Hinshelwood-type rate coefficient relating to TiO$_2$-sensitized primary oxidation
   events in a surface monolayer, i.e., k'_{LH}.
b As per footnote a, but interpreted as a pseudo-equilibrium constant relating to monolayer
   adsorption, i.e., K*.
c Deduced from "best fit" to Equation 3e of data for small extents of equilibrium adsorption
   onto nonilluminated TiO$_2$, i.e., K$_{ads}$.

summarizes the values for k'_{LH} and K*, together with error limits, which were
deduced (as per Equation 2c) from application of this double-reciprocal linearization test to data for initial PCD rates vs C_{eq} obtained from the study of benzyl
alcohol and 4-chlorobenzyl alcohol as the model pollutant during this work. TiO$_2$-
photosensitized decreases in benzyl concentration (–BzOH), as well as increases
in benzaldehyde as the initial product of photodehydrogenation (–H$_2$)*, were
monitored for aqueous suspensions by the HPLC analyses. Consequently, Table
3 contains for the $C_6H_5CH_2OH$-H_2O-TiO$_2$ system two pairs of k_{LH}, plus K* values
derived from double-reciprocal plots of the concentration dependence of those
two processes. Large differences between those pairs of values are apparent,
although they originate from parallel measurements on the same illuminated
system. This lends further weight to doubts expressed by Turchi and Ollis as to
the adequacy of double-reciprocal plots as a sole basis for deducing mechanisms
and kinetic parameters for PCD processes.[21] Similar doubts with respect to their
adequacy as a secure basis for establishing values for adsorption constants are
raised here by the difference between the double-reciprocal-based value of 4080
in Table 3 deduced for a supposed adsorption constant K* for the $C_6H_5CH_2OH$-
H_2O-TiO$_2$ system and an estimate of 320 ± 180 obtained for K$_{ads}$ for the same
system from actual adsorption isotherm data. Within the context of an unmodified
pseudo-first-order k'_{LH} mechanism, Figure 9 further illustrates, for aqueous benzyl
alcohol suspensions this lack of compatibility between the k'_{LH} plus K* parameters
deduced from the double-reciprocal plot and Θ_2 values calculated on the basis of
the K$_{ads}$ values deduced from dark-adsorption measurements. Data on the lower
plot are rates estimated from $k'_{LH} \cdot \Theta_2$. Clearly, they fall much below the rates
actually observed, the principal reason for the discrepancy apparently being that

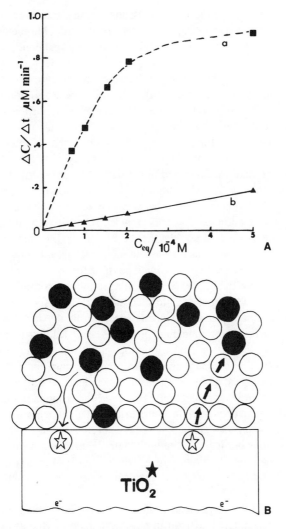

Figure 9. **(A)** Graphical comparison illustrating the large discrepancy between the dependence upon C$_{eq}$ observed experimentally for initial PCD rates in C$_6$H$_5$CH$_2$OH-H$_2$O-TiO$_2$ suspensions (Plot a) and the dependence predicted by inserting into a pseudo-first-order LH expression, *Rate = $k'_{LH} \cdot \Theta_2^m$, the k'_{LH} values obtained from double-reciprocal plots, and the Θ_2^m values known from dark-adsorption studies on this system (Plot b). **(B)** Schematic of possible mechanistic origins of discrepancies illustrated in (A) through involvements in TiO$_2$-sensitized photodegradation events by species not confined to a surface monolayer. The Eley-Rideal-type process to the left-hand side of diagram envisages inward movement of solute from multilayer to become involved in reaction events at the interface; the process on the right-hand side envisages a primary ex-TiO$_2^*$ photooxidizing species diffusing/tunneling outward for reaction events within the solution multilayer. (Solute-Solvent molecules are denoted by filled and open circles, respectively).

the $K_{ads} \sim 322\,M$ value deduced from dark adsorption is much lower than the value $K_{ads}^* \sim 4080\,M$ deduced from slope and intercept of the double-reciprocal plot. One rationalization of such discrepancy would be that the real concentration-induced increase in total number of solute species accessible to oxidizing species photogenerated at the TiO_2-aqueous solution microinterfaces exceeded an apparent surface excess of solute species detected by dark-adsorption measurements. Two classes of process with the potential for selectively enhancing the accessibility between solute species and photogenerated oxidizing species can be envisaged: either photoadsorption, whereby a greater numbers of solute species could be caused to be present in the solution monolayer at the UV illuminated microinterfaces than were detected in the dark,[11a,11b] or contributions to PCD by primary reaction events, which depend upon relocation of solute species, or of photogenerated oxidizing species, between the surface solution monolayer and adjacent solution multilayers, such as depicted schematically in Figure 9B. Contributions of the latter type could either be Eley-Rideal in character,[22] involving inward migration to the surface by solute molecules weakly physically adsorbed in solution multilayers, thereby making it possible for solute species from solution multilayers to contribute to bimolecular encounters and reactions with photogenerated oxidizing species located at the TiO_2 surfaces, or could involve outward movement of primary oxidizing species from the aqueous solution-TiO_2^* interface by diffusion or tunneling processes,[23a,23b] leading to reaction events involving solute species situated within the mixed-solution multilayer region.

As yet, definitive evidence is lacking for or against contributions to TiO_2-sensitized photodegradations by such processes dependent upon surface monolayer-solution multilayer relocations. However, the demonstrations given in the present work of inadequacies in unmodified LH treatments indicate that such possibilities should no longer be ignored for poorly adsorbing pollutants. Indeed, some observations not inconsistent with their occurrence have been made in our laboratories, e.g., the observation of a sizable D_2O-H_2O matrix effect for TiO_2-sensitized photodegradation of poorly adsorbing isopropanol, but not for well-adsorbing pollutants.[10b]

Experimental tests of photoadsorption as a possible contributor to enhanced accessibility are rendered difficult in most systems by virtue of the difficulty of differentiating any such contributions to ΔC^* from parallel contribution by photocatalytic degradation. Ways around this difficulty have not yet been devised for model pollutants of Groups I or II. Cyanuric acid represented, however, an unusually suitable test case for photoadsorption, since work at the University of Turin had demonstrated its remarkable resistance to photocatalytic degradation.[24] Consequently, if any additional decrease in concentration of cyanuric acid were o be detected upon UV illumination of a previously dark-equilibrated cyanuric acid-H_2O-TiO_2 system, it could be attributed to photoadsorption. Results from such an experiment, carried out in the University of Turin, have demonstrated for aliquots withdrawn from the suspension during UV illumination the absence of any measurable decrease in concentration of cyanuric acid relative to aliquots of the suspension withdrawn from the same suspension during the absence of UV illumination. In that system, therefore, no evidence was found for photoadsorption. A similar but less compelling negative result and conclusion emerged from experiments at University College, Cork, Ireland, which sought to detect the persistence of any enhanced capacity of preilluminated H$_2$O-TiO$_2$ suspensions for adsorbing 2-chlorophenol when the latter was injected into the suspension within seconds of the end of illumination.

B4: PCD of Solutes Well Adsorbed onto TiO$_2$ from Acetonitrile Solutions

Evidence was presented in Subsection A4 that switching from H$_2$O to CH$_3$CN as a solvent for benzyl alcohol resulted in greatly enhanced extent of adsorption showing Langmuir-type dependence upon equilibrium concentration of benzyl alcohol. It was of interest, therefore, to check whether better convergence could be achieved between observed and LH-based estimates of the concentration dependence of PCD rates for the better adsorbing C$_6$H$_5$CH$_2$OH-CH$_3$CN-TiO$_2$ suspensions. Other features of note in this system were (1) the apparent lack of processes for formation of surface OH radicals (but see below) or for their escape from the TiO$_2$ surface into near-surface solution layers, and (2) the fact that benzaldehyde was the sole detectable photoproduct whose initial rates of formation exhibited reasonable mass balance (85 ± 15%) with initial rates of PCD of the parent alcohol across the concentration range studied. Plot (a) of Figure 10A summarizes the experimentally determined initial rates of benzaldehyde formation for various C$_{eq}$ values in the 10^{-4} to 2 × 10^{-2} M range. Good lineariza-tion of that data was achieved by double-reciprocal plots, yielding the values k$'_{LH}$ = 1.6 ± 0.3 × 10^{-3} M min^{-1}, and K* = 2.1 ± 0.5 × 10^3 M^{-1}. The divergence of the latter value from K$_{ads}$ = 0.9 ± 0.1 × 10^3 M^{-1} deduced from dark-adsorption data for that system was notably less severe than between values of K* = 4.1 ± 0.4 × 10^3 M^{-1} and K$_{ads}$ ~ 0.3 × 10^3 M^{-1} deduced in corresponding manner for the C$_6$H$_5$CH$_2$OH-H$_2$O-TiO$_2$ system. Not surprisingly then, the calculated concentration dependence of PCD rates arrived at from use of the acetonitrile values in Rate =k$'_{LH}$ · Θ$_2$ results in much better agreement with the experimentally observed trend (cf. plot b with plot a in Figure 10A) than had been possible to achieve in Figure 9A for the corresponding aqueous system. Rates of TiO$_2$-sensitized photodegradation in the C$_6$H$_5$CH$_2$OH-CH$_3$CN-TiO$_2$ system thus approximate more closely to unmodi-fied Langmuir Hinshelwood type behavior than for the other systems studied in this work. Thus, it is tempting to conclude that the predominant contribution to initial PCD rates in that system takes the form of direct interactions of C$_6$H$_5$CH$_2$OH molecules located in the solution monolayer on the TiO$_2$ surface with photogenerated holes, as depicted schematically in Figure 10B.

However, possibilities for involvement of surface-bound OH radicals in the initial stages of TiO$_2$-sensitized photodegradation in the C$_6$H$_5$CH$_2$OH-CH$_3$CN-TiO$_2$ system cannot be excluded in view of the following: (1) concentrations of 4–5

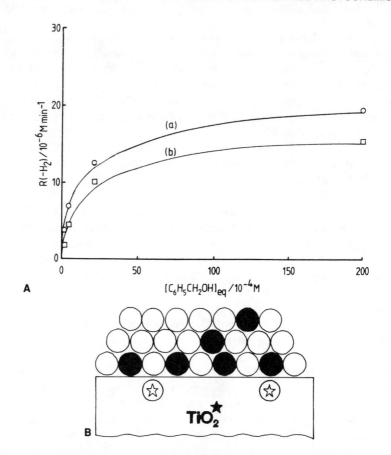

Figure 10. **(A)** Graphical comparison illustrating the considerable agreement achieved between the dependence of initial PCD rates upon C_{eq} observed experimentally for $C_6H_5CH_2OH$-CH_3CN/TiO_2^* suspensions (Plot a) and the dependence predicted (Plot b) by the LH-type rate expression, $\cdot Rate = k'_{LH} \cdot \Theta_2^m$, by employing k'_{LH} values obtained from double-reciprocal plots and Θ_2^m values found from dark-adsorption studies on the same system. **(B)** Schematic representation of an LH-type mechanism making the major contribution to initial TiO_2-sensitized photodegradation rates in $C_6H_5CH_2OH$-CH_3CN-TiO_2^* system: filled circles denote BzOH in the surface-solution monolayer, while open circles denote solvent molecules. The circled asterisk denotes photogenerated holes arrived at the interface to react there with adsorbed solute, either directly or via an OH radical formation from residual surface hydroxyls.

surface hydroxyl groups per square nanometer of TiO_2 surface have been reported for polycrystalline TiO_2 materials after the relatively mild 150°C drying used to precondition our samples.[25] This exceeds the 0.65 nm^{-2} maximum surface density of adsorbent benzyl alcohol deduced from the dark-adsorption studies on $C_6H_5CH_2OH$-CH_3CN-TiO_2 suspensions. Consequently, photogenerated holes reaching the microinterfaces in those systems were statistically more likely to generate OH radicals from OH^- than cation radicals directly from benzyl alcohol. Such surface-bound OH radicals may then either react with adsorbed benzyl alcohol or disappear by electron-hole recombination as per the following process:

$$\left(OH^-\right)_s \underset{e^-}{\overset{h^+}{\rightleftharpoons}} \left(OH\right)_s \xrightarrow{\text{BzOH}} C_6H_5\dot{C}HOH \rightarrow \text{product} \qquad (4)$$

This appears kinetically similar to "geminate recombination" situations considered by Roy et al.[26] and by Noyes[27a-27c] and reported to yield a linear relationship between rate and square root of scavenger concentration. Our data for the concentration dependence of initial $(-H_2)^*$ rates from the $C_6H_5CH_2OH$-CH_3CN-TiO_2 systems were equally well represented by a plot of that form as by a double-reciprocal plot, thereby casting further doubt on the discriminatory power of the latter in mechanistic studies of PCD. (2) The initial PCD rates measured over 20–40 min illumination times corresponded to 5% of $C_o = 10^{-4} M$, but to only 0.1% of $C_o = 10^{-2} M$ initial concentrations of benzyl alcohol. Thus, the possibility could not be excluded, on the basis of those data alone, that Equation 4 merely represented a "photoassisted surface reaction" not satisfying the criteria for truly photocatalyzed heterogeneous processes.[28] Attainment of real photocatalytic character was implied, however, by later observations of TiO$_2$-sensitized photodegradations to greater than 50% decomposition on 0.1 M benzyl alcohol to benzaldehyde in acetonitrile solutions. This was followed during prolonged illuminations using a refractive index detector. However dark-adsorption data of sufficient accuracy to establish reproducible adsorption isotherms extending to those large initial concentrations could not be obtained in this way for the $C_6H_5CH_2OH$-CH_3CN-TiO_2 system.

SUMMARY

A primary experimental concern in this work was to achieve careful characterizations of the extent and nature of equilibrium adsorption of model pollutants onto TiO$_2$ particles from parts per million aqueous solutions. Major interpretative concerns herein are with the degree of correlation attainable between parameters deduced from such adsorption studies and those deduced from measurements on relative efficiencies and solute concentration dependence of TiO$_2$-sensitized PCD of the model pollutants. Aspects of the literature which prompted these concerns were (1) the relative paucity of direct measurements upon equilibrium extent of adsorption in the parts per million range and (2) a widespread assumption that LH mechanisms — featuring bimolecular reaction events between adsorbed species — predominate for TiO$_2$-sensitized PCD degradations, even at these low concentrations.

The following features of results presented in this chapter call into question the adequacy of unmodified LH mechanisms as a secure basis for modeling TiO$_2$-sensitized PCD of model pollutants and its concentration dependence for aqueous solutions at 5–200 ppm. (1) The lack of any direct correlation between measured extents of equilibrium adsorption and initial rates of PCD: rather the observation here of higher rates of PCD for poorly adsorbing phenols and alcohols than for well-adsorbing substituted benzoic acids. (2) In the case of well-adsorbing substituted benzoic acids, a marked contrast was detected between the strong pH dependence of equilibrium dark adsorption and an absence of significant pH effect upon PCD rates. Thus, an order of magnitude decrease in dark adsorption noted at pH 10 relative to that at pH 3.5 was not paralleled by significant changes in initial rates of PCD. (3) A marked lack of agreement — especially for poorly adsorbing pollutants — between experimentally observed dependences of initial PCD rates upon pollutant concentration and dependences predicted on the basis of independently evaluated adsorption parameters and pseudo-first-order LH rate coefficients. Driven by these discrepancies, consideration is given within the chapter to the implication of various experimental observations for the likely involvement of the following modifications or alternatives to LH-type, monolayer-based mechanisms for TiO_2-sensitized PCD:

1. inhibition of PCD efficiency in the case of well-adsorbed pollutants, either via blockage of sites for PCD on the TiO_2 surface or through their action as locations for surface recombinations of photogenerated holes and electrons
2. photoadsorption processes of poorly adsorbing pollutants, causing photostationary extents of adsorption into a surface-solution monolayer to exceed that attained therein by dark adsorption
3. Eley-Rideal-type reaction events, facilitated by inward migration of pollutant species weakly localized in solution multilayers (rather than solely in a solution monolayer) and resulting in their oxidation by primary photooxidizing species photogenerated in the surface solution monolayer
4. occurrence of primary PCD reaction events within near-surface-solution multilayers, facilitated by outward migration of primary photooxidizing species via tunneling or related processes

These considerations lead to the conclusion that neither Steps 1 or 2 are supported by present results, but rather that improved consistency with results observed for weakly adsorbed pollutants — relative to LH-type mechanisms — may be achieved in terms of Steps 3 or 4.

ACKNOWLEDGMENTS

This work was supported by funding provided by the European Commission under Contract No. EV 4V OO68 for which the authors are grateful. We also acknowledge with thanks the many discussions with our co-workers from other laboratories participating in this contract, P. Pichat, E. Pelizzetti, E. Borgarello, K. Hustert, and B. Pouyet.

REFERENCES

1. Matthews, R. W., Environment: photochemical photocatalytic processes. degradation of organic compounds, in *Photochemical Conversion and Storage of Solar Energy*, Pelizzetti, E. and Schiavello, M., Eds., Kluwer, Dordrecht, 1991, 427, and references therein.

2. Al-Sayyed, G., D'Oliveira, J.-C., and Pichat, P., Semiconductor-sensitized photodegradation of 4-chlorophenol in water, *J. Photochem. Photobiol. A: Chem.*, 58, 99, 1991.

3. Terzian, R., Serpone, N., Minero, C., and Pelizzetti, E., Kinetic studies in heterogeneous photocatalysis: Part 5. Photocatalyzed mineralization of cresols in aqueous media with irradiated Titania, *J. Catal.*, 128, 352, 1991, and references therein.

4. Pelizzetti, E., Pramauro, E., Minero, C., Serpone, N., and Borgarello, E., *Photocatalysis and Environment*, Sciavello, M., Ed., NATO ASI Ser. C237, Kluwer, Dordrecht, 1988, 527.

5. Cunningham, J., and Al-Sayyed, G., Factors influencing efficiences of TiO$_2$-sensitized photodegradations. Part 1. Substituted benzoic acids: discrepancies with dark-adsorption parameters, *J. Chem. Soc. Faraday Trans.*, 86, 3935, 1990.

6. Cunningham, J. and Srijaranai, S., Sensitized photo-oxidations of dissolved alcohols in homogeneous and heterogeneous systems. Part 2. TiO$_2$-sensitized photodehydrogenations of benzyl alcohol, *J. Photochem. Photobiol. A: Chem.*, 58, 361, 1991.

7. Hiemenz, P.C., *Principles of Colloid and Surface Chemistry*, 2nd ed., Marcel Dekker, New York, 1986.

8. Adamson, A. A., *Physical Chemistry of Surfaces*, 4th Ed., Wiley, New York, 1982, 369.

9a. Everett, D.H., *Basic Principles of Colloid Science*, Royal Society of Chemistry, London, 1988.

9b. Rouquerol, J. and Sing, K.S.M., Eds., *Adsorption at the Gas Solid and Liquid-Solid Interface*, Elsevier, Amsterdam, 1982, 2.

9c. Physical adsorption in condensed phases, *Faraday Discuss Chem. Soc.*, 59, 9, 1975.

10a. Cunningham, J. and Srijaranai, S., Sensitized photo oxidations of dissolved alcohols in homogeneous and heterogeneous systems. Part 1. Homogeneous photosensitization by uranyl ions, *J. Photochem. Photobiol. A: Chem.*, 55, 219, 1990.

10b. Srijaranai, S., Physical and Chemical Interactions of Dissolved Alcohols with Metal-Oxide Photosensitizers in Homogeneous and Heterogeneous Systems, Ph.D. thesis, National University of Ireland, 1990.

11a. Cunningham, J., Jauch, M., and McNamara, D., Adsorbate activation on metal oxides by single-electron and electron-pair interactions, *Proc. R. Ir. Acad. Sect. B*, 89, 1989, 299.

11b. Kwan, T., Photoadsorption and photodesorption of oxygen on inorganic semiconductors and related photocatalysis, in *Electronic Phenomena, in Chemisorption and Catalysis on Semiconductors*, Hauffe, K. and Wolkenstein, T., Eds., De Gruyter, Berlin, 1969, 184.

12a. Turchi, C. S. and Ollis, D. F., Photocatalytic reactor design: an example of mass-transfer limitations with an immobilised catalyst, *J. Phys. Chem.*, 92, 6852, 1988.

12b. Matthews, R. W., loc. cit., Response to (a), loc. cit., 92, 6853, 1988.

13. Hansen, R. P. and Craig, R. P., The adsorption of aliphatic alcohols and acids from aqueous solutions by non-porous carbons, *J. Phys. Chem.*, 58, 212, 1954.

14. Lamarche, J. M., Foissy, A., Robert, G., Reggiani, J. C., and Bernard, J., Adsorption d'un polymere de l'acide acrylique sur le dioxyde de Titane, in *Adsorption at the Gas-Solid and Liquid-Solid Interface*, Rouquerol, J. and Sing, K. S. W., Eds., Elsevier, Amsterdam, 1982, 117.

15. King, E. J., Acid Base Equilibria, Pergamon Press, Oxford, 1979, 174.

16. Ollis, D. F., Pelizzetti, E., and Serpone, N., Heterogeneous photocatalysis in the environment: application to water purification, in *Photocatalysis*, Serpone, N. and Pelizzetti, E., Eds., Wiley, New York, 1989, 603.

17. Tunesi, S. and Anderson, M., Influence of chemisorption on the photodecomposition of salicyclic acid and related compounds using suspended TiO$_2$ ceramic membranes, *J. Phys. Chem.*, 95, 3399, 1991.

18. Sabate, J., Anderson, M., Kikkawa, H., Edwards, M., and Hill, G. G., A kinetic study of the photocatalyzed degradation of 3-chlorosalicyclic acid over TiO_2 membranes supported on glass, *J. Catal.*, 127, 167, 1991.

19. Deister, U., Warneck, P., and Wurzinger, C., OH radicals generated by NO_3^- photolysis in aqueous solution: competition kinetics and a study of the reaction OH + $CH_2(OH)SO_3^-$, *Ber. Bunsenges. Phys. Chem.*, 94, 594, 1990.

20a. Matthews, R. W., Hydroxylation reactions induced by near-UV photolysis of aqueous TiO_2 supensions, *J. Chem. Soc. Faraday Trans. I*, 80, 457, 1984.

20b. Matthews, R. W. and Sangster, D. F., Measurement by benzoate radiolytic decarboxylation of relative rate constants for hydroxyl radical reactions, *J. Phys. Chem.*, 69, 1938, 1965.

21. Turchi, C. S. and Ollis, D. F., Photocatalytic degradation of organic water contaminants: mechanisms involving hydroxyl/radical attack, *J. Catal.*, 122, 178, 1990.

22. Cunningham, J. and Hodnett, B. K., Kinetic studies of secondary alcohol photooxidation on ZnO and TiO_2 at 348 K, *J. Chem. Soc. Faraday Trans. I*, 77, 2777, 1981.

23a. Bell, R.P., The Tunnel Effect in Chemistry, Chapman & Hall, London, 1980.

23b. Ballard, S. G. and Mauzerall, D., Kinetic evidence for electron tunnelling in solution, in *Tunnelling in Biological Systems*, Chance, B., et al, Eds., Academic Press, New York, 1977, 581.

24. Pelizzetti, E. and Minero, C., unpublished work.

25. Augustynski, J., Aspects of photo-electrochemical and surface behaviour of TiO_2, in *Structure and Bonding*, Monogram No. 69, Springer-Verlag, Berlin, 1988, chap. 1.

26. Roy, J. C., Williams, R. R., and Hamill, W. H., Diffusion kinetics of atom-radical recombinations in radiative neutron capture by halogens in liquid alkyl halides, *J. Am. Chem. Soc.*, 76, 3274, 1954.

27a. Noyes, R. M., Kinetics of competitive processes when reactive fragments are produced in pairs, *J. Am. Chem. Soc.*, 77, 2042, 1955, 27.

27b. Noyes, R. M., Models relating molecular reactivity and diffusion in liquids, *J. Am. Chem. Soc.*, 78, 5485, 1956

27c. Noyes, R. M., Effects of diffusion rates on chemical kinetics, *Prog. React. Kinet.*, 1, 129, 1961.

28. Cunningham, J., Goold, E. L., Hodnett, B. K., Leahy, E. M., and Al-Sayyed, G., Photo assisted gas-solid reactions, photocatalytic processes and end ergonic photo conversions over pure and surface-doped metal oxides, in *Catalysis on the Energy Scene*, Kaliaguine, S. and Makay, A., Eds., Elsevier, Amsterdam, 1984, 283.

CHAPTER 23

Photocatalytic Detoxification of Polluted Aquifers: Novel Catalysts and Solar Applications

Detlef W. Bahnemann, Dirk Bockelmann, Roland Goslich, and Marcus Hilgendorff

INTRODUCTION

Persistent organic chemicals are present as pollutants in wastewater effluents from industrial manufacturers, dry cleaning facilities, or even normal households. They can be found in groundwater wells and surface waters where they have to be removed to achieve drinking water quality.[1,2] Therefore, many processes have been proposed over the years and are currently being employed to destroy these toxins. The so-called photocatalytic detoxification has been discussed as an alternative method for cleanup of polluted water in the scientific literature since 1976.[3] Lately, considerable public attention has been focused on this possibility of combining heterogeneous catalysis with solar technologies to achieve the mineralization of toxins present in water.[4-7] Several reviews have recently been published discussing the underlying reaction mechanisms of photocatalytic detoxification and illustrating examples of successful laboratory and field studies.[8-10] While the overall stoichiometry of most mineralizations appears to be understood,[11] details of the complex reaction mechanism are still not known. Anatase, titanium dioxide, the material with the highest photocatalytic detoxification efficiency, is a wide bandgap semiconductor ($E_g \approx 3.2$ eV).[12] Thus, only light below 400 nm is absorbed and capable of forming the e^-/h^+ pairs[13] which are a prerequisite for the process. Therefore, only the ultraviolet (UV) part, i.e., 5% of the solar energy reaching the surface of the Earth, could be utilized, in principle, when

TiO_2 is the photocatalyst. Hence, it is evident that for solar applications other materials have to be found or developed that exhibit similar efficiencies as anatase TiO_2, but possess spectral properties more closely adapted to the terrestrial solar spectrum. For a solar application of photocatalytic detoxification, it is essential that the incident sunlight is effectively utilized. Therefore, parts of the investigations presented in this chapter concentrate on the synthesis and characterization of photocatalysts that absorb in the visible part of the solar spectrum and simultaneously improve the photocatalytic detoxification properties in this spectral region. The absorption of photons by semiconductors leads to the formation of an equal number of positive and negative charge carriers (e^-/h^+ pairs). While the fate of the hole which induces the desired oxidation process has been studied in detail, most authors did not examine the role of the cathodic process, i.e., the reactions of e_{CB}^-. It is generally assumed that molecular oxygen acts as the oxidant.[8-10] However, hydrogen peroxide (H_2O_2), which should be formed during O_2 reduction, is found only in trace amounts when TiO_2 is used as the photocatalyst.[14] Further reduction of H_2O_2 leads to the formation of ·OH radicals. In fact, it has been shown that the rate of photodegradation can be considerably enhanced when H_2O_2 is used as the oxidant.[15] Separation of the anodic and cathodic process, in principle, is not possible in microheterogeneous photocatalytic systems containing semiconductor particles. Hence, it cannot be decided whether hydroxyl radicals are formed via the oxidation of water or the reduction of molecular oxygen, i.e., whether electrons or holes are more important for the initial step of pollutant degradation which is generally believed to be the reaction of ·OH with the substrate molecule S. Since the efficiency of a complex process is always limited by the slowest reaction step, it is necessary to distinguish between the two possibilities discussed above and to study them separately. In the following, we will therefore also present evidence from photoelectrochemical investigations with separated anode and cathode which have been carried out to further elucidate the underlying reaction mechanisms. Various reactor designs have been tested in laboratory studies where chemical engineering problems characteristic for the different reactor types, such as mass-transfer limitations, have been exploited in detail.[16,17] Here, we will show results obtained with two different solar detoxification reactors at test sites in Almeria, Spain and Campinas, Brazil. Many of the commercially used reactors for photocatalytic cleanup processes, e.g., parabolic trough reactors which have been successfully utilized in solar test fields in Spain and the United States, employ light concentrating systems. Since during the normal use periodical variations of the temperature in the reactor are also encountered, a systematic study of the light intensity and the temperature dependence of photocatalytic mineralizations is required. Here, we present our results of a respective laboratory investigation using titanium dioxide (Degussa P25) as the photocatalyst and dichloroacetic acid as the test compound. Finally, it should be noted that photocatalysis has been compared with other methods for the destruction of toxic chemicals in aqueous solutions on economical[18] and technical[19] grounds. The authors of the respective papers agree that the method bears considerable potential and should be pursued.

EXPERIMENTAL PROCEDURES

The Ti-Fe mixed-oxide colloids with different iron content are prepared as powders, which are stable at room temperature for several months and can be resuspended in water or water-ethanol mixtures yielding transparent colloidal solutions. Details of this preparation have been described previously.[20]

The photocatalytic activity of the Ti-Fe mixed-oxide particles and that of commercially available Degussa P25 titanium dioxide powder was tested with detoxification measurements using dichloroacetic acid (DCA) as the probe molecule. DCA is a relatively strong organic acid ($pK_A = 1.29$) and thus present in its anionic form over the entire pH range studied in this chapter. According to the stoichiometry observed previously,[20] the oxidation of one DCA molecule leads to the formation of one proton. Therefore, we used a pH-stat technique,[21] which allows the *in situ* measurement of H_{aq}^+ formed during the photolysis experiments with extremely high sensitivity. The data from the autotitration system were transferred to a computer, which calculates the concentration of the generated protons from the amount of the added base with respect to the elapsed time. Corrections due to the dissociation equilibria of simultaneously formed H_2CO_3 (as HCO_3^- at $6.3 < pH < 10.3$ and as CO_3^{2-} at pH greater than 10.3) have been considered in the computer code.

The autotitration system (Metrohm) was connected to a ROSS semimicro combination pH electrode 81–15 SC. The titrant solution (0.02 or 0.1 N NaOH) was kept under Ar and calibrated weekly with 0.1 N HCl (Titrisol/Merck). The photochemical reactor was made of quartz glass and filled with a 50-ml colloidal solution which was thermostated, vigorously stirred by a magnetic stirring bar, and continuously purged with molecular oxygen to ensure a constant O_2 concentration throughout the experiment. Illuminations were carried out with a mercury-doped high-pressure xenon lamp (Osram XBO 500 W). Short-wavelength UV radiation was eliminated by a 320-nm bandpass filter, and infrared (IR) radiation was suppressed by a 10-cm water filter. Actinometry was performed using Aberchrome 540[20] in order to determine the light intensity in the wavelength region between 310 and 370 nm. The determination of the light intensity is essential to calculate quantum yields. The quantum yields, i.e., the ratio of reaction rate and corresponding light intensity, have been determined to characterize the efficiency of the degradation process. Monochromatic illuminations were carried out using a 436-nm bandpass filter and focusing the light of the described illumination source onto a 1-cm cell containing 3 ml test solution. A fresh sample was prepared for each experimental point. Actinometry was carried out with the same set-up and the photochemically produced isomer of Aberchrome 540.[22] In the laboratory experiments, each test solution contained 0.5 g/L of the Ti-Fe mixed oxide or the Degussa P25 powder and 2.5 mM DCA at a pH between 2.7 and 2.4. The adjustment of the pH was performed by adding appropriate amounts of 1 M HCl or NaOH. According to the preparation of the Ti-Fe mixed oxides, the colloidal solution contains 2–3 mM chloride. For experiments in the alkaline pH regime (pH ~ 11.3), an appropriate amount of 1 M NaOH was added quickly to the colloidal solution.

A Dionex 4500i ion chromatograph equipped with a HPIC-AS4A separator column was used to determine carbonate, chloride, acetate, glycolate, glyoxalate, and dichloroacetic acid concentrations. Detailed conditions are given in the literature.[23] The Fe(II) concentration was measured spectrophotometrically with ferrocene as the reagent.[24] Absorption spectra of the colloids were recorded using a Bruins Omega 10 UV-Vis spectrophotometer. Chemicals and solvents were of reagent grade and used without further purification. The water employed in all preparations was purified by a Milli Q-RO system (Millipore), resulting in a resistivity greater than 18 MΩcm. Photoelectrochemical experiments were performed using a single crystal rutile TiO_2 electrode (doped with molecular hydrogen for 8 h at 550°C, indium back contact) with a surface area of 1.5 cm^2 at + 1.0 V vs Ag$^+$/AgCl. The counter electrode (Pt) was in a separated dark compartment. Other experimental conditions were 50 mM acetate, pH 7 maintained by pH-stat technique, 0.1 M KCl. Aliquots were taken periodically, diluted 1:5, and analyzed by HPIC as described above.

The experimental set-up of the solar detoxification loop at the Plataforma Solar in Almeria, Spain has been described elsewhere.[25] The solar reactor employed during the experiments at the University of Campinas, Brazil was constructed following a proposal of W. Jardim.[26]

MECHANISMS OF PHOTOCATALYTIC DEGRADATION PROCESSES

Light Intensity and Temperature Dependency

Dichloroacetic acid was used as a model compound in all experiments described in this chapter. Ollis and co-workers have previously postulated a stoichiometry for the photocatalytic destruction of this molecule,[27] which has been supported by our own measurements of the formation of chloride and protons as a function of illumination time.[28] Figure 1 shows that carbonate is indeed formed in the stoichiometric ratio predicted by

$$CHCl_2COO^- + O_2 + 2OH^- \rightarrow 2HCO_3^- + H^+ + 2Cl^- \tag{1}$$

It can be seen in Figure 1 that the concentration of protons formed during the illumination (as measured by the pH-stat method) is indeed equal to one half of the simultaneously formed carbonate concentration (measured by ion chromatography). The former value has been corrected mathematically for the pH change induced by the carbonate protonation equilibria. The irregularities observed in the pH-stat results shown in Figure 1 can be explained by the sluggishness of the pH electrode leading to an overtitration with base. Consequently, the pH climbed

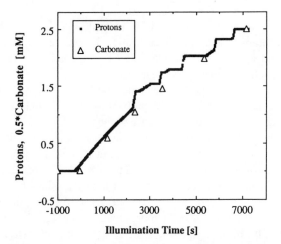

Figure 1. Formation of protons and carbonate as a function of illumination time of an O₂-saturated aqueous suspension containing 0.5 g/L Degussa P25, 24 m*M* dichloroacetate, and 10 m*M* KNO₃ at pH 10.2.

above the predetermined value and no more base was added before enough protons were formed during the subsequent illumination. Figure 1 also illustrates that the concentrations of educts and products of the DCA degradation vary linearly with the illumination time at least during the initial phase of the experiment. Reaction rates, which are given in subsequent figures, therefore have been calculated from the initial slopes of these concentration vs time plots. Figure 2 shows the effect of the light intensity on the rate of the photocatalytic degradation

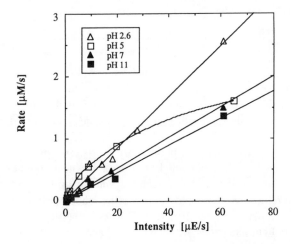

Figure 2. Rate of DCA degradation as a function of illumination intensity at various pH values for O₂-saturated aqueous suspensions containing 0.5 g/L Degussa P25, 24 m*M* dichloroacetate, and 10 m*M* KNO₃.

of dichloroacetate as a function of pH. A linear dependency, i.e., an intensity-independent quantum yield of the mineralization, is observed at pH 2.6, 7, and 11. Only at pH 5 did we find that the rate of DCA degradation increased with the square root of the light intensity. Similar observations have been reported previously when chloroform was used as the model compound.[29] In the latter case, they have been explained by an increased probability of combination of hydroxyl radical intermediates formed on the surface of the titanium dioxide particles used as photocatalysts. When chloroform was the test molecule, this square root dependence was most pronounced at pH 7.2, but it could be observed over the whole pH regime covered in this study ($3.8 \leq pH \leq 8.0$).[29] The present work shows that when illuminations were carried out with light intensities below 1.2×10^{-6} mol photons/L·s, i.e., between 0.015×10^{-6} and 1.1×10^{-6} mol photons/L·s, even at pH 5 the quantum yield of DCA photodegradation no longer depended upon the irradiation intensity. In the range of $2.5–25 \times 10^{-6}$ mol photons/L·s we encountered the highest degradation rates at pH 5 as compared with all other examined pH values. Thus, it appears that a specific mechanistic explanation is required to account for these characteristic properties at pH 5.

We have also performed acid-base titrations to determine the equilibria associated with the protonation and deprotonation of surface-bound hydroxyl groups on TiO_2.[28] With these measurements, it was possible to determine the pH value where $c(TiOH_2^+) = c(TiO^-)$. This so-called *zero proton condition* was found to be $pH_{ZPC} \sim 5$. Thus, at pH 5, most of the surface hydroxyl groups are present as uncharged TiOH groups. In the following, we will offer a possible explanation for the observed square root dependence of the DCA degradation yield at this pH. The reaction of valence-band holes formed during the illumination of the semiconductor particles with these surface states can lead to the generation of surface-bound hydroxyl radicals via

$$TiOH + h_{VB}^+ \rightarrow \left(TiOH_s^{\cdot} \right)^+ \qquad (2)$$

The combination of these ·OH radicals leading to the formation of H_2O_2 is a bimolecular process that can easily explain the observed square root dependence of the degradation yield on the light intensity. A detailed mathematical model, which quantitatively describes this process, has been developed recently by Kormann.[31]

The temperature dependence of photocatalytic processes presents another crucial parameter, the knowledge of which is essential for the design of solar reactors which inevitably will absorb a considerable amount of IR irradiation during operation. Therefore, we measured the rate of the photocatalytic degradation of dichloroacetate at pH 5 between 10 and 90°C. These experiments were carried out at two different light intensities, i.e., at 4.5×10^{-6} and 1.5×10^{-6} mol photons/L·s.

The results of these experiments are illustrated in Figure 3. From this plot, it is evident that the degradation rate increases linearly with the employed light intensity. However, according to the so-called Arrhenius relation

$$k = A\, e^{-E_a/RT} \tag{3}$$

where A is the preexponential factor, E_a is the activation energy [J/mol], R is the gas constant [J/mol*K], T is the temperature [K], and k is the rate constant [M/sec]. The reaction rate should increase linearly with exp($-1/T$). Since in aqueous solution the experimentally accessible temperature range is rather limited, we are not able to decide whether a linear or an exponential fit yields a better fit of the results shown in Figure 3. Therefore, we have also determined the activation energy from a linear regression of plots via log(k) = f(1/T). Even though we would not expect a dependency of the activation energy upon the light intensity, we obtained different values at different illumination intensities. While 13.6 kJ/mol was the result when illuminations were carried out with 4.5 μE/sec, we determined 18.8 kJ/mol for 1.5 μE/sec. This results in a mean activation energy of

$$E_a = 16.2\,[\text{kJ/mol}]$$

Recently, Matthews determined the activation energy for the photocatalytic degradation of salicylic acid using TiO_2 as the photocatalyst and calculated approximately 11 (kJ/mol).[30] Thus, the value obtained for dichloroacetate in the present work appears to be reasonable. A possible explanation for the observed positive activation energy is that an essential reaction step of the complex DCA degradation mechanism is thermally activated. It should be pointed out that the temperature dependence of a photocatalytic detoxification process can even qualitatively be different when other test molecules are employed, e.g., a considerable decrease of the degradation rate of chloroform was observed as the reaction

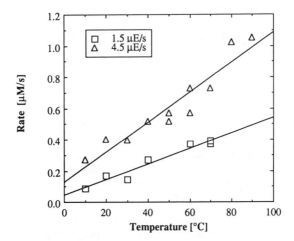

Figure 3. Rate of DCA degradation as a function of reaction temperature at two different illumination intensities for O_2-saturated aqueous suspensions containing 0.5 g/L Degussa P25, 24 mM dichloroacetate, and 10 mM KNO_3 at pH 5.

temperature was raised, especially at high photon fluxes.[10] A detailed study of the temperature dependence of the photocatalytic degradation of a "real" wastewater mixture has, to the best of our knowledge, not yet been published.

Degradation of Acetate: a Photoelectrochemical Study

Over the recent years, we have performed detailed investigations of the mechanism of the photocatalytic degradation of acetate using Degussa P25 titanium dioxide powder as the photocatalyst.[31] The following mechanism seems to describe the observed processes most closely. The acetate molecule is initially attacked by a hydroxyl-like radical that could be either formed via Equation 2, i.e., by the anodic oxidation of surface-bound hydroxyl groups, or through a multistep cathodic process starting with the reduction of molecular oxygen yielding hydrogen peroxide which is further reduced in a Fenton-type reaction (*vide supra*) forming ˙OH and OH⁻. Subsequently, hydrogen abstraction from the methyl group of the acetate molecule results in the formation of a carbon-centered radical:

$$CH_3COO^- + {}^\cdot OH_s \rightarrow {}^\cdot CH_2COO^- + H_2O \qquad (4)$$

Carbon-centered radicals such as those generated in Equation 4 can combine

$$^\cdot CH_2COO^- + {}^\cdot CH_2COO^- \rightarrow \left(CH_2COO\right)_2^{2-} \qquad (5)$$

or, in the presence of molecular oxygen, form peroxy radicals

$$^\cdot CH_2COO^- + O_2 \rightarrow {}^\cdot O_2CH_2COO^- \qquad (6)$$

which may further react to glycolic and glyoxylic acid as the first stable products:

$$^\cdot O_2CH_2COO^- \rightarrow \rightarrow HOCH_2COO^-, \ OCHCOO^- \qquad (7)$$

Indeed, both glycolic and glyoxylic acid are detected when aqueous suspensions of Degussa P25 titanium dioxide powder containing acetate are illuminated with ultra-bandgap light.[31] This finding evinces the intermediacy of hydroxyl radicals in photocatalytic systems containing TiO_2. Until now, however, the question of whether these ˙OH are generated anodically or cathodically, as discussed above, remained unanswered.

In order to understand this important mechanistic aspect of photocatalytic systems, we have performed photoelectrochemical experiments with single crystal (rutile) TiO_2 electrodes as photoanodes and platinum counter electrodes immersed in a second half-cell. The latter was connected only via a salt bridge with the irradiation cell which facilitates equilibration of ionic strength, but did not permit diffusion of acetate or its reaction products. Thus, this set-up ensured that

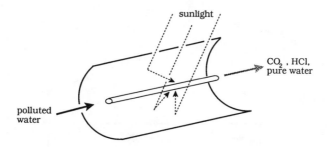

Figure 4. Schematic presentation of a parabolic trough solar detoxification reactor, the Pyrex pipe in the focal plane of the aluminum-coated mirror is filled with an aqueous titanium dioxide (Degussa P25) suspension.

anodic and cathodic processes could be monitored completely independently from each other.

Using this photoelectrochemical cell, we studied the primary products of the photocatalytic degradation of acetate. In the presence of molecular oxygen, we found glycolate as a reaction product (by HPIC) when less than 10% of the originally present acetate had been degraded. Due to the high ionic strength used in these experiments, the glyoxalate peak could not be separated from the broad chloride peak. Hence, glyoxalate could not be detected by this method. Thus, we are confident to draw the following conclusion from these findings. Upon band gap illumination of the TiO_2 catalyst (which can be present either as a particle or as an electrode), electron-hole pairs are formed. While the electrons will most probably be reoxidized by O_2 as oxidant, the detoxification is initiated anodically. Valence-band holes are trapped on the surface of the semiconductor (cf. Equation 2). The resulting surface states attack the pollutant molecule like ˙OH radicals generating predominantly carbon-centered radicals (cf. Equation 4) which in the presence of dioxygen will form peroxy radicals (cf. Equation 6). The latter are the precursors of any stable primary reaction product which will eventually also be oxidized by surface ˙OH_S, leading to a complete mineralization of the pollutant molecule as the final result.

SOLAR APPLICATIONS

The Parabolic Trough Reactor

The experimental set-up of the parabolic trough reactor used for the detoxification studies with the model compound DCA is shown in Figure 4. This reactor type concentrates the photoactive UV part of the solar spectrum by a factor of 30–50. Results of our initial experiments performed at the Plataforma Solar in Almeria, Spain are shown in Figures 5 and 6. It is evident from Figure 5 that an increase of the catalyst concentration results in an increased degradation

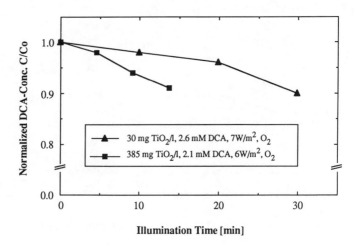

Figure 5. Solar degradation of dichloroacetate using a parabolic trough reactor at PSA, Almeria, Spain, O_2-saturated aqueous suspensions containing Degussa P25.

rate of DCA under otherwise identical conditions. Figure 6 demonstrates that while it seems to have a positive effect on the degradation efficiency when hydrogen peroxide (H_2O_2) is used as the oxidant instead of molecular oxygen (O_2), the combination of both oxidants does not lead to any increase in the DCA oxidation rate as compared with O_2 alone. Further experiments concerning the influence of solar light intensity, DCA concentration, and pH are currently underway and will be published elsewhere.[32]

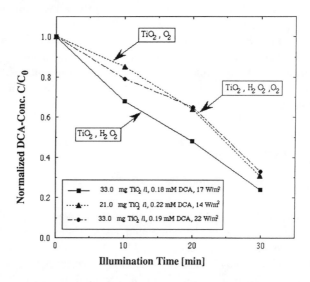

Figure 6. Solar degradation of dichloroacetate using a parabolic trough reactor at PSA, Almeria, Spain, aqueous suspensions containing Degussa P25.

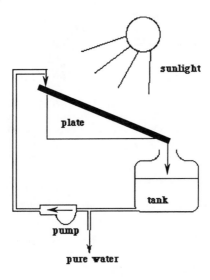

Figure 7. Schematic presentation of a thin-film reactor. Plate A is covered with immobilized titanium dioxide photocatalyst, B is the water reservoir, and C the circulating pump.

The Fixed-Bed Flow Reactor

A thin-film reactor depicted in Figure 7 was constructed and tested at the University of Campinas, Brazil in collaboration with the group of Professor W. Jardim (University of Campinas, Brazil). While pure titanium dioxide (Degussa P25) served as photocatalyst in both solar test systems, it was employed as a suspension in the trough and immobilized on this fixed-bed reactor (details of the immobilization process will be patented and subsequently published elsewhere). Figures 8 and 9 show the results of initial solar detoxification experiments performed with this novel reactor. Figure 8a demonstrates that dichloroacetate is completely degraded after approximately 2 hr in the sun, while almost no DCA loss is observed in the blank (dark) experiment. The simultaneous formation of two chloride ions with every degraded DCA molecule evinces that the stoichiometry given in Equation 1 is indeed obeyed. Figure 8b shows that the degradation kinetics can by nicely described using first-order kinetics. Similar observations have been made during the photocatalytic degradation of other solutes and interpreted by Langmuir-Hinshelwood adsorption phenomena.[8]

While the experimental results depicted in Figure 8 were obtained in a cyclic system (cf. experimental set-up given in Figure 7), Figure 9 illustrates the results of single-path experiments where the contact time of the aqueous DCA solutions with the catalyst was only between 25 and 30 sec. It is evident (filled squares, Figure 9) that the degradation rate increases with increasing solute concentration,

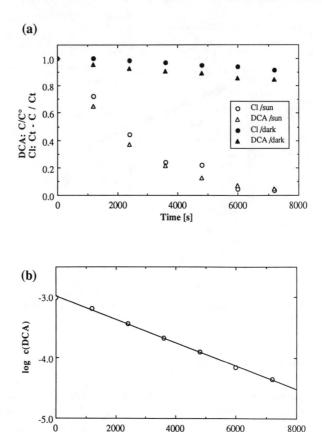

Figure 8. Solar degradation of dichloroacetate using a fixed-bed flow reactor at the University of Campinas (Brazil): aqueous suspensions containing 1 mM DCA, pH 3; flow, 14 ml/min, average light intensity, 25 W/m^2.

evincing that the plateau of the adsorption isotherm has not been reached even at 10 mM DCA. Another way to look at the same set of data is given by the open triangles in Figure 9, which demonstrate that it is indeed possible to achieve degradations of up to 70% in such a single-path experiment. At this point, it is interesting to note that the estimated photon efficiency of this process ($\leq$10% of the incident UV photon flux is utilized under the given experimental condition) is comparable with the efficiency of the parabolic trough reactors. However, since the design of the fixed-bed flow reactor is simpler, the latter type can be realized much cheaper than a trough reactor. Moreover, the immobilization of the catalyst saves an additional separation step which, in practice, will be needed if a suspension reactor is used. Hence, we are optimistic that the novel fixed-bed flow reactor will offer a useful alternative not only for solar applications of the photocatalytic cleanup of polluted aquifers.

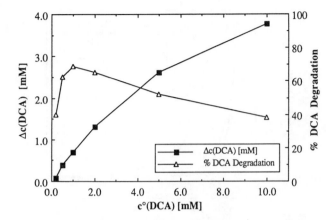

Figure 9. Solar degradation of dichloroacetate using a fixed-bed flow reactor at the University of Campinas (Brazil): aqueous suspensions containing 1 mM DCA, pH 3; flow, 14 ml/min; average light intensity, 25 W/m^2, single-path experiments, contact time 25–30 sec.

TI(IV)-FE(III) MIXED OXIDES: NOVEL PHOTOCATALYSTS?

Recently, we described the preparation of novel Ti(IV)-Fe(III) mixed-oxide particles containing between 0.05 and 50% Fe(III) in their matrix.[20] Besides their interesting photophysical properties, we could demonstrate that these new materials, which are prepared as colloids in aqueous solution and can be stored as stable powders, exhibit considerable photocatalytic activity for the mineralization of our model compound dichloroacetate.[20] We have repeated the synthesis of selected Ti(IV)-Fe(III) mixed oxides that exhibited distinct photocatalytic properties in our previous study. Figure 10 shows the UV/Vis absorption spectra of aqueous colloidal suspensions containing 500 mg/L of the newly synthesized powders with 0, 0.25, 2.5, and 50% Fe(III), respectively. It is obvious that the absorption onset is shifted toward longer wavelengths with an increasing content of Fe(III) in the material. In principle, these particles should be able to utilize parts of the visible solar spectrum when acting as photocatalysts, while pure titanium dioxide only absorbs in the UV region below 400 nm. The photocatalytic activity of the newly synthesized particles was investigated using dichloroacetate as the model compound.

Figure 11 shows the results observed during the illumination of an O_2-saturated colloidal aqueous suspension containing 0.5 g/L Ti(IV)-Fe(III) mixed-oxide powder [50% Fe(III)] at pH 2.5. While DCA is degraded from an initial concentration of 1.2 mM almost to zero within the experimental duration of 2 hr, nearly 0.3 mM Fe(II) are formed simultaneously, indicating a partial reductive dissolution of the catalyst material via

$$\text{Ti}(\text{IV})/\text{Fe}(\text{III})\text{O}_x + e_{CB}^- \rightarrow \text{Ti}(\text{IV})\text{O}_x + \text{Fe}_{aq}^{2+} \qquad (8)$$

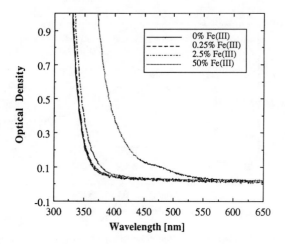

Figure 10. UV/Vis absorption spectra of colloidal aqueous suspensions containing 0.5 g/L Ti(IV)-Fe(III) mixed-oxide powders with various Fe(III) content at pH 2.5.

Alternatively, conduction band electrons (e_{CB}^-) can be scavenged by the molecular oxygen present in solution, leading eventually to the generation of peroxides:[14]

$$O_2 + e_{CB}^- \rightarrow O_2^{\cdot-} \rightarrow \rightarrow H_2O_2 \qquad (9)$$

Hence, the cathodic dissolution of the catalyst should be competing with the desired electron-transfer steps such as the reduction of dioxygen. It is well known

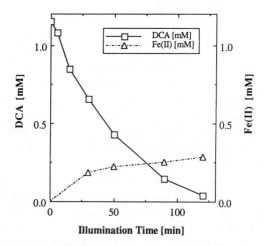

Figure 11. Degradation of dichloroacetate and formation of Fe(II) during the illumination of an O_2-saturated colloidal aqueous suspension containing 0.5 g/L Ti(IV)-Fe(III) mixed-oxide powder (50% Fe(III)) at pH 2.5. Illumination with white light (320-nm filter).

that O_2 is a poor oxidant for conduction-band electrons in most metal oxides.[14] This is readily explained by a rather low-lying conduction-band edge and a one-electron redox potential of O_2 which is even slightly more negative.[14] It has recently been shown that peroxides resulting from the reduction of molecular oxygen are only formed in a reasonable yield when ZnO is employed as the semiconducting metal oxide while only minute yields are observed with TiO_2.[14] In the case of Fe(III) oxides (hematite), no peroxides were detected in analogous experiments, but the catalyst was readily dissolved.[33] It is evident from this discussion that a better electron acceptor than O_2 should be employed to compete more efficiently with the catalyst dissolution. We consider H_2O_2 to be a good candidate for this purpose. Indeed, initial experiments indicate that the Fe(II) release can be suppressed when H_2O_2 is added to the colloidal suspension. However, further investigations are required to decide whether H_2O_2 is competing for the conduction-band electrons via

$$H_2O_2 + e_{CB}^- \rightarrow {}^{\cdot}OH + OH^- \qquad (10)$$

thereby preventing the dissolution of the mixed-oxide photocatalyst, or whether the main process involves the reoxidation of free Fe_{aq}^{2+}, which should eventually also lead to the destruction of the catalyst or at least to the reduction of its photocatalytic activity.

$$Fe_{aq}^{2+} + H_2O_2 \rightarrow Fe_{aq}^{3+} + {}^{\cdot}OH + OH^- \qquad (11)$$

Equation 11 is the well known Fenton reaction which, however, has a rather small rate constant.[34] It is therefore conceivable that scavenging of e_{CB}^- by H_2O_2 via Equation 10 could be rather efficient. A detailed investigation, however, is indicated to differentiate between these mechanisms.

The main incentive to synthesize novel photocatalysts with a hypochromically shifted absorption spectrum came from the lack of photocatalytic activity of pure titanium dioxide above 400 nm. The success of the newly synthesized Ti(IV)-Fe(III) mixed oxides will therefore strongly depend on their ability to act as photocatalysts in the visible part of the solar spectrum. Hence, we have performed illumination experiments using bandpass filters which effectively select the 436-nm emission band of the mercury-doped xenon lamp employed in this study. Figure 12 shows the results of these quasimonochromatic illumination experiments. While at first sight it is evident that the reaction rate is much smaller as compared with the results depicted in Figure 11 (which can be explained by a much smaller photon flux in this "single-wavelength" study), we are confident to report that our newly synthesized Ti(IV)-Fe(III) mixed oxides still exhibit photocatalytic activity above 400 nm. This is true at least for the case of particles containing 50% Fe(III), which is the only material tested under these conditions so far.

The detailed analysis of the data shown in Figure 12 demonstrates that the catalyst dissolution also presents a problem when illuminations are carried out in

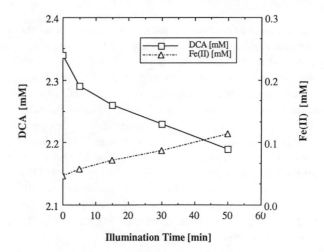

Figure 12. Degradation of dichloroacetate and formation of Fe(II) during the illumination of an O_2-saturated colloidal aqueous suspension containing 0.5 g/L Ti(IV)-Fe(III) mixed-oxide powder (50% Fe(III)) at pH 2.5. "Monochromatic" illumination with bandpass filter at 436 nm (Mercury-doped xenon-lamp).

the visible. Under the experimental conditions employed, the rate of dissolution is approximately half of that of the DCA mineralization. As already discussed for the results shown in Figure 11, the introduction of other oxidants is highly indicated to prevent the dissolution of the catalyst.

SUMMARY AND OUTLOOK

We have shown in this paper that the photocatalytic detoxification of polluted aquifers constitutes a technique that bears a considerable potential for solar as well as for artificial illuminations. Dichloroacetate has been successfully employed as a model compound for laboratory and field studies. It could be shown that its degradation is complete; an activation energy of 16.2 kJ/mol was determined. Except for pH 5, the rate of DCA degradation was found to linearly depend on the illumination intensity over the whole pH-range studied. Results of a photoelectrochemical study have been presented supporting the intermediacy of anodically formed hydroxyl radicals as the key species to explain the high and unspecific oxidation power of photocatalytic systems.

Two different types of solar reactors have been tested in field studies. While they both exhibited an estimated incident photon efficiency of $\leq 10\%$ at their maximum performance, the novel fixed-bed flow reactor offers the advantages of a less expensive construction and reduced running costs as compared with the parabolic trough reactor. Adsorption of the solute onto the catalyst seems to play a major role in explaining variations in the observed DCA degradation rate. Addition of hydrogen peroxide does not improve the degradation efficiency as compared with a system that only uses molecular oxygen as the oxidant. Finally, we have been able to synthesize Ti(IV)-Fe(III) mixed-oxide particles as novel photocatalysts in the form of stable powders which readily dissolve in water yielding transparent colloidal suspensions. These materials exhibit strong absorption bands above 400 nm, i.e., at energies below the bandgap energy of pure titanium dioxide. Moreover, they possess distinct photocatalytic activity, which was demonstrated for the destruction of dichloroacetate. Monochromatic illuminations showed that the photocatalytic action spectrum is indeed extended toward the visible part of the solar spectrum. If the anodic dissolution of the catalyst particles that was observed during these experiments could be suppressed, e.g., by the use of other oxidants such as hydrogen peroxide, we could envisage the newly synthesized material to become a rather promising photocatalyst. For practical purposes, it will eventually be necessary to immobilize these novel catalysts, e.g., on the flat plate of a thin-film reactor, or to prepare larger particle sizes.

We are confident that the photocatalytic detoxification will be a useful method for the detoxification of polluted ground and wastewaters. Extensive laboratory and field studies using various reactor designs have been performed in many laboratories throughout the world, and this technique is now at the stage where the first pilot plants are being constructed and tested. However, several important questions still need to be addressed in laboratory experiments, e.g., the synthesis of more efficient photocatalysts for visible light applications and a detailed chemical analysis to be able to exclude the possible formation of highly toxic residues from the treatment of polluted aquifers. These studies should be carried out in parallel with the initial technical realization of photocatalytic wastewater treatment plants.

ACKNOWLEDGMENTS

Parts of this work were supported by the BMFT (F&E program SOTA, AP 300, Grants 5–395–4387 and 5–395–4484), the Stifterverband für die Deutsche Wissenschaft (Project QF 26), and the C.E.C. (Program "Access to Large-scale Installations," contract no. ERBGE 1*CT0000 19). The inspiring collaboration with Professor W. Jardim (University of Campinas, Brazil) and Mr. J. Blanco (Plataforma Solar in Almeria, Spain) is gratefully acknowledged. The authors wish to thank Ms. R. F. P. Nogueira (University of Campinas, Brazil), Mr. K. Wischnewski (Universität Clausthal), Ms. K. Wolff, (ISFH) and Mr. G. Raabe (ISFH) for performing parts of the measurements.

REFERENCES

1. Häfner, M., Wasser ist kein Naturprodukt mehr, *Natur*, 10, 20, 1989.
2. Dieter, H. H., Gebt dem Grenzwert eine Chance! Pestizide im Trinkwasser, *Wechselwirkung*, 43, 4, 1989.
3. Carey, J. H., Lawrence, J., and Tosine, H. M., Photodechlorination of PCB's in the presence of titanium dioxide in aqueous suspensions, *Bull. Environ. Contam. Toxicol.*, 16, 697, 1976.
4. Robin, E., Sunlight found to help break down organic pollutants, *San Francisco Examiner*, May 20, D-3, 1989.
5. Fettwell, J., Semiconductors help the sun to clean water, *New Sci.*, June 10, 36, 1989.
6. Thornton, J., Sunlight destroys hazardous waste, *SERI S&T*, in review, 8, 1989.
7. Hecht, J., Sunlight gives toxic waste a tanning, *New Sci.*, April 14, 28, 1990.
8. Serpone, N. and Pelizzetti, E., *Photocatalysis: Fundamentals and Applications*, John Wiley & Sons, New York, 1989.
9. Pelizzetti, E., Minero, C., and Maurino, C., The role of colloidal particles in the photodegradation of organic compounds of environmental concern in aquatic systems, *Adv. Colloid Interface Sci.*, 32, 271, 1990.
10. Bahnemann, D., Bockelmann, D., and Goslich, R., Mechanistic studies of water detoxification in illuminated TiO_2 suspensions, *Sol. Energy Mater.*, 24, 564, 1991.
11. Kormann, C., Synthesis and Characterization of Quantum Size Metal Oxide Colloidal Particles. Photocatalytic Peroxide Formation on ZnO and TiO_2, Ph.D. thesis, California Institute of Technology, Pasadena, 1989.
12. Gerischer, H., Solar photoelectrolysis with semiconductor electrodes, *Top. Appl. Phys.*, 31, 115, 1979.
13. Kormann, C., Bahnemann, D. W., and Hoffmann, M. R., Preparation and characterization of quantum size titanium dioxide (TiO_2), *J. Phys. Chem.*, 92, 5196, 1988.
14. Kormann, C., Bahnemann, D. W., and Hoffmann, M. R., Peroxide production on illuminated suspensions of TiO_2, ZnO, and desert sands, *Environ. Sci. Technol.*, 22, 798, 1988.
15. Tanaka, K., Hisanaga, T., and Harada, K., Photocatalytic degradation of organohalide compounds in semiconductor suspensions with added hydrogen peroxide, *New J. Chem.*, 13, 5, 1989.
16. Turchi, C. S. and Ollis, D. F., Photocatalytic reactor design: an example of mass-transfer limitations with an immobilized catalyst, *J. Phys. Chem.*, 92, 6852, 1988.
17. Matthews, R. W., Response to the comment, photocatalytic reactor design: an example of mass-transfer limitations with an immobilized catalyst, *J. Phys. Chem.*, 92, 6853, 1988.
18. Ollis, D. F., Process economics for water purification: a comparative assessment, in *Photocatalysis and Environment: Trends and Applications, NATO-ASI Ser. C237*, Schiavello, M., Ed., Kluwer, Dordrecht, 1988, 663.
19. Matthews, R. W., A comparison between ultraviolet illuminated TiO_2 and ^{60}Co gamma rays for the destruction of organic impurities in water, *Appl. Radiat. Isot.*, 37, 1247, 1986.
20. Bahnemann, D. W., Ultra-small metal oxide particles: preparation, photophysical characterization and photocatalytic properties, *Isr. J. Chem.*, 33, 115, 1993.
21. Bockelmann, D., Photokatalytischer Abbau Halogenierter Kohlenwasserstoffe an Halbleiteroberflächen, Diplomarbeit, TU Clausthal, 1989.
22. Heller, H. G. and Langan, J. R., A new reusable chemical actinometer, *EPA Newslett.*, 71, 1981.
23. Weiß, J., Einführung in die Ionenchromatographie: Grundlagen, Instrumentation und Anwendung, Teil 1, *Chem. Labor Betr.*, 34, 293, 1983.
24. Lange, B. and Vejdelek, Z. J., *Photometrische Analyse*, VCH Publishers, New York, 1987, 102.
25. Sanchez, M., Activities Performed in 1991, Access to Large-Scale Scientific Installations Program, Technical Report, PSA-TR-01/92, 1992.

26. Bockelmann, D., Goslich, R., Hilgendorff, M., Weichgrebe, D., and Bahnemann, D., Solarchemische Detoxifizierung von Abwässern: Laborstudien und Anwendungen, *8th International Solar Forum/Berlin,* Energie und unsere Umwelt, Hohmann, A. and Hohmann, H. H., Eds., DGS-Sonnenenergie Verlags-GmbH, München, 1992, 1340.

27. Ollis, D. F., Hsiao, C.-Y., Budiman, L., and Lee, C.-L., Heterogeneous photoassisted catalysis: conversions of perchloroethylene, dichloroethane, chloroacetic acids, and chlorobenzenes, *J. Catal.*, 88, 89, 1984.

28. Hilgendorff, M., Mechanistische Untersuchungen zum photokatalytischen Abbau von Dichloressigsäure, Diplomarbeit, Universität Hannover, 1991.

29. Kormann, C., Bahnemann, D. W., and Hoffmann, M. R., Photolysis of chloroform and other organic molecules in aqueous TiO_2 suspensions, *Environ. Sci. Technol.*, 25, 494, 1991.

30. Matthews, R. W., Photooxidation of organic impurities in water using thin films of titanium dioxide, *J. Phys. Chem.*, 91, 3328, 1987.

31. Wolff, K., Bockelmann, D., and Bahnemann, D., Mechanistic aspects of chemical transformations in photocatalytic systems, *Symposium on Electronic and Ionic Properties of Silver Halides, Common Trends with Photocatalysis,* Levy, B., Ed., IS&T, Springfield, VA, 1991, 259.

32. Bahnemann, D. W., Bockelmann, D., Goslich, R., and Weichgrebe, D., The solar degradation of dichloroacetate: experiments with the parabolic trough reactor, *Environ. Sci. Technol.*, to be published.

33. Kormann, C., Bahnemann, D. W., and Hoffmann, M. R., Environmental photochemistry: is iron oxide (hematite) an active photocatalyst? A comparative study: α-Fe_2O_3, ZnO, TiO_2, *J. Photochem. Photobiol., A: Chem.* 48, 161, 1989.

34. Walling, C., Fenton's reagent revisited, *Acc. Chem. Res.*, 12, 125, 1975.

CHAPTER 24

Heterogeneous Photocatalysis: Use in Water Treatment and Involvement in Atmospheric Chemistry

Chantal Guillard, Laurence Amalric, Jean-Christophe D'Oliveira, Hervé Delprat, Can Hoang-Van, and Pierre Pichat

INTRODUCTION

Heterogeneous photocatalysis (HP) involves redox reactions initiated by the optical excitation of a semiconductor. This excitation results from the absorption of photons of sufficient energy to provoke the transfer of electrons from the valence band to the conduction band.[1] In this paper, the role of HP in the environment is considered from two viewpoints. First, HP is at the basis of a new technology for water treatment, and a study of the destruction of some organic micropollutants is presented. Second, the importance of HP in tropospheric chemistry needs to be assessed, and laboratory studies in this direction are included.

Over the last few years, the potentialities of HP to destroy organic pollutants in water have been explored.[2,3] Pilot apparatuses have been tested,[4,5] and small units have even been commercialized.[6] Nevertheless, fundamental research is still required, particularly for a better understanding of the basic mechanisms. The chief purpose of the present work is to demonstrate further the generality of the HP method of water treatment by examining and comparing the degradation of monoaromatic pollutants with different substituents. To complete our previous studies,[7-11] we have chosen 2,4-dichlorophenoxyethanoic acid (2,4-D), which is a common herbicide (lethal concentration $LC_{50} = 75$ mg L^{-1} for the perch and the roach; lethal dose LD_{50} for the rat between 0.1 and 0.375 g kg^{-1}) and dimethoxybenzenes (DMBs) ($LD_{50} = 0.9$ g kg^{-1} for the rat for the ortho and meta isomers). Kinetics of formation and destruction of the intermediates and the associated pathways have been studied in detail, in particular to determine whether persisting products are generated. Such studies are useful to increase our knowledge in order to predict, from the structural characteristics of a given pollutant, by which pathways and how easily this pollutant will be degraded by HP. An earlier study[12a,12b] has proposed degradation pathways for the direct photolysis of 2,4-D at 254 nm in water; humic acids are formed. 2,4,5-Trichlorophenoxyethanoic acid is mineralized by HP, but Cl$^-$ ions are not as easily released as from 2,4,5-trichlorophenol; a series of intermediates have been identified, but their temporal variations have not been determined.[13] Direct photolysis of this herbicide at 254 nm also leads to humic substances.[14] The photocatalytic transformation of 1,4-DMB to 4-methoxyphenol and other unspecified hydroxylated products in aqueous suspensions of Pt/TiO$_2$ or WO$_3$ has been reported.[15]

On the other hand, among the enormous amounts of inorganic, water-insoluble, solid particles transported in the atmosphere,[16-20] 20–30% are metal oxides. Although silicoaluminate insulators predominate, semiconductors are also present, principally ferric oxide (1–16% of total oxides). Isolated TiO$_2$ is a less abundant component (around 1%);[21] however, since titanium is the seventh element in the Earth's crust, this dioxide can be encountered as an impurity in other oxides or can form titanates, especially ilmenite (FeTiO$_3$), which is a semiconductor. These particulate materials corresponds to huge surface areas because of their small mean size. Consequently, HP phenomena susceptible to occur on these semiconducting particles cannot be *a priori* neglected in atmospheric chemistry.[16-20]

This paper includes two laboratory studies whose aim is to assess the role of certain tropospheric, water-insoluble, inorganic particles in the fate of organic micropollutants. The first study concerns the phototransformations of naphthalene (NPH), a representative of the important category of polycyclic aromatic hydrocarbon pollutants, adsorbed on synthetic TiO$_2$ and Fe$_2$O$_3$ as well as on natural or collected atmospheric particles. Since, under tropospheric conditions, liquid water droplets can encapsulate water-insoluble particles onto which organics are adsorbed, the second study deals with the phototransformations of ethanal dissolved in water containing TiO$_2$ or Fe$_2$O$_3$. Ethanal has been chosen because it is a key compound in atmospheric chemistry, emitted both from anthropogenic and natural sources, and also formed by photochemical reactions involving OH radicals.[20]

EXPERIMENTAL

The organic compounds were of the highest purity commercially available and were employed as received. TiO$_2$ was Degussa P25 (mainly anatase, 50 m^2 g^{-1}); muscovite (2 m^2 g^{-1}), a phyllosilicate, contained ca. 1 wt% TiO$_2$; the fly ash sample (1 m^{22} g^{-1}) was collected near a coal-fired power plant and contained ca. 3.2 wt% Fe$_2$O$_3$ and insulating oxides such as CaO, MgO, Al$_2$O$_3$, and SiO$_2$. The Fe$_2$O$_3$ samples were either commercial (Merck, hematite, 4 m^2 g^{-1}) or prepared in the laboratory (1) from $Fe(NO_3)_3$ and NH_4OH at pH 8, with calcination at 393 K (80% goethite, 20% hematite, 180 $m^2 g^{-1}$) or 773 K (hematite, 19 $m^2 g^{-1}$) for 15 hr or (2) from $NH_4Fe(SO_4)_2 \cdot 12 H_2O$ and urea at pH 2.2, with calcination at 773 K for 3 hr; the resulting solid (25 $m^2 g^{-1}$) released sulfate and ferric ions when dispersed in water.

Irradiation was provided by a Philips HPK 125-W high-pressure mercury lamp through a circulating-water cell and filters (Pyrex or Corning 0.52). The radiant flux ϕ entering the photoreactor was measured with a power meter (U.D.T., Model 21 A) calibrated against a microcalorimeter. In the case of the gas-phase photoreactor, ϕ was 22 mW cm^{-2} at $\lambda > 340$ nm, which for TiO_2 corresponded to ca. 7.7×10^{16} potentially absorbable photons per second. For the aqueous suspensions, the numbers of photons per second potentially absorbable by TiO_2 were about 6.7×10^{17} ($\lambda > 290$ nm) or 1.9×10^{17} ($\lambda > 340$ nm). These values were calculated as indicated in Reference 7.

To study the phototransformations occurring in aqueous semiconductor suspensions, a cylindrical batch photoreactor with a bottom optical window ca. 4 cm in diameter was used as previously.[7,8] The photoreactor employed for the studies of the transformations of NPH was also cylindrical with an optical window ca. 4.7 cm in diameter. It contained a fixed bed of solid particles spread on an inert porous fiberglass membrane perpendicular to the ultraviolet (UV) beam. In the dynamic mode, the gas mixture was flowed through this layer of particles.

The aqueous suspensions had a volume of 20 cm^3 and contained 50 mg of TiO_2 or 70 mg of Fe_2O_3. The concentrations of chemical compounds were (in millimolar) 0.362 (80 ppm) for 2,4-D, 0.145 (20 ppm) for DMBs, and 0.455 (20 ppm) or 1.14 (50 ppm) for ethanal, unless otherwise indicated. In the case of DMBs, the initial pH was adjusted to 3 with HNO_3, as it has been observed that nitrate ions do not affect the TiO_2-photosensitized degradation of monoaromatics such as monochlorophenols, benzamide, and nitrobenzene.[9a,9b]

Analyses were made by GC (catharometer, FID, or MS detectors) or by HPLC (fixed-wavelength UV, photodiode array, or conductimetric detectors). To measure the kinetic evolution of CO_2, the reaction mixture was flushed in the dark by an O_2 flow, every 30 min during 5 min.

RESULTS AND DISCUSSION

Use of Heterogeneous Photocatalysis in Water Treatment

Kinetics of the Disappearance of 2,4-D and DMBs by TiO_2 Photosensitization or Direct Photolysis

Figures 1 and 2 show that the disappearance of 2,4-D or 1,2-DMB is much more rapid with than without TiO_2 at $\lambda > 340$ nm and even at $\lambda > 290$ nm. In this latter wavelength range and under the conditions used, the initial rate of disappearance

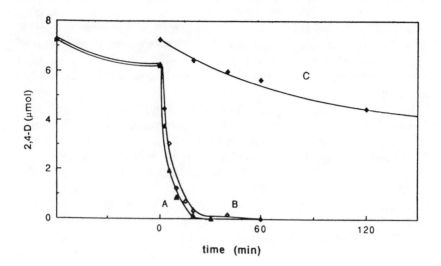

Figure 1. Kinetics of the disappearance of 2,4-D in water (A) with TiO$_2$ at λ > 290 nm, (B) with TiO$_2$ at λ > 340 nm, and (C) without TiO$_2$ at λ > 290 nm.

of 2,4-D was multiplied by ca. 12 in the presence of TiO$_2$, and the half-life, $t_{1/2}$, of the pollutant shortened by a factor of ca. 60. Similarly, the dividing factors of $t_{1/2}$ were approximately 8, 35, or 115 for the para-, ortho-, or meta-DMB isomers, respectively (the former value caused by the absorption maximum of 1,4-DMB at 285 nm). The values confirm the efficiency of HP as compared with direct photolysis in natural media, even though the estimated lower limit of the integral quantum yield of the photocatalytic disappearance of 2,4-D is only ca. 0.03 for the domain corresponding to the removal of half the initial amount of 2,4-D.

The apparent first-order rate constants of the photocatalytic elimination of DMBs at λ > 340 nm were approximately 0.087, 0.069, and 0.023 min^{-1} for

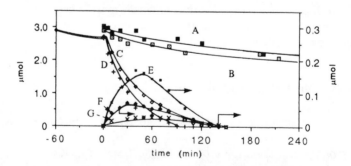

Figure 2. Kinetics of the disappearance of 1,2-DMB in water and resulting appearance/ disappearance of some intermediate aromatic products: (A) without TiO$_2$, λ > 340 nm; (B) without TiO$_2$, λ > 290 nm; (C) with TiO$_2$, λ > 340 nm; (D) with TiO$_2$, λ > 290 nm; (E) 2-methoxyphenol, (F) 2,3-dimethoxyphenol, and (G) 3,4-dimethoxyphenol variations under C conditions. Note that the right-hand scale is expanded ten times.

1,3-DMB, 1,4-DMB, and 1,2-DMB, respectively, under our conditions. This order of reactivity was consistent with the order of the extent of adsorption of these compounds measured in the dark. The higher reactivity of 1,3-DMB could also indicate that the rate-determining step was the attack of an electrophilic species.[22,23]

Although the kinetics of disappearance of 2,4-D cannot be rigorously compared to those of DMBs because of the distinct initial pH used (4.2 for 2,4-D and 3 for DMBs), it nevertheless appears that 2,4-D disappears at least as rapidly as 1,3-DMB, despite a weaker electron availability over the nucleus as a result of the effects of the substituents. This observation can be explained by the transformations of the –OCH$_2$COOH chain (see next section and Scheme 1) induced by the presence of the carboxylic group which can undergo the photo-Kolbe reaction.[24-30] The relatively rapid elimination of the carboxylic group is supported by (1) the fact that 2,4-dichloromethoxybenzene formed from 2,4-D (Scheme 1) is less rapidly eliminated than 2,4-D (Figure 3) and (2) the initially faster formation of CO$_2$ from the herbicide than from DMBs (see the paragraph entitled "mineralization").

Degradation Pathways

The intermediate products shown in Schemes 1 and 2 were identified by the analytical means indicated and comparison with commercial compounds whenever possible. Some pathways were supported by control experiments on the degradation of 2,4-dichlorophenol, chlorohydroquinone, 2-methoxyphenol, 2,3-dimethoxy-phenol, 3,4-dimethoxyphenol, and hydroquinone. The primary intermediates of DMBs (Scheme 2) corresponded to the hydroxylation of the aromatic nucleus with the expected ortho and para orientations due to the methoxy groups.

Similarly, the primary intermediates of 2,4-dichlorophenol (Scheme 1), the main intermediate of 2,4-D (Figure 3), also corresponded to ring hydroxylation with the expected orientations. Transformations of the aliphatic chain of 2,4-D give rise to 2,4-dichloromethoxybenzene, 2,4-dichlorophenyl formate (inferred from the mass spectrum), and formate. These intermediates could be formed by the decarboxylation reaction already evoked:[24-30]

$$R - OCH_2 - COO^- + p^+ \rightarrow R - OCH_2^\circ + CO_2$$

$$R - OCH_2^\circ + O_2 \rightarrow R - OCH_2OO^\circ \rightarrow R - OCHO$$

$$R - OCH_2^\circ + H^\circ \rightarrow R - OCH_3$$

5,7-Dichloro-2-oxo-1,4-benzodioxane (DOBD) was supposed to be formed according to the mass spectrum. This compound might result from the hydroxylation of 2,4-D at the ortho position with respect to the side chain (the corresponding intermediate was not detected) and subsequent formation of a lactone. Since esterification is hindered in aqueous medium, DOBD would be formed during the GC-MS analysis.[12b] These results are in general agreement with those of Reference 13.

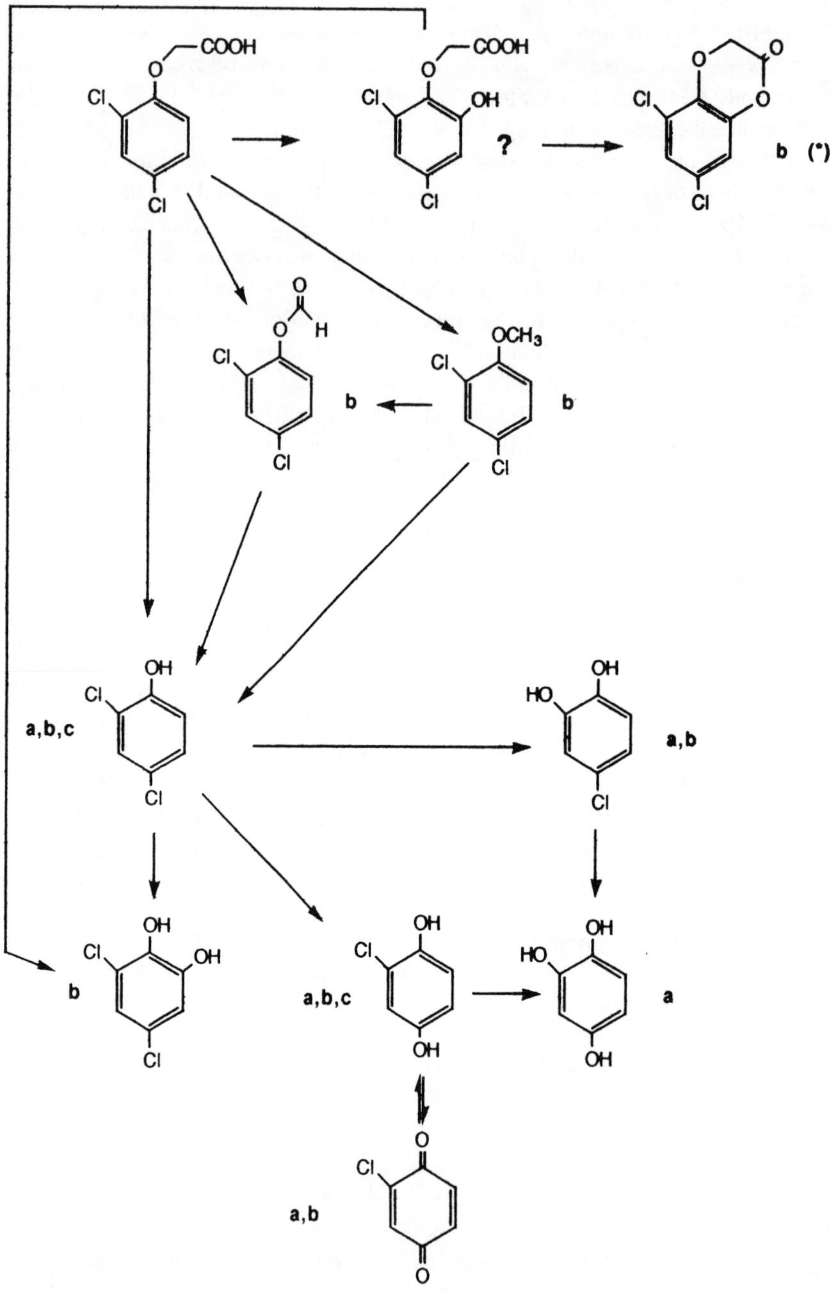

Scheme 1. Aromatic intermediates of the photocatalytic degradation of 2,4-D in water identified by (a) HPLC-UV, (b) GC-MS, (c) HPLC-UV spectrum, and suggested pathways; (?) nonidentified; (*) likely formed during the GC-MS analysis.

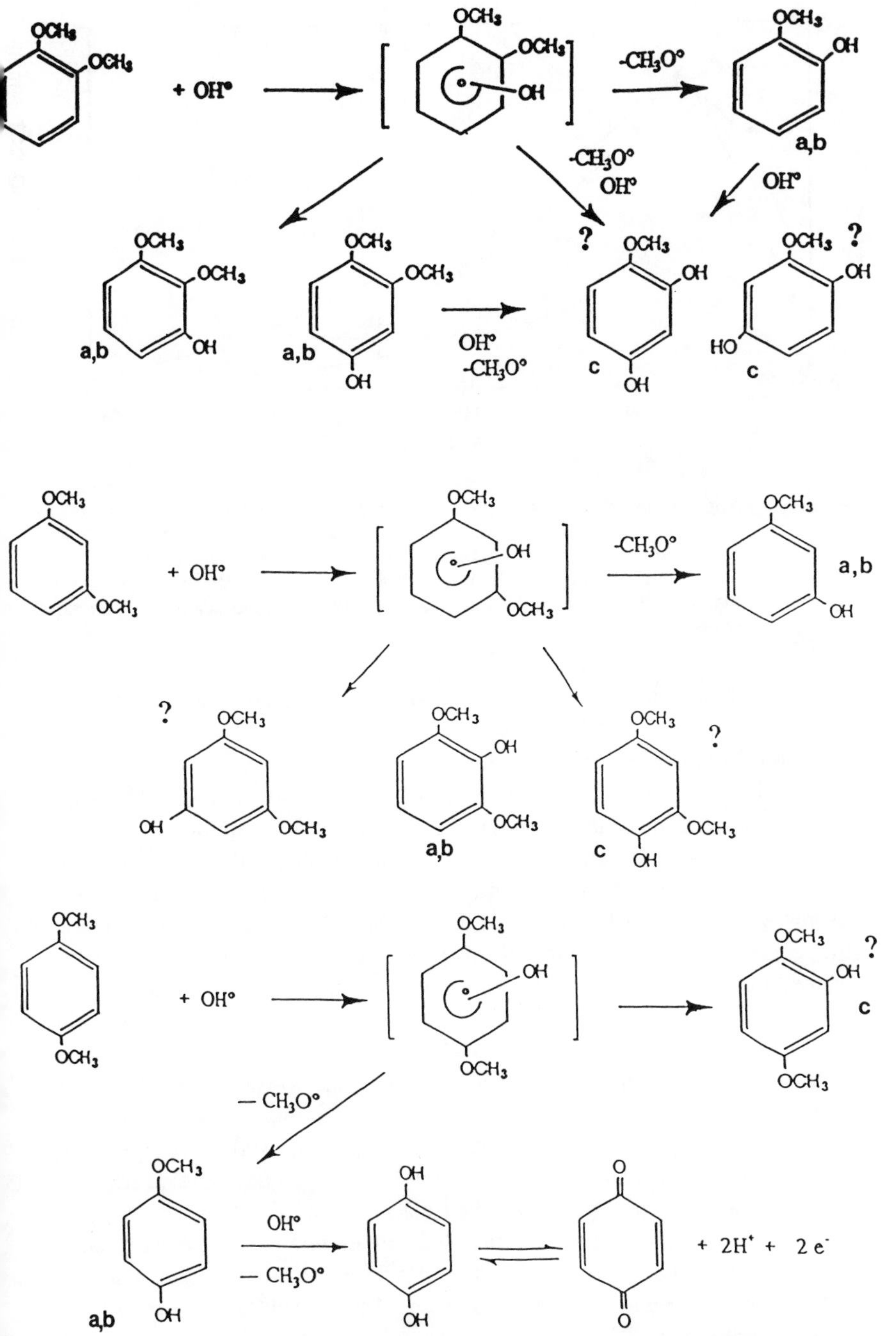

Scheme 2. Aromatic intermediates of the photocatalytic degradation of DMBs in water identified by (a) HPLC-UV spectrum, (b) GC-MS [for the compounds indicated by (c) no authentic sample was available], (?) nonidentified, and suggested pathways.

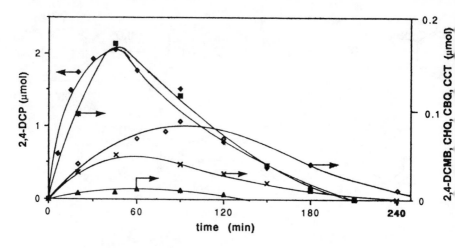

Figure 3. Appearance/disappearance of the main aromatic intermediates of the photocatalytic degradation of 2,4-D (0.362 mM = 80 ppm in water) at λ > 340 nm. Solid diamonds are 2,4-dichlorophenol; squares are 2,4-dichloromethoxybenzene; open diamonds are chlorohydroquinone; crosses are chlorobenzoquinone; and triangles are 4-chlorocatechol. Note that the right-hand scale is expanded 12 times.

Kinetics of Appearance and Disappearance of Intermediates

Figure 3 shows the kinetic variations in the amounts of five of the intermediates of the photocatalytic degradation of 2,4-D. Only 2,4-dichlorophenol could be plotted on the same scale as 2,4-D; the maximum concentrations of the other intermediates were much lower. All these intermediates disappeared within a period of time longer than that required for the disappearance of 2,4-D, which expresses that the transformations of the aliphatic chain were chiefly the cause of the relatively rapid elimination of this herbicide. The concentration of formate, formed from the aliphatic chain, reached its maximum about the time dearomatization was achieved. To eliminate formate it took about two and one half times longer than for the aromatics under our conditions.

Comparison of Figures 3 and 4 confirmed the lower efficiency of direct photolysis with respect to photocatalysis. Furthermore, the distribution of the main intermediates was not the same; 2,4-dichloromethoxybenzene did not form to any appreciable extent by direct photolysis, as would be expected if the decarboxylation were a photocatalyzed reaction. The Cl atom next to the OH group of 2,4-dichlorophenol was eliminated preferentially to the other Cl atom in the absence of TiO$_2$ (or, alternatively, 4-chlorocatechol was less stable than chlorohydroquinone in the presence of TiO$_2$). These results are in agreement with previous reports.[12a,12b]

The temporal variations in the amounts of the aromatic intermediates of the photocatalytic degradation of DMBs indicated that at their maximum

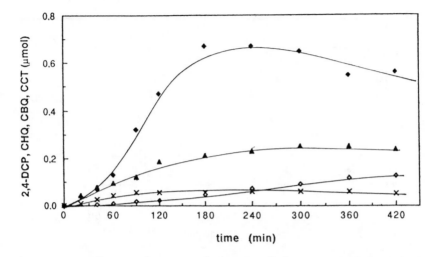

Figure 4. As Figure 3, but with direct photolysis at $\lambda > 290$ nm.

concentrations these intermediates corresponded to only a few percent of degraded DMB and they disappeared within the same time range as did DMB (Figure 2). Control experiments confirmed that they were indeed eliminated at similar rates under identical conditions. Methoxyphenols prevailed over dimethoxyphenols (Figure 2 and Scheme 2). As is usual, the carbon balance was established only at the end, but not in the course of the degradation because of analytical difficulties. The intermediates produced by direct photolysis were different and have not been identified.

Mineralization

By the time that all the chlorinated aromatics formed from 2,4-D had disappeared, according to the GC-MS analyses, still ca. 35% of the expected Cl atoms had not been released as chloride. This observation tended to indicate that unidentified chlorinated aliphatics would be formed. Total dechlorination eventually was achieved, but it required substantially longer than when 2,4-dichlorophenol was the primary pollutant, perhaps due to the steps necessary to form this compound from 2,4-D (Scheme 1).

The formation of CO_2 was initially faster from 2,4-D than from DMBs, consistent with the mineralization of the acetic moiety. In both cases, total mineralization was obtained, but it necessitated an irradiation time markedly longer than did the dearomatization (as previously reported for other aromatic pollutants).[2,3,8-11,13] On the other hand, it also meant that humic substances were not formed from 1,2,4-trihydroxybenzene (Scheme 1) as in natural media.[12,14] Nevertheless, the accuracy of the measurements of CO_2 evolution was insufficient to be definitely sure of their absence.

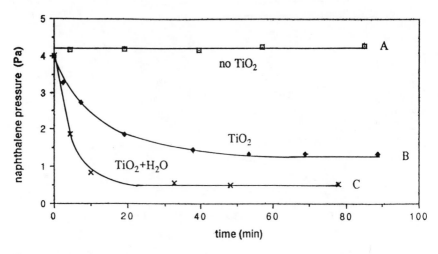

Figure 5. Kinetics of the disappearance of NPH as a function of irradiation time at λ > 340 nm in the differential flow reactor. Curve A is without TiO_2, dry air; curve B is with TiO_2, dry air; and curve C is with TiO_2, air, and H_2O vapor (1.7 kPa).

Involvement of Heterogeneous Photocatalysis in Atmospheric Chemistry

Phototransformations of Naphthalene

Disappearance of Naphthalene Adsorbed on TiO_2.   The amount of adsorbed NPH at equilibrium on the TiO_2 sample we used was 0.63 molecule per square nanometer, which corresponded to a surface coverage of ca. 50%, assuming a cross-sectional area of 0.7 nm^2 for the NPH molecule compared to the value of 0.4 nm^2 assigned to the benzene molecule.[31]

Results of the photodegradation of NPH (4.4 Pa in a flow of air) are presented in Figure 5. Direct photolysis of NPH was negligible at λ > 340 nm under a radiant flux of ca. 22 mW cm^{-2}, as shown by the absence of change in its partial pressure as a function of the irradiation time (Figure 5, curve A). From Figure 5, curve B, a steady-state conversion of 66% in the presence of TiO_2 could be deduced, which corresponded to a rate of 0.4 molecule per square nanometer per hour. The influence of water was obvious from comparison of curves C and B (Figure 5). Indeed the steady state was reached more rapidly in presence of water, an important observation for atmospheric chemistry. The photodegradation level at the steady state also was higher and, accordingly, the corresponding rate of the NPH disappearance was 0.52 instead of 0.4 molecule per square nanometer per hour. The increase in the NPH reactivity in the presence of humidity could be correlated to the properties of the TiO_2-UV system to degrade organics in liquid water as was illustrated in the first part of this chapter.

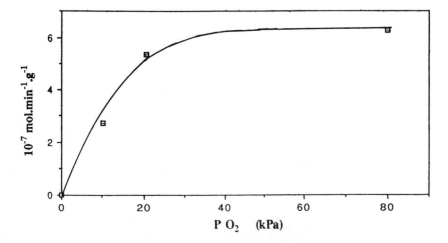

Figure 6. Influence of oxygen partial pressure on the stationary-state rate of NPH disappearance at λ > 340 nm in the presence of TiO$_2$ in the differential flow reactor.

Experiments carried out with various O$_2$ pressures in a flow of N$_2$ showed that the photodegradation of NPH on TiO$_2$ required the presence of O$_2$. The reaction order with respect to O$_2$ was zero above ca. 20 kPa (Figure 6), which indicated that the surface of TiO$_2$ was fully covered by active oxygen species under dry tropospheric conditions.

The disappearance of NPH, previously adsorbed onto TiO$_2$ in the dark, was also determined in a dry-air-filled static reactor as a function of the irradiation time. The initial rate was about 0.9 molecule per square nanometer per hour and, within 2 hr, 90% of NPH was transformed. The kinetics of disappearance of NPH under these conditions corresponded to an apparent first-order law, at least up to an illumination time of ca. 50 min. Blank experiments showed the absence of effect of ultrasonication.

Identification of Intermediate Products of Naphthalene.   No product was detected in the gas phase by on-line GC analysis. Therefore, the reactor was used under static conditions to accumulate the products. After a given irradiation time, the reactor was purged by a N$_2$ flow, and the organic compounds were trapped in CH$_2$Cl$_2$ at 288 K and finally concentrated by evaporating a part of this solvent. The compounds remaining adsorbed on TiO$_2$ were extracted ultrasonically by CH$_3$CN. They were analyzed by GC-MS and, in some cases, HPLC-photodiode array.

The intermediate products identified by GC-MS are presented in Scheme 3. These and others have also been recently found by use of a smog chamber.[32] The formation of products analogous to phthalic acid has been observed in a study of the photooxidation of substituted NPHs in acetonitrile-TiO$_2$ suspensions.[33] As

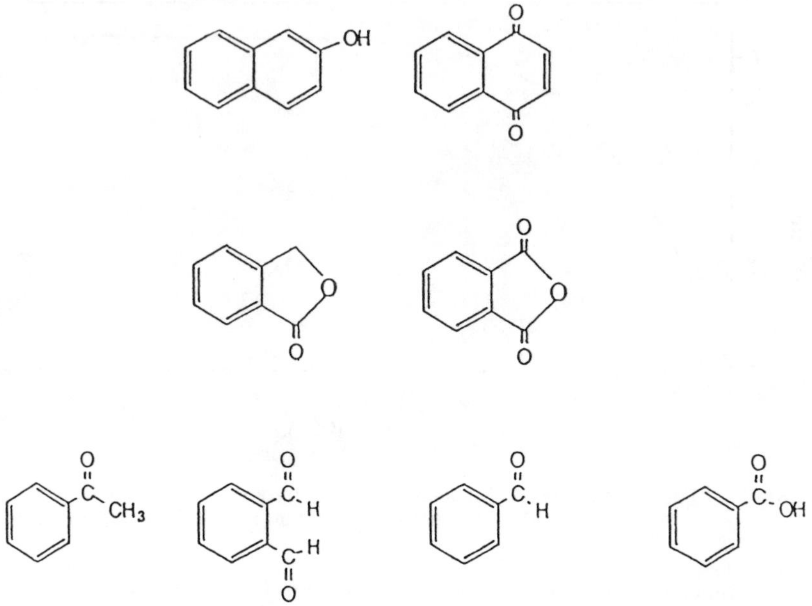

Scheme 3.   Intermediate products of the photodegradation ($\lambda > 340$ nm) of NPH adsorbed on TiO$_2$ in the presence of dry air in the static reactor.

expected, the less volatile intermediates were detected only in the adsorbed phase and the more volatile in the gas phase. Only a few products, apparently the less volatile, such as phthalide, 1,4-naphthoquinone, [2H]-1-benzopyran-2-one, and 2-naphthol were still detected on TiO$_2$ after irradiating for 8 hr. HPLC analysis confirmed the presence of benzaldehyde and acetophenone. Also, an HPLC peak whose retention time corresponded to that of 1,2-naphthoquinone was observed; however, since the associated UV spectrum was not perfectly identical to that of this compound, its formation was uncertain.

As a result of the number of intermediates and the analytical difficulties, only qualitative results could be obtained for the appearance and disappearance of the intermediate products, with the exception of 1,4-naphthoquinone. The kinetic variations in the adsorbed amounts of this compound are shown in Figure 7. The maximum corresponded to ca. 4% of the amount of NPH adsorbed on TiO$_2$, in the dark, before starting the photodegradation experiment in the static reactor. Beyond this maximum, the degradation of 1,4-naphthoquinone appeared to be very slow, since this compound was detectable even after 8 hr. A mechanical mixture of 1,4-naphthoquinone and TiO$_2$, irradiated under the same conditions as with NPH adsorbed on TiO$_2$, yielded benzoic acid and also phthalic anhydride, which might be due to the dehydration of phthalic acid in the mass analyzer.

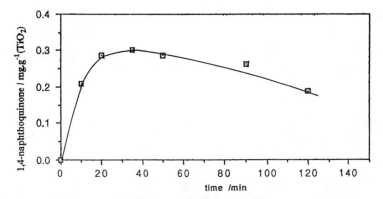

Figure 7.   Amounts of 1,4-naphthoquinone extracted from TiO_2 initially saturated with NPH vs irradiation time (λ > 340 nm, static reactor).

Phototransformations of Naphthalene Over Other Inorganic Particles.   The phototransformations of NPH were studied in the dynamic reactor containing three other inorganic solids that can be found in the air: ferric oxide, muscovite, and a collected fly ash sample.

The rate of the photodegradation of NPH over the fly ash sample was less than that previously observed over TiO_2, although its value per unit surface area was far from negligible (0.19 molecule per square nanometer per hour); the fly ash could significantly contribute to the degradation of NPH. With Fe_2O_3 (80% goethite and 20% hematite) and muscovite particles, no photodegradation rate could be measured. However, products of NPH degradation were detected in the acetonitrile extract of the solids. The distribution of these products as a function of irradiation time appeared somewhat different from that observed with the TiO_2 sample, but because of the very small amounts, it was difficult to be more specific. The observations only showed that these particulate materials exhibit a certain activity that cannot be *a priori* ignored. At this point, we cannot determine whether the activities of muscovite and the fly ash sample arose from their TiO_2 or Fe_2O_3 content or from another origin.

Phototransformations of Ethanal in the Presence of Liquid Water

At pH 5.5, no formation of CO_2 from ethanal was observed in homogeneous aqueous solution. By contrast, CO_2 was evolved in the presence of TiO_2, and the amount corresponded to complete mineralization within experimental accuracy (Figure 8). Doping (0.8 at.%) of TiO_2 with Cr^{3+} yielded a solid with almost the same absorption properties in the UV spectral region[34] as undoped TiO_2 and was found to suppress the degradation of ethanal to CO_2, as would be expected if only HP steps are involved; Cr^{3+} ions act as recombination centers for the photoproduced charges.[34] These results demonstrated the efficiency of HP on TiO_2 compared with

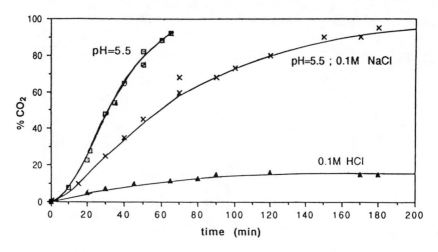

Figure 8. Kinetics of the formation of CO_2 from ethanal in liquid water containing TiO_2.

direct photolysis and photolysis in the adsorbed phase, in line with the results presented in the first part of this chapter. Ethanoic acid was formed as an intermediate (Figure 9). The absence of a correct carbon balance between CH_3CHO, CH_3COO^-, and CO_2 in the course of the degradation was indicative of the existence of another or other intermediate(s).

The effect of chloride ions, which are present in rain droplets, was determined by measuring the evolution of CO_2 in the presence of 0.1 M HCl (pH 1) or NaCl (pH 5.5). Figure 8 shows the dramatic influence of HCl; the formation rate of CO_2 was considerably decreased, and the amount evolved seemed to level off at ca.

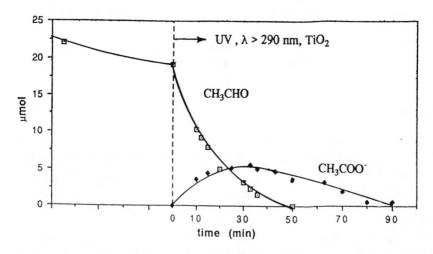

Figure 9. Kinetics of the disappearance of ethanal and the resulting appearance/disappearance of acetate in liquid water containing TiO_2 at pH 5.5.

20% of the maximum possible value. The effect of NaCl at pH 5.5 was much less marked, and the complete transformation of CH_3CHO into CO_2 was achieved. It is understandable that the inhibition by chloride was increased at a pH much lower than the point of zero charge of the TiO_2 (6.3 pH units),[35] since the formation of $Ti-OH_2^+ \dots Cl^-$ species is favored, which maintains a higher number of Cl^- ions closer to the surface. Also, at pH 1, acetate ions did not exist, and the reaction pathway involving the direct capture of photoproduced holes by these anions was suppressed.[24–30]

By replacing TiO_2 by Fe_2O_3, evolution of CO_2 from ethanal was observed only in the case of the sample prepared from ferric sulfate and urea, which released ferric and sulfate ions in the suspension. Further studies are needed to discriminate between the role of Fe_2O_3 and ferric ions.

CONCLUSIONS

Heterogeneous Photocatalysis in Water Treatment

The results presented here corroborate the efficiency of the TiO_2-UV method to destroy monoaromatic pollutants in water. The existence of different intermediates or different intermediate distribution confirms that the mechanisms are distinct from those of direct photolysis. The photocatalytic degradation pathways correspond to electrophilic attack on the aromatic ring, with the expected orientations induced by the substituents, and hydroxylated compounds are formed. Since the adsorption on TiO_2 is weak, it is not surprising that it does not affect these orientations. Initial activities seem to be related both to the extent of adsorption of the pollutant and the electron density on the ring.[9a] In the case of 2,4-D, decarboxylation of the aliphatic chain occurs relatively easily. The resulting hydroxylated aromatics are unstable, and, within experimental accuracy, complete mineralization seems to be achieved for the pollutants studied. All these observations are of interest in predicting the degradability of compounds of a given structure by HP.

Involvement of Heterogeneous Photocatalysis in Atmospheric Chemistry

The fact that numerous compounds are formed from NPH adsorbed on particulate materials that are found in the troposphere illustrates the interest of laboratory studies on the heterogeneous phototransformations of organic pollutants in air.[16-20,32,36] These transformations may generate compounds that are harmful and/or odoriferous and add their noxious effects to those produced by NO_x and the associated nitrate and hydroxy radicals which are homogeneously produced in the atmosphere. Quantitative data required for modeling are difficult to obtain; in this study, the disappearance rate of NPH per unit of adsorbent surface was determined only for the most active solids. Also, there is still a lack of information on the structure of tropospheric particulate matter to allow one to derive general conclusions on the role of photocatalytic phenomena in atmospheric chemistry. Finally, the study on ethanal, which is similar to the investigations on water treatment by HP insofar as TiO_2 is considered, shows that heterogeneous transformations of carbonyl compounds in rain droplets can also intervene in atmospheric chemistry whenever active semiconducting particles are encapsulated there.

ACKNOWLEDGMENTS

The authors are indebted to Professor N. Serpone (Canada) for his helpful contribution to the research on DMBs. Financial support from the CEC (Contract STEP-CT-90–0106-C) and from the EUREKA-EUROTRAC-HALIPP Program (Contracts 89T0159, 89W0081, and 90304) is gratefully acknowledged.

REFERENCES

1. Pichat, P. and Fox, M. A., Photocatalysis on semiconductors, in *Photoinduced Electron Transfer. Part D*, Fox, M. A. and Chanon, M., Eds., Elsevier, Amsterdam, 1988, 291–302.
2. Ollis, D. F., Pelizzetti, E., and Serpone, N., Heterogeneous photocatalysis in the environment: application to water purification, in *Photocatalysis*, Serpone, N. and Pelizzetti, E., Eds., Wiley, New York, 1989, 603–637, and references therein.
3. Bahnemann, D., Cunningham, J., Fox, M. A., Pelizzetti, E., Pichat, P., and Serpone, N., Photocatalytic treatment of waters, in *Aquatic and Surface Chemistry*, Helz, G. R., Zepp, R. G., and Crosby, D. G., Eds., Lewis Publishers, Boca Raton, FL, 1993, chap. 21.
4. ENIRICERCHE, SCICOL, San Donato, Milanese, Italy.
5. Pacheco, J. E. and Tyner, C. E., Enhancement of processes for solar photocatalytic detoxification of water, in *Proc. 1990 ASME Solar Energy Div., Int. Solar Energy Conf.*, Miami, Florida.
6. Nutech Environmental, London, Ontario, Canada.
7. D'Oliveira, J.-C., Al-Sayyed, G., and Pichat, P., Photodegradation of 2- and 3-chlorophenol in TiO_2 aqueous suspensions, *Environ. Sci. Technol.*, 24, 990, 1990.
8. Al-Sayyed, G., D'Oliveira, J.-C., and Pichat, P., Semiconductor sensitized photodegradation of 4-chlorophenol in water, *J. Photochem. Photobiol. A: Chem.*, 58, 99, 1991.
9a. D'Oliveira, J.-C., Maillard, C., Guillard, C., Al-Sayyed, G., and Pichat, P., Decontamination of aromatics-containing water by heterogeneous photocatalysis, in *Proceedings of the International Conference on Innovation, Industrial Progress and Environment*, M.C.I., Paris, 1991, 421–431.
9b. D'Oliveira, J.-C., Guillard, C., Maillard, C., and Pichat, P., Photocatalytic destruction of hazardous chlorine-or nitrogen-containing aromatics in water, *J. Environ. Sci. Health A*, 28, 941, 1993.
10. Maillard, C., Guillard, C., and Pichat, P., Comparative effects of the TiO_2-UV, H_2O_2-UV, H_2O_2/Fe^{2+} systems on the disappearance rate of benzamide or 4-hydroxybenzamide in water, *Chemosphere*, 24, 1085, 1992.

11. Maillard, C., Guillard, C., Pichat, P., and Fox, M. A., Photodegradation of benzamide in TiO_2 aqueous suspensions, *New J. Chem.*, 16, 821, 1993.

12a. Crosby, D. G. and Tutass, H. O., Photodecomposition of 2,4-dichlorophenoxyacetic acid, *J. Agric. Food Chem.*, 14, 597, 1966.

12b. Zepp, R. G., Wolfe, N. L., Gordon, J. A., and Baughman, G. L., Dynamics of 2,4-D esters in surface waters: hydrolysis, photolysis and vaporization, *Environ. Sci. Technol.*, 9, 1144, 1975.

13. Barbeni, M., Morello, M., Pramauro, E., Pelizzetti, E., Vincenti, M., Borgarello, E., and Serpone, N., Sunlight photodegradation of 2,4,5-trichlorophenoxyacetic acid and 2,4,5-trichlorophenol on TiO_2. Identification of intermediates and degradation pathway, *Chemosphere*, 16, 1165, 1987.

14. Crosby, D. G. and Wong, A. S., Photodecomposition of 2,4,5-trichlorophenoxyacetic acid (2,4,5-T) in water, *J. Agric. Food Chem.*, 21, 1052, 1973.

15. Maldotti, A., Amadelli, R., Bartocci, C., and Carassiti, V., Photo-oxidative cyanation of aromatics on semiconductor powder suspensions. I: oxidation processes involving radical species, *J. Photochem. Photobiol. A: Chem.*, 53, 263, 1990.

16. Güsten, H., Photocatalytic degradation of atmospheric pollutants on the surface of metal oxides, in *Chemistry of Multiphase Atmospheric Systems*, Jaeschke, W., Ed., Springer-Verlag, Berlin, 1986, 567–591.

17. Zetzsch, C., Simulation of atmospheric photochemistry in the presence of solid airborne aerosols, *DECHEMA Monogr.*, 104, 1987, 187.

18. Warneck, P., The atmospheric aerosol, in *Chemistry of the Natural Atmosphere*, Academic Press, New York, 1988, 279–373.

19. Takeuchi, K. and Ibusuki, T., Heterogeneous photochemical reactions and processes in the troposphere, in *Encyclopedia of Environmental Control Technology*, Cheremisonoff, P. N., Ed., Gulf Publishing Co., Houston, TX, 1989, 279–326.

20. Isidorov, V. A., *Organic Chemistry of the Earth's Atmosphere*, Springer-Verlag, Berlin, 1990.

21. Fisher, G. L., Chang, D. P. Y., and Brummer, M., Fly ash collected from electrostatic precipitators, *Science*, 192, 553, 1976.

22. Norman, R. O. C. and Thomas, C. B., Reactions of lead tetra-acetate. Part XIX. Oxidation of the dimethoxybenzenes: the behavior of lead tetra-acetate as an ambident electrophile, *J. Chem. Soc. B*, 3, 421, 1970.

23. Liang, J.-J. and Foote, C. S., Electron-transfer photooxygenation 8. Dicyanoanthracene-sensitized photooxidation of ortho-dimethoxybenzene, *Tetrahedron Lett.*, 23, 3039, 1982.

24. Bideau, M., Claudel, B., and Otterbein, M., Photocatalysis of formic acid oxidation by oxygen in an aqueous medium, *J. Photochem.*, 14, 291, 1980.

25. Yoneyama, H., Takao, Y., Tamura, H., and Bard, A. J., Factors influencing product distribution in photocatalytic decomposition of aqueous acetic acid on platinized TiO_2, *J. Phys. Chem.*, 87, 1417, 1983.

26. Chum, H. L., Ratchiff, M., Posey, F. L., Turner, J. A., and Nozik, A. J., Photoelectrochemistry of levulinic acid on undoped platinized n-TiO_2 powders, *J. Phys. Chem.*, 87, 3089, 1983.

27. Bideau, M., Claudel, B., Faure, L., and Rachimoellah, M., Homogeneous and heterogeneous photoreactions of decomposition and oxidation of carboxylic acids, *J. Photochem.*, 39, 107, 1987.

28. Chemseddine, A. and Boehm, H. P., A study of the primary step in the photochemical degradation of acetic acid and chloroacetic acids on a TiO_2 photocatalyst, *J. Mol. Catal.*, 60, 295, 1990.

29. Bideau, M., Claudel, B., Faure, L., and Kazouan, H., The photooxidation of acetic acid by oxygen in the presence of titanium dioxide and dissolved copper ions, *J. Photochem. Photobiol. A: Chem.*, 61, 269, 1991.

30. Mao, Y., Schöneich, C., and Asmus, K.-D., Identification of organic acids and other intermediates in oxidative degradation of chlorinated ethanes on TiO_2 surfaces en route to mineralization. A combined photocatalytic and radiation chemical study, *J. Phys. Chem.*, 95, 10080, 1991.

31. Gregg, S. J. and Sing, K. S. W., *Adsorption, Surface Area and Porosimetry*, Academic Press, London, 1967, 80.

32. Foster, P., Jacob, V., Laffond, M., and Perraud, R., Etude en enceinte de simulation de la photodégradation du naphtalène en présence de particules, in *Physico-Chemical Behavior of Atmospheric Pollutants*, Restelli, G. and Angeletti, G., Eds., Kluwer, Dordrecht, 1990, 289–299.

33. Fox, M. A., Chen, C. C., and Younathan, J. N. N., Oxidative cleavage of substituted naphthalenes induced by irradiated semiconductor powders, *J. Org. Chem.*, 49, 1969, 1984.

34. Herrmann, J.-M., Disdier, J., and Pichat, P., Effect of chromium doping on the electrical and catalytic properties of powder titania under UV and visible illumination, *Chem. Phys. Lett.*, 108, 618, 1984.

35. Jaffrezic-Renault, N., Pichat, P., Foissy, A., and Mercier, R., Study of the effect of deposited Pt particles on the surface charge of TiO_2 aqueous suspensions by potentiometry, electrophoresis, and labeled ion adsorption, *J. Phys. Chem.*, 90, 2733, 1986.

36a. Casado, J., Herrmann, J.-M., and Pichat, P., Phototransformation of o-xylene over atmospheric solid aerosols in the presence of O_2 and H_2O, in *Physico-Chemical Behavior of Atmospheric Pollutants*, Kluwer, Dordrecht, 1990, 283–288.

36b. Guillard, C., Delprat, H., Hoang-Van, C., and Pichat, P., Laboratory study of the rates and products of the phototransformations of naphthalene adsorbed on samples of titanium dioxide, ferric oxide, muscovite and fly ash, *J. Atmos. Chem.*, 16, 47, 1993.

CHAPTER 25

Photocatalyzed Destruction of Water Contaminants: Mineralization of Aquatic Creosote Phenolics and Creosote by Irradiated Particulates of the White Paint Pigment Titania

Nick Serpone, Rita Terzian, Darren Lawless, Anne-Marie Pelletier, Claudio Minero, and Ezio Pelizzetti

INTRODUCTION

Exposure of marine and freshwaters, which contain complex mixtures of dissolved organic and inorganic substrates and biological and colloidal materials, to natural sunlight can lead to a variety of (photo)chemical and physical phenomena.[1] Such physical phenomena as adsorption and solubilization of substrates onto and by particulate matter present in these waters leads to a varied and often enhanced photoreactivity.[2]

Heterogeneous photocatalysis has been rapidly growing as a technology based on irradiation of a photocatalyst, a semiconductor (e.g., TiO_2, ZnO, CdS), which produces electrons in the conduction band (e^-) and holes in the valence band (h^+), following which rapid migration to the particulate surface leads to redox chemistry. In aqueous media, h^+ are thought to oxidize surface OH^- groups and water (Equations 1 and 2) to yield surface-bound $OH^\cdot$ radicals[3a,3b] believed to cause the oxidation and the ultimate mineralization of various classes of organics to carbon dioxide.

$$OH^- + h^+ \rightarrow OH^\cdot \tag{1}$$

$$H_2O + h^+ \rightarrow OH^\cdot + H^+ \tag{2}$$

In this respect, heterogeneous photocatalysis, like the closely related photooxidation processes that utilize (1) light and ozone, (2) light and hydrogen peroxide, or (3) light plus ozone and peroxide, is a member of the family of processes that have become known as *advanced oxidation processes* (AOPs). Like these processes, much of the research in photocatalysis in industrialized countries is driven by legislation that encourages water purification (decontamination, detoxification, decolorization, deodorization) and simultaneous contaminant destruction by means other than by the traditional methods of air stripping and adsorption on activated charcoal. Whereas light alone or oxidant alone (H_2O_2, O_3) lead to partial destruction of contaminants, the simultaneous use of light (sunlight may do) and an oxidant (O_3, H_2O_2, or O_2 with a photocatalyst) completely mineralizes organic carbon to carbon dioxide. This is a principal advantage of AOP approaches to water treatment.

Utilization of this emerging technology to catalyze the photomineralization of a large and varied number of organics and the photoreduction of precious and toxic metals, listed as *priority pollutants* by the U.S. Environmental Protection Agency, has recently been reviewed by us[4-6] and others.[7,8a,8b] The complete mineralization of simple and complex halogenated derivatives of alkanes, alkenes, carboxylic acids, and aromatics has been amply demonstrated repeatedly in various laboratories.[4,5,7,9] So far, only CCl_4[2] and the *s*-triazine herbicides (e.g., atrazine)[4] resist total mineralization.

As part of our continuing systematic fundamental[3,10a,10b,11] and applied[4,12-14] studies into this developing technology, we have recently examined the complete destruction of aquatic phenolics, which represent approximately 10 wt% of the organics found in coal tar creosote from creosote-contaminated sites[15] (Figure 1); there are other components present in creosote together with several polycyclic aromatic hydrocarbons (~85 wt%) and heterocyclics (~5 wt%). We have focused our attention particularly on the phenolics, since these are highly water soluble[15] and therefore more likely to be found in aquatic environments. We earlier reported on the photomineralization of pentachlorophenol,[16] phenol,[10a,10b,17] and cresols,[12] which together with the more recently examined xylenols and 2,3,5-trimethylphenol,[13] is one of several aspects of our work.

Creosote and pentachlorophenol (PCP) are probably the most widely used pesticides in the wood-preserving industry, which account for an annual consumption of approximately 45,000 metric tons in the United States. Between 840 and 1530 dry metric tons of hazardous waste sludge is generated by these wood-treatment facilities.[15] Invariably, through accidental spills, misuse, and inadequate disposal, creosote leachates contaminate the aquatic environments. Two studies[18,19] have examined actual PCP- and creosote-contaminated sites at Pensacola, FL (American Creosote Works, Inc.) and at St.Louis Park, MN; in both cases, the findings of significant concentrations of methane and fatty acids showed that anaerobic degradation occurred in the aquifers.

Near-ultraviolet (UV) irradiation of creosote leachates in the presence of TiO_2 can rapidly and safely eliminate (mineralize) these hazardous substances in air-equilibrated aqueous media. The present remediation techniques use biotic and

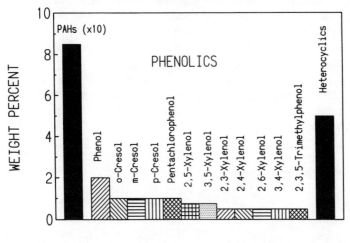

Figure 1. Histograms showing the relative concentrations (weight percent, wt%) of the various components present in coal tar creosote.

abiotic (volatilization, leaching, and direct photolysis) processes to restore contaminated sites with perhaps *in situ* biodegradation as the major method. However, the latter is not without its limitations:[15] (1) the pollutant(s) needs to be in a chemical state that is conducive to microbial utilization; (2) excess organic carbon in the feed robs the microbial population of oxygen and essential inorganic nutrients, thereby limiting its activity; and (3) successful bioremediation necessitates the presence of acclimatized microbial population capable of degrading the pollutant(s). As a case in point, studies by Crawford and Mohn[20] indicate that bioremediation of a PCP-contaminated site takes several months to bring [PCP] from 298 down to 58 ppm, although acclimatization does accelerate the process. The specificity of a microbe to degrade a given organic pollutant is not a limitation in heterogeneous photocatalysis, since the ·OH radical is indiscriminate in its attack of organics.

Actual water-treatment challenges will necessarily be multicomponent. First, as is the case for coal tar creosote, a process must treat several components that constitute the feed. Second, in the course of mineralization of any organic contaminant, oxidation will logically implicate a series of intermediate products of progressively higher oxygen to carbon ratios on the way to carbon dioxide. Thus, demonstrating the formation and elimination of these intermediates is also important to show complete removal of undesirable compounds. This also holds true for the other homogeneous AOP processes.

To illustrate the photocatalytic process, we consider below one of the xylenols (2,3-dimethylphenol). Light alone and the semiconductor catalyst (TiO_2) alone have little, if any, effect on the transformation of 2,3-xylenol. However, taken together (i.e., light plus titania), they lead to a relatively rapid transformation of the compound into intermediate products, which ultimately are also decomposed pathologically to the end product carbon dioxide. The formation of stoichiometric quantities of CO_2 takes longer than the disappearance of the xylenol. This is because intermediates form that also must be degraded. Hence, although the process feed is only one component, mineralization invariably involves multicomponent systems. To demonstrate the practicality of heterogeneous photocatalysis, we have examined the destruction of a solution 360 ppm of creosote in water.

EXPERIMENTAL SECTION

Materials

The individual phenolic substrates were purchased from Aldrich (purity $\leq$ 99%) and used without further treatment. TiO_2 was Degussa P25 (mostly anatase, BET 55 m^2/g, a generous gift from Degussa Canada, Ltd.). Creosote was Armor Coat commercial domestic grade purchased from a hardware store (wood preservative liquid, 100% creosote guaranteed). Water was doubly distilled throughout.

Procedures

Stock solutions of 2,3-xylenol (20 ppm, pH 3 with HCl) and creosote (360 ppm, distilled water) were prepared by weighing the required amount and diluted to 250 mL. The catalyst loading was 2 g/L of TiO_2. The mobile phase for the high-pressure liquid chromatography (HPLC) analyses was HPLC-grade methanol to water (50:50). Irradiation at wavelengths $\lambda \geq 320$ nm was carried out on either 25-mL or 50-mL samples contained in a Pyrex glass reactor with a 900-W mercury-xenon lamp equipped with a water filter to filter out infrared (IR) radiation. HPLC analyses were carried out on filtered (MSI nylon 66 filter, 0.22-μm pore size) 3-mL aliquots using a Waters Associates HPLC chromatograph (Model 501) with UV detection at 214 and 254 nm; the column was a Whatman reverse-phase C18 (partisil-10, ODS-3). The temporal evolution of CO_2 from the mineralization of creosote (100 ppm, 25 mL sample, oxygen-saturated suspension) was followed by sampling the head-space volume above the solution using a GOW-MAC gas chromatograph (GOW-MAC Instrument Co., Bridgewater, NJ) with a Porapak-N column and a TCD detector; helium was the carrier gas. The percent carbon present in a given sample of creosote was determined by the S.I.R.U. laboratory of this department (we thank Mr. B. Patterson for the courtesy). UV-visible (UV-Vis) absorption and/or diffuse reflectance spectra were recorded on a Schimadzu 265 SP UV-Vis spectrophotometer.

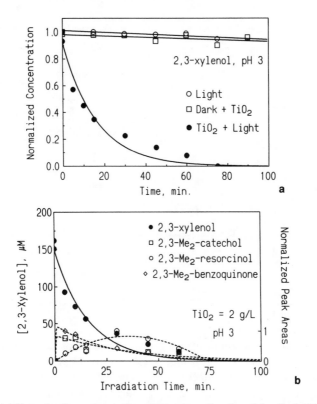

Figure 2. (a) Plots showing concentration changes as a function of time in the mineralization of 2,3-xylenol (20 ppm) in air-equilibrated aqueous solutions at pH 3 (with HCl) for light alone (direct photolysis), TiO_2 in the dark, and light plus TiO_2. For other details, see text. (b) Plots showing the changes in the concentrations of 2,3-xylenol (20 ppm) as a function of irradiation time, together with the formation and subsequent degradation of three aromatic intermediates (2,3-dimethylcatechol; 2,3-dimethylresorcinol; 2,3-dimethylbenzoquinone) during the temporal course of the photocatalyzed mineralization. Trace quantities of 2,3-dimethylhydroquinone are not shown.

RESULTS AND DISCUSSION

Photomineralization of 2,3-Xylenol

Irradiation of air-equilibrated aqueous suspensions containing TiO_2 particulates and 20 ppm of the xylenol leads to the disappearance of the latter in a relatively short time (~75 min; Figure 2) via good apparent first-order kinetics, $k_{deg} = 0.063 \pm 0.009$ min^{-1}. Light alone (direct photolysis) and TiO_2 alone in the dark (adsorption of xylenol, approximately 7%) have no consequence on the degradation of the xylenol. However, light and the catalyst together show a marked effect, indicating the necessity of TiO_2. Degradation of 2,3-xylenol is

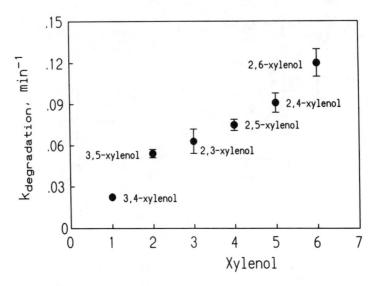

Figure 3. Graph comparing the rates of photocatalyzed degradation of all six xylenols examined.

followed by the formation of hydroxylated intermediates: 2,3-dimethylcatechol ($k_f = 0.03 \pm 0.01$ min^{-1}; $k_{deg} \sim 3.7$ min^{-1}), 2,3-dimethylresorcinol ($k_f \sim 0.007$ min^{-1}; degradation via zero-order kinetics), 2,3-dimethylbenzoquinone ($k_f = 0.039 \pm 0.009$ min^{-1}; $k_{deg} \sim 3.2$ min^{-1}), (Figure 2b), together with traces of 2,3-dimethyl-hydroquinone. No other intermediate was detected by HPLC.

The photomineralization of the remaining five xylenols (see Figure 3) under otherwise identical conditions has also been examined; a full account will be published elsewhere.[13] In every instance, degradation follows reasonably good apparent first-order kinetics. The relevant rate constants are depicted in Figure 3, which compares the relative degradation of the initial xylenols. They are shown as increasing rates of degradation. It is evident that the 3,x-xylenols degrade more slowly than the 2,x-xylenols, indicating a substituent effect on the rates. The related cresols (methylphenols) degrade more slowly by nearly a factor of 5 under similar but not identical conditions. By contrast, the 2,3,5-trimethylphenol degrades faster than the xylenols under identical conditions; $k_{deg} = 0.14 \pm 0.002$ min^{-1}.[13]

Photomineralization of Creosote

Addition of 100 mg of TiO$_2$ to a 50-mL sample of creosote leads to an immediate adsorption of the water-insoluble creosote components (oily droplets, mostly polycyclic aromatic hydrocarbons) onto the TiO$_2$ particulates that turned from white to an intense dark-beige color. The water-soluble phenolics remain largely in solution. Following irradiation of the air-equilibrated suspension, the temporal course of the degradation of these phenolics was monitored by absorption spectroscopy (Figure 4). For the first 60 min of irradiation, there are no

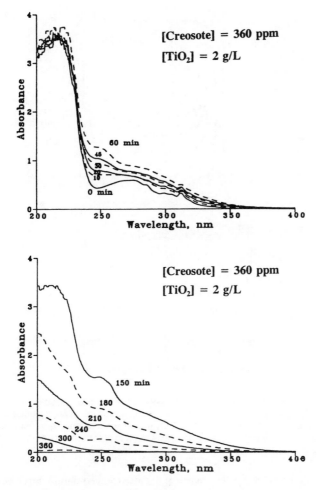

Figure 4. Absorption spectra of the filtrates containing water-soluble components of creosote during the photocatalyzed degradation of 360 ppm of creosote under air-equilibrated conditions (see text).

changes at wavelengths below 240 nm, but the spectra show greater absorptivity in the range 240–350 nm with increasing irradiation time; maximum absorption is reached after ca. 1 hr, following which a dramatic absorption decrease is evident throughout the wavelength range 200–400 nm from 150 to 360 min. After 6 hr of irradiation, there are no more (water-soluble) aromatic species evident in solution. The changes in the concentration of solution substrates were also followed by HPLC methods. Figure 5 illustrates the chromatograms as intensity vs retention time vs irradiation time. The changes observed here parallel those seen in Figure 4. After 6 hr, the chromatogram is identical to a water blank that contained TiO_2 but no creosote, thereby confirming the total disappearance of the water-soluble creosote components.

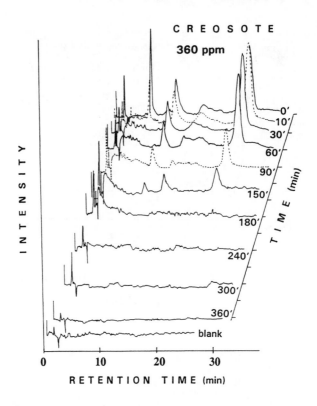

Figure 5. High-pressure liquid chromatograms of the initial water-soluble substances in creosote and intermediates following light irradiation of an aqueous suspension containing the photocatalyst TiO_2 and creosote under air-equilibrated conditions. The samples used were those used to record the absorption spectra (see Figure 4).

During irradiation, the catalyst particulates also showed significant color changes. The diffuse reflectance spectra of the particulates, illustrated in Figure 6, were taken after filtration on the MSI nylon membrane; the reflectance spectrum of pure TiO_2 particulates (blank) is also shown for comparison. The spectral features below 400 nm are due to TiO_2 particulates; the region above 400 nm shows increasing broad absorption to 850 nm up to 90 min of irradiation. These spectral features are reminiscent of surface complexes[14] formed between Ti(IV) on the catalyst particle surface and the various coordinating organic species. Prolonged irradiation (time greater than 1.5 hr) leads to absorption decrease until, after approximately 15 hr of irradiation, the appearance (white color) of the TiO_2 particulates has been restored.

Figure 7 summarizes the spectral changes as absorbance (at 215 nm in Figure 4) or reflectance (at 450 nm in Figure 6) vs irradiation time. Here also, there are relatively no changes at 215 nm until after ca. 3 hr of irradiation. At this point, all the water-soluble species *degrade collectively* via excellent first-order kinetics

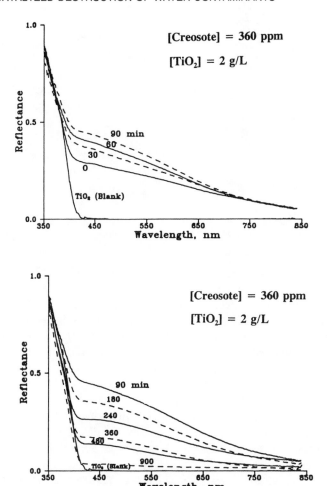

Figure 6. Diffuse reflectance spectra of the TiO$_2$ particulates after filtration of the irradiated aliquots taken during the photocatalyzed degradation of creosote.

(solid line) after an "induction" period of approximately 3 hr: k_{deg} = 0.0185 ± 0.0004 min^{-1}, $t_{1/2}$ = 36 min; similar kinetics were obtained at 250 nm (0.0183 ± 0.0009 min^{-1}; $t_{1/2}$ = 38 min). The species that are adsorbed on the catalyst surface also *decay collectively* via good first-order kinetics (dashed line) at 450 nm after an "induction" period of approximately 2 hr: k_{deg} = 0.0032 ± 0.0002 min^{-1}, $t_{1/2}$ = 216 min. We suspect that the slower degrading substances are the polycyclic aromatic hydrocarbons (PAHs) and other water-insoluble substrates. We estimate from these kinetics that after 24 hr of irradiation the concentrations of the observable species is reduced from 360 ppm to 6 ppb. Any attempt at describing the events occurring during the "induction period" would be too speculative at this

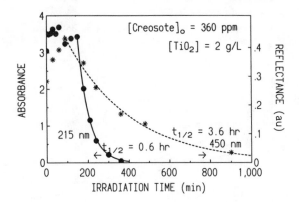

Figure 7. Kinetic analyses of the changes in concentrations (as absorbance) from the data in Figures 4 and 6 at the wavelengths indicated.

time, owing to the complexity of the system examined. Such effort will have to await further studies with an analytical methodology that will identify the various intermediate products involved along the course of the mineralization.

As emphasized earlier, total mineralization necessitates demonstration of stoichiometric evolution of the ultimate oxidation product, CO_2. Experimental difficulties (air-equilibrated conditions in a sealed reactor, time of irradiation needed to mineralize 360 ppm, and not least the suspension running out of needed oxygen) suggested we demonstrate total oxidation using an initial concentration of creosote of 100 ppm and a saturated oxygen atmosphere in a septum-sealed Pyrex glass reactor. The results are graphically illustrated in Figure 8 in terms of micromoles of CO_2 evolved vs irradiation time. The data were fitted (solid line) to double-exponential growth kinetics: $k_{CO2}^{(1)} = 0.042 \pm 0.008$ min^{-1}, $t_{1/2}^{(1)} = 17$ min and $k_{CO2}^{(2)} = 0.006 \pm 0.002$ min^{-1}, $t_{1/2}^{(2)} = 113$ min. The former correspond to CO_2 formation from the *faster* degraded water-soluble components, while the latter kinetics correspond to CO_2 formation from the *slower* degraded substances (the PAHs). The dashed line denotes the expected stoichiometric quantity of carbon dioxide, while the dotted curves represent the two growth kinetic components. Under these conditions, total mineralization of 100 ppm creosote is achieved in approximately 6–7 hr.

CONCLUDING REMARKS

The general ability of photocatalysis to mineralize a host of water contaminants, as demonstrated here for creosote and its various components (see also Reference 4), and to serve as a photochemical process basis for metals recovery[6] has been established. In the near future, researchers/entrepeneurs who wish to commercialize **heterogeneous photocatalysis** will need to design photoreactors and improve the light-absorbing characteristics of the catalyst(s) to increase the efficiencies of light utilization. As well, there will be a strong need to demonstrate

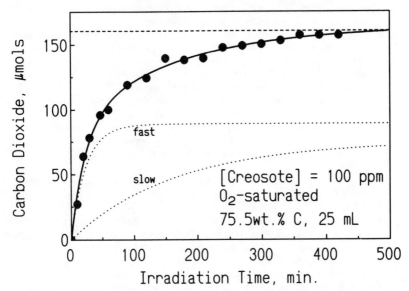

Figure 8. Plots showing the double exponential evolution of stoichiometric quantities of carbon dioxide formed from the total photomineralization of 100 ppm of creosote under a saturated oxygen atmosphere. See text for additional details.

economic feasibility if this emerging technology, and the related AOPs are to compete with presently used so-called "best" technologies.

ACKNOWLEDGMENTS

Our work is supported by NSERC Canada, CNR Rome, and by NATO (Grant No. CRG 890746).

REFERENCES

1. Zika, R. G. and Cooper, W. J., *Photochemistry of Environmental Aquatic Systems*, ACS Symposium Ser. No. 327, American Chemical Society, Washington, D.C., 1987.
2. Thomas, J. K., *The Chemistry of Excitation at Interfaces,* Advances in Chemistry Ser. No. 181, American Chemical Society, Washington, D.C., 1984.
3a. Lawless, D., Serpone, N., and Meisel, D., The role of OH radicals and trapped holes in photocatalysis: a pulse radiolysis study, *J. Phys. Chem.*, 95, 5166, 1991.
3b. Serpone, N., Lawless, D., Terzian, R., and Meisel, D., Redox mechanisms in heterogeneous photocatalysis. The case of holes vs OH radical oxidation and free vs surface-bound OH radical oxidation processes, in *Electrochemistry in Microheterogeneous Fluids*, McKay, R. and Texter, J., Eds., VCH Publishers, New York, 1992.
4. Ollis, D. F., Pelizzetti, E., and Serpone, N., Photocatalyzed destruction of water contaminants, *Environ. Sci. Technol.*, 25, 1522, 1991.

5. Ollis, D. F., Pelizzetti, E., and Serpone, N., Heterogeneous photocatalysis in the environment: Application to water purification, in *Photocatalysis. Fundamentals and Applications*, Serpone, N. and Pelizzetti, E., Eds., Wiley, New York, 1989, chap. 18.

6. Serpone, N., Lawless, D., Terzian, R., Minero, C., and Pelizzetti, E., Photochemical transformation and disposal of inorganic substances in the environment: hydrogen sulfide, cyanide and strategic and toxic metals, in *Photochemical Conversion and Storage of Solar Energy*, Pelizzetti, E. and Schiavello, M., Eds., Kluwer, Dordrecht, 1991, 451–475.

7. Matthews, R. W., Environment: photochemical and photocatalytic process. Degradation of organic compounds, in *Photochemical Conversion and Storage of Solar Energy*, Pelizzetti, E. and Schiavello, M., Eds., Kluwer, Dordrecht, 1991, 427–449.

8. Ollis, D. F., Contaminant degradation in water, *Environ. Sci. Technol.*, 19, 480, 1985.

9. Chemseddine, A. and Boehm, H. P., A study of the primary step in the photochemical degradation of acetic acid and chloroacetic acids on a TiO$_2$ photocatalyst, *J. Mol. Catal.*, 60, 295, 1990.

10a. Al-Ekabi, H. and Serpone, N., Kinetic studies in heterogeneous photocatalysis. I. Degradation of chlorinated phenols in aerated aqueous solutions over TiO$_2$ supported on a glass matrix, *J. Phys. Chem.*, 92, 5726, 1988.

10b. Serpone, N., Terzian, R., Minero, C., and Pelizzetti, E., Kinetic studies in heterogeneous photocatalyzed oxidation of organics over TiO$_2$ irradiated with AMI simulated sunlight: oxidation of phenols, cresols and fluorophenols, in *Photosensitive Metal-Organic Systems*, Kutal, C. and Serpone, N., Eds., Advances in Chemistry Ser., Vol. 238, American Chemical Society, Washington, D.C., 1993, 281-314.

11. Terzian, R., Serpone, N., Draper, B., Fox, M. A., and Pelizzetti, E., Pulse radiolysis of pentahalophenols: formation, kinetics, and properties of pentahalophenoxyl and dihydroxypentahalocyclohexadienyl radicals, *Langmuir*, 7, 3081, 1991.

12. Terzian, R., Serpone, N., Minero, C., and Pelizzetti, E., Photocatalyzed mineralization of cresols in aqueous media with irradiated titania, *J. Catal.*, 128, 352, 1991.

13. Terzian, R., Serpone, N., and Pelizzetti, E., Photocatalyzed degradation of xylenols and trimethylphenols, in preparation.

14. Terzian, R., Serpone, N., Minero, C., and Pelizzetti, E., Kinetic studies in heterogeneous photocatalysis 4: the photomineralization of a hydroquinone and a catechol, *J. Photochem. Photobiol. A: Chem.*, 55, 243, 1990.

15. Mueller, J. G., Chapman, P. J., and Pritchard, P. H., Creosote-contaminated sites: their potential for bioremediation, *Environ. Sci. Technol.*, 23, 1197, 1989.

16. Barbeni, M., Pramauro, E., Pelizzetti, E., Borgarello, E., and Serpone, N., Photodegradation of pentachlorophenol by visible light using a semiconductor catalyst, *Chemosphere*, 14, 195, 1985.

17. Okamoto, K., Yamamoto, Y., Tanaka, H., and Tanaka, M., Heterogeneous photocatalytic decomposition of phenol over TiO$_2$ powder, *Bull. Chem. Soc. Jpn.*, 58, 2015, 1985.

18a. Goerlitz, D. F., Troutman, D. E., Godsy, E. M., and Franks, B. J., Migration of wood-preserving chemicals in contaminated groundwater in a sand aquifer at Pensacola, Florida, *Environ. Sci. Technol.*, 19, 955, 1985.

18b. Troutman, D. E., Godsy, E. M., Goerlitz, D. F., and Ehrlich, G. G., Phenolic Contamination in the Sand-and-Gravel Aquifer from a Surface Impoundment of Wood Treatment Wastes, Pensacola, Florida, Report 84–4230, U.S. Geological Survey Water-Resources Investigations, Tallahassee, FL, 1984.

19. Ehrlich, G. G., Goerlitz, D. F., Godsy, E. M., and Hult, M. F., Degradation of phenolic contaminants in ground water by anaerobic bacteria: St. Louis Park, Minnesota, *Ground Water*, 20, 703, 1982.

20. Crawford, R. L. and Mohn, W. W., Microbial removal of pentachlorophenol from soil using a Flavobacterium, *Enzyme Microb. Technol.*, 7, 617, 1985.

CHAPTER 26

Mechanistic Studies of Chloro- and Nitrophenolic Degradation on Semiconductor Surfaces

K. A. Gray, P. Kamat, U. Stafford, and M. Dieckmann

INTRODUCTION

The photocatalytic degradation of a wide range of organic compounds using semiconductors such as TiO_2 and ZnO has been reported in the literature.[1] For a variety of substituted phenolic compounds, complete mineralization has been observed to occur via an array of identified intermediates and to require the presence of oxygen.[2,3] In general, it is commonly believed that photodegradation in semiconductor systems is mediated by the generation of the hydroxyl-free radical.[4] Although a number of reaction pathways have been proposed, the actual mechanism of degradation of phenolic compounds by semiconductors is not known and remains a controversial topic.

The purpose of this research is to elucidate the mechanistic details of chlorophenol and nitrophenol degradation on TiO_2 by identifying charge-transfer species, short-lived transients, and longer-lived intermediates using a number of spectroscopic techniques. One objective of this paper is to present the results of an *in situ* FTIR technique employed to probe the adsorption and photocatalytic degradation of 4-chlorophenol (4-CP) on TiO_2 particles in a gas-solid system. The second objective is to discuss the photodegradation of nitrophenols on a TiO_2 surface that is based on the principle of visible light sensitization by colored compounds. The findings of these two studies are significant in that they provide the fundamental basis for the development of efficient technologies to destroy hazardous chemicals.

0-87371-871-2/94/$0.00+$.50

CHLOROPHENOL DEGRADATION

A novel use of diffuse reflectance FTIR spectroscopy has allowed us to study the photocatalytic transformation of 4-CP adsorbed to the dry surface of TiO_2 (Degussa P25). It is possible using this technique to probe surface reactions in the absence of solution in order to distinguish the mode of adsorption, to identify primary intermediates, and to kinetically study surface phenomena as a function of various reaction components.[5] Sample preparation, the photoreactor, and the FTIR analysis are explained elsewhere.[6]

In Figure 1, the FTIR spectra are shown for (a) TiO_2, (b and c) 4-CP adsorbed onto TiO_2, and (d) 4-CP in a condensed phase. Peak positions and identities are also included in the figure. Between 1600–1000 cm^{-1}, no peaks are observed for TiO_2, although below 1000 cm^{-1} it absorbs strongly. At low levels of 4-CP adsorption to the TiO_2 surface (a), major changes in the spectrum relative to that of the condensed phase (d) are seen. There are shifts in the aromatic C=C and C–O stretches, disappearance of the O–H bending peaks, and decrease in the C–H bending peaks. For TiO_2 saturated with 4-CP (c), some of the details observed in the (d) spectrum reappear, but the significant shifts seen in spectrum (b) remain. In addition, the peak at 1488 cm^{-1} splits to produce a C=C stretch in the same position as in the condensed phase. Comparison of these spectra has been interpreted to indicate that 4-CP is attached to the surface by a strong phenolate link (absence of O–H and shift in C–O peaks) and that with adsorption the aromatic ring loses some degree of freedom (decrease in relative heights of C=C stretches). In addition, shifts in peak positions indicate changes in the internal molecular bond energies. At high levels of adsorption, a condensed phase of 4-CP begins to appear on the surface (split aromatic peak), and under vacuum can be easily removed. This removable portion of 4-CP is considered to be physisorbed. That which remains and produces the FTIR spectra shown in (b) is chemisorbed. This chemisorbed layer has been measured gravimetrically and corresponds to a single monolayer.[6]

Similar adsorption and FTIR experiments have been conducted with other preparations of TiO_2.[7] The amount of 4-CP adsorbed per surface area (BET determination) was measured gravimetrically for Aldrich TiO_2 (anatase and rutile forms) and for a reduced Degussa P25 TiO_2. In each case, the chemisorbed quantity was less than that measured for Degussa P25 TiO_2. Furthermore, interpretation of the FTIR spectra illustrate that 4-CP was adsorbed less strongly to these other surfaces. These findings suggest that the physicochemical properties of a semiconductor surface are important, and the affinity of a compound for the surface may be an important factor in determining photocatalytic efficiency.

Diffuse reflectance FTIR was used to monitor the photodegradation of a thin layer of TiO_2 powder saturated with 4-CP and irradiated with ultraviolet (UV) light. In Figure 2, the spectra of 4-CP/titania powder are illustrated for three irradiation times (t = 0, 2, and 20 hr). This reaction was conducted in an atmosphere of water-saturated oxygen. The spectral data are presented in Kubelka-Munk units, which are a function of sample reflectance and are linear with respect

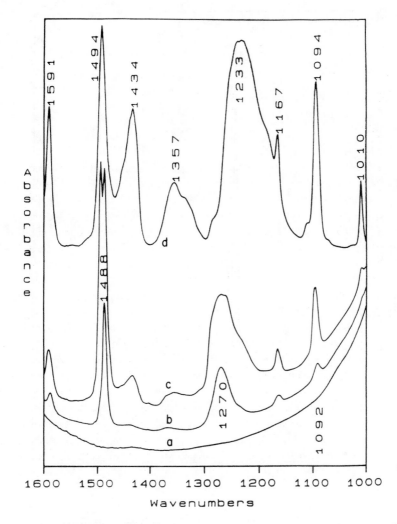

Figure 1.   FTIR spectra of TiO₂ **(a)** prior to exposure to 4-CP vapor at 25°C, **(b)** after 30 min **(c)** after 7 hr. Spectrum **(d)** shows condensed phase of 4-CP which is included for reference and has the following peak identities: 1591 and 1494 — C=C stretch; 1434 and 1357 — O–H bend; 1233 — C–O stretch; 1167 and 1094 — C-H bend; 1010 — C-H bend.

to surface concentration for diffuse reflectance spectra. As the reaction progressed, the spectrum changed, indicating the transformation of 4-CP to hydroquinone (HQ). Product identification was based on a comparison of the spectrum at 2 hr to that of pure HQ adsorbed onto TiO₂. Hydroquinone has also been identified as the predominant intermediate detected in aqueous slurry studies.[8,9] After 20 hr of irradiation, HQ disappeared and the remaining material corresponded to carbonates.[10] By determining the areas beneath representative peaks, FTIR data can be used to evaluate the kinetic behavior of the degradation process.

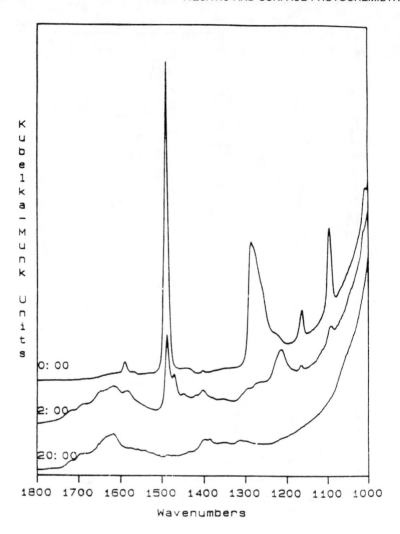

Figure 2. Photocatalytic degradation of 4-CP adsorbed on TiO$_2$ powder monitored using diffuse reflectance FTIR spectroscopy. The samples were analyzed following UV irradiation for 0, 2, and 20 hr.

Control reactions have been conducted that demonstrate that the reaction will not take place in the absence of light or TiO$_2$. The reaction has been followed under a number of conditions in order to investigate the roles of oxygen and water in the mechanism of 4-CP degradation.[7,11] The rate of 4-CP disappearance and HQ appearance and gradual disappearance were measured under UV light and various oxygen conditions. Under dry oxygen conditions, 4-CP was degraded to HQ, which also gradually disappeared. These rates increased with increasing water vapor pressure and were found to be greatest under a saturated oxygen atmosphere. Under conditions of UV irradiation and a hydrated N$_2$ atmosphere (no

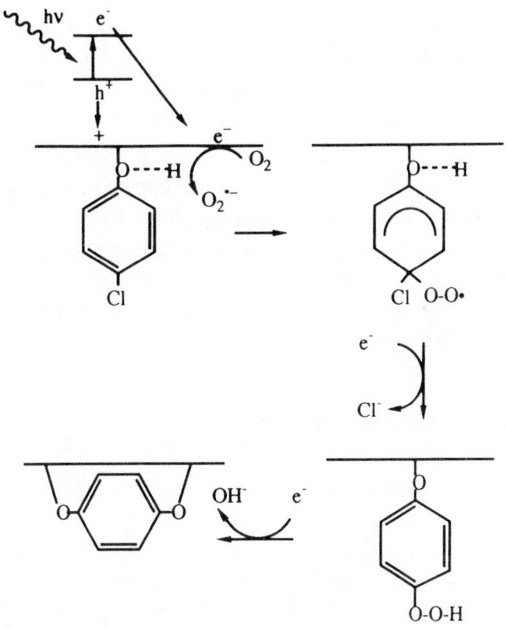

Figure 3. Possible reaction scheme for the reductive dechlorination of 4-CP to hydro-quinone under dry oxygen and UV irradiation on TiO_2 powder.

oxygen), 4-CP degradation to HQ was observed, but HQ did not undergo photodegradation.

The fact that degradation occurred under dry conditions suggests that the reactions can be mediated through pathways other than those involving the hydroxyl-free radical, such as direct interaction between the organic compound and photoinduced charge carriers at the semiconductor surface or by reaction with oxygen in either its molecular or reduced form.[12,13] It seems more probable that under dry oxygen conditions HQ evolves through the reductive dechlorination of 4-CP by the superoxide radical, O_2^- in a possible reaction scheme shown in Figure 3. Increasing amounts of water vapor would cause a shift in the equilibria of activated oxygen species, a greater ease of protonation, and the formation of hydroxyl-free radical by the oxidation of water at the positive hole, all of which would account for the dramatic increase in the rate and extent of reaction.

Partial degradation of 4-CP to HQ occurred with UV light under hydrated N_2 conditions, but complete mineralization failed to occur. Although it is generally thought that photocatalytic reactions do not proceed in the absence of oxygen because it is required to sweep electrons to prevent charge recombination, in this case 4-CP appears capable of serving as an electron acceptor. The reaction stops, however, presumably because HQ is unable to serve as a suitable electron acceptor.

Using diffuse reflectance FTIR, an *in situ* study of photodegradation on semiconductor surfaces has been performed. The nature and extent of surface adsorption can be distinguished with this technique, and it is suggested that the physicochemical properties of the surface and its affinity for an adsorbate may control the efficiency of semiconductor photocatalysis. Reaction kinetics and products can be followed with FTIR and based on these results oxygen was found to play an intrinsic role in the complete destruction of 4-CP. In this system, 4-CP degradation occurs on the TiO_2 surface, and 4-CP appears to be capable of various surface interactions such as direct charge transfer at the semiconductor surface. One of the primary advantages of using this FTIR technique in powder systems is that solvent effects can be controlled and individual reaction steps revealed. The focus of future work will be to determine the relationship between photocatalysis in powder systems and aqueous slurry systems.

NITROPHENOL DEGRADATION

Normally, photocatalysis in semiconductor systems requires the use of sufficiently energetic light in order to initiate the charge separation. Generation of UV light can be costly, and it is likely that wider applications would be found for photocatalytic degradation if the range of semiconductor activity could be extended to visible light. It is well established that dyes such as rose bengal when excited with visible light can inject an electron into the conduction band of a semiconductor undergoing degradation in the process.[14] The focus of this work was to determine if colored pollutants interact with semiconductors in a similar way.

Diffuse reflectance studies have been performed to investigate the photodegradation of nitrophenols on various oxide surfaces (i.e., Al_2O_3, SiO_2, TiO_2, etc.). In Figure 4, results are shown for 2,5-dinitrophenol (2,5-DNP) coated on various solid powders and irradiated with a halogen light source. This graph illustrates that significant disappearance of the parent compound occurred only on TiO_2 powders. In Figure 5, the effect of surface coverage on the rate of degradation is illustrated. For a particular amount of surface, a calculated amount of 4-nitrophenol (4-NP) was coated onto the particles using rotary evaporation of the solvent. Normalized Kubelka-Munk units at 410 nm are plotted as a function of visible light irradiation time. Very little degradation of 4-NP was observed at a degree of surface coverage equaling four times a monolayer. A high degree of surface coverage prohibits direct interaction (charge injection) between excited 4-NP and the semiconductor surface. The highest degradation rate was observed at the quarter monolayer level, because at low surface coverage most molecules of 4-NP are in direct contact with the surface and will undergo degradation with excitation and charge injection. Similar experiments were conducted with an alumina surface, and no degradation was observed at any level of coverage. These experiments demonstrate that the intrinsic properties of the support material control the photochemistry of nitrophenols and that degradation does not occur by direct photolysis of adsorbed species. The position and number of nitro substitutions on the aromatic ring has been shown to also influence the rate of degradation on semiconductors.[15]

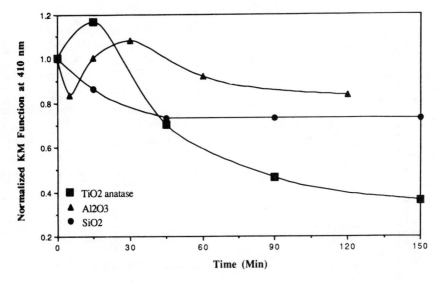

Figure 4. Photodegradation of 2,5-DNP on various support materials. Diffuse reflectance measured at 410 nm and expressed in normalized Kubelka-Munk units.

In order to prove that the observed results were not due to excitation of the semiconductor, a 385-nm filter was placed on the light source. There was little difference between the rate of degradation for the filtered and unfiltered light sources, demonstrating that charge transfer was initiated by visible light excitation of the adsorbed 4-NP. Degradation produced with a UV source is compared to that produced with visible light in Figure 6. The disappearance of 4-NP as measured by the Kubelka-Munk function at 410 nm appeared to be more rapid with visible light irradiation. This is not entirely accurate, though, because UV irradiation produced a product with an absorbance maximum around 450 nm, which interfered with monitoring at 410 nm (Figure 7). The product produced with UV irradiation resisted further degradation and was not observed with visible irradiation.

The acidic and basic forms of nitrophenol have very different absorbance spectra. The absorbance maximum of the acid is around 320 nm, and with increasing pH it shifts to a maximum around 400 nm. Furthermore, the basic form has a larger extinction coefficient, which accounts for the higher rate of disappearance observed with basic 4-NP than with the acid form. Different products are also observed with the photodegradations of these different forms.

Reactions were conducted and monitored in a closed cell in order to assess the role of oxygen for this process. Similar degradation rates were observed in a vacuum and in dry air. With the addition of moisture, however, the rate increased. Laser flash photolysis experiments are currently underway to characterize the 4-NP short-lived transient. No evidence has been found for the existence of a singlet (lack of fluorescence upon excitation). Results from picosecond laser flash photolysis suggest the presence of a triplet excited state

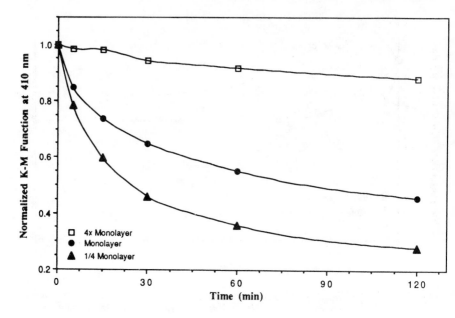

Figure 5.   Dependence of 4-NP photodegradation on surface coverage of TiO_2.

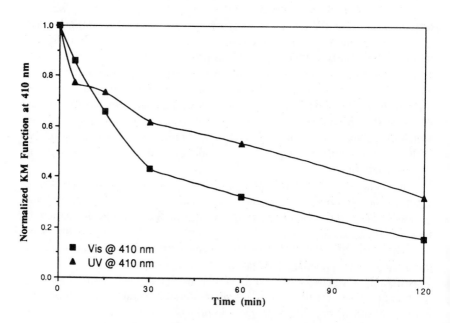

Figure 6.   Photodegradation of 4-NP on TiO_2 with visible and UV irradiation (diffuse reflectance monitored at 410 nm).

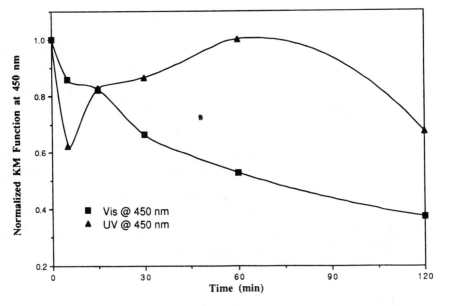

Figure 7. Product formation with UV irradiation of 4-NP as illustrated by absorbance at 450 nm and compared to visible light irradiation.

which absorbs at 470 nm. This excited state displays strong solvent effects, is extremely short-lived in water (decays within 500 ps), and the lifetimes are increased in ethanol and benzene.

A variety of products has been identified for the interaction of nitrophenols with various free radicals or excited states.[16-18] Based on these results and preliminary data from pulse radiolysis, it seems likely that one mechanism of 4-NP degradation may involve denitration due to photoinduced hydrolysis. At the present time, product identification is underway.

Visible light excitation of nitrophenols will cause degradation of the compound adsorbed to semiconductor surfaces. The process appears to take place in a manner similar to that observed with such dye sensitizers as rose bengal. Visible light irradiation produces an excited state of the colored compound. Charge injection into the conduction band of the semiconductor will occur when direct interaction between the excited state and the surface occurs. Subsequent degradation of the parent compound is observed. For nitrophenols, a number of factors influence the rate of degradation: position and number of nitrosubstitutions on the aromatic ring, degree of surface coverage, and acid-base characteristics. Different rates and products are observed when UV irradiation is used to promote degradation. This indicates that the reaction is taking place via different mechanisms. Work continues to identify the products and characterize the short-lived transients of the reaction.

REFERENCES

1. Ollis, D. F., Pelizzetti, E., and Serpone, N., Heterogeneous photocatalysis in the environment: application to water purification, *Photocatalysis: Fundamentals and Applications*, Serpone, N. and Pelizzetti, E., Eds., 1989, 603–637.

2. Al-Ekabi, H., Serpone, N., Pelizzetti, E., Minero, C., Fox, M. A., and Draper, R. A., Kinetic studies in heterogeneous photocatalysis, *Langmuir,* 5, 250, 1989.

3. D'Oliveira, J.-C., Al-Sayyed, G., and Pichat, P., Photodegradation of 2- and 3-chlorophenol in TiO$_2$ aqueous suspensions, *Environ. Sci. Technol.,* 24, 990, 1990.

4. Turchi, C. S. and Ollis, D. F., Photocatalytic degradation of organic water contaminants: mechanisms involving hydroxyl radical attack, *J. Catal.,* 122, 178, 1990.

5. Kung, K.-H. and McBride, M. B., Bonding of chlorophenols on iron and aluminum oxides, *Environ. Sci. Technol.,* 25, 702, 1991.

6. Stafford, U., Gray, K. A., Kamat, P. V., and Varma, A., An *in-situ* diffuse reflectance FTIR investigation of photocatalytic degradation of 4-chlorophenol on a TiO$_2$ powder surface, *Chem. Phys. Lett.,* 205, 55, 1993.

7. Stafford, U., unpublished results (1992).

8. Al-Ekabi, H. and Serpone, N., Photocatalytic degradation of chlorinated phenols in aqueuos solutions over TiO$_2$ supported on a glass, *J. Phys. Chem.,* 92, 5726, 1988.

9. Al-Sayyed, G., D'Oliveira, J.-C., and Pichat, P., Semiconductor-sensitized photodegradation of 4-chlorophenol in water, *J. Photochem. Photobiol. A: Chem.,* 58, 99, 1991.

10. Miller, F. A. and Wilkins, C. H., Infrared spectra and characteristic frequencies of inorganic ions. Their use in quantitative analysis, *Anal. Chem.,* 24, 1253, 1952.

11. Gray, K. A., Stafford, U., Dieckmann, M. S., and Kamat, P., Mechanistic studies in TiO$_2$ systems: photoctalytic degradation of chloro- and nitrophenols, in *Proceedings of the First International Conference on TiO$_2$ Photocatalytic Purification and Treatment of Water and Air*, Ollis, D. and Al-Ekabi, H., Eds., Elsevier, Amsterdam, 1993.

12. Lawless, D., Serpone, N., and Meisel, D., Role of OH radicals and trapped holes in photocatalysis. A pulse radiolysis study, *J. Phys. Chem.,* 95, 5166, 1991.

13. Sheldon, R. A. and Kochi, J. K., *Metal-Catalyzed Oxidations of Organic Compounds,* Academic Press, New York, 1981.

14. Gopidas, K. R. and Kamat, P. V., Photochemistry on surfaces. 4. Influence of support material on the photochemistry of an adsorbed dye, *J. Phys. Chem.,* 93, 6428, 1989.

15. Dieckmann, M., Gray, K. A., and Kamat, P. V., Photocatalyzed degradation of adsorbed nitrophenolic compounds on semiconductor surfaces, *Water Sci. Technol.,* 25, 277, 1992.

16. Kotronarou, A., Mills, G., and Hoffmann, M. R., Ultrasonic irradiation of p-nitrophenol in aqueous solution, *J. Phys. Chem.,* 95, 3630, 1991.

17. Alif, A., Boule, P., and LeMaire, J., Photochemistry and environment. XII. Phototransformation of 3-nitrophenol in aqueous solution, *J. Photochem. Photobiol.,* 50, 331, 1990.

18. O'Neill, P., Steenken, S., van der Linde, H., and Schulte-Frohlinde, D., Reaction of OH radicals with nitrophenols in aqueous solutions, *Radiat., Phys. Chem.,* 12, 13, 1978.

CHAPTER 27

Spin-Trap EPR Studies of Intermediates Involved in Photodegradation Reactions on TiO₂: Is the Process Heterogeneous or Homogeneous?

Lizhong Sun, K.-Michael Schindler, Aitken R. Hoy, and James R. Bolton

INTRODUCTION

Photodegradation, using small semiconductor particles as a photocatalyst, is an efficient and practical method to convert organic pollutants in water into harmless substances. For example, the mineralization of 4-chlorophenol proceeds via the reaction

$$C_6H_5OCl(aq) + 13/2\ O_2 \xrightarrow[\text{TiO}_2]{h\nu} 6CO_2 + HCl + 2H_2O \qquad (1)$$

The irradiation of TiO₂ particles with light of energy larger than the bandgap (3.2 eV) creates electron-hole pairs. These transfer to the surface of the particles and there react with hydroxyl groups on the surface to produce ·OH radicals. The ·OH radicals may react either with adsorbed pollutant (heterogeneous reaction) or desorb into solution and react there with solution phase pollutant molecules (homogeneous reaction). In either case, the pollutant is decomposed.

Carey et al.[1] were the first to find that irradiation of TiO₂ suspensions with 365-nm light brings about the complete degradation (mineralization) of chloro-organic molecules. Following this initial report, there have been many papers[2-3] suggesting the usefulness of UV-irradiated TiO₂ for the activation of stable organic molecules in aquatic systems. Some investigators[4-6] have proposed

0-87371-871-2/94/$0.00+$.50

409

kinetic models for the photodegradation process and have advanced some possible schemes to explain the mechanism of the photocatalytic process. In spite of these explorations, the detailed mechanism of the photodegradation process is still not clear.

In this research, we have used electron paramagnetic resonance (EPR) spectroscopy with the spin-trap 5,5-dimethyl-1-pyrroline N-oxide (DMPO) (see below) to examine the nature of the photocatalytic process and, particularly, to examine whether or not the process is heterogeneous. We have determined the initial growth rates of the DMPO-OH spin-adduct electron paramagnetic resonance (EPR) signal in systems with different scavengers within a competitive reaction scheme. The kinetic behavior of truly homogeneous systems was compared with that of the TiO_2 systems to establish the photocatalytic pathway.

In past investigations, we have obtained poorly reproducible results,[7] which we ascribed to a variable particle-size distribution. Thus, in this investigation, we have used a centrifugation technique to obtain a reproducible and narrow size distribution for the TiO_2 particles.

EXPERIMENTAL

Electron Paramagnetic Resonance and Photolysis Apparatus

Electron Paramagnetic Resonance spectra were obtained using a Bruker Model ESP300 Electron Paramagnetic Resonance Spectrometer. Photolysis was carried out using a water-filtered 150-W xenon-mercury UV lamp (Oriel Corporation of America) fixed 80 cm from the flat cell in the EPR cavity. It is important to carry out all comparative experiments on the same day without changing the configuration of the lamp assembly.

Preparation of TiO_2 Suspensions

A suspension of TiO_2 was prepared by adding commercial TiO_2 powder (anatase) (Aldrich Chemical Company, Inc.) to 5.0 mM phosphate buffer. The suspension was sonicated for 1 hr (Model 8845-4 ultrasonic cleaner, Cole-Parmer Instrument Company) to break the larger particles into smaller ones. The suspension was then centrifuged (Superspeed Centrifuge Sorvall Inc.) to narrow the particle-size distribution.

The particle size in the final suspension was found to be in the range 90–160 nm [scanning electronic microscopy (SEM)]. The stability and reproducibility of the suspension were confirmed by determining absorption [ultraviolet-visible (UV-Vis) spectrophotometer, Hewlett Packard model 8450A] and EPR spectra.

The concentration of the buffer was varied within a range of 0.0–5.0 mM and was found to have no effect on the measurements.

Langmuir Isotherms

Langmuir isotherms were determined to assess the extent of adsorption of the spin trap and some of the scavengers on TiO₂ particles. A suspension of TiO₂ with the spin trap or scavenger was shaken at frequent intervals over a period of minutes and then centrifuged. The supernatant in the upper part of the centrifugation tube was then extracted to measure the concentration of the adsorbent in liquid solution using UV-Vis spectrophotometry. The difference of the concentration of adsorbent in the solution before and after the adsorption onto the TiO₂ particles was determined. The fraction θ of surface covered is the ratio of the concentration difference at a given concentration to that where the surface is saturated (i.e., at very high adsorbent concentration).

Spin-Trapping Technique

Electron paramagnetic resonance spectroscopy is a technique that is sensitive to the presence of molecules with unpaired electrons ($<10^{-9}$ M free radicals can be detected). However, the radical intermediates produced in the TiO₂ photocatalysis reaction are usually too short-lived to be detected directly by EPR. This problem can be solved by the use of the spin-trapping technique in which compounds called spin traps are used to convert reactive radicals into relatively stable radicals (spin adducts), which can be then detected by EPR.

$$R^{\cdot} + \text{spin trap} \rightarrow \text{spin adduct} \tag{2}$$

The spin trap used in this experiment was DMPO (Aldrich, stored at $-20°C$). The spin-trapping reaction is

$$\tag{3}$$

The spin-adduct DMPO, produced when $\cdot$OH radicals react with DMPO, has a lifetime of more than 30 min in our system.

The initial growth rate of the DMPO-OH spin adduct was used as a convenient measure of relative reaction rates. The magnetic field was fixed at the second peak of the first-derivative EPR signal of the DMPO-OH spin adduct (Figure 1a). The initial growth rate (Figure 1b) was obtained using a linear least squares fitting method.

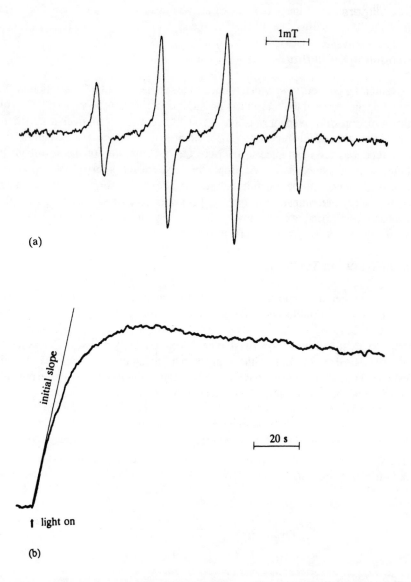

Figure 1. Electron paramagnetic resonance (EPR) signals of the DMPO-OH spin adduct in a TiO$_2$ suspension with 2 mM DMPO. **(a)** first-derivative EPR spectrum of DMPO-OH; $a^H = a^N = 1.49$ mT, sweep width: 9.0 mT, modulation frequency: 100 kHz, modulation amplitude: 0.214 mT, time constant: 20.48 ms, conversion time: 5.12 ms, microwave frequency: 9.74 GHz, microwave power: 20.0 mW, and number of scans: 10. **(b)** The growth curve of DMPO-OH at the second peak of (a), center field: 347.55 mT, sweep width: 0.00 mT, time constant: 655.36 ms, and conversion time: 163.84 ms.

Scavengers

These scavengers were tested: formate (sodium formate from Fisher Scientific Company), phenol (Caledon Laboratory, Ltd.), and 2-propanol (Fisher Scientific Company). The concentrations used were such that direct photolysis of the scavenger by the UV light can be ignored.

Competitive Reaction Experiment

The detection of the DMPO-OH spin adduct by EPR does not necessarily prove the presence or intermediacy of ·OH radicals in the photocatalytic reaction, as there are other known reactions, not involving ·OH radicals, that may lead to the same spin adduct. For example, Finkelstein et al. have proposed the following mechanism:[8]

$$\tag{4}$$

In our case, this mechanism must be insignificant, as we detect no signal on photolysis of DMPO alone. We have measured the initial rate of production of the DMPO-OH spin adduct. However, this rate is markedly dependent on light intensity, fraction of light absorbed, surface activity, etc. Thus, the absolute initial rate is not very useful. One common technique for solving such ambiguities is the use of a competitive reaction scheme. Since both scavengers and spin traps can react with ·OH radicals, the addition of a scavenger to a solution containing spin traps forces the two species to compete for the scarce ·OH radicals. Hence, the rate of production for the ·OH spin adduct should decrease as the scavenger concentration increases.

The formate ion is a convenient scavenger, in that it is known to react with ·OH radicals to produce CO_2^- radicals which are also trapped by DMPO. With H_2O_2 as a source of ·OH radicals, the reactions are as follows:

$$H_2O_2 \xrightarrow{\ hv\ } 2\ ^{\cdot}OH \tag{5a}$$

$$^{\cdot}OH + HCO_2^- \longrightarrow H_2O + CO_2^{-\cdot} \tag{5b}$$

$$^{\cdot}OH + DMPO \longrightarrow DMPO\text{-}OH \tag{5c}$$

$$CO_2^{-\cdot} + DMPO \longrightarrow DMPO\text{-}CO_2^- \tag{5d}$$

The DMPO-CO$_2$ spin adduct has a spectrum easily distinguished from that of DMPO-OH.

The observation of an increase in the DMPO-CO_2 signal, along with a corresponding decrease in the DMPO-OH signal, as the concentration of formate is increased in the system, indicates that the spin trap and scavenger are competing for the same species.

RESULTS AND DISCUSSION

Assessment of Extent of Adsorption of the Spin Trap and Scavengers on TiO_2 Particles

Standard Langmuir adsorption isotherms for DMPO and the scavenger formate were determined in the dark for TiO_2 suspensions. Linear relations between the reciprocal of the fraction of the surface covered by adsorbents $1/\theta$ and the reciprocals of the equilibrium concentrations of adsorbents in the solution $1/c_{eq}$ were obtained for these adsorbents. The following adsorption equilibrium constants were found: $K_{DMPO} = 0.084 \pm 0.006$ mM^{-1} and $K_{formate} = 0.036 \pm 0.004$ mM^{-1}. In the case of phenol, we have not found any significant change of phenol concentration in the solution arising from adsorption. This means that $K_{phenol} <$ 0.01 mM^{-1}.

The above results imply that only weak adsorption of the three substances takes place on the TiO_2 particles. Thus, the adsorption behavior of these scavengers on TiO_2 does not cause any significant change in the concentration of scavenger or the DMPO spin trap in the solution. That is, we are in the weak adsorption region where the surface coverage is linear in the concentration of all adsorbents in solution.

The TiO_2 Photocatalysis Pathway

As mentioned above, using TiO_2 as a photocatalyst, the photodegradation reaction takes place when surface-generated ·OH radicals react with pollutant molecules. But we are not sure if the active agents, ·OH radicals, react with adsorbed pollutants on the surface of the particles in a *heterogeneous reaction* or desorb from the surface and react with free pollutants in solution in a *homogeneous reaction*.

The spin-trapping/EPR technique, within a competitive reaction scheme, has been used to investigate this point. Two systems were investigated: one containing H_2O_2 as a source of ·OH radicals, DMPO as the spin trap and a scavenger, and the other containing a TiO_2 suspension, DMPO and a scavenger. For both systems, the concentration of scavenger was varied, while keeping all other concentrations fixed. Since the H_2O_2 system is known to produce free ·OH radicals on UV illumination and its reaction is homogeneous, it can be used as a "standard." If the reaction in the TiO_2 system is also homogeneous, i.e., if ·OH radicals generated by holes desorb into solution and react there with pollutants, the kinetic behavior

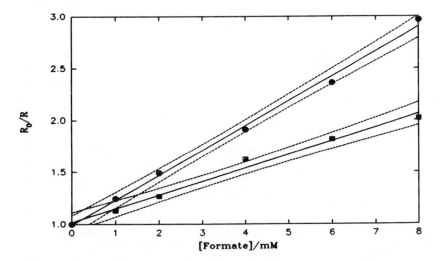

Figure 2. Competitive reaction of formate as scavenger. R is the initial growth rate of DMPO-OH spin adduct in the solution with scavenger, and R$_0$ is the initial growth rate in the absence of the scavenger; -●-: H$_2$O$_2$ system, [H$_2$O$_2$] = 4.0 mM, and [DMPO]= 5.0 mM; -■-: TiO$_2$ system, [DMPO] = 5.0 mM. The confidence level for both lines is 95%.

of each system *must* be the same as that of the standard. The reason is that once in solution after a few diffusion steps, the ˙OH radicals lose all "memory" of how they were created. Any significant deviation in the kinetics of the TiO$_2$ system from that of the standard then constitutes proof that at least some of the ˙OH radicals react heterogeneously on the surface.

The effects of the scavenger concentration on the competition reaction (i.e., the initial rate R) for both systems with different scavengers were examined. Results with formate as a scavenger are shown in Figure 2. Both systems exhibit a linear relation between R$_0$/R and the formate concentration, where R$_0$ is the initial rate of the system in the absence of scavenger. There is a marked difference between the slopes of the two lines. The Sigma Plot program was used to establish 95% confidence limits about the regression lines. Since the two lines are not even close to overlapping at this level of confidence, we are virtually certain that the behavior of the two systems is significantly different. Similar results were obtained with phenol and 2-propanol as scavengers (Figures 3 and 4, respectively).

These results indicate that the reaction pathway of ˙OH radicals in TiO$_2$ suspensions is quite different from that of H$_2$O$_2$ system, which is well known as a homogeneous process. We thus conclude that the photocatalysis pathway in TiO$_2$ is at least partly heterogenous.

Based on our experimental results, we propose that the photodegradation pathway in TiO$_2$ suspensions is as follows. When light with energy larger than the bandgap of TiO$_2$ irradiates the particles in solution, induced electron-hole pairs in the particles transfer to the surface of the semiconductor and react there with

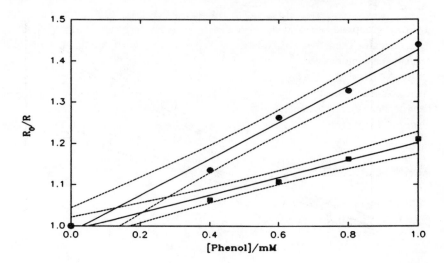

Figure 3. Competitive reaction of phenol as scavenger. R is the initial growth rate of
DMPO-OH spin adduct in the solution with scavenger, and R_0 is the initial growth
rate in the absence of the scavenger. -●-: H_2O_2 system, $[H_2O_2]$ = 4.0 mM,
[DMPO] = 6.0 mM; -■-: TiO_2 system, [DMPO] = 6.0 mM. The confidence level
for both lines is 95%.

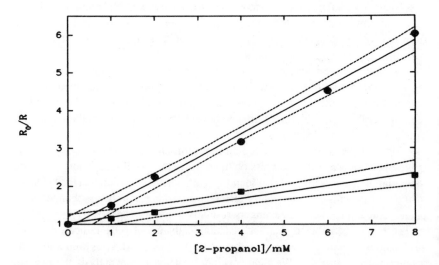

Figure 4. Competitive reaction of 2-propanol as scavenger. R is the initial growth rate of
DMPO-OH spin adduct in the colution with scavenger, and R_0 is the initial growth
rate in the absence of the scavenger. -●-: H_2O_2 system, $[H_2O_2]$ = 4.0 mM and
[DMPO] = 2.0 mM; -■-: TiO_2 system, [DMPO] = 2.0 mM. The confidence level
for both lines is 95%.

hydroxyl groups adsorbed on the surface of the particles to produce surface-bound ·OH radicals. Then the ·OH radicals react with adsorbed pollutant heterogeneously, causing the decomposition of the pollutant.

An extension of this work will focus on the study of the kinetic behavior of the "heterogeneous" catalytic process. It should be possible to obtain a table of heterogeneous rate constants that should be valuable in understanding the mechanism by which scavengers/pollutants are removed on TiO_2 particles.

ACKNOWLEDGMENTS

We thank the Ontario Ministry of the Environment of Canada for an operating grant to support this work.

REFERENCES

1. Carey, H. J., Lawrence, J., and Tosine, H. M., Photodechlorization of PCB's in the presence of titanium dioxide in aqueous suspensions, *Bull. Environ. Contam. Toxicol.*, 16, 697, 1976.
2. Pelizzetti, E., The role of colloidal particles in the photodegradation of organic compounds of environmental concern in aquatic systems, *Adv. Colloid Interface Sci.*, 2, 271, 1990.
3. Anpo, M., Photocatalysis on small particle TiO₂ catalysis. Reaction intermediates and reaction mechanisms, *Res. Chem. Intermed.*, 11, 67, 1989.
4. Gopidas, K. R. and Kamat, P. V., Photoelectrochemistry in particulate systems, 11: reduction of phenosafranin dye in colloidal TiO₂ and CdS suspensions, *Langmuir*, 5, 22, 1989.
5. Terian, R., Serpone, N., Pelizzetti, E., and Hidaka, H., Kinetic studies in heterogeneous photocatalysis 4. The photomineralization of a hydroquinone and a catechol, *J. Photochem. Photobiol. A: Chem.*, 55, 243, 1990.
6. Matthews, R. W., Photocatalytic oxidation and adsorption of methylene blue on thin films of near-ultraviolet-illuminated TiO₂, *J. Chem. Soc. Faraday Trans. I*, 85, 1291, 1989.
7. Bolton, J. R., Mechanism of the photochemical degradation of organic pollutants on TiO₂, in *Proceedings of Symposium on Advanced Oxidation Processes in Contaminated Water and Air of Pollutants*, Halevy, M., Ed., Wastewater Technology Centre, Burlington, Ontario, Canada, 1990.
8. Finkelstein, E., Rosen, G. M., and Rauckman, E. J., Spin trapping of superoxide and hydroxyl radical: practicle aspects, *Arch. Biochem. Biophys.*, 200, 1, 1980.

CHAPTER **28**

Phenol and Nitrophenol Photodegradation Using Aqueous TiO$_2$ Dispersions

Antonino Sclafani, Leonardo Palmisano, and Mario Schiavello

INTRODUCTION

This chapter collects data from various studies on the photodegradation of phenol and of nitrophenol isomers in aqueous oxygenated dispersions of TiO$_2$ that was obtained from various sources and preparations.[1-8]

The main scope of these studies was to investigate the influence of the following parameters on the photoreactivity: initial pH of the dispersion; oxygen partial pressure; addition of such ions as Cl$^-$, SO$_4^{2-}$, Ag$^+$, etc. and addition of H$_2$O$_2$; alternative preparation methods of TiO$_2$; polymorphic form of TiO$_2$ (anatase or rutile); or their mixture; etc.

The theme, which links these studies, is the attempt to find correlations among the above parameters, some pertaining to the reaction conditions, some to the features of the photocatalysts, and some to the level of photoreactivity.

EXPERIMENTAL

Apparatus and Procedure

A detailed description has been reported in previous papers.[1-8] In this chapter, the essentials will be given. Two different types of photoreactors were used for the photodegradation runs. The first type, which consisted of Pyrex glass flasks placed in a Solarbox (Co.Fo.Me.Gra., Milan, Italy), was used for runs aimed only at checking the photoactivity of various powders. These batch reactors were magnetically stirred. They contained 50 ML of an aqueous dispersion in which the concentration of the catalyst and the organic compounds were 1 and 0.1 g · L^{-1}, respectively. The lamp used was a 1500-W xenon lamp (Philips XOP 15-OF). The temperature inside the Solarbox reached 313 K. The majority of runs lasted 1 hr. Some of them, when the complete disappearance of the organic compounds was followed, lasted 4–6 hr. Oxygen was bubbled into the system before starting the runs and every hour, as the complete mineralization of the organic compounds was achieved only in the presence of this gas.[1,8,9a,9b]

The second type of photoreactor was a biannular Pyrex batch photoreactor containing 1.5 L of aqueous dispersion. It was mainly used when the kinetics of the process were investigated or mass balances were determined by measuring evolved carbon dioxide. The concentrations of the photocatalyst and the organic compounds were varied in the 0.2–2 and 0.03–0.3 g · L^{-1} ranges, respectively. The photoreactors were provided with a number of ports in their upper section; 500-W or 1000-W high- or medium-pressure mercury lamps (Helios Italquartz) were positioned within the inner part of the photoreactors. The photoreactivity experiments were performed at 300 K. During the runs, pure oxygen or helium-oxygen mixtures were continuously bubbled into the dispersions. The system was magnetically stirred, and samples for analysis were withdrawn every 15 min.

The quantitative determination of phenol was performed by a standard colorimetric method.[10] For the nitrophenols, the absorption at 280, 275, and 315 nm was measured for 2-, 3-, and 4-nitrophenol, respectively. Nitrates and nitrites, the products of complete oxidation of nitrophenols, were determined by the brucine standard colorimetric method.[10] The photoactivity was shown as a reaction rate (in molecules nm^{-2} · s^{-1} or mg · L^{-1} · h^{-1} · m^{-2}) or as a rate constant (expressed according to the experimental rate law).

Catalysts

Commercial specimens (all reagent grade BDH GPR, Merck Optipur, Tioxide 1601/1 and 1601/2, Montedison B5, Degussa P25, and Carlo Erba RPE) of "Rome prepared" (hp) titania (anatase or rutile) and TiO$_2$ powders were used.

Two series of differently prepared TiO$_2$ hp specimens were investigated. The first one was obtained by reacting an aqueous solution of TiCl$_3$ (15 wt%, Carlo Erba) with an aqueous solution of ammonia (25 wt%, Merck). The solids were separated by filtration, washed repeatedly with bidistilled water, and dried in air at 393 K for 24 hr. Subsequently, they were divided into several portions. Each portion underwent a different thermal treatment in a range of temperatures from 473 K up to 1073 K and for times ranging from 3 to 192 hr.

The second series of TiO$_2$ hp was obtained by reacting TiCl$_4$ (Carlo Erba) with pure water at about 278 K and adjusting with NaOH the initially very low pH value up to 4.5. The subsequent preparation procedure was the same as for TiO$_2$ hp prepared by the previous method.

RESULTS AND DISCUSSION

Only a few results, which are relevant to the scope of this paper, are here presented and discussed.

Rutile vs Anatase Photoactivity

The following results should be considered. The commercial TiO$_2$ specimens, consisting of the anatase phase, included the Degussa P25 specimen (anatase 80%, rutile 20%). They were all found to be active, with levels of activity ranging from 0.45 to 4.4 mg · L^{-1} · hr^{-1} · m^{-2}, whereas the commercial TiO$_2$ specimens consisting of rutile were completely inactive.[1]

For the hp specimens, it was found that (1) the specimens containing only anatase exhibited a great variety of photoactivity; (2) the specimens, fired at relatively high temperature (=1073 K) and containing only the rutile phase, were completely inactive; and (3) the specimens formed by mixtures of anatase and rutile, with various ratios, were found to be as active or more active as those containing only anatase. It is worth reporting that a specimen containing only rutile, fired 3 hr at 973 K, showed an activity of 3.1 mg · L^{-1} · hr^{-1} · m^{-2}, while the same specimen fired 24 hr at the same temperature was found to be completely inactive.[4]

Experiments carried out as previously described, but with the addition of Ag$^+$ ions in the aqueous dispersion, showed that both rutile and anatase were photoactive in the presence of He or O$_2$. In the presence of helium, anatase was slightly more active than rutile. In the presence of O$_2$, the photoactivity increased, and both modifications exhibited almost the same photoactivity.[4,7]

This body of results appears to indicate that the anatase modification is always photoactive for the studied reactions, showing a spread of activity according to the source and to the preparation methods, while rutile is inactive or active according to the methods of its preparation and to the experimental conditions.

From the great range of activity of the anatase specimens, it is easy to think that the photoactivity of a semiconductor is due to an interplay of several factors, as, for instance, the textural and the surface acid-base properties, the capacity of absorbing photons and reagents, the lifetime of the e-h formed pairs, etc. These factors are very dependent on the preparation methods, and therefore, a range of activity is expected when specimens of various origin and preparations are investigated. As for the inactivity/activity of rutile in comparison to the activity of anatase, the results suggest some lines of explanation. The results of the home-made specimens suggest that the hydroxylation-dehydroxylation surface equilibrium plays an important role in determining the level of photoactivity. The behavior of the rutile modifications prepared at high temperature (1073 K) and especially that prepared at a lower temperature (973 K), but for 24 hr instead of 3 hr, is a clear indication that an irreversible dehydroxylation occurs, as is also known from other studies.[11-13]

The drastic decrease of surface OH_s groups has a detrimental effect on the charge separation process (OH_s, together with the anionic organic species is a trap for holes, h^+). Therefore, the activity on the dehydroxylated specimens is decreased. Moreover, the decrease of OH_s groups reduces the adsorption of O_2 and of organic molecules.[1,14,15]

In the photooxidation of organic molecules, the likely predominant steps for charge transfer are described in the following equations (Equations 1–3). For the hole, we have

$$OH^- + h^+ \rightarrow OH^\cdot \tag{1}$$

$$RO^- + h^+ \rightarrow RO^\cdot \tag{2}$$

where RO^- is $C_6H_5O^-$ or $O_2NC_6H_4O^-$. For the electrons, when O_2 is present, we have

$$O_{2(ads)} + e^- \rightarrow O_{2(ads)}^- \tag{3}$$

These processes are possible on both modifications of TiO_2 on thermodynamic grounds, but the oxygen adsorption is scarce on the surface of rutile for the reasons given above and Equation 3 does not significantly occur.

When Ag^+ is present in aqueous dispersion, the charge-trapping step for electrons is

$$Ag_{(ads)}^+ + e^- \rightarrow Ag_{(ads)} \tag{4}$$

This step predominates both when He or O_2 are bubbled through rutile suspensions. Therefore, the addition of Ag^+ allows, for the rutile, the occurrence of Equation 4 and therefore the charge-separation process, with a rate comparable to that of anatase.

In conclusion, the results show two possible reasons for the inactivity of rutile: an irreversible surface dehydroxylation and difficulty in performing the electron-trapping step with O_2.

The Effect of pH

It was found that the photoreactivity was pH dependent. Namely, a maximum of reactivity was found around pH 3. The reactivity steadily increased in alkaline medium, and, for pH higher than 13, it was found higher than at pH 3.[1,3]

The enhancement of photoreactivity in alkaline medium is due to several factors: (1) charge separation via the reaction of OH^- and h^+ is favored in basic medium; (2) adsorption of oxygen on the surface of TiO_2 is enhanced when the presence of the OH^- group is increased;[1,13-16] (3) the conduction band becomes more negative ($E_{cb} = E_{cb}^\circ - 0.059$ pH), favoring the reduction processes; and (4) in alkaline medium, the anionic species of the organic molecules are prevalent, while the undissociated forms are prevalent in acidic medium. By reflectance spectroscopy, it has been shown that the anionic species are adsorbed on the surface of TiO$_2$ much more than the undissociated ones.[1,3,8]

The Influence of the Concentration of the Reactants and the Photocatalyst

Given below are some of the results to be considered. The maximum photo-activity was found under these conditions: 80–100 mg $\cdot$ L^{-1} ($\approx 0.8 \cdot 10^{-3}$–$10^{-3}$ M) of phenol, 1 g $\cdot$ L^{-1} of TiO$_2$ and 0.6 atm of O$_2$. Moreover, an inhibiting effect of the initial concentration of the organic compounds on the photodegradation rate was clearly evident. Two different types of sites on the surface of TiO$_2$ were hypothesized: one able to adsorb the organic species and the other able to adsorb O$_2$ species in competition with Cl$^-$ ions (see next paragraph). Thus, the reaction rate was written following the Langmuir-Hinshelwood theory as

$$r = k'' \Theta_{O_2} \Theta_{R-OH} \tag{5}$$

where k'' is the second-order rate constant for the surface decomposition of the organic pollutant under investigation, and Θ_{O_2} and Θ_{R-OH} are the fractional site coverages for oxygen and for the organic pollutant.

By considering that Θ_{O_2} is equal to one, when the dispersion is saturated with O$_2$ and taking into account the adsorption constant of the organic pollutant (K_{R-OH}), the previous equation becomes

$$r = \frac{k'' K_{R-OH} [R - OH]}{1 + K_{R-OH} [R - OH]_0}$$

where $[R-OH]_0$ is the initial concentration, and $[R–OH]$ is the concentration corresponding to the same time at which r is calculated.

Similar kinetics were found for phenol and for the nitrophenol isomers. (The numerical values of k'' and K_{R-OH} are different for the different compounds, and, of course, they depend also on the geometry of the photoreactor and the power of the lamp.)

The Influence of the Addition of Cl$^-$, SO$_4^{2-}$. NO$_3^-$ Ions, and of H$_2$O$_2$

Only chloride ions, among the various added anions, were found to be significantly detrimental for photoreactivity when their concentration was higher than 3–4% w/w. In order to explain the results obtained in this study, one can invoke a different extent of interaction between the various ions and the TiO$_2$ surface. The photoreactivity results suggest that an interaction does not occur, or if it does occur, it has no, or scarcely any, influence on photoreactivity, at least in the range of concentrations used in this study, except for the case of addition of Cl⁻. It is known that Cl⁻ ions are preferentially adsorbed,[15] and the decrease of OH⁻, as previously said several times, has the effect of depressing the charge separation step and thus decreasing the photoreactivity.

As for the addition of H_2O_2, a thorough study was performed by considering the features of the homogeneous and heterogeneous photodecompositions of pure H_2O_2, pure C_6H_5OH, and various mixtures in the presence of He or O_2. It is worth noting that the homogeneous process in all the experimental conditions investigated did not allow a complete mineralization of phenol, but only its partial oxidation.

The most important result was that the maximum in photoactivity for the decomposition of phenol was found in the heterogeneous system and when H_2O_2 and O_2 were simultaneously present. It was also observed that photoreactivity was higher when H_2O_2 was present in the dispersion without O_2 in comparison to that found when O_2 was bubbled into the dispersion without H_2O_2. For the case when H_2O_2 is present alone or with O_2, it was found that the rate law is similar to that previously described.

The following mechanism was invoked to explain the beneficial influence of the addition of H_2O_2 both in the presence and in the absence of O_2, confining the discussion to the heterogeneous regime. As said before, at least two types of sites can be hypothesized to exist on the surface of TiO_2: one type able to adsorb the organic species and the other able to adsorb O_2 and H_2O_2 in competition. When the three species are simultaneously present, charge separation occurs involving adsorbed oxygen species as electron traps and adsorbed H_2O_2 species both as hole and electron traps.[6] As a consequence of these processes, radical species, which are powerful oxidants, are formed. Thus, the oxidation process may be envisaged to occur according to two parallel pathways: one involving the radical species produced by the electron-trapping process that involves reduced oxygen species and the other by the radical species produced by H_2O_2 photodecomposition. This hypothesis has been checked by determining the rate constants, which can be written following the above mechanism.

ACKNOWLEDGMENTS

The authors wish to thank Ministero dell'Università e della Ricerca Scientifica e Tecnologica (Rome) and CNR (Rome) for financially supporting this work.

REFERENCES

1. Augugliaro, V., Palmisano, L., Sclafani, A., Minero, C., and Pelizzetti, E., Photocatalytic degradation of phenol in aqueous titanium dioxide dispersion, *Toxicol. Environ. Chem.* 16, 89, 1988.

2. Augugliaro, V., Davi, E., Palmisano, L., Schiavello, M., and Sclafani, A., The photocatalytic degradation of phenol in aqueous titanium dioxide dispersions: the influence of hydrogen peroxide, in *Environmental Contamination*, Orio, A. A., Ed., CEP Consultants, Ltd., Edinburgh, 1988, 206.

3. Palmisano, L., Augugliaro, V., Schiavello, M., and Sclafani, A., Influence of acid-base properties on photocatalytic and photochemical processes, *J. Mol. Catal.*, 56, 284, 1989.

4. Sclafani, A., Palmisano, L., and Schiavello, M., The influence of the preparation methods of TiO$_2$ on the photocatalytic degradation of phenol in aqueous dispersions, *J. Phys. Chem.*, 94, 829, 1990.

5. Sclafani, A., Palmisano, L., and Davi, E., Photocatalytic degradation of phenol by TiO$_2$ aqueous dispersions: rutile and anatase activity, *New J. Chem.*, 14, 265, 1990.

6. Augugliaro, V., Davi, V., Palmisano, L., Schiavello, M., and Sclafani, A., Influence of hydrogen peroxide on the kinetics of phenol photodegradation in aqueous titanium dioxide dispersion, *Appl. Catal.*, 65, 101, 1990.

7. Sclafani, A., Palmisano, L., and Davi, E., Photocatalytic degradation of phenol in aqueous polycrystalline TiO$_2$ dispersions: the influence of Fe^{3+}, Fe^{2+} and Ag$^+$ on the reaction rate, *J. Photochem. Photobiol. A: Chem.*, 56, 113, 1991.

8. Augugliaro, V., Palmisano, L., Schiavello, M., Sclafani, A., Marchese, L., Martra, G., and Miano, F., Photocatalytic degradation of nitrophenols in titanium dioxide aqueous dispersion, *Appl. Catal.*, 69, 323, 1991.

9a. Okamoto, K., Yamamoto, Y., Tanaka, H., Tanaka, M., and Itaya, A., Heterogeneous photocatalytic decomposition of phenol over TiO$_2$ powder, *Bull. Chem. Soc. Jpn.*, 58, 2015, 1985.

9b. Okamoto, K., Yamamoto, Y., Tanaka, H., Tanaka, M., and Itaya, A., Kinetics of heterogeneous photocatalytic decomposition of phenol over anatase TiO$_2$ powder, *Bull. Chem. Soc. Jpn.*, 58, 2023, 1985.

10. Taras, H. J., Greenberg, A. E., Hoak, R. D., and Rand, M. C., Eds., *Standard Methods for the Examination of Water and Wastewater*, 13th ed., American Public Health Association, Washington, D.C., 1971, 520.

11. Bickley, R. I. and Jayanty, R. K., Photo-adsorption and photo-catalysis on titanium dioxide surfaces, *J. Chem. Soc. Faraday Discuss.*, 58, 194, 1974.

12. Morterra, C., An infrared spectroscopic study of anatase properties, *J. Chem. Soc. Faraday Trans. I*, 84, 1617, 1988.

13. Primet, M., Pichat, P., and Mathieu, M. V., Infrared study of the surface of titanium dioxide I. Hydroxyl groups, *J. Phys. Chem.*, 75, 1216, 1971.

14. Bickley, R. I. and Stone, F. S., Photoadsorption and photocatalysis at rutile surfaces. Part I. Photoadsorption of oxygen, *J. Catal.*, 31, 389, 1973.

15. Boonstra, A. H. and Mutsaers, C. A. H. A., Relation between the photoadsorption of oxygen and the number of hydroxyl groups in a titanium dioxide surface, *J. Phys. Chem.*, 79, 1694, 1975.

16. Munuera, G., Gonzalez-Elipe, A. R., Rives-Arnau, V., Navio, A., Malet, P., Soria, J., Conesa, J. C., and Sanz, J., Photoadsorption of oxygen on acid and basic TiO$_2$ surfaces, in *Adsorption and Catalysis on Oxide Surface*, Che, M. and Bond, G. C., Eds., Elsevier, Amsterdam, 1985, 113.

CHAPTER 29

Accelerated Photooxidative Dissolution of Oil Spills

Adam Heller and J. R. Brock

INTRODUCTION

We consider here the application of $n\text{-}TiO_2$ photoassisted oxidation to cleaning up oil spills. Research in the past 15 years has shown that on large bandgap stable, oxide semiconductors, such as $n\text{-}TiO_2$, photoassisted oxidation of organic compounds takes place.[1-16] Photoassisted oxidation involves the reaction sequence

$$hv \rightarrow e^- + h^+ \tag{1}$$

$$h^+ + H_2O \rightarrow \text{·}OH + H^+ \tag{2}$$

$$n \text{·}OH + \text{hydrocarbon} \rightarrow \text{water soluble carbonates and organics,}$$

$$\text{e.g., } HCO_3^- \text{ carbonyl compound,}$$

$$\text{carboxylates, phenols} \tag{3}$$

$$O_2 + 2H^+ + 2e^- \rightarrow H_2O_2 \tag{4}$$

$$2H_2O_2 \rightarrow 2H_2O + O_2 \tag{5}$$

0-87371-871-2/94/$0.00+$.50

In Equation 1, an electron (e^-)-hole (h^+) pair is produced upon absorption of a photon having an energy greater than the bandgap of the photocatalyst. The hole oxidizes water to an ·OH radical and a proton (Equation 2).[17] The ·OH radicals oxidize hydrocarbons and other water-insoluble organic compounds to carbonates and ketones, aldehydes, carboxylic acids, etc., also converted ultimately to HCO_3^- (Equation 3). The electrons remaining on the semiconductor, after the holes are consumed in Equation 2, either recombine with subsequently generated holes, a process resulting in loss in quantum efficiency, or reduce dissolved oxygen (Equation 4). In the presence of an excess of organic material, the quantum efficiency thus depends on the rate of electron stripping by O_2 from the semiconductor particles.

The oxygen-reduction reaction can be slow and thus limit the quantum efficiency. Maintenance of high quantum efficiency requires that the flux of O_2 to the semiconductor particles match the solar ultraviolet (UV) photon flux that they absorb.[18,19] Because the rate of catalytic O_2 reduction on the TiO_2 surfaces depends on the TiO_2 phase and the TiO_2 surface area and surface chemistry, the method of photocatalyst preparation, impurity level, and particle size all affect the quantum efficiency. For this reason, the reported quantum efficiencies diverge, even for particles of the same (anatase or rutile) phase and size. When the O_2-reduction kinetics on the semiconductor surface limits the quantum efficiency, the rate and quantum efficiency can be increased by incorporating traces of an O_2-reduction catalyst, such as Pd^0 or Pt^0, in the TiO_2 surface[20,21] or by reductive-thermal treatment of TiO_2.[22] Whether or not the quantum efficiency is limited by the kinetics of O_2 reduction or by mass transport, it will increase with the surface area of the photocatalyst.

For the rate of the photocatalytic process to be fast, it is essential that part of the TiO_2 catalyst be above the air-oil interface rather than immersed in the optically thick crude oil. Therefore, the catalyst is attached to a buoyant support. Buoyancy of the support is one of the key rate- and efficiency-controlling factors.[23] The buoyant photocatalyst support must not be itself photooxidized. Hollow glass or glass-ceramic microspheres, consisting of manufactured borosilicate glasses or of fly ash-derived aluminosilicate glass ceramics are appropriate supports[24] and are available at low cost. They have, except for their buoyancy, the appearance of fine sand. The manufactured microbeads are engineered to 50, 100, or 200 μm diameter, and their density ranges from 0.05 to 0.8 g cm^{-3}. The fly ash-derived microbeads, known as cenospheres, are 50–200 μm in diameter and their density is near 0.7 g cm^{-3}. The wall thicknesses of the different beads vary from 2 to 20 μm.

Here, we consider the rate and efficiency of catalytic photoassisted oxidation of strongly UV-absorbing crude oil on water, as well as the catalytic photoassisted oxidation of 3-octanol on water. In the UV, 3-octanol is transparent and serves as a model aliphatic compound. We show that TiO_2-coated, buoyant ceramic microspheres photoassist oxidation of both aliphatic and aromatic compounds. Unlike natural noncatalytic photooxidation, that eliminates UV-absorbing aromatic hydrocarbons, but produces high steady-state concentrations of toxic and biodeg-radation retarding phenols, catalytic photoassisted oxidation of aromatics, while producing phenols, readily eliminates these.

METHODS

Activation of Degussa P25 TiO$_2$

Degussa's P25 TiO$_2$, consisting of 35% rutile, 65% anatase particles of ap-proximately 100 nm diameter ,was activated by thermal H$_2$ reduction and HCl etching. Related activation has been observed in approximately 200-nm rutile particles upon 550°C reduction by adsorbed carbonized triethanolamine followed by HCl etching.[22] Although activation improves the photooxidation rate and quantum efficiency when these are photocatalyst surface kinetics limited, it does not increase the rate or efficiency when these are O$_2$ mass transport limited. In our experiments, because the kinetics were often O$_2$ mass transfer limited, the Degussa P25 TiO$_2$ was not activated prior to attachment.

Preparation of Coated Microspheres

Hollow aluminosilicate microspheres (cenospheres) (SLG, PQ Corporation, Valley Forge, PA) were coated by modified versions of the earlier reported high- and low-temperature processes.[16] Beads made by either process had about the same photoactivity. In the modified version of the low-temperature process, 750 mL ethanol, 75 mL methyltrimethoxysilane, and 8 mL water (pH adjusted to 3.0 by adding HCl) were refluxed for 4 hr. Then, 250 g of water-washed and dried cenospheres (100 μm, 0.7 g cm^{-3}) were added and the mixture was refluxed for 2 hr. Next, 100 g of Degussa P25 TiO$_2$ was added and the mixture was refluxed for an additional 2 hr. The volatiles were then evaporated at $135 \pm 15°C$. The resulting solid was loosened and sieved to produce free-flowing coated beads. The beads were etched in 1 M HCl for 30 min, water washed, and dried. The low-temperature process yields beads that are hydrophobic. The earlier de-scribed[16] high-temperature process yields beads that are initially hydrophilic and are made hydrophobic by treatment with the reactive silane Hüls' Glassclad 6C.

Assay of Bead Photoactivity

All measurements were performed at ~ 46 W m^{-2} irradiance by 300–400 nm [365 nm (max.)] light. The rate of O$_2$ consumption was measured in 200-mL water-jacketed, 7-cm diameter isothermal photoreactors. In the octanol experi-ments, 1 g of beads and 5 mL of octanol were applied to 60 mL 0.5 M NaCl in water. After loading, the reactor was flushed with 79% N$_2$–21% O$_2$ and sealed. A water manometer was used to follow the drop in pressure. Because analysis of the gas phase following irradiation showed formation of a significant amount of CO_2, some of the CO_2 was stripped from the gas by attaching an Ascarite II (Thomas Scientific, Swedesboro, NJ) containing tube. It was not ascertained that the gas did not contain residual CO_2. Therefore, the O_2 uptake-based results represent lower limits of the rate and efficiency of photoassisted oxidation. The experiments on 3-octanol photooxidation were run for 3 hr, both when the liquid was stagnant and when it was swirled by placing the reactor in an orbit shaker operated at 120 rpm.

Experiments on crude oil were performed using Texaco Bruce #1. Four series of experiments were performed. The first of these was aimed at determining whether or not mousse formation — that is, emulsification of the nonvolatile crude residue — can be prevented by oleophilic TiO_2-coated microbeads and whether or not the nonvolatile bead-adsorbed residue is readily photooxidized by UV light. In the second series, the rate of photoassisted oxidation of the crude was determined by measuring the rate of oxygen uptake from pure air (79% nitrogen, 21% oxygen). The O_2 uptake experiments on crude oil were run for 4.5 hr. In the third, oil (2 mL) was aggregated with beads (2 g) on water (200 mL), and exposed for 2000 hr to UV light under stagnant conditions. In the fourth, the oil-bead mass ratio required for aggregation of the oil to a mechanically harvestable and combustible semisolid mass was determined.

The experiments on the photooxidation of the nonvolatile residue involved mixing 1 mL of oil with 3 g of beads and evaporating the volatile fraction by heating to 100°C for 15 hr. The resultant dark oily residue, after being agglomerated by oleophilic photoactive beads, had the appearance and texture of tar-soaked sand. This tar-like sand was placed on 120 mL of a 0.5 M aqueous NaCl solution, agitated by bubbling air, and exposed to near UV light. One sample was exposed for 120 hr and the other for 240 hr. After each experiment, the oil-bead agglomerate was collected, washed to remove the salt, dried at 100°C, weighed, and then photographed.

HPLC Analysis of the Hydrocarbons and Phenols

High-performance liquid chromatography (HPLC) separation of the hydrocarbon fraction was performed on a Supelcosil LC-PAH (C18) column (15 cm × 4.6 mm). The mobile phase was water-acetonitrile mixture 65:35 (v/v) held for 5 min, followed by a linear gradient to 100% acetonitrile over 25 min and holding for 5 min. The flow rate was 1 mL/min, and the detector was set at 254 nm.

HPLC separation of the phenolic fraction was also performed on a Supelcosil LC-PAH (C18) column (15 cm × 4.6 mm). The mobile phases (four 10-min steps, all linear gradients) were gradient (1) 5% acetonitrile vs water to 10% acetonitrile, (2) 10 to 50% acetonitrile, (3) 50 to 70% acetonitrile, and (4) 70 to 100% acetonitrile. The flow rate was 1 mL/min and the detector was set at 214 nm.

RESULTS AND DISCUSSION

Photoassisted Oxidation

Estimate of the Theoretical Solar UV Flux-Limited Rate of Crude Oil Dissolution by Photoassisted Oxidation

To estimate the theoretical photon flux-limited rate of dissolution of oil on water through its catalytic photoassisted oxidation, we assume that one carbon in six needs to be oxidized. We justify this assumption by the increase in solubility when, for example, hexane is oxidized to hexanol or hexanone or dodecane to a diol. Thus, insertion of 1-g atom of oxygen per about 80 g of organic material is assumed to be necessary for its solubilization.

The cross-sectional area of 1 ton of ideally spread 50-µm diameter hollow glass microbeads of 0.4 g cm^{-3} density is 7.5×10^4 m^2. The UV (sub-400 nm) flux of sunlight filtered by one atmospheric mass (AM1) is 35 W m^{-2}. Thus, the UV flux on 1 ton of fully dispersed floating beads is 2.6×10^6 W.

With the UV part of the solar flux having a mean photon energy of about 3.4 eV, the UV solar flux is 8 Einsteins sec^{-1} ton^{-1}. Thus, at the photon flux-controlled theoretical limit and with current doubling, 640 g of oil are solubilized each second by 1 ton of beads. At this rate, the beads solubilize a mass of oil equaling their own in 26 min. However, because about 40% of the UV light is reflected when the beads are optimally TiO$_2$ coated,[23] the theoretical solubilization time cannot be faster than 43 min.

Similar considerations show that beads of 100 µm average diameter and 0.7 g cm^{-3} density could theoretically solubilize a mass of oil equaling their own in 3 hr at AM1 irradiance. We find, in practice, that slow O$_2$ or soluble photoproduct mass transport and UV absorption by crude oil films coating the less than ideally buoyant beads reduce this rate by two orders of magnitude. Nevertheless, even at 1% of the theoretical solar UV photon flux-limited rate, the beads solubilize an oil mass equaling their own in a period between 2 weeks and 2 months. We note that in microspheres the angle of the sun with respect to zenith affects the rate only because sunlight is filtered by a thicker atmosphere in the early morning, in the late afternoon, in the winter, or farther from the equator and not because of the usual angular dependence of the solar irradiance.[23] Thus, the rate of oil cleanup with spherical microbeads is not greatly affected by the time of the day, by the season, or by the latitude, in contrast with the power output of a flat stationary solar cell.

Photoassisted Oxidation of 3-Octanol on n-TiO$_2$-Coated Microspheres

In oxygen uptake experiments with beads made by the high-temperature process, 0.034 equivalents of oxygen, or 0.017 mol of O$_2$, were consumed per

Einstein under stagnant conditions. The uptake increased to 0.062 equivalents or 0.031 mol of O_2 per Einstein when "waves" were produced by circular movement ("swirling") of the photochemical reactor in an orbit shaker at 120 rpm. The increase in efficiency upon bubbling O_2 or shaking suggests that photoassisted oxidation under stagnant conditions can be mass transport limited at approximately 46 W m^{-2} irradiance.

Photoassisted Oxidation of Crude Oil

The measurements were performed on crude oil films in 0.5 M NaCl in water at approximately 46 W m^{-2} irradiance by 300 – 400 nm (365 nm maximum) light. In O_2 uptake measurements, the efficiency increased upon increasing the bead to oil ratio and upon agitation by wave-motion-imitating circular swirling of the fluid in an orbit shaker operated at 120 rpm.

Under stagnant conditions, a high bead to oil mass ratio increased the rate of photooxidation both because more oxygen diffused to the surface of bead-oil aggregates and because the thickness of the UV-absorbing oil films covering the top layer of beads was reduced. At a bead to oil mass ratio of 6.5:1, 0.010 mol of O_2 were consumed per Einstein in experiments with beads made by the high-temperature process. At a 2.3:1 mass ratio, 0.005 mol were consumed; and at 1.2:1 mass ratio, 0.0008 mol were consumed. With "waves," O_2 consumption increased at 2.3:1 bead to oil mass ratio to 0.010 mol O_2 per Einstein. At the measured rate of O_2 consumption, 2.3 tons of the 100-μm 0.7-g cm^{-3} density cenospheres are projected to photosolubilize 1 ton of oil under AM1 sunlight in about 150 hr — that is, in about 2 weeks.

Figure 1 shows the results of an experiment where the nonvolatile residue from 1 mL of crude oil was photocatalytically oxidized on 3 g of beads. In this experiment, the volatile fraction was evaporated by heating to 100°C for 15 hr. The beads used were made by the low-temperature process. After 240 hr of exposure, 70% of the organic matter was lost and the initially tar-rich beads turned into free-flowing, nearly colorless "sand."

The exposure time dependences of the concentrations of phenols produced by natural photooxidation of crude oil and by photocatalytic oxidation are compared in Figure 2. The concentrations of the phenolic intermediates increased initially in both oxidation processes; in the catalytic process, the concentration reached a maximum, then dropped to a lower level. The drop is attributed to continued photoassisted oxidation of the TiO_2-adsorbed compounds.

Upon mixing the beads with oil a semisolid oil-bead mass forms within less than 1 hr. With cenospheres, it is possible to strip from the surface of water an equal mass of oil. This mass can be mechanically harvested or ignited. The hollow ceramic microbeads thermally insulate the adsorbed oil from convective water cooling. As a result, the ignition temperature is easily reached. Thus, thin oil films that otherwise cannot be ignited can be burned after uptake in the oil-bead mass. The oil-bead aggregates break up upon approximately 1 day exposure, i.e., the

Figure 1.   A 240-hr photoassisted oxidation of a nonvolatile crude oil fraction converts the initially tar-soaked oil-bead mass into free-flowing and nearly colorless "sand."

photocatalytic beads spread on the surface, carrying the adsorbed oil and accelerating its photooxidation.

Solar Collection Efficiency and Optics of the Catalytic Microbeads

Because crude oil, even at micrometer thickness, absorbs most of the UV flux, only the part of the beads that is above the oil film collects useful light. When 75% of the bead surface is above the oil film, the theoretically feasible solar collection efficiency of UV photons reaches approximately 60% when 60% of the bead's surface is covered with TiO_2 particles. Greater TiO_2 coverage provides little increase in efficiency because the index of refraction of TiO_2 is high and more light is reflected. The efficiency of solar UV collection does not change rapidly with the index of refraction of the oil; with the index of refraction of the glass, i.e., its aluminum to silicon ratio; or with the thickness of the wall of the microbead. Thus, industrial grade heterogeneous microbeads can be used as long as the glass does not absorb in the UV. Significantly, the "cosine effect" that decreases the

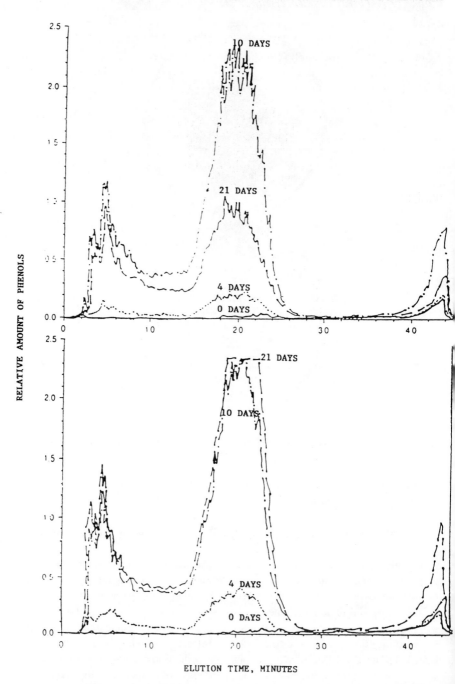

Figure 2. Change in concentration of water-soluble phenolic compounds during photoassisted oxidation of Texaco Bruce #1 crude oil. Bottom: 2 mL of crude oil, no beads: —— 0 days, ····· 4 days, – · – 10 days, and – – 21 days. Top: 2 g of beads coated by high-temperature process with 2 mL of the crude oil: —— 0 days, ····· 4 days, – · –10 days, and – –21 days.

solar irradiance on a flat terrestrial surface with increasing angle relative to the zenith is absent for the microbeads whose surface is spherical. Thus, the irradiance depends on the time of day and day of year primarily because the sunlight is filtered by different atmospheric thicknesses.[23]

ACKNOWLEDGMENTS

This work was supported by the Division of Advanced Energy Projects, Office of Basic Energy Sciences, U.S. Department of Energy, under Grant DE-FG05–90ER12101.

REFERENCES

1. Izumi, I., Dunn, W. W., Wilbourn, K. O., Fan, F. R., and Bard, A. J., Heterogeneous photocatalytic oxidation of hydrocarbons on platinized titanium dioxide powders, *J. Phys. Chem.*, 84, 3207, 1980.
2. Hashimoto, K., Kawai, T., and Sakata, T., Photocatalytic reactions of hydrocarbons and fossil fuels with water, *J. Phys. Chem.*, 88, 4083, 1984.
3. Ollis, D. F. and Turchi, C., Heterogeneous photocatalysis for water purification: contaminant mineralization kinetics and elementary reactor analysis, *Environ. Prog.*, 9, 229, 1990.
4. Turchi, C. S. and Ollis, D. F., Photocatalytic degradation of organic water contaminants: mechanisms involving hydroxyl radical attack, *J. Catal.*, 122, 178, 1990.
5. Matthews, R. W., Carbon dioxide formation from organic solutes in aqueous suspensions of ultraviolet-irradiated titanium dioxide. Effect of solute concentration, *Aust. J. Chem.*, 40, 667, 1987.
6. Matthews, R. W., Kinetics and photocatalytic oxidation of organic solutes over titanium dioxide, *J. Catal.*, 111, 264, 1988.
7. Matthews, R. W., Purification of water with near-UV illuminated suspensions of titanium dioxide, *Water Res.*, 24, 653, 1990.
8. Kormann, C., Bahnemann, D. W., and Hoffmann, M. R., Photolysis of chloroform and other organic molecules in aqueous TiO_2 suspensions, *Environ. Sci. Technol.*, 25, 494, 1991.
9. Al-Ekabi, H. and Serpone, N., Kinetic studies in heterogeneous photocatalysis. 1. Photocatalytic degradation of chlorinated phenols in aerated aqueous solutions over TiO_2 supported on a glass matrix, *J. Phys. Chem.*, 92, 5726, 1988.
10. Pelizzetti, E., Borgarello, M., Minera, C., Pramauro, E., Borgarello, E., and Serpone, N., Photocatalytic degradation of polychlorinated dioxins and polychlorinated biphenyls in aqueous suspensions of semiconductors irradiated with simulated solar light, *Chemosphere*, 17, 499, 1988.
11. Barbeni, M., Morello, M., Pramauro, E., Pelizzetti, E., Vincenti, M., Borgarello, E., and Serpone, N., Sunlight photodegradation of 2,4,5-trichlorophenoxy-acetic acid and 2,4,5,Trichlorophenol on TiO_2. Identification of intermediates and degradation pathway, *Chemosphere*, 16, 1165, 1987.
12. Barbeni, M., Pramauro, E., Pelizzetti, E., Borgarello, E., and Serpone, N., Photodegradation of pentachlorophenol catalyzed by semiconductor particles, *Chemosphere*, 14, 195, 1985.
13. Barbeni, M., Promauro, E., Pelizzetti, E., Borgarello, E., Graetzel, M., and Serpone, N., Photodegradation of 4-chlorophenol catalyzed by titanium dioxide particles, *Nouv. J. Chim.*, 8, 547, 1984.
14. Carey, J. H. and Oliver, B.G., Photochemical treatment of waste water by ultraviolet irradiation of semiconductors, *Water Pollut. Res. J. Canada*, 15, 157, 1980.

15. Fujihira, M., Satoh, Y., and Osa, T., Heterogeneous photocatalytic reactions on semiconductor materials. Part II. Photoelectrochemistry at semiconductor TiO$_2$/insulating aromatic hydrocarbon liquid interface, *J. Electroanal. Chem. Interfacial Electrochem.*, 126, 277, 1981.

16. Jackson, N.B., Wang, C.M., Luo, Z., Schwitzgebel, J., Ekerdt, J. G., Brock, J. R., and Heller, A., Attachment of TiO$_2$ powders to hollow glass microbeads: activity of the TiO$_2$-coated beads in the photoassisted oxidation of ethanol to acetaldehyde, *J. Electrochem. Soc.*, 138, 3660, 1991.

17. Jaeger, C. D. and Bard, A. J., Spin trapping and electron spin resonance detection of radical intermediates in the photodecomposition of water at TiO$_2$ particulate systems, *J. Phys. Chem.*, 83, 3146, 1979.

18. Gerischer, H. and Heller, A., The role of oxygen in photooxidation of organic molecules on semiconductor particles, *J. Phys. Chem.*, 95, 5261, 1991.

19. Gerischer, H. and Heller, A., Photocatalytic oxidation of organic molecules at TiO$_2$ particles by sunlight in aerated water, *J. Electrochem. Soc.*, 139, 113, 1992.

20. Kraeutler, B. and Bard, A. J., Heterogeneous photocatalytic preparation of supported catalysts, *J. Am. Chem. Soc.*, 100, 4317, 1978.

21. Wang, C. M., Heller A., and Gerischer, H., Palladium catalysis of O$_2$ reduction by electrons accumulated on TiO$_2$ particles during photoassisted oxidation of organic compounds, *J. Am. Chem. Soc.*, 114, 5230, 1992.

22. Heller, A., Degani, Y., Johnson, D. W., Jr., and Gallagher, P. K., Controlled suppression or enhancement of the photoactivity of titanium dioxide (rutile) pigment, *J. Phys. Chem.*, 91, 5987, 1987.

23. Rosenberg, I., Brock, J. R., and Heller, A., Collection optics of TiO$_2$ photocatalyst on hollow glass microbeads floating on oil slicks, *J. Phys. Chem.*, 96, 3423, 1992.

24. Heller, A. and Brock, J. R., Materials and Methods for Photocatalyzing Oxidation of Organic Compounds on Water, U.S. Patent No. 4,997,576, March 5, 1991.

CHAPTER 30

Photosensitization of Semiconductor Colloids by Humic Substances

K. Vinodgopal and Prashant V. Kamat

INTRODUCTION

The photocatalyzed degradation of organic environmental pollutants in the presence of large bandgap semiconductors such as TiO_2 and ZnO has become a subject of increasing study over the last decade.[1,2] An important way to extend the response of the semiconductor is by photosensitization. Photosensitization of a stable large bandgap semiconductor is an interesting and useful phenomenon, which extends the semiconductor's absorptive range and thus enables photoelectrochemical reactions under visible light irradiation.[3]

An obvious choice as photosensitizers would be humic substances (HS). The importance of HS lies in their ability to initiate the photochemical transformation of organic compounds in natural water and their eventual degradation.[4] A considerable volume of literature exists on the photochemistry of HS, although knowledge of the primary photochemical processes is limited.[5-7] Laser flash photolysis studies have indicated the formation of three transients during photolysis, viz., the aqueous hydrated electron and transients with radical and triplet properties, respectively. The hydrated electron is believed to result from photoejection of an electron from the excited state of HS. The electron can now back react to form a neutral HS molecule, or electron transfer is now possible to a suitable electron acceptor.

For the first time, we have obtained evidence for a direct charge-transfer interaction between semiconductors such as ZnO or TiO_2 and HS (Suwanee River reference humic and fulvic acids, SHA and SFA) by using the fluorescence emission of the latter as a probe. Our preliminary results that describe the charge-injection mechanism from HS to semiconductor are presented here.

EXPERIMENTAL

Reference Suwanee River fulvic acid (SFA) was obtained from the International Humic Substances Society (Colorado School of Mines, Golden, CO). A colloidal suspension of 0.02 M ZnO in ethanol was prepared by the method described by Spanhel and Anderson[8] with stoichiometric addition of LiOH to the organometallic zinc precursor solution. Since the colloidal ZnO tends to precipitate in aqueous solution, solutions of SFA were prepared in ethanol in the following manner. A concentrated solution of SFA was prepared in doubly distilled water; 50 μL of this stock solution was then diluted with 3 mL of ethanol such that the optical density of the resulting solution was about 0.2 (pathlength 1 cm) at 400 nm. All solutions were deoxygenated by bubbling nitrogen gas through them for a period of 5–15 min.

For the TiO$_2$ studies, the colloids were prepared in aqueous alkaline solution by addition of 200 μL of a 1-M solution of TiCl$_4$ to 25 mL of a 0.02-M NaOH solution. The SFA solutions were prepared in phosphate buffer at pH 12 such that the optical densities were similar to the solutions used in the ZnO case. Details of the experimental set-up, including the fluorescence emission and laser flash photolysis studies, are given elsewhere.[9]

RESULTS AND DISCUSSION

The fluorescence quantum yield of SFA decreased upon successive addition of colloidal ZnO. Data for the SFA-ZnO mixture is shown in Figure 1. The marked decrease in the fluorescence yield was due to the quenching of the excited singlet states of the humic material by the semiconductor colloid. The observed quenching of the excited singlet states of SFA is attributed to electron transfer to the semiconductor metal oxide colloid.

The quenching is due to the strong adsorption of the sensitizer on the semiconductor particles. The participation of ZnO-TiO$_2$ colloid in the quenching of HS emission can be analyzed by considering an equilibrium between adsorbed and unadsorbed molecules of the sensitizer (HS) with an apparent association constant K$_{app}$:

$$\text{Semiconductor colloid} + \text{HS} \rightarrow [\text{Semiconductor} \cdots \text{HS}] \qquad (1)$$

The fluorescence quenching data can be analyzed to yield values for the apparent association constants, as well as the fluorescence quantum yield, of the

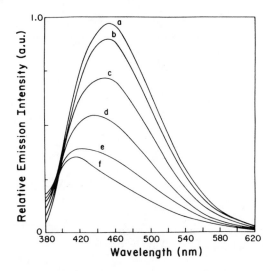

Figure 1. Fluorescence emission spectrum of SFA in ethanol at various concentrations of colloidal ZnO: (a) 0 M, (b) 0.13 mM, (c) 0.16 mM, (d) 0.19 mM, (e) 0.32 mM, and (f) 0.64 mM. Excitation wavelength was 370 nm. (All spectra were corrected for background.) (From Vinodgopal, K. and Kamat, P. V., *Environ. Sci. Technol.*, 26, 1963, 1992. With permission.)

associated (HS-semiconductor) complex. Such analysis of the SFA-ZnO data gives a K_{app} value of 12,000, indicating a strong association between it and the zinc oxide colloids. The fluorescence quantum yield of the associated complex is less than 10^{-3}, suggesting that the quenching of excited fulvic acid by ZnO is highly efficient (71%). As indicated above, such a quenching is indicative of the charge injection into the semiconductor (Equation 3).

$$SFA \rightarrow SFA*$$ (2)

$$SFA * ZnO \rightarrow SFA^{+\cdot} + ZnO(e)$$ (3)

If indeed such a process should occur on the ZnO surface, we should be able to characterize the products by laser flash photolysis measurements.

Laser Flash Photolysis

The transient absorption spectra recorded following the excitation of SFA in the absence of ZnO colloids is shown in Plot a in Figure 2. As shown earlier,[5] the spectrum exhibits absorption bands in the region 400–500 nm and at wavelengths greater than 600 nm. The low-wavelength transient is presumably due to the cation radical SFA$^{+\cdot}$, while the broad absorption at longer wavelengths is due to the solvated electron that is formed as a result of photoionization of SFA:

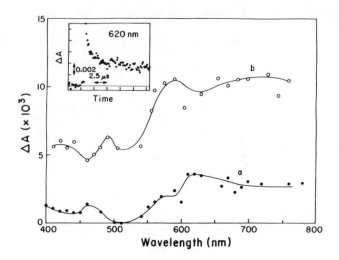

Figure 2. Transient absorption spectrum recorded immediately after 532-nm laser pulse excitation of (a) SFA; (b) with 2 mM ZnO. Inset shows the absorption-time profile at 620 nm in spectrum b(...). (From Vinodgopal, K. and Kamat, P. V., *Environ. Sci. Technol.*, 26, 1963, 1992. With permission.)

$$SFA \rightarrow SFA^{+\cdot} + e^-_{solv} \qquad (4)$$

The transient spectrum recorded in presence of ZnO (Plot b in Figure 2) colloids exhibits similar features, but the yield of the product is considerably higher. In this case, Equation 3 dominates so that the injected electron gets trapped at the ZnO surface. It has been shown earlier[10] that trapped electrons in ZnO colloids also exhibit broad absorption in the red region. Such a trapping of electrons on the semiconductor surface reduces the recombination between the injected charge and the cation radical. This long lived charge at the semiconductor surface is responsible for controlling the photocatalytic properties of the semiconductor. Further support for the charge injection from excited SFA into ZnO colloids was obtained by varying the concentration of ZnO colloid while monitoring the maximum absorption at 620 nm, which was taken as a measure of the net electron-transfer yield in the sensitization of the ZnO colloids. The dependence of ΔA on the ZnO concentration is shown in Figure 3. At low-ZnO concentrations, an increase in the electron-transfer yield was observed as an increasing amount of excited SFA interacted with ZnO colloids. At higher concentrations, the yield reached a plateau, since the charge injection process is limited by the availability of SFA adsorbed onto the ZnO surface. The SFA-TiO$_2$ system gives similar results, and the dependence of ΔA on the TiO$_2$ concentration is shown in Figure 4. Additional mechanistic details will be presented elsewhere.[11]

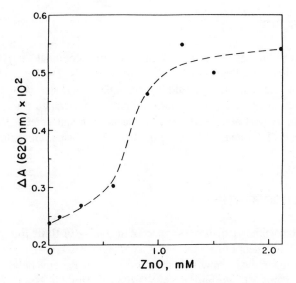

Figure 3. Dependence of the net electron-transfer yield on ZnO concentration. The electron-transfer yield was monitored by the maximum transient absorbance (ΔA) at 620 nm following the 532-nm laser pulse excitation of SFA in ethanol ($\cdots$ –). (From Vinodgopal, K. and Kamat, P. V., *Environ. Sci. Technol.*, 26, 1963, 1992. With permission.)

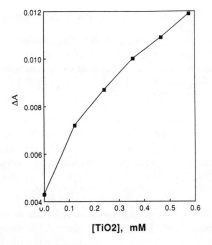

Figure 4. Dependence of the net electron-transfer yield on TiO_2 concentration. The electron-transfer yield was monitored by the maximum transient absorbance (ΔA) at 620 nm following the 532-nm laser pulse excitation of SFA in pH 12 phosphate buffer.

CONCLUSIONS

We have shown that SFA can be used to sensitize colloidal TiO_2 and ZnO by charge injection. The key question then that arises is whether such a charge trapped on the semiconductor would be accessible for charge transfer to another substrate, such as a potential environmental pollutant. Detailed studies to answer this question, as well as to elucidate the mechanism and kinetic details of the sensitization of TiO_2 and ZnO by humic and fulvic acids, are currently being carried out.

ACKNOWLEDGMENTS

P. V. Kamat acknowledges the support of the Office of Basic Energy Sciences of the U.S. Department of Energy, and K. Vinodgopal acknowledges the support of Indiana University Northwest through a Summer Faculty Fellowship and a Grant-In-Aid. This is Contribution No. NDRL-3450 from the Notre Dame Radiation Laboratory, University of Notre Dame, Notre Dame, IN.

REFERENCES

1. Pelizzetti, E. and Schiavello, M., Eds., *Photochemical Conversion and Storage of Solar Energy*, Kluwer, Dordrecht, 1991.
2. Ollis, D. F., Pelizzetti, E., and Serpone, N., Photocatalyzed destruction of water contaminants, *Environ. Sci. Technol.*, 25, 1522, 1991.
3. Kamat, P. V., Chauvet, J., and Fessenden, R. W., Photosensitization of a TiO_2 semiconductor with a chlorophyll analog, *J. Phys. Chem.*, 90, 1389, 1986.
4. Zepp, R. G., *Humic Substances and their Role in the Environment*, Frimmel, F. H. and Christman, R. F., Eds., John Wiley & Sons, New York, 1988, 193–214.
5. Fischer, A. M., Winterle, J. S., and Mill, T., in *Photochemistry of Environmental Aquatic Systems*, Zika, R. G. and Cooper, W. J., Eds., ACS Symposium Ser. No. 327, American Chemical Society, Washington, D.C., 1987, 141–156.
6. Power, J. F., Sharma, D. K., Langford, C., Bonneau, R., and Joussot-Dubien, J. in *Photochemistry of Environmental Aquatic Systems*, Zika, R. G. and Cooper, W. J., Eds., ACS Symposium Ser. No. 327, American Chemical Society, Washington, D.C., 1987, 157–173.
7. Zepp, R. G., Braun, A. M., Hoigne, J., and Leenherr, J. A., Photoproduction of hydrated electrons from natural organic solutes in aquatic environments, *Environ. Sci. Technol.*, 21, 485, 1987.
8. Spanhel, L. and Anderson, M. A., Semiconductor clusters in the sol-gel process: quantized aggregation, gelation and crystal growth in concentrated ZnO colloids, *J. Am. Chem. Soc.*, 113, 2826, 1991.
9. Das, P. K., Encinas, M. V., Small, R. D., Jr., and Scaiano, J. C., Photoenolation of O-alkyl-substituted carbonyl compounds. Use of electron transfer processes to characterize transient intermediates, *J. Am. Chem. Soc.*, 101, 6965, 1979.
10. Kamat, P. V. and Patrick, B., Photophysics and photochemistry of quantized ZnO colloids, *J. Phys. Chem.*, 96, 6829, 1992.
11. Vinodgopal, K. and Kamat, P. V., Environmental photochemistry on surfaces. Charge injection from excited fulvic acid into semiconductor colloids, *Environ. Sci. Technol.*, 26, 1963, 1992.

CHAPTER 31

The Role of Support Material in the Photodegradation of Colored Organic Compounds

Prashant V. Kamat and K. Vinodgopal

INTRODUCTION

In recent years, semiconductor particulate systems have been employed to degrade organic pollutants (e.g., degradation of chlorophenols on TiO_2 surface).[1] In such systems, the semiconductor particles are excited with ultraviolet (UV) or visible light to induce charge separation. The photogenerated holes would then oxidize the pollutant. An alternative approach is to excite the adsorbed organic material and then inject charge from the excited organics into the semiconductor particle.[2] The oxidized form of the organic material can then undergo further degradation. This process, which is commonly referred to as photosensitization, is extensively used in photoelectrochemistry and imaging science. The principle of such a process is described in Figure 1. In this scheme, the colored substrate itself acts as a sensitizer and initiates its own degradation. The advantage of this process is the utilization of visible light for degrading colored pollutants.

Further, the insulating materials such as silica, alumina, and clays are naturally abundant and can provide ordered, two-dimensional environments for effecting and controlling photochemical processes more efficiently than can be attained in homogeneous solutions. However, little effort has been made so far to study the influence of the support material on the photochemical degradation of the adsorbed substrate.

The hazardous nature of polychlorinated and polybrominated dibenzofurans has given rise to several methods to degrade them to less environmentally harmful

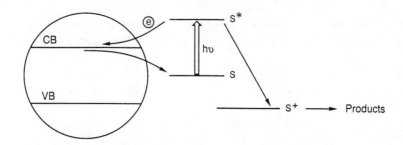

Figure 1. Photosensitization of a semiconductor particle with an organic substrate. The conduction and valence bands of the semiconductor particle are indicated by CB and VB, respectively. The electron-donating energy levels (oxidation potential) of the ground and excited state of the sensitizer are indicated by S and S*. Electron injection from excited sensitizer results in the oxidation of the substrate (S+). (Vinodgopal, K. and Kamat, P. V., *J. Phys. Chem.*, 96, 5053, 1992. With permission.)

products.[3] In the present study, we have used 1,3-diphenylisobenzofuran (DPBF) as a model compound, since it strongly absorbs in the visible and is photochemically reactive in solutions[4-6] and on surfaces.[7,8] Although DPBF itself is not a major chemical pollutant, the study of furan photochemistry on oxide surfaces can yield valuable information regarding the degradation of other colored pollutants with visible light. The mechanistic and kinetic details of DPBF photodegradation on Al_2O_3, TiO_2, and ZnO surfaces are described here.

EXPERIMENTAL SECTION

Sample Preparation

1,3-Diphenylisobenzofuran (DPBF) was obtained from Aldrich and was used as supplied. TiO_2 (particle diameter 30 nm and surface area 50 m^2/g) and Al_2O_3 (particle diameter 20 nm and surface area 100 m^2/g) was obtained from Degussa Corporation. ZnO powder (surface area 1 m^2/g) was obtained from Johnson Mathey Chemicals, Ltd. (Puratronic grade). 1,3-Diphenylisobenzofuran was coated onto the surface of oxides by dispersing the individual oxide particles in acetonitrile and adding a known amount of DPBF solution. After stirring the suspension in dark for 2–4 hr, the solvent was evaporated off. The amount of DPBF required to achieve the necessary coverage was determined from the surface area of the particle and the area occupied by a single DPBF molecule (assumed to be 1 nm^2).

Optical Measurements

The diffuse reflectance absorption spectra of DPBF-coated oxide samples were recorded with a Cary 219 spectrophotometer with a diffuse reflectance attachment (Harrick Scientific). Corrected emission and excitation spectra of the solid samples were measured with an SLM photon counting spectrofluorimeter in a front face configuration. The measurements on evacuated samples were carried out in a vacuum-tight $10 \times 10 \times 40$ mm^3 rectangular quartz cell. Steady-state photolysis was carried out with a collimated light beam from a 250-W halogen lamp.

Diffuse Reflectance Laser Flash Photolysis Experiments

Time-resolved diffuse reflectance laser flash photolysis experiments were carried out with the set-up described earlier.[2] The 532-nm laser pulse (10 mJ, pulse width 6 ns) from a Quanta Ray DCR-1 Nd:YAG laser system was used for the excitation of the sample. A 1000-W xenon lamp was used as the monitoring source. The diffusely reflected monitoring light from the sample was collected and focused onto a monochromator that was fitted to a photomultiplier tube, and the photomultiplier output was input to a Tektronix 7912A digitizer. Before triggering each laser pulse, the cell was shaken to expose a fresh surface for excitation. The principle of diffuse reflectance laser flash photolysis is described by Kessler et al.[9]

RESULTS AND DISCUSSION

Absorption and Emission Characteristics

The diffuse reflectance spectrum of DPBF on a TiO$_2$ surface is shown in Figure 2. DPBF-TiO$_2$ sample absorbs strongly in the visible with a broad maximum at 425 nm. Compared to the solution spectrum, the absorption band on the oxide surface is broad, and this broadening of the spectrum is attributed to the interaction between the support and DPBF.

The emission spectra of degassed samples of DPBF on Al$_2$O$_3$, TiO$_2$, and ZnO are shown in Figure 3. All these samples exhibited emission maxima around 500 nm. Comparison of the relative quantum yields on these particles indicates that the fluorescence yields are significantly lower on a semiconductor (TiO$_2$ and ZnO) surface when compared to the fluorescence yield on an insulator surface (alumina).

Steady-State Photolysis

When air-equilibrated samples were irradiated with visible light, DPBF readily underwent degradation over an oxide surface. The photodegradation occurred in both air-equilibrated and degassed samples over TiO$_2$ and ZnO surfaces (see, for example, Spectrum b in Figure 2). However, on the Al$_2$O$_3$ surface, photodegradation was seen only in air-equilibrated samples. The products formed during photolysis

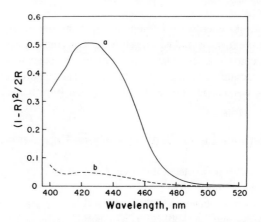

Figure 2. Diffuse reflectance spectra of degassed DPBF coated on TiO_2 particles (0.08 mmol/g support). The spectrum was recorded **(a)** before and **(b)** after steady-state photolysis. (Vinodgopal, K. and Kamat, P. V., *J. Phys. Chem.*, 96, 5053, 1992. With permission.)

exhibited absorption in the 420- to 460-nm region. This weak absorption persisted even when the photolysis was extended for longer duration. The interaction of excited DPBF with ambient oxygen and/or the support material is responsible for its photodegradation.

Furans such as DPBF have been widely used as a probe to detect singlet oxygen in a triplet-triplet sensitization reaction, since they get degraded quantitatively by

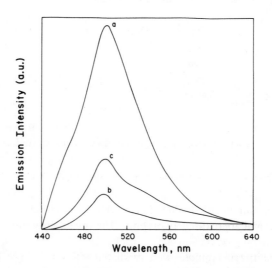

Figure 3. Emission spectra of degassed DPBF coated on **(a)** Al_2O_3 (0.16 mmol/g), **(b)** TiO_2 (0.08 mmol/g) and **(c)** ZnO (0.08 mmol/g). Excitation wavelength was at 380 nm. (Vinodgopal, K. and Kamat, P. V., *J. Phys. Chem.*, 96, 5053, 1992. With permission.)

reacting with singlet oxygen.[10] Direct excitation of DPBF in the presence of oxygen also produces singlet oxygen as both the singlet and triplet excited states of DPBF are capable of transferring energy to O_2.[5,6] The singlet oxygen, thus produced by excitation of DPBF on oxide surfaces, quickly reacts with ground state molecules of DPBF and initiates the degradation. The reaction scheme can be summarized as follows:

$$DPBF + h\nu \rightarrow {}^1DPBF^* \rightarrow {}^3DPBF^* \tag{1}$$

$$^1DPBF^* \left(or \ ^3DPBF^* \right) + O_2 \rightarrow DPBF + {}^1O_2^* \tag{2}$$

$$^1O_2^* + DPBF \rightarrow Products \tag{3}$$

The reaction between $^1O_2^*$ and DPBF (Equation 3) was the major pathway for the degradation of DPBF in air-equilibrated samples. As indicated earlier,[5,6] the primary product of Equation 3 is a peroxy compound, which then can undergo further degradation to yield stable products. In our earlier study of the DPBF-Al_2O_3 system, we have shown that the interaction of DPBF with surface-bound oxygen occurs at a faster rate than the surrounding atmospheric oxygen.[7] However, in the absence of oxygen, a different reaction mechanism initiated the photodegradation on the semiconductor surface, the details of which are discussed below.

Mechanism of Photodegradation on TiO$_2$ and ZnO Surface

It was evident from the emission spectra of DPBF in Figure 3 that the singlet excited state of DPBF is readily quenched on the semiconductor (TiO$_2$ and ZnO) surface. As shown earlier,[2] such a quenching behavior represents charge injection from the singlet excited DPBF into the conduction band of the semiconductor (Equation 4).

$$DPBF^* \left(S_1 \right) + TiO_2 \rightarrow DPBF^{+\cdot} + TiO_2 \left(e \right) \tag{4}$$

The time-resolved transient absorption spectra recorded after 532-nm laser pulse excitation of degassed DPBF-TiO$_2$ sample is shown in Figure 4. The transient absorption spectrum recorded immediately after the laser pulse excitation shows maxima at 510 and approximately 600 nm. The broad absorption in the red and near-infrared (IR) region is characteristic of trapped electrons at the TiO$_2$ surface. The transient absorption in the 510-nm region is attributed to the formation of a cation radical, DPBF$^{+\cdot}$, at the TiO$_2$ surface (Equation 4). The absorption characteristics of DPBF$^{+\cdot}$ have been independently confirmed by pulse radiolysis experiments.[8] The cation radical decayed quickly ($\tau = 15$ μsec) as it underwent irreversible chemical changes. This was evident from the bleaching observed at 450 nm

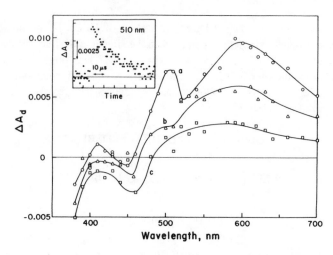

Figure 4.   Time-resolved transient absorption spectra recorded following 532-nm laser pulse excitation of degassed DPBF on TiO_2 (0.08 mmol/g) at time intervals: (a) 2 sec, (b) 25 sec, and (c) 56 sec. The inset shows decay trace at 430 nm. (Vinodgopal, K. and Kamat, P. V., *J. Phys. Chem.*, 96, 5053, 1992. With permission.)

in the transient spectrum recorded at longer times scales. As indicated in the steady-state photolysis experiments, such a bleaching represents irreversible degradation of DPBF.

The results presented in Figure 5 highlight the importance of the interaction between semiconductor and DPBF in promoting the photodegradation. The surface coverage of DPBF on TiO_2 was varied in these experiments by increasing the DPBF concentration on these particles. The half-life of degradation at quarter and one monolayer coverages of DPBF were 0.6 and 1.0 min, respectively. At these submonolayer and monolayer coverages, most of the DPBF molecules are in direct contact with the TiO_2 surface, and when subjected to visible light irradiation they can efficiently participate in the charge-injection process. As a result of this process, complete degradation of DPBF is seen in a matter of a few (5–10) minutes of irradiation. On the contrary, for the samples with four times monolayer coverage, the half-life of DPBF degradation was more than 18 min. Only a partial degradation (~32%) is seen in this sample, since most of the DPBF molecules are not in direct contact with the TiO_2 surface. Thus, the dependence of DPBF degradation on the surface coverage highlights the catalytic role of the semiconductor, which directly participates in the photoinduced degradation of DPBF in the absence of oxygen. As indicated earlier, no such degradation could be observed on a degassed Al_2O_3 surface.

Although the charge injection may also occur in air-equilibrated samples, we cannot sort out the contribution of the two different mechanisms on the degradation

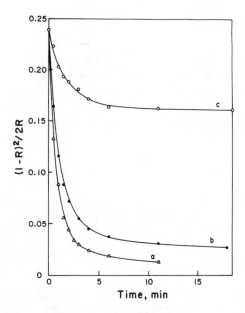

Figure 5. Normalized decay traces representing the degradation of degassed DPBF-TiO$_2$ samples under steady-state irradiation with visible light at different coverages. The DPBF coverage was (a) 0.25, (b) 1.0, and (c) 4.0 times monolayer. (Vinodgopal, K. and Kamat, P. V., *J. Phys. Chem.*, 96, 5053, 1992. With permission.)

of DPBF. Since we could not observe any transient absorption of DPBF$^{+\cdot}$ in air-equilibrated DPBF-TiO$_2$ samples, we consider the charge-injection mechanism to be a minor pathway in these experiments. We propose to analyze the surface photochemical products to resolve this ambiguity.

CONCLUSIONS

The photodegradation of a model furan compound, DPBF, has been carried out on the solid surfaces of metal oxides. In air-equilibrated samples, the photo-degradation is seen as a result of its reaction with singlet oxygen. However, in degassed samples, the photodegradation is seen only when the support material is a semiconductor (e.g., TiO$_2$ and ZnO). The direct participation of the semiconductor surface in the charge-transfer process with the excited DPBF was elucidated by the diffuse reflectance laser flash photolysis studies. The present study highlights the application of surface photochemical processes for degrading environmentally hazardous organic contaminants.

ACKNOWLEDGMENTS

K. Vinodgopal acknowledges the support of Indiana University Northwest through a Summer Faculty Fellowship and a Grant-in-Aid. P. V. Kamat acknowledges the support of the Office of Basic Energy Sciences of the U. S. Department of Energy. This is Contribution No. 3435 from the Notre Dame Radiation Laboratory, University of Notre Dame, Notre Dame, IN.

REFERENCES

1. Ollis, D. F., Pelizzetti, E., and Serpone, N., Photocatalyzed destruction of water contaminants, *Environ. Sci. Technol.*, 25, 1522, 1991, and references cited therein.
2. Gopidas, K. R. and Kamat, P. V., Photochemistry on surfaces. Influence of support material on the photochemistry of adsorbed dye, *J. Phys. Chem.*, 93, 6428, 1989.
3. Tiernan, T. O., Wagel, D. J., Vanness, G. F., Garrett, J. H., Solch, J. G., and Rogers, C., Dechlorination of PCDD and PCDF sorbed on activated carbon using KPEG reagent, *Chemosphere*, 19, 573, 1989.
4. Lala, D., Rabek, J. R., and Ranby, B., The effect of 1,3-diphenylisobenzofuran on the photo-oxidation of cis-polybutadiene, *Eur. Polym. J.*, 16, 735, 1980.
5. Stevens, B., Ors, J. A., and Christy, C. N., Photoperoxidation of unsaturated organic molecules. 19. 1,3 Diphenylisobenzofuran as sensitizer and inhibitor, *J. Phys. Chem.*, 85, 210, 1981.
6. Stevens, B. and Small, R. D., Jr., Singlet oxygen quenching by the triplet state of 1,3 diphenylisobenzofuran, *J. Photochem.*, 9, 233, 1978.
7. Vinodgopal, K. and Kamat, P. V., Photochemistry on surfaces. Photochemical behavior of 1,3-diphenylisobenzofuran over alumina, *J. Photochem. Photobiol. A: Chem.*, 63, 119, 1992.
8. Vinodgopal, K. and Kamat, P. V., Photochemistry on surfaces. Photodegradation of 1,3-diphenylisobenzofuran over metal oxide particles, *J. Phys. Chem.*, 96, 5053, 1992.
9. Kessler, R. W., Krabichler, S., Uhl, S., Oelkrug, D., Hagan, W. D., Hyslop, J., and Wilkinson, F., Transient decay following pulse excitation of diffuse scattering samples, *Opt. Acta*, 30, 1099, 1983.
10. Gorman, A. A. and Rodgers, M. A., Singlet molecular oxygen, *J. Chem. Soc. Rev.*, 10, 205, 1981.

CHAPTER 32

Hydrogen Peroxide in Heterogeneous Photocatalysis

Edward J. Wolfrum and David F. Ollis

INTRODUCTION

There has been much interest in recent years in so-called "Advanced Oxidation Processes" (AOPs) for the oxidation of organic pollutants in groundwater.[1,2] The formal definition of an AOP is a process in which the hydroxyl radical (OH·) is considered to be the primary oxidant species. Examples of AOPs include ozonation, O_3/H_2O_2, UV/H_2O_2, and heterogeneous photocatalysis.

Heterogeneous photocatalysis has shown considerable promise as a degradation technique for a broad range of organic pollutants, both in aqueous systems and in the gas phase. The preferred catalyst, TiO_2, is inexpensive, nonhazardous, and widely available. The oxidants O_2 and near-ultraviolet (UV) light are similarly regarded as nonhazardous. The near-UV activation allows solar-driven photocatalysis, and semiconductor oxides in the environment have been suggested to play a role in the environmental fate of some pollutants.

General descriptions of heterogeneous photocatalysis exist in several excellent review articles,[3-7] as well as a comprehensive review of the subject in a recent book.[8] A brief summary is presented here only for the sake of completeness. Absorption of a photon (hv) by a semiconductor photocatalyst results in the promotion of an electron e⁻ from the valence band to the conduction band, leaving behind a hole h⁺ in the valence band:

$$TiO_2 + h\nu \rightarrow h_{VB}^+ + e_{CB}^-$$ (1)

These electron-hole pairs, called "excitons," can recombine in the bulk catalyst or diffuse to the catalyst particle surface and react with species adsorbed there. In an oxygenated aqueous system, the photogenerated holes are assumed to react with either hydroxyl anions or organic species, while the electrons react with molecular oxygen.

$$h_{VB}^+ + OH^- \rightarrow OH^· \tag{2}$$

$$h_{VB}^+ + R \rightarrow R^{+·} \tag{3}$$

$$e_{CB}^- + O_2 \rightarrow O_2^{-·} \tag{4}$$

Thus, the h_{VB}^+ and $OH^·$ radicals are the primary oxidant species in photocatalysis, while molecular oxygen serves only to scavenge the photoproduced electrons in order to prevent electron-hole recombination (O_2^- may play a subsequent role as well). Whether the radical, $OH^·$ (Equation 2), or the hole, h_{VB}^+ (Equation 3), is the primary reactive species or even whether the two species can be distinguished definitively in an aqueous system is still in question. For example, photocatalytic oxidation of aromatics leads to the corresponding hydroxylated intermediates,[7,9,10] and several electron paramagnetic resonance (EPR) investigations of photocatalysis have found the spin-trap adduct of the $OH^·$ radical to be the primary species formed.[11-14] These results support the importance of Equation 2. In contrast, Draper and Fox[15] investigated the photocatalytic oxidation of a variety of aromatic compounds in aqueous systems using diffuse reflectance flash photolysis and found evidence of direct electron-transfer oxidation as indicated by the formation of cation radicals, not hydroxyl radical addition to the aromatic compounds; whether electron transfer was mediated by $OH^·$ or whether it occurred directly from the organic to a photogenerated hole on the catalyst particle surface was not known.

H_2O_2 IN PHOTOCATALYSIS

Proposed Mechanisms

Combinations of AOPs often produce significant increases in organic degradation (e.g., $UV/O_3/H_2O_2$ vs UV/O_3 or UV/H_2O_2[1]). Consequently, there has been interest in adding H_2O_2 to heterogeneous photocatalytic systems in order to enhance the degradation rate of organic pollutants. Most literature exploring the effect of added hydrogen peroxide (H_2O_2) in heterogeneous photocatalysis find that the presence of H_2O_2 increases the rate of degradation of organic species. However, there are some reports to the contrary. While there is a significant amount of rate data regarding this issue, the mechanisms by which the simultaneous presence of H_2O_2 and an illuminated photocatalyst produces the measured rate effect is still unclear.

As the primary oxidant species in AOPs are hydroxyl radicals, their production and subsequent consumption are of critical importance. In the UV-H_2O_2 process, photolysis of H_2O_2 produces two hydroxyl radicals per molecule of H_2O_2:

$$H_2O_2 + h\nu \rightarrow 2OH^{\cdot} \tag{5}$$

In heterogeneous photocatalysis, hydroxyl radicals are produced via Equation 2. Addition of H_2O_2 may provide a secondary, homogeneous source of $OH^{\cdot}$ radicals via Equation 5. Another possible source is the reduction of H_2O_2 by photogenerated electrons:

$$H_2O_2 + 2e_{CB}^- + 2H^+ \rightarrow 2H_2O \tag{6}$$

$$H_2O_2 + e_{CB}^- \rightarrow OH^{\cdot} + OH^- \tag{7}$$

Like O_2, H_2O_2 can scavenge photogenerated electrons and be reduced either to water (Equation 6) or to another $OH^{\cdot}$ (Equation 7). Both these pathways decrease the rate of electron-hole recombination, increasing the steady-state hole concentration and in turn increasing the rate of organic degradation. Small (micromolar) amounts of H_2O_2 are known to be produced during photocatalysis with CdS, ZnO, and TiO_2 photocatalysts in aqueous systems as an intermediate in the reduction of O_2 to H_2O.[16-18] Thus, the role of H_2O_2 as an electron scavenger seems well established, and H_2O_2 may simply serve as either an electron scavenger in the absence of O_2 or a more efficient one with O_2, which would account for its beneficial effect.

In contrast to these reductive mechanisms, oxidation of H_2O_2 by an $OH^{\cdot}$ radical or a photogenerated hole is possible.

$$H_2O_2 + OH^{\cdot}/h_{VB}^+ \rightarrow HO_2^{\cdot} + H_2O/H^+ \tag{8}$$

The perhydroxyl radical formed by Equation 8 is significantly less reactive with respect to organics than the $OH^{\cdot}$ radical.[19,20] By competing with organic pollutants for either $OH^{\cdot}$ radicals or h_{VB}^+, H_2O_2 can decrease the rate of organic degradation. Rate inhibition has been observed in AOPs by the presence of multiple reactants,[9,21] reaction intermediates,[22] and inorganic scavenger species[1,2,23] and has been either demonstrated or theorized to result from competition with the organic solute for $OH^{\cdot}$ radicals; where H_2O_2 addition has a deleterious effect on the rate of organic degradation, the cause may be competition for $OH^{\cdot}$ radicals that would otherwise oxidize the organic pollutant.

Thus, when H_2O_2 reacts with the photogenerated holes or $OH^{\cdot}$ radicals (the oxidative pathway), H_2O_2 addition should have a negative effect on degradation rates, while when H_2O_2 reacts with the photogenerated electrons (the reductive pathway) a positive rate effect of H_2O_2 addition is expected.

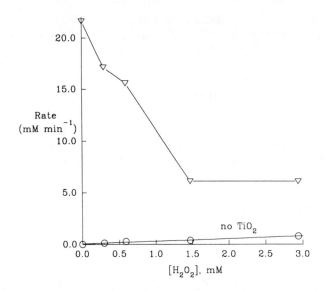

Figure 1. Rate of photocatalytic degradation of PCE as measured by chloride production
vs $[H_2O_2]_o$ in the presence ($\triangledown$) and absence ($\bigcirc$) of TiO_2. $[PCE]_o = 0.3$ mM,
$[TiO_2] = 1.0$ g/L. Homogeneous photolysis of H_2O_2 provides a negligible contribution to degradation rate (see text). Data from McGeever.[24] (Glaze, W. H., et al.,
Treatment of hazardous waste chemicals using advanced oxidation processes,
in Tenth International Ozone Symposium, Monaco, 1991. With permission.)

LITERATURE REVIEW

For H_2O_2 addition to illuminated aqueous slurries of rutile TiO_2 and perchloroethylene (PCE), McGeever[24] found that in all cases H_2O_2 addition decreased the
degradation rate of PCE, as measured by chloride ion production. Representative
data are shown in Figure 1. The top curve in this figure shows the rate of chloride
release vs the initial H_2O_2 concentration in the presence of catalyst, while the
bottom curve shows the same rate in the absence of catalyst, demonstrating that
the contribution to chloride-ion release by homogeneous photolysis is negligible;
thus, the interaction between TiO_2 and H_2O_2 may be purely chemical. McGeever
postulated that the cause of the rate decrease with added H_2O_2 was the formation
of surface peroxy-titanium complexes that were irreversibly blocking catalytic
sites.

Brown and Darwent[25] investigated the oxidation of methyl orange (MO) dye
by colloidal TiO_2 at pH 11.2 and found that addition of H_2O_2 decreased the rate
of dye oxidation. The rate of dye oxidation was independent of [MO] in the
absence of H_2O_2, but in the presence of H_2O_2 plots of inverse rate vs $[MO]^{-1}$ were
linear, with both the slope and the y-intercept increasing with increasing $[H_2O_2]$
(Figure 2). The authors suggested that these results indicated that H_2O_2 is reacting

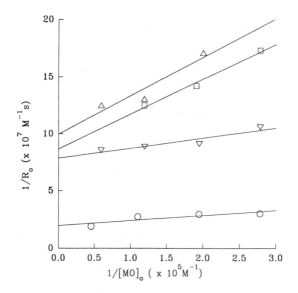

Figure 2. Inverse rate vs inverse $[MO]_o$ with $[H_2O_2]_o = 0.0$ (○), 45 (▽) , 84 (□), and 120 μM (△). Difference in slopes and intercepts of the lines shown suggest that MO and H_2O_2 are reacting with different oxidant species (see text). Data from Brown and Darwent.[25] (Peyton, G. R., *Significance and Treatment of Volatile Organic Compounds in Water Supplies*, Lewis Publishers, Chelsea, MI, 1991, 313. With permission.)

not with the OH· radical, but instead with a precursor to this species, assumed to be the photogenerated hole h_{CB}^+. However, the form of their derived kinetic equations require only that H_2O_2 and MO react with different species; their data do not rule out MO reaction with h_{CB}^+ and H_2O_2 reaction with OH·. Regardless, these researchers saw no rate increase upon H_2O_2 addition under any conditions, and a decrease was typical (Figure 2).

In a series of papers, Tanaka et al.,[26,27] Hisanaga et al.,[28] and Harada et al.[29] investigated the effect of H_2O_2 addition on the photocatalytic degradation of organochlorine compounds and organophosphorus insecticides using several different reactor assemblies with a variety of light sources and filters. Addition of low-H_2O_2 concentrations increased the degradation rate in all but two cases, while high concentrations reduced the rate; an optimum H_2O_2 concentration was demonstrated, as illustrated with data in Figure 3. However, at any H_2O_2 concentration, the degradation rate is greater with added H_2O_2 than without. The two exceptions noted above were for carbon tetrachloride and 1,1,1-trichloroethane. The researchers measured the photocatalyzed half-life, $t_{1/2}$, of these compounds with TiO$_2$ alone, platinized TiO$_2$ (Pt-TiO$_2$), and TiO$_2$ with H_2O_2. For 1,1,1-trichloroethane, $t_{1/2} = 125$ min for TiO$_2$ alone and 150 and 140 min for Pt-TiO$_2$ and TiO$_2$ with H_2O_2, respectively. Carbon tetrachloride was significantly less reactive than any other compound studied, with a $t_{1/2}$ value of 480 min under all three

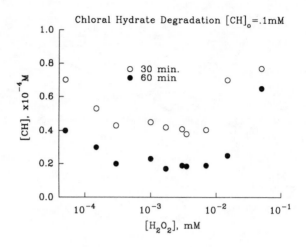

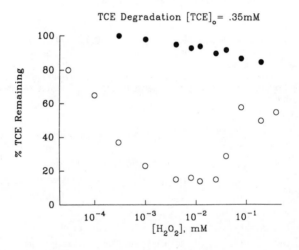

Figure 3.   (Top) Chloral hydrate concentration after 30 (○) and 60 (●) min illumination in the presence of 3.0 g/L TiO_2. (Bottom) Percent of TCE remaining after 10 min of illumination with (○) and without (●) 4.6 g/L TiO_2. In both cases, researchers found an optimum H_2O_2 concentration (see text). Data from Tanaka et al.,[26,27] Hisanaga et al.,[28] and Harada et al.[29] (Schiavello, M., Ed., *Photocatalysis and Environment: Fundamentals and Applications*, Kluwer, Dordrecht, 1987. With permission.)

reaction conditions. Thus, the chemical nature of the pollutant was shown to have an influence on the effect of added H_2O_2.

Sclafani et al.[30] investigated the effect of H_2O_2 on the photocatalytic degradation of phenol by both anatase and rutile TiO_2. They found that with anatase TiO_2, addition of either Ag^+ or H_2O_2 to a reaction mixture sparged with helium greatly increased the rate, but that Ag^+ or H_2O_2 addition to an O_2-sparged system caused

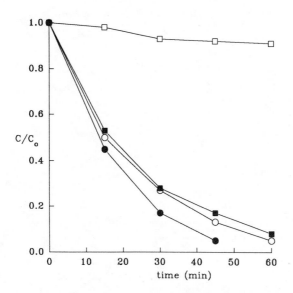

Figure 4. C/C_o vs time for rutile TiO_2 in the absence ($\square$) and presence ($\blacksquare$) of H_2O_2 and for anatase TiO_2 in the absence ($\bigcirc$) and presence ($\bullet$) of H_2O_2 in O_2-sparged slurries. $[Phenol]_o$ = 1.2 mM, $[H_2O_2]_0$ = 24 mM, and $[TiO_2]$ = 1.0 g/L. Addition of H_2O_2 dramatically increased the rate of phenol oxidation for rutile but not anatase TiO_2 (see text). Data from Sclafani et al.[30]

only a slight increase in rate. With rutile TiO_2 in either O_2- or He-sparged slurries, phenol degradation was observed only in the presence of either H_2O_2 or Ag^+ ions. Since it is known that Ag^+ scavenges electrons and is reduced to metallic Ag on the catalyst particle surface,[31,32] these results indicate that H_2O_2 is also serving as an electron scavenger. Figure 4 shows the difference in effect of adding H_2O_2 to O_2-sparged anatase and rutile TiO_2 slurries. Hydrogen peroxide greatly increases the rate for rutile, while only slightly increasing the rate for anatase.

Augugliaro et al.[33] later examined the H_2O_2-TiO_2 issue in more detail by investigating the photocatalytic oxidation of phenol by anatase TiO_2 in either He- or O_2-sparged systems. When the sparging gas was helium, they showed that H_2O_2 dramatically increased the rate of phenol oxidation from a negligible value, but when O_2 was used, only a modest rate increase was seen. Langmuir-Hinshelwood kinetics for H_2O_2 loss were seen in the He-sparged system, while simple zero-order kinetics were seen under O_2-sparging. The researchers suggested that H_2O_2 competed with O_2 for photogenerated electrons at the catalyst particle surface. Figure 5 shows how the apparent first-order rate constant for phenol oxidation varies with H_2O_2 for the O_2- and He-sparged systems.

Wei et al.[34] investigated the effect of Cu^{2+} and Fe^{3+} ions on the rate of phenol oxidation in O_2-sparged slurries of anatase TiO_2. They found that addition of either metal ion to TiO_2 slurries containing H_2O_2 increased the rate of phenol oxidation, but that in the absence of H_2O_2 the rate of phenol oxidation was

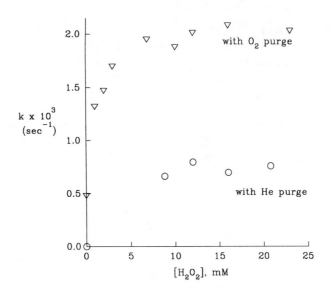

Figure 5. Rate of phenol oxidation vs $[H_2O_2]$ with He purging ($\bigcirc$) and O_2 purging ($\triangledown$). $[Phenol]_0 = 1.06$ mM, and $[TiO_2] = 1.0$ g/L. Phenol degradation rate is significantly faster with O_2 rather than He sparging (see text). Data from Augugliaro et al.[33]

decreased by Cu^{2+} addition and increased by Fe^{3+}. They did not report the effect of added H_2O_2 in the absence of any transition-metal ions, which would have served as a "baseline" for the H_2O_2-M^{n+} data.

Using a photoelectrochemical slurry cell, Peterson et al.[35] investigated the influence of H_2O_2 on photocatalysis. They measured photocurrents produced during illumination of the slurry in the presence and absence of H_2O_2. The measured photocurrent was the sum of an anodic current due to photoproduced electrons and a cathodic current due to photoproduced holes. In the absence of H_2O_2, the net steady-state photocurrent was anodic. When anatase TiO_2 was used, addition of 5 mM H_2O_2 quenched the anodic photocurrent (i.e., electrons reduced H_2O_2 before reaching the anode), leaving only a net cathodic photocurrent. Increasing the $[H_2O_2]$ to 10 mM decreased this steady-state cathodic photocurrent. With rutile TiO_2, the steady-state photocurrent in the absence of H_2O_2 was found to be significantly lower than that for anatase, and addition of H_2O_2 always caused a net cathodic photocurrent, which became more cathodic as the H_2O_2 increased. The authors suggested that these results indicated that with anatase TiO_2, H_2O_2 can scavenge electrons (quenching the anodic current), but that if an excess of H_2O_2 was added, it can also scavenge holes (or equivalently OH^- radicals), decreasing the cathodic photocurrent. With rutile TiO_2, the latter process appeared not to occur. Thus, addition of H_2O_2 to rutile TiO_2 slurries would always be beneficial, while for anatase TiO_2, an optimum H_2O_2 concentration should exist.

Chemseddine and Boehm[36] examined the photocatalytic oxidation of acetic acid (AA), monochloroacetic acid (MCA), dichloroacetic acid (DCA), and trichloroacetic acid (TCA) in aqueous systems. In all cases, the molar ratio (S) of CO_2 formed to acid originally present was always less than its stoichiometric value of 2.0. This ratio was approximately 1.2 for AA, MCA, and DC, while for TCA $S \approx 0.2$. Upon addition of H_2O_2 (0.19 M H_2O_2 in a 0.2 mM MCA solution), the rate of CO_2 production decreased at short times, but S increased from 1.0 to 1.5. Thus, while H_2O_2 decreased the rate of solute oxidation, it increased the overall solute conversion.

A U.S. patent was awarded to Lichtin et al.[37] for a process in which the presence of H_2O_2 and a transition-metal oxide photocatalyst (TiO_2 was the preferred catalyst) could completely oxidize liquid-phase organic pollutants when the volume fraction of water in the reaction mixture was as low as 0.5%. We have been unable to determine the reaction mechanism(s) for this system from the experimental details available in the patent application.

Gratzel et al.[38] investigated the effect of the inorganic oxidizing agents H_2O_2, peroxydisulfate ($S_2O_8^{2-}$), periodate (IO_4^-), and bromate (BrO_3^-) on the heterogeneous photocatalytic degradation of organophosphorus compounds. All four oxidants greatly enhanced the organic degradation rate, and the authors attributed this to both a scavenging of the photogenerated electrons, thereby reducing the rate of electron-hole recombination, as well as to direct attack on the organic species by the inorganic oxidants themselves.

Pelizzetti et al.[39] later examined the effect of $S_2O_8^{2-}$, IO_4^-, H_2O_2, and chlorate (ClO_3^-) on the photocatalytic oxidation of 2-chlorophenol, 2,7-dichlorodibenzo-dioxin, and atrazine. The former two oxidants were found to be quite effective in increasing the organic degradation rate, while the latter two were found to have only marginal effectiveness. The researchers used the same catalyst (Degussa P25) as Gratzel et al.,[38] although catalyst loading, reactor geometry, and other reaction conditions differed.

CURRENT INVESTIGATIONS

We have been investigating the role of the wavelength of the incident light on the H_2O_2-TiO_2 issue,[40] and this section presents a brief summary of our results using Degussa P25 TiO_2. The reactor used in this study was a quartz annular reactor assembly similar to those used previously.[41,42] Only a single bulb was used, either nUV ($\lambda = 366$ nm) or UV ($\lambda = 254$ nm), and it was placed in the center line of the annulus. The outer surface of the reactor was wrapped with aluminum foil to provide reflection. Our results on the photocatalytic oxidation of aqueous 80 μM benzene solutions are summarized in Figures 6–9. To allow comparison of the different lamps, we report the effective quantum yield, Φ, defined as the volumetric rate of benzene disappearance (moles per liter per second) divided by the volumetric rate of photon entry into the reactor (Einstein per liter per second).

The effect of wavelength of the incident illumination on the optimum TiO_2 concentration [TiO_2] has not been previously examined. Figure 6 shows the

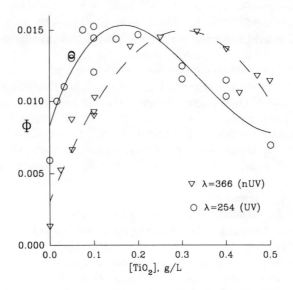

Figure 6. Variation of effective quantum yield Φ with [TiO₂] for heterogeneous photocatalysis with UV and nUV illumination. [Benzene]₀ ≈ 80μM for all runs.

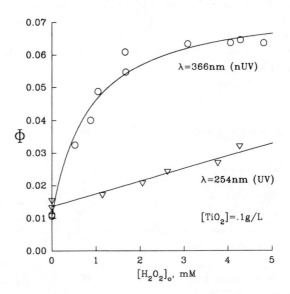

Figure 7. Variation of effective quantum yield Φ with [H₂O₂] for heterogeneous photocatalysis with UV and nUV illumination. [TiO₂] = 0.1g/L, and [Benzene]₀ ≈ 80 μM.

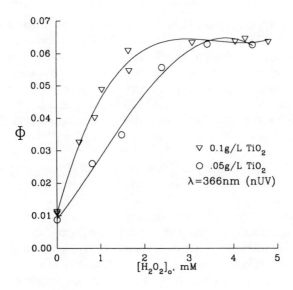

Figure 8. Variation of effective quantum yield Φ with $[H_2O_2]$ for heterogeneous photocatalysis with nUV illumination. $[Benzene]_o \approx 80\ \mu M$.

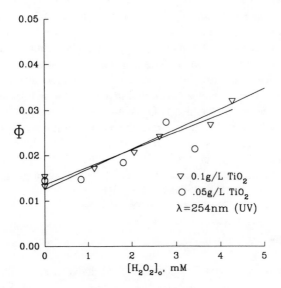

Figure 9. Variation of effective quantum yield Φ with $[H_2O_2]$ for heterogeneous photocatalysis with UV illumination. $[Benzene]_o \approx 80\ \mu M$.

variation in effective quantum yield Φ with [TiO_2] at two different illumination wavelengths, $\lambda = 254$ nm and $\lambda = 366$ nm. With UV illumination, the optimum [TiO_2] is ≈ 0.1 g/L, and Φ drops rapidly with higher [TiO_2]. With nUV illumination, the optimum concentration is ≈ 0.35 g/L, and the decrease in Φ with still greater [TiO_2] is not nearly as dramatic as with UV illumination. Previous work in our laboratory[41,42] found the rate exhibited a broad maximum at [TiO_2] ≈ 1 g/L with seven bulbs (one center, six peripheral) in the reactor assembly. Other groups have reported either a maxima in a plot of rate vs [TiO_2][43,44] or a plateau value where further increases in catalyst loading did not change the reaction rate.[45,46]

The effect of H_2O_2 on the effective quantum yield of photocatalyzed benzene degradation is summarized in Figures 7–9. Figure 7 shows the variation of Φ with [H_2O_2] under both UV and nUV illumination at [TiO_2] = 0.1 g/L. Hydrogen peroxide significantly increases the photocatalyzed reaction rate under nUV illumination, but has a less positive effect with UV illumination. The difference in the curve shapes, as well as the different magnitudes of influence, suggests that there is an important, previously unreported wavelength influence on the mechanism for H_2O_2-photocatalyst interaction. As the extinction coefficient for H_2O_2 decreases by two orders of magnitude when the wavelength increases from 254 to 366 nm,[47] H_2O_2 can compete with TiO_2 for UV photons, while with near-UV illumination TiO_2 is assumed to be the only absorbing species.

The appreciable sensitivity of Φ vs [H_2O_2] under nUV illumination for two values of [TiO_2] is shown in Figure 8. Both curves reach the same maximum effective quantum yield, but in the presence of more TiO_2 this maximum is reached at a lower H_2O_2 concentration. In contrast, the corresponding variation of Φ with [H_2O_2] under UV illumination for the same two values of [TiO_2] (Figure 9); thus, with UV illumination, no TiO_2 concentration effect is observed.

Our experimental results indicate that two chemically identical systems, differing only in the wavelength of incident illumination, show significant differences regarding the rate effect of added H_2O_2. While both UV- and nUV-illuminated, aerated systems show an increase in rate, the increase is greater for nUV than for UV illumination, and the shape of the curves describing rate vs [H_2O_2] show significant differences. To understand these results, we are currently developing a mathematical model which takes into account the light absorption properties of both H_2O_2 and TiO_2.

DISCUSSION OF RESULTS

The literature and laboratory[40] results cited above demonstrate the several variables that an exact mechanism for H_2O_2 will need to include. The majority of the literature cited in this work indicates that H_2O_2 addition has a positive effect on the rate of organic solute degradation, particularly in the absence of molecular oxygen,[30,33] the standard electron acceptor. Since the necessity of molecular oxygen in bulk solution for the complete mineralization of organic species is well known, it may be that H_2O_2 often serves simply as a source of active oxygen in an otherwise O_2-deficient system.

The ability of added species, both organic[9,21,22] and inorganic,[1,2,33] to scavenge OH· radicals and thereby decrease the rate of pollutant degradation is well established. The results of Peterson et al.,[35] and Tanaka et al.[26,27] Hisanaga et al.,[28] and Harada et al.[29] which indicate that excessive H_2O_2 addition diminishes its beneficial effect support the idea of H_2O_2 as an OH· radical scavenger.

The intriguing data of Hisanaga et al.[28] for organochlorine degradation show that the chemical nature of the organic pollutant is also important in determining the effect of H_2O_2 addition. This effect may be associated with whether pollutant attack is initiated by oxidation (h_{VB}^+ or OH·) or by reduction (e_{CB}^-).

Our results[40] show that in two chemically identical systems, the wavelength of the incident light is a critical parameter in determining the amount of H_2O_2 rate enhancement, suggesting the possibility of photon competition between H_2O_2 and TiO_2. The difference in sensitivity of the degradation rate to both [TiO_2] and [H_2O_2] at different incident illumination wavelengths indicates that a full kinetic model of H_2O_2 rate influence will have to include homogeneous and heterogeneous absorption and thus photon pathlength and light scattering.

ACKNOWLEDGMENT

E. J. Wolfrum is pleased to acknowledge an Electronic Materials Fellowship from the National Science Foundation for the 1991–1992 academic year.

REFERENCES

1. Glaze, W. H., Kang, J. W., and Zigler, S. S., Treatment of hazardous waste chemicals using advanced oxidation processes, in Tenth International Ozone Symposium, Monaco, 1991.
2. Peyton, G. R., Oxidative treatment methods for removal of organic compounds from drinking water supplies, in *Significance and Treatment of Volatile Organic Compounds in Water Supplies,* Ram, N. M., Christman, R. F., and Cantor, K. P., Eds., Lewis Publishers, Chelsea, MI, 1991, 313.
3. Schiavello, M., Ed., *Photocatalysis and Environment: Fundamentals and Applications*, Kluwer, Dordrecht, 1987.
4. Pelizzetti, E., Minero, C., and Maurino, V., The role of colloidal particles in the photodegradation of organic compounds of environmental concern in aquatic systems, *Adv. Colloid Interface Sci.*, 32, 271, 1990.
5. Serpone, N., *Solar Photochemistry and Heterogeneous Photocatalysis: A Convenient and Practical Utilization of Sunlight Photons*, Elsevier, Amsterdam, 1989, 297.
6. Fox, M. A., Photoinduced electron transfer, *J. Photochem. Photobiol. A: Chem.*, 52, 617, 1990.
7. Matthews, R. W., Environment: photochemical and photocatalytic process. Degradation of organic compounds, in *Photochemical Conversion and Storage of Solar Energy*, Pelizzetti, E. and Schiavello, M., Eds, Kluwer, Dordrecht, 1991, 427.
8. Serpone, N. and Pelizzetti, E., Eds., *Photocatalysis: Fundamentals and Applications*, John Wiley & Sons, New York, 1989.

9. Turchi, C. S. and Ollis, D. F., Mixed reactant photocatalysis: intermediates and mutual rate inhibition, *J. Catal.*, 119, 483, 1989.

10. Turchi, C. S. and Ollis, D. F., Photocatalytic degradation of organic water contaminants: mechanisms involving hydroxyl radical attack, *J. Catal.*, 122, 178, 1990.

11. Ceresa, E. M., Burlamacchi, L., and Visca, M., An ESR study on the photoreactivity of TiO$_2$ pigments, *J. Mater. Sci.*, 18, 289, 1983.

12. Greenwald, R. A., Ed., *CRC Handbook of Methods for Oxygen Radical Research*, CRC Press, Boca Raton, FL, 1985, 151.

13. Harbour, J. R., Tromp, J., and Hair, M. L., Photogeneration of H$_2$O$_2$ in aqueous TiO$_2$ dispersions, *Can. J. Chem.*, 63, 204, 1985.

14. Maldotti, A., Amadelli, R., and Carassiti, V., An EWR spin-trapping investigation of azide oxidation on TiO$_2$ powder suspensions, *Can. J. Chem.*, 66, 76, 1988.

15. Draper, R. B. and Fox, M. A., Titanium dioxide photosensitized reactions studies by diffuse reflectance flash photolysis in aqueous suspensions of TiO$_2$ powder, *Langmuir*, 6, 1296, 1990.

16. Harbour, J. R. and Hair, M. L., Superoxide generation in the photolysis of aqueous cadmium sulfide dispersions. Detection by spin trapping, *J. Phys. Chem.*, 81, 1791, 1977.

17. Harbour, J. R. and Hair, M. L., Radical intermediates in the photosynthetic generation of H$_2$O$_2$ with aqueous ZnO dispersions, *J. Phys. Chem.*, 83, 652, 1979.

18. Kormann, C., Bahnemann, D. W., and Hoffmann, M. R., Photocatalytic production of H$_2$O$_2$ and organic peroxides in aqueous suspensions of TiO$_2$, ZnO, and desert sand, *Environ. Sci. Technol.*, 22, 798, 1988.

19. Buxton, G. V., Greenstock, C. L., Helman, W. P., and Ross, A. B., Critical review of rate constants for reactions of hydrated electrons, hydrogen atoms and hydroxyl radicals (OH$^\cdot$) in aqueous solution, *J. Phys. Chem. Ref. Data*, 17, 513, 1988.

20. Bielski, B. H. J., Cabelli, D. E., Arudi, R. L., and Ross, A. B., Reactivity of HO$_2$/O$_2^-$ radicals in aqueous solution, *J. Phys. Chem. Ref. Data*, 14, 1041, 1985.

21. Sundstrom, D. W., Weir, B. A., and Redig, K. A., Destruction of mixtures of pollutants by UV-catalyzed oxidation with H$_2$O$_2$, in *Emerging Technologies in Hazardous Waste Management*, Tedder, D. W. and Pohland, F. G., Eds., ACS Symposium Ser. No. 422, American Chemical Society, Washington, D. C., 1990, 67.

22. Wolfrum, E. J. and Ollis, D. F., Modeling the UV/H$_2$O$_2$ process: competitive light absorption and competitive reaction by H$_2$O$_2$, organic reactants, and intermediates, *Environ. Sci. Technol.*, 1992, submitted.

23. Abdullah, M., Low, G.K.-C., and Matthews, R. W., Effects of common inorganic anions on rates of photocatalytic oxidation of organic carbon over illuminated titanium dioxide, *J. Phys. Chem.*, 94, 6820, 1990.

24. McGeever, C. E., Effect of Hydrogen Peroxide on Photocatalysis of Perchloroethylene in Aqueous Suspensions of Titanium Dioxide, Master's thesis, University of California at Davis, 1983.

25. Brown, G. T. and Darwent, J. R., Methyl orange as a probe for photooxidation reactions of colloidal TiO$_2$, *J. Phys. Chem.*, 88, 4955, 1984.

26. Tanaka, K., Hisanaga, T., and Harada, K., Efficient photocatalytic degradation of chloral hydrate in aqueous semiconductor suspensions, *J. Photochem. Photobiol. A: Chem.*, 48, 155, 1989.

27. Tanaka, K., Hisanaga, T., and Harada, K., Photocatalytic degradation of organohalide compounds in semiconductor suspensions with added hydrogen peroxide, *Nouv. J. Chem.*, 13, 5, 1989.

28. Hisanaga, T., Harada, K., and Tanaka, K., Photocatalytic degradation of organochlorine compounds in suspended TiO$_2$, *J. Photochem. Photobiol. A: Chem.*, 54, 113, 1990.

29. Harada, K., Hisanaga, T., and Tanaka, K., Photocatalytic degradation of organophosphorus insecticides in aqueous semiconductor suspensions, *Water Res.*, 24, 1415, 1990.

30. Sclafani, A., Palmisano, L., and Davi, E., Photocatalytic degradation of phenol by TiO$_2$ aqueous dispersions: rutile and anatase activity, *Nouv. J. Chem.* 14, 265, 1990.

31. Ohtani, B., Okugawa, Y., Nishimoto, S., and Kagiya, T., Photocatalytic activity of TiO$_2$ powders suspended in AgNO$_3$ solution. Correlation with pH-dependent surface structures, *J. Phys. Chem.*, 91, 3550, 1988.

32. Ohtani, B., Kakimoto, M., Miyadzu, H., Nishimoto, S., and Kagiya, T., Effect of surface-adsorbed 2-propanol on the photocatalytic reduction of silver and/or nitrate ions in acidic TiO$_2$ suspension, *J. Phys. Chem.,* 92, 5773, 1988.

33. Augugliaro, V., Davi, E., Schiavello, M., and Sclafani, A., Influence of H$_2$O$_2$ on the kinetics of phenol photodegradation in aqueous TiO$_2$ dispersion, *Appl. Catal.*, 65, 101, 1990.

34. Wei, T.-Y., Wang, Y.-Y., and Wan, C.-C., Photocatalytic oxidation of phenol in the presence of H$_2$O$_2$ and TiO$_2$ powders, *J. Photochem. Photobiol. A: Chem.*, 55, 115, 1990.

35. Peterson, M. W., Turner, J. A., and Nozik, A. J., Mechanistic studies of the photocatalytic behavior of TiO$_2$ particles in a photoelectrochemical slurry cell and the relevance of photodetoxification reactions, *J. Phys. Chem.*, 95, 221, 1990.

36. Chemseddine, A. and Boehm, H. P., A study of the primary step in the photochemical degradation of acetic acid and chloroacetic acid on a TiO$_2$ photocatalyst, *J. Mol. Catal.*, 60, 295, 1990.

37. Lichtin, N. N., DiMauro, T. M., and Svrluga, R. C., Catalytic Process for Degradation of Organic Materials in Aqueous and Organic Fluids to Produce Environmentally Compatible Products, U. S. Patent No. 4,861,484, 1989.

38. Gratzel, C. K., Jirousek, M., and Gratzel, M., Decomposition of organophosphorus compounds on photocatalyzed TiO$_2$ surfaces, *J. Mol Catal.*, 60, 375, 1990.

39. Pelizzetti, E., Carlin, V., Minero, C., and Gratzel, M., Enhancement of the rate of photocatalytic degradation on TiO$_2$ of 2-chlorophenol, 2,7-dichlorodibenzodioxin, and atrazine by inorganic oxidizing species, *J. Photochem. Photobiol. A: Chem.*, 15, 351, 1991.

40. Wolfrum, E. J., Mechanistic Investigations of Heterogeneous Photocatalysis, Ph.D. thesis, North Carolina State University, Raleigh, 1992.

41. Sczechowski, J. G., The Role of Dissolved Oxygen in the Photocatalytic Decomposition of Dilute Aqueous Solutions of Chlorinated Hydrocarbons, Master's thesis, North Carolina State University, Raleigh, 1987.

42. Turchi, C. S., Heterogeneous Photocatalytic Degradation of Organic Water Contaminants: Kinetics and Hydroxyl Radical Mechanisms, Ph.D. thesis, North Carolina State University, Raleigh, 1990.

43. Ayoub, P. M., A Transport Reactor for Photocatalytic Reactions, Ph.D. thesis, Northwestern University, Raleigh, 1990.

44. Augugliaro, V., Palmisano, L., Sclafani, A., Minero, C., and Pelizzetti, E., Photocatalytic degradation of phenol in aqueous TiO$_2$ dispersions, *Toxicol. Environ. Chem.*, 16, 89, 1988.

45. Matthews, R. W., Letter to the editor. Photocatalytic oxidation of chlorobenzene in aqueous suspensions of titanium dioxide, *J. Catal.,* 97, 565, 1986.

46. Minero, C., Aliberti, C., Pelizzetti, E., Terzian, R., and Serpone, N., Kinetic studies in heterogeneous photocatalysis (6) AM1 simulated sunlight photodegradation over titania in aqueous media: a first case of fluorinated aromatics and identification of intermediates, *Langmuir*, 7, 928, 1991.

47. FMC Corporation, *H$_2$O$_2$ Physical Properties Data Book*, Becco Chemical Division, Buffalo, NY, 1955.

CHAPTER 33

Homogeneous Photodegradation of Pollutants in Contaminated Water: An Introduction

James R. Bolton and Stephen R. Cater

INTRODUCTION

As society becomes more complex and the population grows, pollutants, coming from a variety of industrial and agricultural activities, are contaminating our water and air to an unacceptable level. It has now been recognized in some countries that the most effective way to deal with these problems is to remove and destroy these pollutants at the source, before they are discharged into the environment. Unfortunately, we have a legacy of decades of neglect; already many streams, rivers, lakes, and groundwaters, as well as urban air, are polluted to the point that they must be cleaned up. This problem will be with us for many decades.

The remediation of industrial wastewaters, polluted groundwaters, and contaminated air has traditionally relied on two methods: treatment with activated carbon (for water and air) and air stripping (for volatile organics in water). Recently, catalytic combustion has been employed for contaminated air. However, the first two methods only transfer the pollutants from one phase to another, and ultimately there is still a disposal problem. Thus, in recent years, a number of new technologies, called Advanced Oxidation Processes (AOP), have emerged that are capable of converting the pollutants into harmless chemicals.[1-3] They are called oxidation processes because they promote reactions that bring about a nearly complete oxidation (or mineralization) of the pollutants to yield HCO_3^-, H_2O, and a small amount of acid from any halogens, nitrogen, or sulfur present. For example, 2,4-dichlorophenol ($C_6H_4OCl_2$) reacts according to

$$C_6H_4OCl_2 + 5\,H_2O + 6\,O_2 \longrightarrow 8\,H^+ + 6\,HCO_3^- + 2\,Cl^- \qquad (1)$$

0-87371-871-2/94/$0.00+$.50

Almost all reactions of this type are exothermic (i.e., energy releasing) and thermodynamically spontaneous; however, they are usually kinetically slow in the absence of initiators.

The AOP technologies almost all rely on the generation of very reactive free radicals, such as the hydroxyl ($^{.}$OH) radical, to function as initiators. Four major approaches to AOP are under development at present:

1. *Homogeneous photolysis* (UV/H_2O_2 and UV/O_3) — These processes employ UV photolysis of H_2O_2 and/or O_3 and other additives in homogeneous solution to generate $^{.}$OH and other radicals.

2. *Dark homogeneous oxidation* — These processes do not employ UV light; they usually involve the use of Fenton's reaction,[4,5] ozone at high pH,[3,6] and ozone/peroxide.[3,7]

3. *Heterogeneous photolysis* (UV/TiO_2)[8] — Here, solid particles of the semiconductor TiO_2 absorb UV light and generate $^{.}$OH and other radicals in reactions on the surface of the particles. The process is heterogeneous because there are two active phases, solid and liquid.

4. *Radiolysis*[9] — A source of high-energy radiation (e.g., a linear accelerator) is used to irradiate the wastewater. $^{.}$OH, H$^{.}$ atoms, hydrated electrons, and other radicals are generated in the radiolysis of water.

In this chapter, we shall only focus on the first topic, namely processes that involve homogeneous photolysis. The review by Cunningham et al.[10] in this book covers the third topic. We shall also focus primarily on applications in water treatment, since this area is much more developed than the photochemical treatment of contaminated air.

Homogeneous photolysis involves the use of very powerful (up to 60 kW) UV lamps to irradiate contaminated waters containing additives, such as H_2O_2, to generate reactive radical initiators, such as the hydroxyl radical or hydrated electrons. The photochemical reactions take place primarily in homogeneous aqueous solution.

The range of contaminated waters that may be treated with homogeneous photolysis systems is quite broad, including drinking water, groundwater, industrial process water, marine tank ballast water, process water in offshore drilling rigs, etc. The degree of treatment required varies from complete mineralization to partial treatment, followed by some other process, such as biodegradation. Every application is different with its own required treatment strategy. This is why treatment tests on sample waters are essential in the design of viable UV-treatment processes.

HISTORICAL PERSPECTIVE

The use of homogeneous photodegradation to treat contaminated water dates back to the early 1970s. Prengle et al.[11] and Garrison et al.[12] at Houston Research, Inc. investigated the use of UV/ozone for the treatment of organic pollutants. Hoigné and Bader[6] around the same time were investigating the decomposition of ozone in aqueous solution for the oxidation of organic compounds. They identified hydroxyl radicals as the important oxidizing species in the process.

The photolysis of hydrogen peroxide to produce hydroxyl radicals has been known for decades and is summarized in the work of Baxendale and Wilson[13] and by Skuratova.[14] However, these and earlier reports did not deal with the application of UV/peroxide for destruction of organic compounds in water. This application can be attributed first to Koubek[15] in 1975 who applied the use of UV/peroxide for lowering the chemical oxygen demand (COD) in wastewater. The earliest published account of the use of UV light for organic compound degradation was the work in 1968 by Bulla and Edgerley[16] who investigated the direct UV photodegradation of chlorinated pesticides. Throughout the 1980s, several papers on all aspects of photodegradation of organic pollutants in water were published. These covered the use of ozone,[3,17,18] as well as applications with the photolysis of hydrogen peroxide.[19-22] Hoigné and Bader[6] were the first to recognize that the hydroxyl radical is the principal radical species and a powerful oxidizing species in the oxidation of organic compounds with ozone. For further reading on the historical aspects of other forms of advanced oxidation, the readers are also directed to review the articles in References 1 to 3 and by Zeff.[23]

BASIC CONCEPTS

For the benefit of readers who have a limited background in photochemistry, we present here a brief outline of some of the basic concepts of photochemistry.

Photon Energies and Bond Energies

In order to effect photochemistry, photons of light must be absorbed. The energy of a photon is given by

$$U = h\nu = hc/\lambda \tag{2}$$

where h is Planck's constant, c is the speed of light, and ν and λ are the frequency and wavelength of the light. For a bond to be broken in a molecule, U must be greater than the bond energy of that bond.

Absorbance, Extinction Coefficients, and Absorption Coefficients

When light of wavelength λ enters a medium, its spectral irradiance E_λ(W m^{-2} nm^{-1})* is attenuated according to the Beer-Lambert law, which is given in two forms, one for the gas phase and the other for the liquid phase:

---

*Spectral irradiance is defined as the light power (watts) incident on 1 m^2 of surface in a 1-nm band of wavelengths.

$$\ln\left(E_\lambda^o/E_\lambda^\ell\right) = \alpha_\lambda\, p_i\, \ell \qquad\qquad \text{(gas phase)} \qquad\qquad (3a)$$

$$\log\left(E_\lambda^o/E_\lambda^\ell\right) = \varepsilon_\lambda\, c_i\, \ell \qquad\qquad \text{(liquid phase)} \qquad\qquad (3b)$$

E_λ^o and E_λ^ℓ are the spectral irradiances incident and at a di_tance ℓ into the medium, α_λ is the *absorption coefficient* (cm^{-1} atm^{-1}), p_i is the partial pressure (atm) of component i, ε_λ is the *decadic extinction coefficient* (M^{-1} cm^{-1}), and c_i is the concentration (M) of component i. The absorbance A_λ (formerly called *optical density*) at wavelength λ is the product $\varepsilon_\lambda c_i \ell$. At 298 K, the decadic extinction coefficient and the absorption coefficient are related by

$$\varepsilon_\lambda/M^{-1}cm^{-1} = 10.619\, \alpha_\lambda/cm^{-1}\, atm^{-1} \qquad\qquad (4)$$

Quantum Yield

The photolysis quantum yield ϕ is defined as the number of molecules of target compound that decompose in the photolysis reaction divided by the number of photons of light absorbed by the compound, as determined in a fixed period of time. Normally, unity is the maximum quantum yield possible; however, if the photolysis reaction initiates a chain reaction, the quantum yield can be considerably greater than unity.

Cage Effects in Solution

When a molecule decomposes photolytically in the gas phase, the fragments fly apart with no restrictions. However, in a condensed medium, such as an aqueous solution, "cage" effects generally reduce photolysis quantum yields. For example, if a molecule AB photodissociates in solution,

$$AB \underset{}{\overset{h\nu}{\rightleftharpoons}} \left(A^\cdot ... B^\cdot\right) \longrightarrow A^\cdot + B^\cdot \qquad\qquad (5)$$

there is a significant probability for the recombination of the primary radicals within the solvent cage. The reason for this is that the rate of diffusion of the primary radicals out of the cage is often much smaller than the rate of recombination. This is why quantum yields are reduced in the liquid phase. For example, the quantum yield for the photolysis of H_2O_2 in aqueous solution is only about half of its value in the vapor phase.

ABSORPTION SPECTRA AND PHOTOLYSIS OF O_2, O_3, H_2O, AND H_2O_2

Dioxygen, ozone, water, and hydrogen peroxide are important components in water or air.* Figure 1 shows their vapor phase absorption spectra in the usual

---

*Nitrogen is also a major component of air, but it does not absorb light until below 125 nm and thus is transparent to all normal sources of UV light.

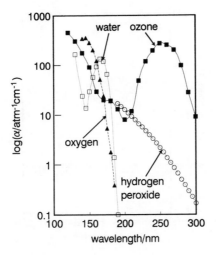

Figure 1. Absorption spectra (on a log scale) of O_2, H_2O, O_3, and H_2O_2 in the gas phase. Data taken from Reference 50.

gas-phase units of α (cm^{-1} atm^{-1}). The spectra of O_3 and H_2O_2 in the gas phase and aqueous solution are shown in Figures 2 and 3. Although the aqueous solution spectra are similar to the corresponding spectra in the gas phase, there are significant differences, with the extinction coefficients being generally larger in aqueous solution than in the gas phase.

Table 1 gives the principal photolysis reactions of these components and the corresponding quantum yields. The threshold wavelength is that corresponding to

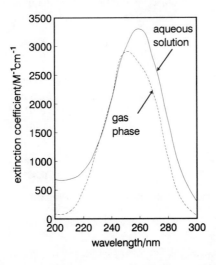

Figure 2. Absorption spectra of ozone in the gas phase and in aqueous solution. Gas-phase data are from Reference 50; aqueous-solution data are from the work of Taube.[51]

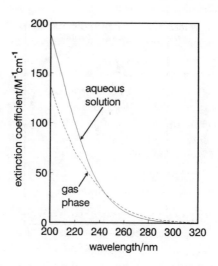

Figure 3. Absorption spectra of hydrogen peroxide in the gas phase and in aqueous solution. Gas-phase data are from Reference 50; aqueous-solution data were obtained in this laboratory. The extinction coefficient at 254 nm (18.7 M^{-1} cm^{-1}) agrees with that determined by Weeks and Matheson.[52]

Table 1. Primary Photochemical Processes of Simple Molecules in the Gas Phase

Absorber	Primary Products	Threshold Wavelength[a] (λ_D/nm)	Absorption Threshold[b] (λ_{abs}/nm)	Quantum Yield	Remarks
O_2	$2\ \cdot O(^3P)$	242	190	1.0[c]	Below 176 nm, one of the O atoms may be formed as $O(^1D)$
O_3	$O_2 + \cdot O(^1D)$	410	330	0.99[d] 0.61[e]	In the gas phase at 14°C in water (λ = 254 nm); the $\cdot O(^1D)$ reacts with H_2O to form H_2O_2
H_2O	$H\cdot + \cdot OH$	242	190	1.0[c]	In the gas phase
H_2O_2	$2\ \cdot OH$	561	310	1.0[c]	In the gas phase; ind. of λ from 200–350 nm
				0.50[f]	At 254 nm in water
				0.30[g]	At 313 nm in water

[a] Defined as the maximum wavelength for which the photon energy matches the bond energy of the bond that breaks in the photolysis reaction.
[b] Defined as the wavelength where the absorption coefficient α is about 0.1 (ε about 1).
[c] Data taken from Reference 50.
[d] Data taken from Reference 53.
[e] Data taken from Reference 51.
[f] Data taken from References 24 and 52.
[g] Data taken from Reference 54.

the bond energy of the weakest bond in the molecule; however, the molecule must absorb if it is to dissociate. Column 4 gives the absorption threshold. For example, O_3 can theoretically dissociate to yield excited $\cdot O(^1D)$ atoms* at wavelengths as long as 410 nm; however, O_3 does not absorb significantly until below 330 nm. This absorption is very important in the atmosphere, since the absorption of the UV component of sunlight by O_3 in the stratosphere is responsible for shielding us from harmful solar radiation below 300 nm.

Note that the quantum yields for the photodissociation of both H_2O_2 and O_3 are much smaller in aqueous solution than in the gas phase. This arises from the cage effect in solution, as discussed above.

RADICAL INTERMEDIATES FORMED IN PHOTOLYSIS REACTIONS

As seen in Table 1, the dissociation reactions of O_3, O_2, H_2O, and H_2O_2 produce very reactive radical intermediates, such as $\cdot OH$, $O\cdot$, and $H\cdot$. Oxygen atoms react very rapidly with water to form $\cdot OH$ radicals. In addition, certain solutes, such as $Fe(CN)_6^{4-}$, can undergo photoionization to produce hydrated electrons (e_{aq}^-).

$$Fe(CN)_6^{4-} \xrightarrow{h\nu} Fe(CN)_6^{3-} + e_{aq}^- \tag{6}$$

Hydrated electrons react rapidly with oxygen to produce $\cdot O_2^-$ (in basic solution) and $HO_2\cdot$ in acid solution. They can also react directly with pollutants (see below).

Table 2 lists several properties of these reactive radical intermediates. Note that e_{aq}^- and $H\cdot$ are strong reducing agents and $\cdot OH$ and $\cdot O^-$ are strong oxidizing agents. $HO_2\cdot$ and $\cdot O_2^-$ are only mild oxidizing or reducing agents and are found to have much slower rate constants than the other species for reaction with a variety of substrates. The reducing radicals (e_{aq}^- and $H\cdot$) can be converted into oxidizing radicals by the use of N_2O (Equations 7 and 8), reactions that occur close to the diffusion controlled limit ($\sim 1 \times 10^{10}$ M^{-1} s^{-1}).

$$e_{aq}^- + H^+ + N_2O \longrightarrow N_2 + \cdot OH \tag{7}$$

$$H\cdot + N_2O \longrightarrow N_2 + \cdot OH \tag{8}$$

$\cdot OH$ radicals can react with a variety of substrates, often to produce other reactive radical intermediates. For example, the $\cdot CO_2^-$ and $\cdot CO_3^-$ radicals are produced by reactions with formate or carbonate. These are generally reducing radicals, although not as strong reducing agents as e_{aq}^- and $H\cdot$.

---

*At wavelengths as long as 1180 nm, O_3 can photodissociate to give ground state $\cdot O(^3P)$ atoms; however, the $\cdot O(^3P)$ atoms are relatively unreactive and most recombine with O_2 to reform O_3.

Table 2. Properties of Some Aqueous Radicals[a]

Property	Radical					
	H·	e_{aq}^-	·OH	·O⁻	HO_2	O_2^-
Absorption maximum/nm	188	715	230	240	225	245
$\varepsilon/M^{-1}\,cm^{-1}$	1620	18500	530	370	1050[b]	2350
pK_a	9.6		11.9		4.8	
$E°/V$	−2.1	−2.9	1.77	1.64	−0.3	−0.33
Diffusion coefficient/10^{-5} cm² sec⁻¹	8	4.90	2.3		1.5	
$\Delta G_f°/kJ\,mol^{-1}$	222	276	13	96		
$\Delta H_f°/kJ\,mol^{-1}$	213	277	−7	38	138	80
$S_{298}°/J\,mol^{-1}\,K^{-1}$	38	13	96			
$\Delta G(hyd)/kJ\,mol^{-1}$	222	276	13	96		
$\Delta H(hyd)/kJ\,mol^{-1}$	−3	−167	−370			395

[a]   Data taken from References 29 and 55.
[b]   Measured in the gas phase. Value is the average of several measurements — see
      review by Lightfoot et al.[56]

Table 3 summarizes rate constants for several radical intermediates often found
in UV-irradiated aqueous solutions. Note that some of these rate constants are
considerably greater than the diffusion-controlled limit (ca. $1 \times 10^{10}\ M^{-1}\ s^{-1}$). The
reason probably lies in the possibility that species such as e_{aq}^-, H·, ·OH, and ·O⁻
can move through water without moving molecules; all that is required is a simple
bond rearrangement.

UV TREATMENT PROCESSES

The use of UV light for the photodegradation of pollutants in water can be
classified into three principal areas:

1. photooxidation — light-driven oxidation processes principally initiated by
    hydroxyl radicals
2. photoreduction — light-driven reduction processes principally initiated by
    hydrated electrons
3. direct photodegradation — processes where degradation proceeds following
    direct excitation of the pollutant by UV light

Electrical Energy Per Order

The evaluation of the efficiency of photodegradation treatment processes is
difficult because the reaction rate depends on many factors (quantum yield, light
intensity, pathlength, pollutant concentration, etc.). Thus, there is need for a
"figure of merit" that can be used to assess the relative performance of each

Table 3. Rate Constants for Reaction of e_{aq}^-, H·, ·OH, and ·O⁻ Radicals with Various Compounds[a]

Substrate	k/10⁹ M⁻¹ s⁻¹			
	e_{aq}^-	H·	·OH	·O⁻
H_2O_2	11	0.090	0.027	—
O_2	19	21	8	3.6(11)
O_3	36(9)	38(2)	0.11	—
HCO_3^-	<0.001	4.4e–5	8.5e–3	—
CO_3^{2-}	3.9e–4(11)	—	3.0	—
CN·	—	—	7.6	0.26(14)
Fe^{2+}	~0.16	7.5e–3	0.43(3)	3.8(5)
CH_3OH	<1e–5	2.6e–3	0.97	0.75(14)
HCOH	~0.01	5.0e–5(1)	~1(1)	—
HCOOH	0.14(5)	4.4e–4	0.13(1)	—
C_2H_5OH	—	0.017(1)	1.9	1.2(13)
CH_3COOH	0.20	9.8e–5	0.016	—
CH_3COCH_3	6.5	2.6e–3	0.11	—
CH_3CN	0.037(12)	3.6e–3(2)	0.022	0.21(14)
CH_2Cl_2	6.3	4e–4	0.058(10)	—
$CHCl_3$	30	0.011	~0.005(5.7)	—
$CHCl=CCl_2$	19	—	4.2	—
$ClCH_2COOH$	6.9	7.2e–6(2)	0.043(1)	—
Diethyl ether	—	—	3.6	0.95(13)
Benzene	0.009(10)	0.91	7.8	—
Toluene	0.014(10)	2.6(3)	3.0(3)	2.1(13)
Phenol	0.02	1.7	6.6	—
Benzophenone	10	6.1	8.8	—
Benzaldehyde	—	1.4(1)	4.4(9)	—
Benzoic acid	7.1(3.5)	0.92	4.3	—
Benzoate	3.2	1.1	5.9	0.04(14)
Chlorobenzene	0.50	1.4(1)	5.5	—
Benzoquinone	23	8.3(2)	1.2	—
Nitrobenzene	37	1.0(1)	3.9	—
Benzonitrile	19	0.68	4.4	0.07(14)

(Rate constants are for pH ≈ 7, unless otherwise indicated by a number in parentheses after the rate constant value.)

[a] Compiled from the RCDC Bibliographic Database, Radiation Chemistry Data Center (RCDC), Radiation Laboratory, University of Notre Dame, Notre Dame, IN 46556.

system. Since the rate of photodegradation of a pollutant usually follows first-order kinetics, every decade in reduction of the pollutant concentration requires the same time and hence the same amount of UV irradiation. As a figure of merit, we have proposed the *electrical energy per order* (EE/O), defined as the electrical energy (in kilowatt hours) required to reduce the concentration of a pollutant by one order of magnitude in 1000 U.S. gallons (3785 L) of water.* The EE/O value may be calculated from the equations

*This concept was introduced by Mr. Keith Bircher of Solarchem Environmental Systems, Markham, Ontario, Canada.

$$EE/O = \frac{P \times (t/60) \times 3785}{V \times \log[c_i/c_f]} \qquad \text{(batch experiments)} \qquad (9a)$$

$$EE/O = \frac{P \times 1000 \text{ gal}}{R \times 60 \times \log[c_i/c_f]} \qquad \text{(flow through)} \qquad (9b)$$

where P is the lamp power (in kilowatts), t is the irradiation time (in minutes), V is the reactor volume (in liters), R is the flow rate (in gallons per minute), and c_i and c_f are the initial and final concentrations over the irradiation time. The EE/O values are related to the first-order rate constant k_1 (per minute) by

$$EE/O = \frac{145.25 \, P}{V \times k_1} \qquad (10)$$

Note that the EE/O value is independent of light intensity, flow rate, and reactor volume as long as the assumption of first-order kinetics holds. The EE/O values of ten or less are considered very favorable.

It should be noted that the EE/O values can be quite matrix dependent. If the wastewater has a significant UV absorption (as is often the case), part of the UV light will be absorbed by nonactive components in the wastewater and not by H_2O_2. This will result in a higher EE/O value. Also, if the wastewater contains fine particles that absorb UV light, a similar increase in EE/O values should be expected. Finally, if the wastewater has a high COD or contains components, such as carbonate, that compete for ·OH radicals, the EE/O values will increase.

Photooxidation

Photooxidation involves the use of UV light plus an oxidant to generate hydroxyl radicals. The hydroxyl radicals then attack the organic pollutants to initiate oxidation. The oxidants act as the source of the hydroxyl radical; there are two major oxidants used: hydrogen peroxide and ozone.

Hydrogen Peroxide

As noted in Figures 1 and 3, hydrogen peroxide absorbs fairly weakly in the UV region with increasing absorption as the wavelength decreases. At 254 nm, the molar extinction coefficient (ε_λ) is 18 M^{-1} cm^{-1}, whereas at 200 nm ε_λ is 190 M^{-1} cm^{-1}. The primary process for absorption of light below 365 nm is dissociation to give two hydroxyl radicals (Equation 11).*

---

*A comprehensive review of H_2O_2 photolysis reactions has recently appeared.[25]

$$H_2O_2 \xrightarrow{\ h\nu\ } 2\ ^{\cdot}OH \tag{11}$$

In the gas phase, the quantum yield for Equation 11 is 1.0, whereas in aqueous solution cage recombination reduces the quantum yield to 0.50.[13,25,26] The use of hydrogen peroxide is now very common for the treatment of contaminated water due to several practical advantages:

1. The peroxide is available as an easily handled solution that can be metered into the aqueous wastewater to give a wide range of accurate, reproducible concentrations.
2. There are no air emissions. This can be a significant problem with the treatment of volatile organics with UV/ozone.
3. Hydroxyl radicals are generated from hydrogen peroxide with a high quantum yield and good electrical efficiency

The major drawback to the use of hydrogen peroxide is the relatively low molar extinction coefficient, which means that in waters with high inherent UV absorption the fraction of light absorbed by the hydrogen peroxide can be low unless prohibitively large concentrations are used. This increases the costs for the treatment of the water.

Ozone

Ozone is generated as a gas in air or oxygen in concentrations generally ranging from 1 to 8% (v/v). Ozone has a strong absorption band centered at 260 nm with an $\varepsilon_\lambda = 3000$ M^{-1} cm^{-1}. Absorption of light by ozone at this wavelength leads to a quantitative formation of hydrogen peroxide[18,27] (Equations 12 and 13).*

$$O_3 \xrightarrow{\ h\nu\ } O\left(^1D\right) + O_2 \tag{12}$$

$$O\left(^1D\right) + H_2O \longrightarrow H_2O_2 \tag{13}$$

Hydroxyl radicals are then formed by reaction of ozone with the conjugate base of hydrogen peroxide (Equations 14–17).

$$H_2O_2 + H_2O \longrightarrow HO_2^- + H_3O^+ \tag{14}$$

$$HO_2^- + O_3 \longrightarrow O_3^- + HO_2^{\cdot} \tag{15}$$

$$O_3^- + H_2O \longrightarrow HO_3^{\cdot} + OH^- \tag{16}$$

$$HO_3^{\cdot} \longrightarrow\ ^{\cdot}OH + O_2 \tag{17}$$

---

*In aqueous solution, Equation 13 initially generates two ·OH radicals; however, these are created very close to each other, and most of these pairs recombine to form H_2O_2.

Since the net result of ozone photolysis is the conversion of ozone into hydrogen peroxide, UV/ozone appears to be only a rather expensive method to make hydrogen peroxide. However, there are other oxidation-related processes occurring in solution, such as photolysis of the generated hydrogen peroxide and photodegradation of pollutants, as well as direct reaction of ozone with the pollutant.

The optimum conditions for using UV/peroxide and UV/ozone are different and dependent on the wastewater. The photodissociation of hydrogen peroxide is independent of pH and temperature. However, the efficiency of the use of the generated hydroxyl radicals is pH dependent. In waters of high carbonate and bicarbonate alkalinity, there is a competition for the hydroxyl radical between the pollutant and the carbonate/bicarbonate.[7,28] Carbonate is a much more efficient scavenger of hydroxyl radicals due to its higher rate constant with the hydroxyl radical ($k = 3.9 \times 10^8$ M^{-1} s^{-1})[29] than is bicarbonate ($k = 8.5 \times 10^6$ M^{-1} s^{-1}).[29] UV-peroxide treatment is therefore more efficient at lower pH values. Reactions with ozone tend to be more effective at generating hydroxyl radicals at higher pH values (Equations 14–17), so in waters of high alkalinity the hydroxyl radicals are not efficiently utilized.

Ozone can have advantages in waters of high inherent UV absorbance, since the reaction of ozone with hydrogen peroxide does not rely on light. There have been theoretical comparisons of the relative efficiencies of UV/ozone vs UV/peroxide.[3] In these studies, the method of optimization has been based on using the same molar concentrations of ozone or hydrogen peroxide. In practice, this is rarely done, since it is not possible to achieve a high concentration of ozone in water, but virtually any concentration of hydrogen peroxide can be introduced. We prefer to base a comparison on the amount of energy required to generate a mole of hydroxyl radicals under optimum conditions for each type of treatment. In this case, the amount of hydrogen peroxide added is sufficient to capture at least 90% of the incoming photons below 280 nm. Assuming a 30% efficiency of generating photons below 300 nm and an ozone generation efficiency of 10 kWh per pound of ozone from air,[30] the following energy requirements are found:

Treatment	Energy per mole of ˙OH
Ozone/peroxide	1.05 kWh
UV/peroxide	0.54 kWh

Under optimum conditions then, the generation of hydroxyl radicals from UV/peroxide is more energy efficient. As the fraction of light absorbed by the peroxide drops, the energy required to generate hydroxyl radicals increases proportionately. Thus, in waters of high inherent UV absorption, the relative energy requirements can reverse, making the ozone-peroxide system more energy efficient.

The optimum treatment to use can vary for each particular wastewater. In some applications, ozone may make the most practical sense, while in others peroxide is the oxidant of choice. When a gas-phase oxidant, i.e., ozone, is added to water containing volatile organics, air stripping of these components can occur. If measures are not taken to prevent and/or allow for air stripping, a low EE/O will be the apparent (and incorrect) result of a UV-ozone treatment. Most waters must be tested prior to design of a treatment system to ensure that there are no interfering species in the matrix that have not been accounted for.

Ozone and hydrogen peroxide are not the only oxidants that have been used. In some cases, HOCl has been effective. Also some companies claim to have proprietary additives that can decrease the EE/O values by as much as a factor of 5 in specific treatment cases.

Photoreduction

Some pollutants, such as chloroalkanes (e.g., chloroform), treat very slowly in AOP systems. The reason can be seen in Table 3, where it is apparent that these compounds have very small rate constants for reaction with $\cdot$OH radicals. Note, however, that they have very high rate constants with the hydrated electron e_{aq}^-, which is one of the strongest reducing agents known.* There are several known photochemical reactions by which the hydrated electron can be generated in water (e.g., Equation 8).

The hydrated electron acts principally as an agent to convert organic halogen atoms into inorganic halide ions. For example, chloroform reacts according to

$$CHCl_3 + e_{aq}^- \longrightarrow \cdot CHCl_2 + Cl^- \tag{18}$$

Recently, Bolton et al.[31] reported that a UV photoreduction treatment of chloroform (and other chloroalkanes) proceeds at rates up to four times faster than is possible with a UV-H_2O_2 treatment.

Direct Photodegradation

Some pollutants are able to dissociate in the presence of UV light only. For this to happen, the pollutant must absorb light that the lamp emits and have a reasonable quantum yield of photodissociation. Organic pollutants absorb light over a wide range of wavelengths, but generally absorb more strongly to lower wavelengths especially below 250 nm. In addition, the quantum yield of photodissociation tends to increase at lower wavelengths, since the photon energy is increasing; this increases the chance of photofragments escaping the solvent cage.[32] One example of a pollutant that can be successfully photodegraded is N-nitrosodimethylamine (NDMA), which has a strong absorption band at 227 nm ($\varepsilon_\lambda = 7000$ M^{-1} cm^{-1}) and a photodissociation quantum yield of 0.13 at pH 7 and 0.25 at pH 2.[33] Other candidate pollutants are halogenated compounds such as Freons® and iron cyanide complexes, which react slowly with $\cdot$OH. The net

---

*The reducing power of the hydrated electron is comparable to metallic potassium (see Table 2).

chemical result of photodissociation is usually oxidation, since the free radicals generated from the dissociation can react with dissolved oxygen in the water. In practice, the range of wastewaters that can be successfully treated by UV alone is very limited. Most contaminated waters contain a variety of pollutants that must be destroyed, and the chances of all of these pollutants being able to photodegrade in a cost effective manner is slight. It will almost always be advantageous to add an oxidant for the treatment.

MECHANISM STUDIES

Most applications of photodegradation involve the use of advanced oxidation. The mechanism of advanced oxidation is complex and has not been elucidated for many compounds. The exact reaction mechanism depends on a number of factors, including other organic and inorganic species present, their concentration, and the solution pH. Almost always the mechanism involves several steps and intermediates, even when the contaminant is a simple molecule, such as chloroform.

The initial attack of the hydroxyl radical gives either an addition or a hydrogen abstraction; aromatic and olefinic compounds give addition, whereas saturated organics give hydrogen abstraction. The resulting radical can react with dissolved oxygen, thus beginning a series of oxidative reactions leading to eventual formation of bicarbonate and water. In general, if chlorine or other halogens are present, then these are converted to inorganic halide ions early in the reaction so that quantitative formation of halide is observed suggesting very little or no chlorinated organic intermediates are formed.[22,34]

Most mechanistic investigations have involved the use of ozone. There have also been studies using UV/peroxide and radiolysis. There is considerable mechanistic overlap because all these processes involve hydroxyl radicals as the principal oxidative species. Mechanistic work has included studies on the photooxidation of methanol,[35] chloroform,[36] benzene,[37] phenol,[38-40] and 2,4-dinitrotoluene.[41] We shall consider the postulated mechanisms of oxidation of methanol and benzene to highlight the features of the chemistry of advanced oxidation.

In the case of methanol, the proposed mechanism[35] involves an initial attack of the hydroxyl radical on methanol to abstract a hydrogen atom to give the hydroxymethyl radical ($^{\cdot}CH_2OH$), which then reacts with oxygen to form formaldehyde. Hydroxyl radical attack on formaldehyde leads to production of formic acid and then to carbon dioxide. The postulated reaction scheme is shown in Scheme 1.

In the case of benzene, the postulated mechanism is much more complex.[37] Here, the initial attack by the hydroxyl radical results in addition to the aromatic ring. The initial product is phenol, which is then successively oxidized to 1,2- and 1,4-dihydroxybenzene (catechol and hydroquinone). These can be oxidized to the corresponding quinones, or further addition of hydroxyl radical gives 1,2,3- and 1,2,4-trihydroxybenzene. At this stage, the aromatic ring is cleaved on further

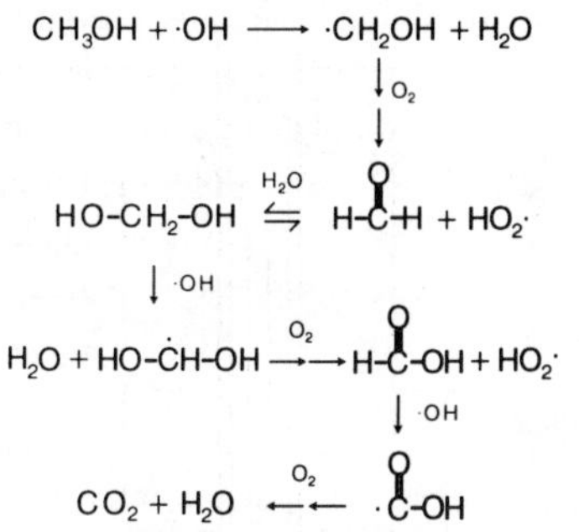

$$CH_3OH + \cdot OH \longrightarrow \cdot CH_2OH + H_2O$$

$$\downarrow O_2$$

$$HO\text{-}CH_2\text{-}OH \overset{H_2O}{\rightleftharpoons} H\text{-}\overset{O}{\underset{}{C}}\text{-}H + HO_2\cdot$$

$$\downarrow \cdot OH$$

$$H_2O + HO\text{-}\overset{\cdot}{C}H\text{-}OH \overset{O_2}{\longrightarrow} H\text{-}\overset{O}{\underset{}{C}}\text{-}OH + HO_2\cdot$$

$$\downarrow \cdot OH$$

$$CO_2 + H_2O \overset{O_2}{\longleftarrow} \cdot \overset{O}{\underset{}{C}}\text{-}OH$$

Scheme 1.

oxidation to yield unsaturated six-carbon aldehydes and acids, such as muconaldehyde or muconic acid. These are then oxidized to smaller aldehydes and acids, such as glyoxal, oxalic acid, maleinaldehyde, and maleic acid. Subsequent oxidation leads to formic acid and carbon dioxide.

While the mechanism for benzene is much more complex than that for methanol, the same general pattern of oxidation is observed. That is,

organic pollutant $\longrightarrow$ aldehydes $\longrightarrow$ carboxylic acids $\longrightarrow$ carbon dioxide

Although it is possible to drive the reaction all the way to carbon dioxide, this is not always necessary. Generally, the toxicity of the treated water is less than the untreated water, since the process converts toxic chlorinated and aromatic compounds into biodegradable species such as aldehydes and carboxylic acids. Complete TOC reduction is possible,[19] but not always cost effective. The ultimate fate of the carbon is likely to be bicarbonate at the pH levels normally carried out.

Lipczynska-Kochany and Bolton[42-44] have utilized a flash photolysis technique with HPLC detection of products to follow the reaction sequence in the photolysis of 4-chlorophenol in aqueous solution. Under direct photolysis conditions,[42,43] they find that 4-chlorophenol photolyses almost entirely to form p-benzoquinone, which then further photolyses to form hydroquinone and 2-hydroxy-p-benzoquinone. Under conditions where H_2O_2 is added so that almost all the UV light is absorbed by H_2O_2, the principal intermediate is 4-chlorocatechol, with a minor amount of hydroquinone. The latter products are those expected from the attack of $\cdot OH$ radicals on 4-chlorophenol.

In the photoreduction treatment of chloroform, Bolton et al.[31] have shown that the $\cdot CHCl_2$ radicals, formed in Equation 18, dimerize to form 1,1,2,2-tetrachloroethane, which also reacts with e_{aq}^- to generate Cl^- ions. Other intermediates found were trichloroethylene and cis- and $trans$-1,2-dichloroethylene. Eventually, almost all the organic chlorine is converted to chloride, since analysis indicated that over 95% of the original organic chlorine can be accounted for as chloride.

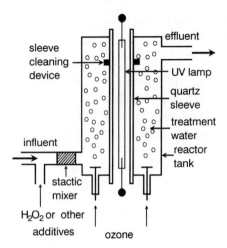

Figure 4. Diagram of a UV reactor for homogeneous photodegradation treatments.

HO$_2^-$ radicals are often formed as intermediates in AOP reactions, but react rather slowly with most substrates. Fortunately, HO$_2^-$ photolyses ($\varepsilon_\lambda = 1300\ M^{-1}$ cm^{-1} at 230 nm)[45] to give ·OH radicals with a high quantum yield.[46] Thus, if the UV lamp has a strong output in this wavelength region, the relatively unreactive HO$_2^-$ radicals can be converted into very reactive ·OH radicals.

THE REACTOR

In a typical application, the UV reactor is comprised of an inlet with provision for addition of hydrogen peroxide or other additives, a chamber which houses the lamps protected in quartz sleeves, possible oxidant mixing elements, and an outlet. A stylized schematic of a reactor is shown in Figure 4. In this case, a reactor employing vertically mounted lamps is shown, although lamps can also be mounted horizontally.

It is important that the chemicals, such as hydrogen peroxide, be well mixed prior to entrance to the reactor. In flow-through applications, the amount of treatment can be dramatically decreased if the peroxide is not adequately mixed. Mixing is also important inside the reactor, especially in waters of high inherent UV absorption. In this case, all the light can be absorbed in a fraction of a centimeter, so it is important to exchange the treated water nearest the quartz with untreated water nearest the reactor wall.

The pathlength between the quartz sleeve and the reactor wall should be sized to allow for a reasonable fraction (>0.9) of the light to be absorbed. Allowing the UV light to hit the outside wall results in a wastage of light. Most commercial systems are constructed of stainless steel, which is a poor reflector of UV light. It is not practical or cost effective to use reflective walls, since it is less expensive to increase the pathlength and fouling of the reflectors from continuous use would erase any benefits from reflection of light.

It is essential that the quartz sleeve remain clean during operation. Fouling of the sleeve will block the transmittance of light into the water and hence slow down or even stop the treatment. The length of time before the sleeve fouls is dependent on the application, but can vary anywhere from 5 min to 2 weeks.

The reactor design must be different to accommodate the use of ozone. An ozone contactor is required. This is frequently a sparging device, although these devices are prone to clogging during continuous use. Provision must be made to allow for adequate contact of the ozone with the water. This can be a contact tower or a baffle tank arrangement. In addition, the offgas must be separated from the water and treated to destroy any residual ozone.

UV LIGHT SOURCES

This section deals with a discussion of UV lamps for use in water treatment. We begin with an introduction to the type of UV light required and then discuss different types of lamps that can be employed in advanced oxidation technology.

Type of UV Light

In photochemical experiments, the rate of the reaction is equal to the light absorbed (E_a) by the photoreactive species multiplied by the quantum yield for the particular process under question.* The amount of light absorbed depends on the total incident light irradiance, and the amount of that light absorbed by the species, according to Equation 3b.

In order to maximize the rate of reaction, there must be a good wavelength overlap between the emission of the lamp and the absorbance of the absorbing species. In practice, most organic compounds, as well as hydrogen peroxide, absorb more strongly to lower wavelengths. Figure 5 presents the absorption spectrum for selected pollutants in water and clearly shows the trend to increased absorption at lower wavelengths. Thus, it is desirable to have a UV lamp that has its output in the same region where pollutants absorb light, i.e., less than 300 nm. Some compounds, such as polycyclic aromatic hydrocarbons (PAHs), do have significant absorption above 300 nm, but generally absorption of light at these high wavelengths does not lead to significant photodegradation.

Below 200 nm, the absorption of water begins to come into play, and at 185 nm all of the light is absorbed by a few micrometers of water. Thus, it generally can be stated that for homogeneous photochemical destruction of waterborne

---

*Since the spectral irradiance E_λ and the quantum yield ϕ_λ may vary with wavelength, the reaction rate must be computed in small wavelength intervals and then integrated over the wavelength range in which photolysis is significant.

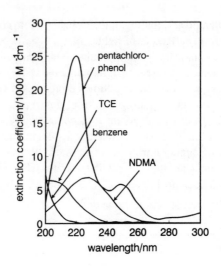

Figure 5. UV spectra of some common pollutants.

pollutants, light in the range of 200–300 nm is required. We must now consider the type of lamps available that can deliver light in this region.

Lamp Types

For a more complete discussion of photochemical lamps, see References 47 and 48. There are four types of lamps that are available with significant output between 200 nm and 300 nm. These are

- low-pressure mercury lamps
- medium-pressure mercury lamps
- pulsed xenon flashlamps
- proprietary lamps

Low-Pressure Mercury Lamps

Low-pressure mercury lamps operate at low pressures (~10^{-2} torr) and temperatures. The operating temperature is typically 40°C. These lamps have low power inputs in the range of 20–120 W. They generally have long lifetimes in the range of 4,000–10,000 hr. About 90% of the light output is centered around 254 nm, with an electrical efficiency of around 30%. There is also a significant output at 185 nm, although this light would be completely absorbed by a thin film of water. Since almost all of the output is around 254 nm, the absorbing species must have significant light absorption at this wavelength. Ozone does absorb very strongly at 254 nm, and hence these lamps have found wide use in UV-ozone systems. Hydrogen peroxide absorbs only weakly at 254 nm; however, it is possible to absorb significant fractions of the light by hydrogen peroxide by ensuring that the hydrogen peroxide concentration is high enough. Low-pressure lamps are not particularly effective for treating refractory compounds such as chlorinated alkanes which require light of wavelengths below 240 nm to photodissociate. This is one drawback in the use of low-pressure lamps. The major drawback is the low power available, which means that a very large number of lamps would be required to treat water at reasonable flow rates. Not only does this increase the size of the system, but also the capital cost and lamp replacement costs.

Medium-Pressure Mercury Lamps

Medium-pressure mercury lamps operate with more mercury in the bulb so they produce higher pressures and temperatures than low-pressure lamps. The bulb temperature is typically in the range of 400–600°C and even higher. The lamp lifetime is shorter than with low pressure lamps, in the range of 3000–4000 hr. These lamps emit light over a wide wavelength range, including the UV, visible, and IR regions. The efficiency in the UV range 200–300 nm is in the range of 5–20%, which is lower than for low-pressure lamps and would reflect in higher electrical power operating costs. Most medium-pressure lamps do not have significant emission below 230 nm. The lamp powers available are quite varied, from 100 to 60,000 W. Thus, use of these lamps result in the need for less lamps than for low-pressure lamp systems, but with higher electrical costs.

Pulsed Xenon Flashlamps

These lamps operate differently than the mercury vapor lamps. Instead of being continuous sources, the lamps are pulsed rapidly, which involves applying a short intense burst of energy to the lamp followed by an off time. By increasing the current density to the lamp, the amount of UV light generated can be increased. With the current state of the flashlamp technology, a theoretical maximum of 28% of the lamp output can be in the range of 200–300 nm for a plasma temperature of approximately 15,000 K. Practical efficiencies, however, are around 20%. Assuming a maximum power supply efficiency of 90% gives a total output efficiency of 18%, which is a slight improvement over the typical medium pressure mercury lamps. The xenon flashlamps also produce more output below 240 nm than do the medium-pressure mercury lamps. The biggest drawback to their use is a short lamp lifetime. Since the lamps must operate at very high current densities to give the approximately 20% UV-C output, this strains the lamps and reduces the lifetime to the range of 50–500 hr. The short lifetime makes these lamps quite impractical.

Proprietary Lamps

Some commercial advanced oxidation suppliers use proprietary lamps, which are improved light sources based on medium-pressure mercury lamps. Some lamps are operated at higher power densities than are the conventional medium-pressure lamps. This results in a smaller lamp size for the same power input and also gives an improved efficiency and spectral emission. The operating temperatures of the lamps are in the range of 700–1000°C. The lamps can have UV-C efficiencies of approximately 30% and emit strongly below 240 nm. Lamp lifetimes are usually in excess of 3000 hr.

SOME GENERAL REMARKS

Effect of Pollutant Concentrations

There are two factors that come into play as the concentration of the pollutant varies. The disappearance of the pollutant frequently follows first-order kinetics. This means that it takes as much energy to go from 100 to 1 ppm as it does to go from 100 to 1 ppb. Thus, the amount of pollutant destroyed per unit energy input is greater at higher concentrations. The second factor is that the EE/O does tend to decrease at lower initial pollutant concentrations. Pollutants with more carbon atoms tend to give the largest differences as the initial concentration is varied. The likely reason for this is that by-products absorb light and act as a filter.

Deep-Cleaning Requirements

It is usually possible to reduce the pollutant concentrations to any desired level, since photochemical reactions generally follow first-order kinetics. The cost of treatment increases, however, with the degree of removal required. In fact, as noted above, the EE/O decreases at lower concentrations so the treatment rate constant actually increases at lower concentrations. Thus, the technology is able to be applied to ever more stringent environmental regulations as they are enforced.

Economic Considerations

When considering the use of advanced oxidation, a full economic analysis should be taken. This includes the capital costs and operating and maintenance costs. Operating costs include the costs of electrical energy, oxidants and other chemical costs, and lamp replacement costs. In general, for treatment of pollutants in the low parts per million or lower range, the electrical energy costs are the largest component, whereas at higher concentrations the costs of the oxidant can dominate. Frequently, the operating and capital costs can be significantly lowered by the addition of proprietary additives or catalysts.

Sample Preservation

When carrying out tests using advanced oxidation, it is important to adequately preserve the samples to ensure that no background reactions, due to the presence of any residual hydrogen peroxide, occur before analysis. Unfortunately, there does not seem to be any sufficient literature protocol for practitioners to follow. Based on our experience, we feel the following procedure will result in representative samples:

- adjust sample pH to 7 ± 1
- destroy residual hydrogen peroxide with addition of sodium sulfite (1:1 molar ratio)
- store samples at 4°C

The sulfite can be added as a 10% solution, or alternatively the pure solid can be weighed directly into the sample container prior to addition of the sample. Once the peroxide is destroyed, then the sample can be stored at acidic pH if there is a concern about biological decomposition. Under no circumstances do we recommend that the only sample preservation is adjustment to acidic pH and cold storage. If there is any iron in the water, then the samples will likely decompose due to Fenton's reaction. It is better to leave the samples at neutral pH than to acidify if any residual hydrogen peroxide is not destroyed. Alternatively, we do not recommend the sometimes-used procedure where an excess of ferrous ammonium sulfate is added at acidic pH to destroy residual peroxide, since we have observed[49] that Fenton's decomposition of the samples occurs under these conditions.

CONCLUSIONS

The AOP technologies offer virtually complete destruction of almost all organic pollutants in water and air. While several AOP systems have been installed and are successfully treating wastewaters, the AOP technologies are still in the early stages of development and implementation. Only the homogeneous photolysis processes have been fully implemented commercially, and even in that case only a small fraction of potential sites are employing these processes. Nevertheless, in important application niches, the AOP technologies are proving to be cost effective and often are the only method available to achieve complete destruction of the contaminants.

On the research side, we still know only a small fraction of the information necessary to determine the mechanism and pathways of the complex reactions taking place in these systems. There is thus a great need for fundamental photochemical studies to elucidate these mechanisms, to develop new photochemical and other degradative processes, and to understand and remove kinetic barriers. It is interesting that the vast majority of published photochemical papers are on systems in nonaqueous solution. An increased emphasis on fundamental photochemical studies in aqueous solution would be very helpful.

ACKNOWLEDGMENTS

We are grateful to our colleagues at Solarchem: Keith Bircher, Adele Buckley, Peter O'Connor, Sam Stevens, and Ali Safarzadeh-Amiri for their helpful comments.

REFERENCES

1. Braun, A. M., Progress in the applications of photochemical conversion and storage, in *Photochemical Conversion and Storage of Solar Energy*, Schiavello, M. and Pelizzetti, E., Eds., Kluwer, Dordrecht, 1991.
2. Halevy, M., Ed., *Proceedings of a Symposium on Advanced Oxidation Processes for the Treatment of Contaminated Water and Air*, Wastewater Technology Centre, Burlington, Ontario, Canada, 1990.
3. Glaze, W. H., Kang, J.-W., and Chapin, D. H., The chemistry of water treatment processes involving ozone, hydrogen peroxide and ultraviolet radiation, *Ozone Sci. Eng.*, 9, 335, 1987.
4. Fenton, H. J. H., Oxidation of tartaric acid in the presence of iron, *J. Chem. Soc.*, 65, 899, 1894.
5. Eisenhauer, H. R., Oxidation of phenolic wastes, *J. Water Pollut. Control Fed.*, 36, 1116, 1964.
6. Hoigné, J. and Bader, H., Ozonation of water: role of hydroxyl radicals as oxidizing intermediates, *Science*, 190, 782, 1975.
7. Paillard, H., Brunet, R., and Dore, M., Optimal conditions for applying an ozone — hydrogen peroxide oxidizing system, *Water Res.*, 22, 91, 1988.
8. Serpone, N. and Pelizzetti, E., Eds., *Photocatalysis: Fundamentals and Applications*, Wiley-Interscience, New York, 1989.
9. Getoff, N., Radiation- and photoinduced degradation of pollutants in water. A comparative study, *Radiat. Phys. Chem.*, 37, 673, 1991.
10. Cunningham, J., Pelizzetti, E., Pichat, P., Serpone, N., Fox, M. A., and Bahnemann, D., Photocatalytic treatment of waters, in *Aquatic and Surface Photochemistry*, Helz, G. R., Zepp, R. G., and Crosby, D. G., Eds., Lewis Publishers, Chelsea, MI, 1993, chap. 21.
11. Prengle, H. W., Jr., Mauk, C. E., Legan, R. W., and Hewes, C. G., III, Ozone/UV process effective wastewater treatment, *Hydrocarbon Process.*, 54, 82, 1975.
12. Garrison, R. L., Mauk, C. E., and Prengle, H. W., Jr., Advanced ozone-oxidation system for complexed cyanides, in *1st International Symposium on Ozone for Water and Wastewater Treatment*, International Ozone Institute, Watersberg, CN, 1975, 551.
13. Baxendale, J. H. and Wilson, J. A., The photolysis of hydrogen peroxide at high light intensities, *Trans. Faraday Soc.*, 53, 344, 1957.
14. Skuratova, S. I., The generation and identification of hydroxy-radicals in aqueous solution, *Russ. J. Phys. Chem.*, 50, 179, 1976.
15. Koubek, E., Photochemically induced oxidation of refractory organics with hydrogen peroxide, *Ind. Eng. Chem. Process Des. Dev.*, 14, 348, 1975.
16. Bulla, C. D., III and Edgerley, E., Jr., Photochemical degradation of refractory organic compounds, *J. Water Pollut. Control Fed.*, 40, 547, 1968.
17. Peyton, G. R., Huang, F. Y., Burleson, J. L., and Glaze, W. H., Destruction of pollutants in water with ozone in combination with ultraviolet radiation. 1. General principles and oxidation of tetrachloroethylene, *Environ. Sci. Technol.*, 16, 448, 1982.
18. Peyton, G. R. and Glaze, W. H., Mechanism of photolytic ozonation, in *Photochemistry of Environmental Aquatic Systems*, Zika, R. G. and Cooper, W. J., Eds., ACS Symposium Ser. No. 327, American Chemical Society, Washington, D.C., 1987, chap. 6.
19. Malaiyandi, M., Sadar, M. H., Lee, P., and O'Grady, R., Removal of organics in water using hydrogen peroxide in presence of ultraviolet light, *Water Res.*, 14, 1131, 1980.
20. Mansour, M., Parlar, H., and Korte, F., Removal of pollutants from the aquatic environment by photooxidation, *Stud. Environ. Sci.*, 23, 457, 1984.

21. Ho, P. C., Photooxidation of 2,4-dinitrotoluene in aqueous solution in the presence of hydrogen peroxide, *Environ. Sci. Technol.*, 20, 260, 1986.

22. Sundstrom, D. W., Klei, H. E., Nalette, T. A., Reidy, D. J., and Weir, B. A., Destruction of halogenated aliphatics by ultraviolet catalyzed oxidation with hydrogen peroxide, *Haz. Waste Haz. Mater.*, 3, 101, 1986.

23. Zeff, J. D., UV-OX process for the effective removal of organics in wastewaters, in Water-1976: II. Biological wastewater treatment, *AIChE Symp. Ser.*, 73, 206, 1976.

24. Hunt, J. P. and Taube, H., The photochemical decomposition of hydrogen peroxide. Quantum yields, tracer and fractionation effects, *J. Am. Chem. Soc.*, 74, 5999, 1952.

25. Luňák, S. and Sedlák, P., Photoinitiated reactions of hydrogen peroxide in the liquid phase, *J. Photochem. Photobiol.*, 68, 1, 1992.

26. Weeks, J. L. and Matheson, M. S., The primary quantum yield of hydrogen peroxide decomposition, *J. Am. Chem. Soc.*, 78, 1273, 1956.

27. Taube, H., Photochemical reactions of ozone in solution, *Trans. Faraday Soc.*, 53, 656, 1957.

28. Staehelin, J. and Hoigné, J., Decomposition of ozone in water in the presence of organic solutes acting as promotors and inhibitors of radical chain reactions, *Environ, Sci. Technol.*, 19, 1206, 1985.

29. Buxton, G. V., Greenstock, C. L., Helman, W. P., and Ross, A. B., Critical review of rate constants for reactions of hydrated electrons, hydrogen atoms and hydroxyl radicals in aqueous solution, *J. Phys. Chem. Ref. Data*, 17, 513, 1988.

30. Elsenhaus, K. H., A comparison of ozone generation from air and from oxygen from an economic, operational and safety point of view, in *Proceedings of the 10th Ozone World Congress*, Vol. 1, International Ozone Association, 1991, 65.

31. Bolton, J. R., Cater, S. R., and Safarzadeh-Amiri, A., The use of reduction reactions in the photodegradation of organic pollutants in waste streams, paper presented at the Symposium on Environmental Aspects of Surface and Aquatic Photochemistry, American Chemical Society Conference, San Francisco, CA, April, 1992.

32. Wayne, R. P., *Principles and Applications of Photochemistry*, Oxford Science Publications, 1988, 55.

33. Ho, T.-F. L. and Bolton, J. R., Photodegradation of N-nitrosodimethylamine in aqueous solution: mechanism and quantum yield measurements, unpublished work.

34. Cooper, W., Nickelsen, M., Waite, T., and Kurucz, C., High-energy electron beam irradiation: an advanced oxidation process for the treatment of aqueous based organic hazardous wastes, in *Proceedings of a Symposium on Advanced Oxidation Processes for the Treatment of Contaminated Water and Air*, Halevy, M., Ed., Wastewater Technology Centre, Burlington, Canada, 1990.

35. Peyton, G. R., Smith, M. A., and Peyton, B. M., Photolytic Ozonation for Protection and Rehabilitation of Ground-Water Resources: A Mechanistic Study, Report 87–206, University of Illinois Water Resources Center, 1987.

36. Getoff, N., Advancements of radiation induced degradation of pollutants in drinking and waste water, *Appl. Radiat. Isot.*, 40, 585, 1989.

37. Srinivasan, T. K. K., Balakrishnan, I., and Reddy, M. P., On the nature of the products of radiolysis of aerated aqueous solutions of benzene, *J. Phys. Chem.*, 73, 2071, 1969.

38. Yamamoto, Y., Niki, E., Shiokawa, H., and Kamiya Y., Ozonation of organic compounds. 2. Ozonation of phenol in water, *J. Org. Chem.*, 44, 2137, 1979.

39. Otake, T., Tone, S., Kono, K., and Nakao, K., Photo-oxidation of phenols with ozone, *J. Chem. Eng. Jpn.*, 12, 289, 1979.

40. Ho, P. C., Evaluation of ultraviolet light/oxidizing agent as a means for the degradation of toxic organic chemicals in aqueous solution, in *Management of Hazardous and Toxic Wastes in Process Industries*, Kdaczkowski, S. T. and Crittenden, B. D., Eds., Elsevier Applied Science, 1987, 563.

41. Ho, P. C., Photooxidation of 2,4-dinitrotoluene in aqueous solution in the presence of hydrogen peroxide, *Environ. Sci. Technol.*, 20, 260, 1986.

42. Lipczynska-Kochany, E. and Bolton, J. R., Flash photolysis/HPLC method applied to the study of photodegradation reactions, *J. Chem. Soc. Chem. Commun.*, 1596, 1990.

43. Lipczynska-Kochany, E. and Bolton, J. R., Flash photolysis/HPLC method applied to the study of photodegradation reactions: applications to 4-Chlorophenol in aerated aqueous solution, *J. Photochem. Photobiol.*, 58, 315, 1991.

44. Lipczynska-Kochany, E. and Bolton, J. R., Flash photolysis/HPLC applications II: direct photolysis vs. hydrogen peroxide mediated photodegradation of 4-chlorophenol as studied by a flash photolysis/HPLC technique, *Environ. Sci. Technol.*, 26, 259, 1992.

45. Getoff, N. and Prucha, M., Spectroscopic and kinetic characteristics of HO_2^- and O_2^- species studied by pulsed radiolysis, *Z. Naturforsch.*, 38a, 589, 1983.

46. Lee, M., Observation of $O(^1D)$ produced from the photodissociation of HO_2 at 193 and 248 nm, *J. Chem. Phys.*, 76, 4909, 1982.

47. Calvert, J. G. and Pitts, J. N., Jr., *Photochemistry*, John Wiley & Sons, New York, 1966.

48. Phillips, R., *Sources and Applications of Ultraviolet Radiation*, Academic Press, New York, 1983.

49. Cater, S. R., unpublished work.

50. Baulch, D. L., Cox, R. A., Crutzen, P. J., Hampson, R. F., Jr., Kerr, J. A., Troe, J., and Watson, R. T., Evaluated kinetic and photochemical data for atmospheric chemistry: supplement I. CODATA Task Group on Chemical Kinetics, *J. Phys. Chem. Ref. Data*, 11, 327, 1982.

51. Taube, H., Photochemical reactions of ozone in solution, *Trans. Faraday Soc.*, 53, 656, 1957.

52. Weeks, J. L. and Matheson, M. S., The primary quantum yield of hydrogen peroxide decomposition, *J. Am. Chem. Soc.*, 78, 1273, 1956.

53. Amimoto, S. I., Force, A. P., and Wisenfeld, J. R., Ozone photochemistry: production and deactivation of $O(^1D)$ following photolysis at 248 nm, *Chem. Phys. Lett.*, 60, 40, 1978.

54. Dainton, F. S., The primary quantum yield in the photolysis of hydrogen peroxide at 3130 Å and the primary radical yield in the X- and γ-radiolysis of water, *J. Am. Chem. Soc.*, 78, 1278, 1956.

55. Spinks, J. W. T. and Woods, R. J., *An Introduction to Radiation Chemistry*, Wiley-Interscience, New York, 1990, 278.

56. Lightfoot, P. D., Cox, R. A., Crowley, J. N., Destriau, M., Hayman, G. D., Jenkin, M. E., Moorgat, G. K., and Zabel, F., Organic peroxy radicals, kinetics, spectroscopy and tropospheric chemistry, *Atmos. Environ.*, 26A, 1805, 1992.

CHAPTER **34**

Determination of Rate Constants for Reactive Intermediates in the Aqueous Photodegradation of Pollutants Using a Spin-Trap/EPR Method

Aitken R. Hoy and James R. Bolton

INTRODUCTION

There has been considerable interest in recent years in the photochemical removal of pollutants from water. Most commercial processes (called advanced oxidation processes, AOP) are based on the photochemical generation of hydroxyl radicals (from the photolysis of H_2O_2 or O_3), which then attack and oxidize the pollutants. However, there are some refractory pollutants, such as halogenated alkanes, that are slow to react. Recently,[1] it has been found that photochemically generated hydrated electrons are effective in treating these compounds. Since this is now a commercial process, it is important to measure the rate constants for the reaction of the hydrated electrons e_{aq}^- with various pollutant molecules under different conditions. For example, for halogenated alkanes RX, the reaction of interest is

$$e_{aq}^- + RX \longrightarrow R^\cdot + X^- \qquad (1)$$

Traditionally, the reaction rates of hydrated electrons have been measured using pulse radiolysis. In this technique, the electrons are produced by a beam of high energy particles depositing energy into the solution. The electrons quickly thermalize within a solvent cage. This species is called the *hydrated electron* and may be detected by its strong absorption in the red. Rates of reaction are deduced by measuring the decay rate of the hydrated-electron absorption as a function of

0-87371-871-2/94/$0.00+$.50

491

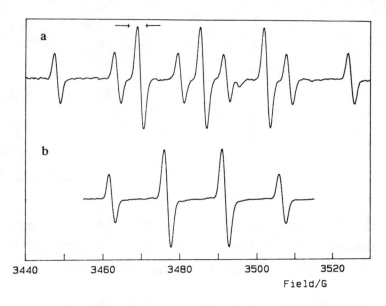

Figure 1. **(a)** The EPR first-derivative spectrum of the spin-adduct DMPO-H. The peak
used for monitoring the DMPO-H concentration is indicated by arrows. **(b)** The
EPR first-derivative spectrum of the spin-adduct DMPO-OH.

reactant concentration. Almost all available reaction rates have been measured by
this decay rate technique.[2]

In our method, hydrated electrons are produced by the UV photolysis of
ferrocyanide.[3]

$$Fe(CN)_6^{4-} \longrightarrow Fe(CN)_6^{3-} + e_{aq}^- \qquad (2)$$

The hydrated electrons are detected indirectly using an electron paramagnetic
resonance (EPR) spectrometer in conjunction with the spin-trap 5,5-dimethyl-1-
pyrroline N-oxide (DMPO). DMPO reacts with hydrated electrons in the presence
of water to form the spin-adduct DMPO-H, whose nine-line spectrum (Figure 1a)
is easily monitored, even in the presence of the hydroxyl radical adduct (DMPO-
OH) (Figure 1b) which is often produced along with DMPO-H. The rate constant
for a given pollutant molecule is deduced by measuring the initial rate of production of DMPO-H as a function of pollutant concentration. This approach mini-
mizes the effects of secondary reactions. Since the pollutant and DMPO compete
for the scarce e_{aq}^- species, the initial rate of production of DHPO-H decreases as
the pollutant concentration increases (Figure 2).

EXPERIMENTAL

Experiments were performed on a Bruker ESP 300 EPR spectrometer using a
150-W water-filtered xenon-mercury arc lamp for photolysis. Solutions were

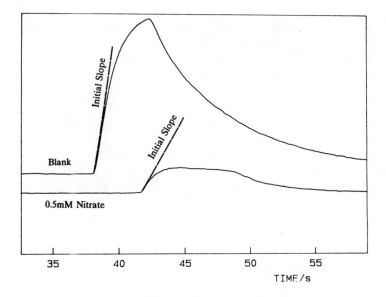

Figure 2. Time evolution of DMPO-H signal. The upper curve has no scavenger, and the lower curve has 0.5 mM nitrate as scavenger. The initial slopes are indicated schematically.

introduced into the quartz EPR cell inside the cavity using a peristaltic pump. All solutions were approximately 1.0 mM in potassium ferrocyanide, 0.5 mM in DMPO, and 1.0 mM in phosphate buffer, while pollutant/calibrant concentrations were in the range 0.1–2.0 mM. In some cases, small quantities of ethanol (which does not react with e_{aq}^-) were added to the solution to help dissolve the pollutant. All experiments reported here were performed at room temperature and at pH 7.0.

The concentration of DMPO-H was monitored by tuning the spectrometer to the peak of a convenient spectral line (see Figure 1a) and recording the signal amplitude before and during irradiation. These time scans were recorded digitally at 82 msec/channel and later analyzed by computer to extract the initial slope. The background level was estimated by averaging all points before the start of illumination (at least 50). The next 20 points were fitted to a quadratic and the initial slope determined at the point where the quadratic intersected the background. Subsequently, the first point of the quadratic was omitted, and the next 20 points were used in the fit to give a new value of the initial slope. This procedure was continued until a maximum in the initial slope was found. This algorithm can be rationalized as follows. There is a region of maximum slope in the rising portion of the signal vs time plot. This corresponds to the region where the effects of the finite (though small) instrumental time constant at short times is offset by the effects of finite DMPO-H lifetime at longer times. This maximum slope is then extrapolated to time zero with a correction for the curvature of the plot.

An experimental run typically involved five or six repeated measurements at each of five different scavenger concentrations. In addition, five or six repeats at zero scavenger concentration (blank) were measured at the start and end of each run to monitor lamp stability and changes in alignment. Thus, a typical run produced about 40 values for the initial slope and corresponding scavenger concentrations. Runs in which the initial and final blanks disagreed were discarded. Reproducibility, as measured by the standard deviation of repeated initial slopes, was approximately 4% of the mean for blanks (strongest signals) and approximately 10% of the mean for the highest scavenger concentrations (weakest signals).

For each scavenger, at least three experimental runs were performed using freshly prepared solutions. Scavenger concentrations were adjusted so that the range of initial slopes was approximately the same for all scavengers. Typically, this range was a factor of 3–4 between the blanks and the highest scavenger concentrations.

KINETIC MODEL AND ANALYSIS

We have analyzed our results within the following kinetic model:

$$Fe(CN)_6^{4-} \xrightarrow{\ h\nu\ } Fe(CN)_6^{3-} + e_{aq}^- \qquad \beta \ (3)$$

$$e_{aq}^- + H_2O + DMPO \longrightarrow DMPO\text{-}H + OH^- \qquad k_2 \ (4)$$

$$e_{aq}^- + P \longrightarrow Products \qquad k_3 \ (5)$$

$$e_{aq}^- + O_2 \longrightarrow O_2^{\cdot -} \qquad k_4 \ (6)$$

where β is the rate of production of e_{aq}^- and depends on both the ultraviolet (UV) flux and on the ferrocyanide concentration, neither of which were strictly controlled from run to run. The reaction with oxygen is included because most of our results were obtained in air-saturated water. Technical difficulties precluded the use of air-free solution at this time.

On applying the steady-state approximation to the above scheme, we find that

$$\left[e_{aq}^-\right] = \beta / \left(k_2[DMPO] + k_3[P] + k_4[O_2]\right) \qquad (7)$$

The rate of formation of DMPO-H (R) then becomes

$$R = d[DMPO\text{-}H]/dt = \beta\, k_2[DMPO] / \left(k_2[DMPO] + k_3[P] + k_4[O_2]\right) \qquad (8)$$

which can be rearranged to give

$$R = \beta\left(k_2[\text{DMPO}]/k_{\text{eff}}\right)\Big/\left\{1 + \left(k_3/k_{\text{eff}}\right)[P]\right\} \qquad (9)$$

or

$$R = \beta'\Big/\left\{1 + \left(k_3/k_{\text{eff}}\right)[P]\right\} \qquad (10)$$

where $k_{\text{eff}} = k_2$ [DMPO] $+ k_4$ [O_2], and $\beta' = \beta$ (k_2 [DMPO]/k_{eff}). Since the same concentration of DMPO (0.5 mM) was used in all experiments and the concentration of oxygen in air-saturated water is also constant at a given temperature, k_{eff} is a constant. Therefore, the final equation for R depends on two parameters, β', which is a constant for any one experimental run but varies from run to run, and (k_3/k_{eff}), which is a constant for any particular pollutant.

A nonlinear least squares computer program was used to fit the experimental data using Equation 10. The input data consisted of the concentrations and corresponding initial slopes for all repeats at all concentrations (~40 values/run). A complete data set consisted of N (N $\geq$ 3) independent runs (i.e., ~40 N initial slopes), while the output consisted of N values of β' plus one value of k_3/k_{eff}. The values of β' produced are of little absolute significance, as they depend on the unknown absolute sensitivity of the EPR spectrometer. However, in cases where it was known that the system had not been adjusted between runs, the repeatability of the initial slope of blanks was of the order of 5%. Unfortunately, this condition could not always be guaranteed and the value of β' for each run was left as a floating parameter.

The above fitting procedure was also applied to individual runs as a test of the run to run consistency of the values of k_3/k_{eff}. It was found that individual runs agreed at the combined 3σ level and also agreed with the overall fit at this level. It should also be noted that Equation 10 may also be cast in the linear form

$$1/R = \left\{1 + \left(k_3/k_{\text{eff}}\right)[P]\right\}\Big/\beta' \qquad (11)$$

When the data are plotted this way, the plots are linear; however, this introduces the need for properly weighted data. Thus, we prefer the nonlinear form used in Figure 3.

RESULTS

The raw output of our computer analysis consists of the "relative rate constant" k_3/k_{eff} along with related statistics. In order to convert this to a useful k_3, we require a value for k_{eff}. Although the rate constant k_4 and the concentration of oxygen in air-saturated water are well known,[2,4] the rate constant k_2 is not well established.[2] Therefore, we decided to calibrate our method against two much-studied scavengers; the nitrate ion and nitrobenzene. The results of this calibration are shown in

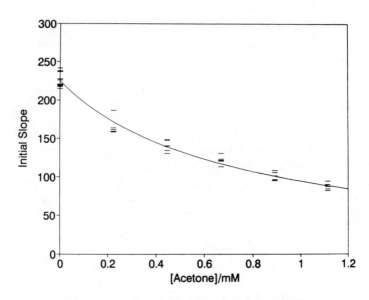

Figure 3. A typical data set and corresponding best-fit line from one run with acetone as scavenger. The nonlinear form (Equation 10) is displayed.

Table 1. The two values for k_{eff} are not only mutually consistent, but they are also in the middle of the range calculated from the literature values of k_2 and k_4.

Using the calibrated value of k_{eff}, we were then able to determine the rate constant for several pollutant molecules. These results are presented in Table 2.

DISCUSSION

The agreement between our results and the literature is erratic. However, there is no evidence of any trend as we appear to be above the literature value as often as we are below. The literature values themselves may not be entirely trustworthy. Most have been determined only once, and for the few that have

Table 1. Calibration of the Spin-Trap Method

	$k_3/10^{10}\ M^{-1}s^{-1}$ [a]	$(k_3/k_{eff})/mM^{-1}$ [b]	$k_{eff}/10^7\ s^{-1}$
NO_3^-	0.965 ± 0.079	0.699 ± 0.017	1.38 ± 0.11
Nitrobenzene	3.91 ± 0.24	3.27 ± 0.09	1.20 ± 0.08
		$k_{eff} = 1.28 \pm 0.10\ 10^7\,s^{-1}$ [c]	
		$k_{eff} = k_2[DMPO] + k_4[O_2] = (1.0\ to\ 1.5) \times 10^7\,s^{-1}$ [d]	

[a] Average of values in Reference 2. Errors (1σ) from distribution of values.
[b] Statistical errors (1σ) from present work are given.
[c] Final calibrated value for k_{eff}.
[d] Range of k_{eff} using values from References 2 and 4.

Table 2. Rate Constants for the Reaction of the Hydrated Electron with Various Pollutants

Compound	This Work[a]	Literature	Ref.
		$k/10^{10}\ M^{-1}\ s^{-1}$	
Dichloromethane	0.60 ± 0.02	0.63	2
Trichloromethane	0.84 ± 0.03	3.0	2
Tetrachloromethane	0.69 ± 0.04	1.3, 1.6, 3.0	2, 2, 5
1,2-Dichloroethane	0.76 ± 0.04	0.29	6
1,1,1-Trichloroethane	0.68 ± 0.04	1.4	6
1,2-Dibromoethane	1.41 ± 0.04	1.2	6
Trichloroethylene	0.80 ± 0.04	1.9	2
1,2-Dichlorobenzene	0.81 ± 0.04	0.47	2
Bromobenzene	0.39 ± 0.05	1.0, 0.43	2, 2
2,4-Dichlorophenol	0.05 ± 0.04	—	
Acetone	1.62 ± 0.03	0.65 ± 0.09[a,b]	2
Benzaldehyde	4.46 ± 0.15	—	
Benzonitrile	2.53 ± 0.09	1.9	2

[a] Statistical errors (1σ) are given. [b] Average of six determinations.

multiple determinations the variability is often large. However, acetone is a case where this comment does not apply. The value listed in Table 2 is the average of six determinations. Our value lies substantially outside the literature range.

Since all the literature values quoted derive from pulse radiolysis decay rate studies, any systematic problems with that method would likely show up in this comparison. We are not, however, suggesting at this time that the discrepancy is entirely in the literature. There are possible problems in our method that require further study. For instance, the reaction of DMPO with e_{aq}^- in the presence of water may well be a multistep process. If DMPO⁻ is the first product, a reasonable but untested hypothesis, then we have implicitly assumed that the reaction of DMPO⁻ to form DMPO-H is fast compared to its rate of reaction with other scavengers. In evidence, we can say that we see no sign of nonlinearity in any inverse plots (see Equation 11). This is rather strong evidence, since almost any alternative reaction scheme would give curved inverse plots. Clearly, further investigation of these possibilities is required before our results may be considered reliable.

CONCLUSIONS

We have developed and calibrated a new method of measuring the rate constants for the reactions of the hydrated electron with other molecules and have used this technique to measure rate constants for some important pollutants. However, further work is needed to investigate the discrepancies between some of our results and the literature.

ACKNOWLEDGMENTS

This work was supported by Research Grant No. E560G from the Ontario Ministry of the Environment for which we are extremely grateful.

REFERENCES

1. Bolton, J. R., Cater, S. R., and Safarzadeh-Amiri, A., Use of reduction reactions in the photodegradation of organic pollutants in waste streams, Symposium on Environmental Aspects of Surface and Aquatic Chemistry, American Chemical Society Conference, San Francisco, April, 1992.
2. Buxton, G. V., Greenstock, C. L., Helman, W. P., and Ross, A. B., Critical review of rate constants for reactions of hydrated electrons, hydrogen atoms and hydroxyl radicals ($^{\cdot}OH/^{\cdot}O^{-}$) in aqueous solution, *J. Phys. Chem. Ref. Data*, 17, 513, 1988.
3. Shirom, M. and Stein, G., Excited state chemistry of the ferrocyanide ion in aqueous solution I. Formation of the hydrated electron, *J. Chem. Phys.*, 55, 3372, 1971.
4. Murov, L., *Handbook of Photochemistry,* Marcel Dekker, New York, 1973, 89.
5. Hart, E. J., Gordon, S., and Thomas, J. K., Rate constants for hydrated electron reactions with organic compounds, *J. Phys. Chem.*, 68, 1271, 1964.
6. Lal, M., Schöneich, C., Mönig, J., and Asmus K.-D., Rate constants for the reactions of halogenated radicals, *Int. J. Radiat. Biol.*, 54, 773, 1988.

CHAPTER 35

Reaction Pathways in Advanced Oxidation Processes

C. C. David Yao and Theodore Mill

INTRODUCTION

Because of the many possible reactions in each Advanced Oxidation Process (AOP), simple kinetic evaluations generally will not distinguish among possible oxidation pathways nor the effects of external variables such as pH, humic acid, or HCO_3^- concentration on the reaction rates and efficiencies. Kinetic models, which can simulate kinetic features of AOPs, are useful for optimization and for predicting the efficiencies of AOPs.

This chapter summarizes experiments and modeling conducted with several different AOPs in which ultraviolet (UV) light with hydrogen peroxide, ozone, or titanium oxide have been used to oxidize butyrate (B) and propionate (P) ions at pH 7 or 2.2. Three major objectives of this study are to characterize the principal oxidant(s) in AOPs through relative reactivity measurements; to develop detailed kinetic models which accurately describe rates of loss of oxidants, B and P; and the effect of system variables on rates and efficiencies.

RELATIVE REACTIVITY MEASUREMENTS AND OXIDANTS

A way to characterize the major oxidant(s) is to measure the relativity ratio of the oxidant toward different compounds (Equations 1 and 2). For a specific oxidant, the loss rate ratio should be the same as the ratio of the rate constants for the compounds (k_A/k_B).[1] It is unlikely that two different oxidants will have the same ratio of rate constants toward the selected probe chemicals.

$$\text{OX} + \text{A} \longrightarrow \text{OXIDATION PRODUCTS} \left(k_A \right) \qquad (1)$$

$$\text{OX} + \text{B} \longrightarrow \text{OXIDATION PRODUCTS} \left(k_B \right) \qquad (2)$$

$$k_A / k_B = \ln \left([A_o] / [A_t] \right) / \ln \left([B_o] / [B_t] \right)$$

Characterization of Oxidants in AOPs

Butyrate (B) and propionate (P) anions were chosen as kinetic probes for AOP experiments because they are analytically detectable at 6-μM concentrations, they have no UV spectra above 210 nm, they produce no halogen atoms on oxidation, and they have well-known reactivities toward HO· which gives a relative reactivity ratio of 2.4 for the anions ($k_B = 2.0 \times 10^9$ and $k_p = 8.2 \times 10^8\ M^{-1}\ s^{-1}$).[2] Moreover, B and P will have negligible reactivity toward ozone and secondary radicals such as $HO_2^·$ and $RO_2^·$.

H₂O₂-UV System

The H_2O_2-UV system generates HO· by photolyzing the peroxide HO-OH bond with UV light below 300 nm (Equation 3). A low-pressure mercury lamp was filtered to remove light below 250 nm used in our experimental system. All experiments were conducted with 100 μM peroxide in a batch reactor. Experiments were conducted with 6 μM each of B and P at pH 7 with the H_2O_2-UV, where the relative reactivity of B and P was found to be 2.2, nearly the same as the calculated ratio (Figure 1). Since the H_2O_2-UV system is a well-known primary producer of HO·,[3] this experiment checks the independent measurements of k_{ox} for the B and P system.[2]

$$H_2O_2 + UV \longrightarrow 2HO· \qquad (3)$$

$$HO· + B(P) \longrightarrow H_2O + B·(P·) \qquad (4)$$

$$B·(P·) + O_2 \longrightarrow BO_2^· (PO_2^·) \equiv RO_2^· \qquad (5)$$

$$HO·\ \text{or}\ RO_2^· + H_2O_2 \longrightarrow H_2O\ \text{or}\ RO_2H + HO_2^· \qquad (6)$$

$$HO· + HCO_3^- \longrightarrow H_2O + CO_3^{-·} \qquad (7)$$

Secondary oxidants such as $HO_2^·$ or $RO_2^·$ are unreactive with B or P and do not contribute to the relative reactivity ratio.[4] We estimate that for ·CO_3^- to play

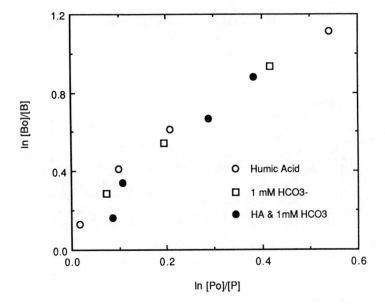

Figure 1. Reactivity ratio plot for oxidation of propionate (P) and butyrate (B) ions by H_2O_2 in the presence and absence of humic acid (HA) and HCO_3^-.

a significant role, its rate of oxidation of B and P ($R(\cdot CO_3^-)$) must be greater than 10% of $R(HO\cdot)$). However, the reactivity ratio of $CO_3^-/HO\cdot$ toward acetate ion is less than 1×10^{-5}, and the steady-state concentration ratio of $\cdot CO_3^-/HO\cdot$ is less than 10^{4}.[5] The similarities in relative reactivities of B and P in the presence and absence of $\cdot CO_3^{2-}$ and humic acid indicate that $HO\cdot$ is the dominant oxidant despite the presence of $\cdot CO_3^-$ as an additional oxidant.

Ozone Systems

Two kinds of ozone systems were studied: ozone in the dark and in the light.

Ozone decomposes in the dark by a chain reaction initiated by HO^- and propagated by $\cdot O_2^-$. The major products are oxygen and hydrogen peroxide.[6,7] Peroxide (via HO_2^-) continues the chain. Thus, any model developed to describe ozone decomposition must include the O_3-H_2O_2 system as well.

Photolysis of ozone produces $O(^1D)$ oxygen atom, which either inserts in water to form H_2O_2 or forms $HO\cdot$ which rapidly decomposes ozone, but can be scavenged by B and P. A simplified O_3-H_2O_2-UV model is presented in Equations 8–12.

$$O_3 + UV \longrightarrow O_2 + O \tag{8}$$

$$O + H_2O \longrightarrow H_2O_2 \tag{9}$$

$$O + H_2O \longrightarrow 2\,HO\cdot \tag{10}$$

$$HO^{\cdot} + O_3 \longrightarrow \longrightarrow 2O_2 + H^+ \tag{11}$$

$$HO^{\cdot} + B + O_2 \longrightarrow \longrightarrow H_2O + BO_2^{\cdot} \tag{12}$$

TiO$_2$/UV System

The oxidation of organic chemicals on illuminated TiO$_2$ is well documented according to Equations 13–18.[8,9] Ultraviolet photons (<360 nm) promote electrons from the TiO$_2$ valence band to the conduction band creating electron-hole pairs. The hole and electron can either recombine or diffuse to the surface where holes and electrons react with surface-adsorbed species, producing HO$^{\cdot}$ radical and superoxide ions. HO$^{\cdot}$ is believed to be responsible for the oxidation of organic molecules.

$$TiO_2 + h\nu \longrightarrow e^- + h^+ \tag{13}$$

$$h^+ + OH^- \text{ (sur.)} \longrightarrow HO^{\cdot} \tag{14}$$

$$O_2 + e^- \longrightarrow O_2^{-\cdot} \tag{15}$$

$$^{\cdot}O_2^- + HO_2^{\cdot} \longrightarrow HO_2^- + O_2 \tag{16}$$

$$HO_2^- + H^+ \longrightarrow H_2O_2 \tag{17}$$

$$HO^{\cdot} + B + O_2 \longrightarrow \longrightarrow H_2O + BO_2^{\cdot} \tag{18}$$

Figure 2 shows results from reactivity ratio measurements with B and P at pH 7.0 in six different AOP systems. The relative reactivity of B and P was found to be 2.2 in all AOPs, the same as the calculated ratio based on HO$^{\cdot}$ in the literature. The constancy of the reactivity ratio in these systems points to HO$^{\cdot}$ as the dominant oxidant in all AOPs.

To confirm this conclusion, experiments with B and P were also performed at pH 2.2. The reactivity ratio for butyric and propionic acids (BH and PH) in the H$_2$O$_2$-UV and O$_3$-UV systems are shown in Figure 3. Ratios in Figure 3 are close to 3.5, the same as the ratio of reactivities of BH and PH towards HO$^{\cdot}$ measured independently at pH 2.2.[2] Again, we find that HO$^{\cdot}$ is the principal oxidant in AOPs.

COMPARISON OF EXPERIMENTS WITH THE MODEL

We developed a H$_2$O$_2$-UV kinetic model with 45 reactions and an O$_3$-UV model with 84 reactions to evaluate the role of HO$^{\cdot}$ in the AOPs.[10] All models

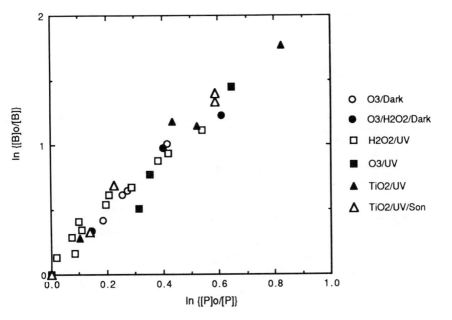

Figure 2.   Reactivity ratio plot for 6 μM propionate and butyrate ions in different oxidation systems at pH 7.0. TiO$_2$/UV/Son refers to the TiO$_2$-UV system with sonication.

were developed with Acuchem software.[11] Each model includes four sections: initiation, oxidation, termination steps, and protonation-deprotonation steps for radicals and ionic species.

H$_2$O$_2$/UV System

To evaluate the role of HO$^\cdot$ in the H$_2$O$_2$-UV system, we conducted experiments in the absence and presence of 6 μM each B and P and with different concentrations of HCO$_3^-$ and humic acid (HA). The scavenging rate for HO$^\cdot$ in this concentration range of HCO$_3^-$ and HA should be as large as possible, while the concentrations of B and P should be as low as possible in order not to affect the $[\text{HO}]_{ss}$ in the system.

Figure 4 shows that the measured and calculated peroxide losses in MQ water alone (100 μM HCO$_3^-$) or with different amounts of bicarbonate ion up to 15 mM are in excellent agreement. The interpretation of the lack of effect of bicarbonate ion on both model and experiments is that although high concentrations of HCO$_3^-$ scavenge nearly all HO$^\cdot$ and prevent its attack on H$_2$O$_2$, $^\cdot$CO$_3^-$ radicals themselves oxidize H$_2$O$_2$ efficiently, resulting in the same net loss of H$_2$O$_2$. Even 100 mM HCO$_3^-$ has no significant effect on the loss rate of H$_2$O$_2$.

In the presence of 5 ppm HA, the calculated results also agree with the experimental ones.[10]

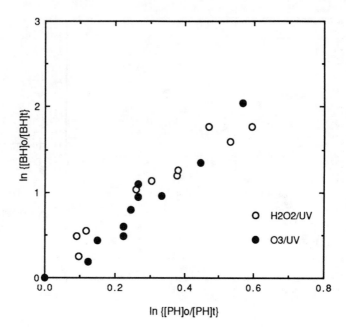

Figure 3. Reactivity ratio plot for 6 μM propionic and butyric acids in H_2O_2-UV and O_3-UV systems at pH 2.2.

O_3 System

The ozone system includes three types of initiation steps: initiation by HO^-, initiation by HO_2^-, and initiation by light. Each initiation step will generate $HO^\cdot$ with its propagation reactions. To evaluate their relative importance, we first performed experiments to understand the role of light in the generation of $HO^\cdot$. Then we compare the relative rates of these three initiation steps in the next section.

In the O_3-UV system, ozone is photolyzed to O atoms, and O atoms react with water in two ways: one is to generate hydrogen peroxide directly (Equation 9) and the other is to generate $HO^\cdot$ (Equation 10). To evaluate their relative importance, we photolyzed ozone at pH 2.2, where the ozone-HO_2^- chain reactions are unimportant and measured the peroxide formed.[10] To confirm the importance of the ozone-HO_2^- chain reactions, we also photolyzed ozone with H_2O_2 at pH 7.0 in the presence of 6 μM B and P. To evaluate the ozone model, the results from models under the same conditions were compared with those from the experiments in both cases.

Figure 5 shows the calculated peroxide increases, assuming that 0, 1, 3, 5, and 10% of atomic O reacts to give free $HO^\cdot$ in the model in the light and in the absence of BH and PH at pH 2.2. The average of triplicate measured results indicates that no more than 5% atomic O gives free $HO^\cdot$. This result is included in the model with 6 μM BH and PH at pH 2.2 as shown in Figure 6 and indicates

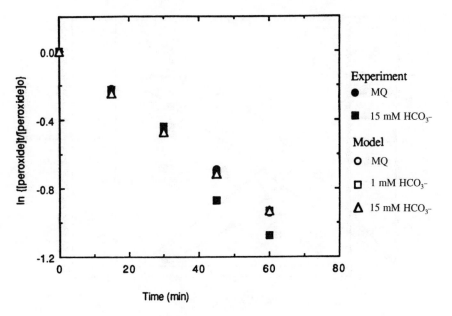

Figure 4. The measured hydrogen peroxide loss in the absence of butyrate and propionate ions in MQ water with and without HCO_3^- and HA.

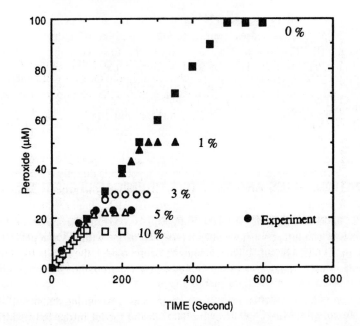

Figure 5. The calculated peroxide increases in the light and in the absence of BH and PH for different percentage of direct generation of HO radical from 100 μM ozone at pH 2.2.

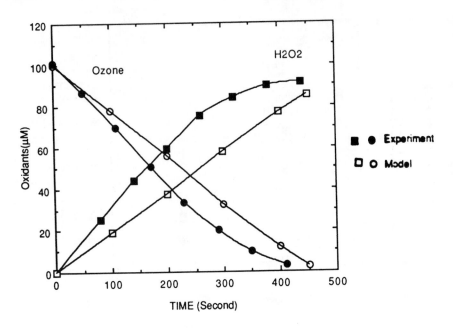

Figure 6. The measured and calculated losses of ozone and the increases of hydrogen peroxide in the O_3-UV system in the presence of 6 μM BH and PH at pH 2.2.

that the ozone model can reasonably predict the changes of peroxide and ozone at pH 2.2. Peroxide increases from 20 to 90% based on ozone lost because of 6 μM PH and BH. This dramatic increase must be due to scavenging of 80% of HO$^\cdot$ by BH and PH. Thus, the HO$^\cdot$-O_3 chain leading to oxygen was cut.

To further evaluate the ozone model, we compared calculated and measured ozone and peroxide changes at pH 7 where peroxide is not stable. The results, shown in Figure 7, are in good agreement. More important, the model correctly describes the sharp drop in peroxide production when the pH changes from 2 to 7.

OXIDATION RATES AND RELATIVE CHEMICAL EFFICIENCIES

Our experimental system used a low-pressure mercury lamp filtered to remove light below 250 nm. All experiments were conducted with 100 μM peroxide or ozone in a batch reactor. Figure 8 indicates the relative measured loss rates of oxidants in various systems in the absence of B and P. The H_2O_2-UV system is much less reactive in UV light than the O_3-UV system. This is because the UV absorption of H_2O_2 is weak with $\varepsilon_{254} = 19\ M^{-1}\ cm^{-1}$, while the UV absorption of O_3 is strong with $\varepsilon_{254} = 2850\ M^{-1}\ cm\ sec^{-1}$. Loss of ozone at high pH (pH 9.0) in the dark is only two times slower than in the light at low pH. This implies that the ozone dark system may be characterized by a faster loss rate than the light system when the pH value of the dark system is higher than 9. To compare the dark

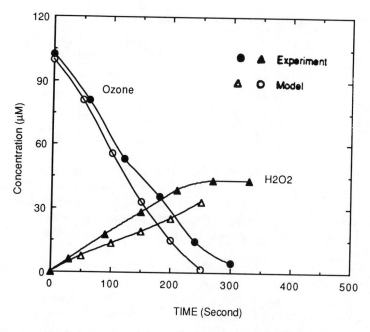

Figure 7. The measured and calculated ozone losses and peroxide increases in an O₃-H₂O₂-UV system in the presence of 6 μM BH and PH at pH 7.0.

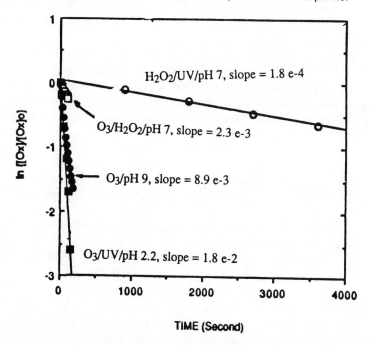

Figure 8. Comparisons of measured loss rates of oxidants in different AOPs in the absence of BH and pH.

Table 1.   Zero-Order Rates of Oxidation of Butyrate in AOPs

AOP	pH	Measured Rate ($\mu M^{-1} \times 10^2$)	Calculated Rate ($\mu M\ s^{-1} \times 10^2$)
O_3-UV	2.2	1.0	0.6
H_2O_2-UV	2.2	0.6	0.9
O_3-Dark	7.0	0.09	0.12
O_3-H_2O_2	7.0	2.5	3.1
O_3-H_2O_2-UV	7.0	3.9	3.9

systems of ozone alone and the O_3-H_2O_2 system, the latter system with 10 μM peroxide at pH 7 has a loss rate only about four times slower than the ozone alone system at pH 9, while ozone alone at pH 7 has a rate about 100 times slower than that at pH 9. These results suggest that a small amount of peroxide at pH 7 can adjust the loss rate of ozone to compete with that in the system at higher pH in the absence of peroxide.

Table 1 summarizes the measured and calculated zero-order oxidation rates of butyrate in the AOPs. The calculated rates among the AOPs agree with the experiments within a factor of 2. The O_3-H_2O_2 system in the light at pH 7.0 has the highest oxidation rate, but only 20 to 40% higher than the O_3-H_2O_2 system in the dark. This difference is due to photolysis of ozone and peroxide. No simple ozone-UV system can be observed at pH greater than 4 without a contribution from the O_3-H_2O_2 system because the O_3-UV system produces mostly H_2O_2 in the initial reaction in water. Among all AOPs tested, the O_3-dark system at pH 7.0 has the lowest oxidation rate.

The O_3-H_2O_2-UV system includes contributions from the O_3-HO^-, O_3-HO_2O_2, O_3-UV, and the H_2O_2-UV systems. The rates of photolytic generation of $HO^{\cdot}$ from ozone or peroxide are independent of pH, but do depend on the concentration of oxidant. Scavenging of $HO^{\cdot}$ by organics may be pH dependent (e.g., butyrate ion is more reactive than butyric acid) and is certainly concentration dependent.

Thus, the oxidation rate of the O_3-H_2O_2-UV system at pH 7 is nearly the sum of oxidation rates of the O_3-UV system at pH 2 (purely photolysis of ozone), the O_3-dark system at pH 7 (contribution of $HO^{\cdot}$), the O_3-H_2O_2 system at pH 7 (contribution of $HO^{\cdot}$), and about 10% of the rate of the H_2O_2-UV system at pH 2. Table 2 lists the measured efficiencies of AOPs. It shows that H_2O_2-UV system is the most efficient system, but that the other four AOPs have similar efficiencies.

Tables 1 and 2 list rate constants and efficiencies for five AOPs. From Table 1, we learn that the rate of loss of ozone in the O_3-H_2O_2 system in light and dark have very similar values. From Table 2, we see that the efficiencies of these two systems are nearly identical. The H_2O_2-UV system is the most efficient one among AOPs, but with a relatively slow oxidation rate. The selection and optimization of an AOP for a specific treatment depends on many factors that affect cost efficiency, but not chemical efficiency.

Table 2.   Efficiencies of AOPs[a]

AOP	pH	Measured[b] E, %	Ratio of Measured Rates
O_3-UV	2.2	5.0	1.0
H_2O_2-UV	2.2	50	0.58
O_3-Dark	7.0	13.8	0.087
O_3-H_2O_2	7.0	8.3	2.5
O_3-H_2O_2-UV	7.0	8.6	3.9

[a] During oxidation of 5 μM butyrate.
[b] E = ΔButyrate/(Δ added oxidant).

SELECTION AND OPTIMIZATION OF AOPs

Factors in Selection of Advanced Oxidation Processes

The choice of an oxidation system for a specific treatment depends on the cost of equipment, operation, and maintenance. We evaluated the operating expenses for each AOP in terms of a unit dollars per kilogallon for the same treated water. The variables affecting the operating costs include water characteristics, treatment process design, and operation. Variables related to water characteristics include the type and concentration of inorganic and organic contaminants, light transmittance of the water, and type and concentration of dissolved solids (if any). The variables directly related to treatment process design and operation are the UV dosage, the oxidant dosage, pH, temperature, the volume of water treatment per unit time, and the residence time of the water in the reactor. Lamp maintenance and operating costs are fixed once the type and the number of lamps is decided. The oxidant dosage, pH, and temperature can be varied according to the needs of the particular system. Operating expenses are based on the efficiencies of using oxidant(s) and UV light (if any).

UV Lamps

UV light plays a critical role in the formation of HO· and in direct photolysis of dissolved organic compounds. Some organic molecules such as chloroform or methylene chloride will react more rapidly with HO· if first they are partly oxidized to formaldehyde or formic acid by direct absorption of UV light.[2] The amount of energy absorbed by the compounds and by oxidants is related to the intensity of the UV light, the absorbance coefficients, contaminant and oxidant concentrations, and pathlength.[12] Therefore, oxidation times can be greatly reduced by increasing the intensity of UV light, using short wavelengths (<300 nm), long pathlengths, and high concentration of oxidants. However, if the formation rate of HO· is too fast, some energy may be wasted by recombination reactions of 2HO· to reform H_2O_2. Additionally, the efficiency in generating photon is also a very important factor in cost efficiency.

Defining the Chemical Efficiency of Oxidants

AOPs generate and consume HO˙ by different paths. The chemical efficiency of AOP oxidants may be defined as in Equation 19.

$$E(\%) = 100\,\Delta[\text{Probe chemical}]\big/\big(\Delta[\text{ozone}] + \Delta[H_2O_2]\big) \qquad (19)$$

The consumption of ozone and peroxide, changed into dollars, is

$$E(\$) = \Delta[\text{Probe chemical}]\big/(\$\ \text{of oxidants}) \qquad (20)$$

To compare the efficiencies among the AOP systems, three parameters need to be considered: (1) loss of added oxidant; (2) formation of oxidant, if any; and (3) the loss of substrate. In ozone systems (dark and light), peroxide, either added to the system or generated from ozone, plays an important role in controlling the efficiencies of the system under different conditions.

An Example in Selection and Optimization of AOPs

A typical question to be asked before using any aqueous treatment is, "What is the cheapest way to reduce the concentration level of the contaminant to the acceptable level?". Here, we present a procedure to select conditions to reduce a 10-ppm butyrate (100 μM), 6-ppm HCO_3^- (100 μM) solution at pH 9 to 100 ppb butyrate within 2 min in each batch.

Considerations for Residence Time

Among the AOPs the H_2O_2-UV system is the slowest (Table 1), and the HO˙ generation rate should be independent of pH.[3] The number of lamps needed to meet the requirement, calculated from the oxidation rate with a single lamp, is about 35 and is too large and makes the oxidation process with H_2O_2 and UV impractical.

Chemical Efficiency and UV Light

Table 3 lists the reaction rates of for reaction of ozone with important species in AOPs based on published data or our own experiments. At pH 9 with 10 μM peroxide, 98% of consumed ozone reacts with peroxide anions (1×10^{-1} sec^{-1}) instead of with photons (2.2×10^{-3} sec^{-1}) to generate HO˙. This implies that the UV lamp is unnecessary in this ozone system. Figure 9 shows that the comparison of calculated butyrate loss rates in the O_3-H_2O_2 systems with 1 mM O_3 and 0.1 mM butyrate at pH 9. It is not a surprise that the oxidation rate of butyrate is doubled when the initial peroxide concentration is changed from 50 μM to 100 μM. The unexpected result is that in the ozone-dark system the efficiency is not sacrificed

Table 3. Reactions of Ozone with Important Species in AOPs[a]

Species	Conc (M)	k (sec^{-1})
HO_2^-	2×10^{-8}	1×10^{-1} (10 μM, pH 9)
$h\nu$(UV lamp)		2.2×10^{-3b}
HO^-	1×10^{-5}	7×10^{-4}
$HO\cdot$	10^{-11}–10^{-14}	1.1×10^{-3}–1.1×10^{-6}
HCO_3^-	1×10^{-4}	$<1 \times 10^{-6}$
B	100×10^{-6}	6×10^{-7}

[a] References 2 and 6.
[b] Experimental value for ozone loss with our lamp.

when the oxidation rate is increased (Figure 10). This is due to the fact that only approximately 2% of HO˙ will be scavenged by peroxide in the system.

CONCLUSIONS

In this study, we have shown that HO˙ is the major oxidant in all of the AOPs we examined. This means that the choice of an AOP for a specific treatment system is based on issues of efficiency in generating HO˙ with the different chemical and photochemical systems. The most efficient system for generating

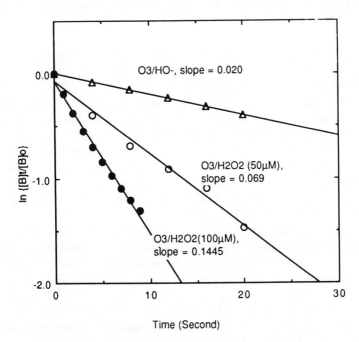

Figure 9. Comparisons of calculated butyrate loss rates in the O_3-H_2O_2 systems with 1 mM O_3 and 0.1 mM butyrate at pH 9.0.

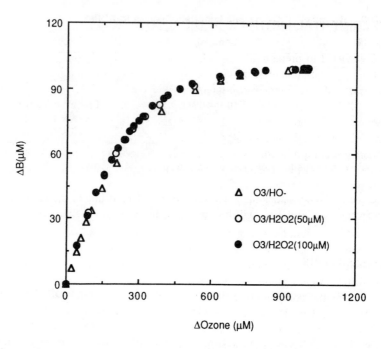

Figure 10. Efficiency comparisons of the O_3-H_2O_2-dark system with different concentrations of H_2O_2.

HO⁻ is the H_2O_2-UV system. However, this AOP is too slow to use with high-flow systems where short residence times require high rates of HO⁻ generation.

Ozone AOPs use 250-nm UV light very efficiently, but only 5% of the photolyzed ozone forms HO⁻. This means that only above pH 7, where HO_2^- is available, is the ozone AOP an efficient source of HO⁻. This also means that above pH 7, the ozone-UV AOP has no advantage over O_3-H_2O_2-dark system for generating HO⁻ and oxidizing organics. The efficiency for generating HO⁻ in the O_3-H_2O_2-dark system is unaffected by adding more H_2O_2 to speed up the reaction; thus, this AOP can both make and use HO⁻ rapidly and efficiently for oxidations. Physical mixing of H_2O_2 with ozone is the chief limitation on the amount of H_2O_2 that can be added to speed up this reaction.

Kinetic models to describe the time dependence of the AOP reactions have been developed for peroxide and ozone AOPs and are shown to be reliable predictors of the rates and concentrations of the oxidants and organic compounds in the systems. The models provide a valuable tool for optimizing conditions and for selecting and efficiently using a specific AOP with a specific hazardous waste stream. Although the models did not include any direct reaction of ozone with organic compounds, these reactions can be included where they are important. Current limitations on use of these kinetic models include effects of humic acids on rates of oxidation of organic compounds and possible complications associated with oxidations of halogenated compounds.

EXPERIMENTAL METHODS AND CALCULATIONS

Water used to prepare all solutions was from a Milli-Q purification system (Millipore Corp.), featuring reverse osmosis, activated carbon, ion exchange, and 0.2-μM membrane filters. Buffer solutions of the stock 0.5 M phosphate for pH 7.0 and 2.7 were prepared by mixing 0.5 M phosphate solution at different pHs.

Humic acid solution was prepared by dissolving 10 g of Aldrich humic acid in 1 L of 0.1 M NaOH, neutralizing with H_3PO_4, filter-sterilizing through 0.2-μM membranes, and pre-aging in sunlight for 3 days in summer and 4 days in other seasons. The humic acid solution had an absorbance of about 0.20 cm^{-1} at 260 nm at about 5 mg/L.

Ozone solutions were prepared using a 0.5·L·min^{-1} oxygen feed into a Welsbach Model T-408 ozone generator. The gas stream first passed through pH 6 phosphate buffer (to remove nitrogen oxides) and then into distilled water cooled in ice. Continuous gas flow maintained an ozone concentration of about 100 μM.

Solutions of organic compounds were prepared by weight by rapidly stirring with Milli-Q water an amount calculated to give, at most, half their water solubility. The organic compounds, propionic, butyric, and hexanoic acids (Aldrich), were all reagent grade. Stock solutions were used within 7 days.

Titanium sulfate solution was prepared by diluting the reagent grade solution of 15% w/v $Ti(SO_4)_2$ with 23% sulfuric acid (BDH Chemicals, Ltd.) and adding concentrated H_2SO_4 to make a solution 12.5 mM $Ti(SO_4)_2$ with approximately 3N sulfuric acid. Albone 35 35% w/w H_2O_2 was from DuPont, Wilmington, DE. Perchloric acid (70–72%, J. T. Baker Chemical Co.) was used for preparing pH 2.2 solutions.

UV light was supplied by a 25 × 0.5-cm low-pressure mercury vapor penlight (UV Products, Santa Clara, CA), with greater than 85% output at 254 nm. The lamp was immersed in a filter solution of 0.4 M nickel sulfate and 0.6 M cobalt sulfate, placed between the lamp and the reaction solutions to remove light below 220 nm and above 380 nm. A fresh solution was used for each photolysis, and the penlight was warmed up more than 10 min before exposing the solution.

Ozone was monitored in a 1-cm cuvette at 260 nm. At pH $\geq$ 9, ozone is not stable in solution, and reactions were initiated by injecting buffer into a 1-cm cuvette containing ozone solution, followed by rapid mixing. Ozone is stable below pH 7 in solution, and reactions were initiated by injecting buffer solution into 1 L of 100 μM ozone water. After mixing, samples were withdrawn at regular intervals and transferred into a 1-cm cuvette for UV analysis.

When the solution had too great an absorbance to allow direct measurement of ozone by UV, as in the presence of humic acid, the ozone loss was followed by the indigo method.[1] In such cases, large (1000-mL) reaction vessels were used to limit volatilization of ozone into the head space created by sample removal. After mixing the reactants, samples were withdrawn at regular time intervals and transferred to vials containing indigo solution and 30 μL of 85% w/w phosphoric acid. The vials were immediately mixed, and ozone was monitored by bleaching of indigo absorbance at 596 nm.

Experimental rate constants (k_{exp}) for ozone consumption were evaluated from the absorbance data (A_t) assuming pseudo-first-order kinetics:

$$\ln \frac{A_o - A_\infty}{A_t - A_\infty} = k_{exp}t \tag{21}$$

In the direct UV method, k_{exp} was evaluated automatically by the HP8950 computer system controlling the spectrometer. However, for slow reactions that were stopped well before the consumption of ozone went to completion, the computer failed to give accurate results because of larger errors in estimating the final absorbance of the solutions. In these cases, we used sufficient nitrogen gas to purge out the remaining ozone, and then the background absorbance corresponding to infinite reaction time was measured.

Hydrogen peroxide was monitored in a 1-cm cuvette at 420 nm using the titanium complex method.[13] Even though ozone had no significant effect on the measurement, ozone-free samples were preferred. Samples were withdrawn at regular intervals and equally divided into two vials. Stock $Ti(SO_4)_2$ solution was added to one sample and MQ water was added to the other, both in the ratio of 1:10 (v/v). After mixing, the samples were transferred to 1-cm cuvettes and measured. The concentration was calculated using a calibration curve determined from known concentrations of hydrogen peroxide.

Probe chemicals B and P were measured by gas chromatography (GC) using a flame ionization detector. At regular intervals, fixed amounts (10 ml) of ozone-organic reaction mixture were pipetted into vials and purged with nitrogen to remove remaining ozone. Each sample was purged about 30 sec with no significant loss of probe chemicals. Mixtures were adjusted to pH greater than 11 by adding $1N$ NaOH solution and evaporated to dryness. Residues were dissolved in 0.5 mL solution containing 120 μM internal standard of hexanoate ion acidified to pH less than 2 with 4 M HCl solution and chromatographed.

Acids were measured with a Varian 3700 GC equipped with an AT-1000 10 m × 0.54 mm i.d. Megabore column with a 1.2-μm film thickness and a flame ionization detector. An HP8452A Diode Array Spectrophotometer with an HP89500A UV/Vis ChemStation was used for absorption measurements of ozone and hydrogen peroxide.

Acuchem[11] is a computer program for solving the system of differential equations describing the temporal behavior of homogeneous multicomponent chemical reactions. Acuplot is a program for the output file and the graphics. Both programs were supplied by National Institute of Standards and Technology (Gaithersburg, MD) and run on an IBM personal computer with a math coprocessor, display, EPSON MX-100III printer, and a 200-MB hard disk (Procom Technology).

ACKNOWLEDGMENT

We thank Dr. W. R. Haag for his contribution to the study of the H_2O_2-UV system. The work described here was supported by the U.S. Air Force Civil Engineering Laboratory, Tyndall AFB, Florida through Contract F08635-90-0061.

REFERENCES

1. Bader, H. and Hoigné, J., Determination of ozone in water by the indigo method, *Water. Res.*, 15, 449, 1981.
2. Buxton, G. V., Greenstock, C. L., Helman, W. P., and Ross, A. B., Critical review of rate constants for reactions of hydrated electrons, hydrogen atoms, and hydroxyl radicals (HO/O^-) in aqueous solution, *J. Phys. Chem. Ref. Data*, 17, 513, 1988.
3. Calvert, J. G. and Pitts, J. N., *Photochemistry*, John Wiley & Sons, New York, 1967, 5–352.
4. Bielski, B. H. J., Cabelli, D. E., and Arudi, R. L., Reactivity of HO_2/O_2^- radicals in aqueous solution, *J. Phys. Chem. Ref. Data*, 14, 1041, 1985.
5. Neta, P., Huie, R. E., and Ross, A. B., Rate constants for reactions of inorganic radicals in aqueous solution, *J. Phys. Chem. Ref. Data*, 17, 1027–1284, 1988.
6. Peyton, G. R. and Glaze, W. H., Mechanism of photolytic ozonation, in *Photochemistry of Environmental Aquatic Systems*, Zika, R. G. and Cooper, W. J., Eds., ACS Symposium Ser., No. 327, American Chemical Society, Washington, D.C., 1987, 76.
7. Peyton, G. R. and Glaze, W. H., Destruction of pollutants in water with ozone in combination with ultraviolet radiation. 3. Photolysis of aqueous ozone, *Environ. Sci. Technol.*, 22, 761–767, 1988.
8. Peterson, M. W., Turner, J. A., and Nozik, A. J., Mechanistic studies of the photocatalytic behavior of TiO_2 particles in a photoelectrochemical slurry and the relevance to photodetoxification reactions, *J. Phys. Chem.*, 95, 221, 1991.
9. Turchi, C. S. and Ollis, D. F., Photocatalytic degradation of organic water contaminants: mechanisms involving hydroxyl radical attack, *J. Catal.*, 122, 178, 1990.
10. Mill, T., Haag, W. R., and Yao, C. C. D, Kinetic features of advanced oxidation processes for treating aqueous chemical mixtures, in Chemical Oxidation Technology For The Nineties, 2nd International Symposium, Nashville, Tennessee, 1992.
11. Braun, W., Herron, J. T., and Kahaner, D. K., Acuchem: A computer program for modeling complex chemical reaction systems, *Int. J. Chem. Kinet.*, 20, 51, 1988.
12. Zepp, R. G. and Cline, D. M., Rates of direct photolysis in aquatic environments, *Environ. Sci. Technol.*, 11, 357, 1977.
13. Satterfield, C. N. and Bonnell, A. H., Interferences in the titanium sulfate method for hydrogen peroxide, *Anal. Chem.*, 27, 1174, 1955.

CHAPTER 36

Photooxidation of Organic Compounds in Water and Air Using Low-Wavelength Pulsed Xenon Lamps

Werner R. Haag

INTRODUCTION

This study examines the use of pulsed xenon lamps for the photooxidation of organic compounds in water and air. Ultraviolet light (UV)-induced and other radical oxidation processes (ROPs) are currently of interest for the onsite destruction of organic contaminants in environmental media. Compared to phase-transfer processes like stripping and adsorption, destructive processes are attractive because they mitigate any future toxicity or liability issues. They are also useful for compounds like p-dioxane or vinyl chloride that, in any case, cannot readily be removed from water by stripping or from air by adsorption, respectively. The most common UV oxidation systems currently use titania catalysts, hydrogen peroxide, or ozone in combination with low- and medium-pressure mercury lamps having UV output either peaking at 254 nm or broad-band emission from 200 to 300 nm.[1] Hydrogen peroxide is usually the preferred HO· radical source because ozone is more expensive, less soluble in water, and requires off-gas control.

This chapter compares the spectra and energy efficiencies of current lamp systems with those of pulsed xenon lamps that have a peak emission near 230 nm. We summarize the usefulness of the pulsed lamps for direct photolyses of chlorinated volatile organic compounds (VOCs) in addition to H_2O_2-induced photolyses in terms of rates, quantum efficiencies, and products and compare the costs of UV photons with those of chemical reagents commonly used for treatment. Finally, we compare the advantages and disadvantages of carrying out the photooxidation processes in air vs water.

0-87371-871-2/94/$0.00+$.50

517

METHODS

Relative lamp emission spectra were measured on an Oriel InstaSpec III 1024 diode array detector calibrated with a Molectron J25 pyroelectric calorimeter. Relative intensities were placed on an absolute basis using hydrogen peroxide actinometry[2] in a reactor scrupulously cleaned of organic residues.

Gas-phase photooxidation methods were described previously.[3] Briefly, the reactor was a fan-mixed, 208-L, cylindrical steel vessel with a single 6-in. flashlamp inserted in the middle through the side, giving light paths of 25–35 cm. All measurements were performed at 300–340 K and 1 atm and normalized using CCl_4 actinometry assuming a quantum yield of unity.[4] Samples were removed by syringe and analyzed by capillary gas chromatography with either electrolytic conductivity detection or photoionization detection (HP 5890).

Aqueous-phase reactions were conducted in a 9.0-L cylindrical aluminum vessel 25 cm in diameter and 20 cm in height. Two 3-in., 1380-W pulsed xenon lamps were contained in a 7-cm diameter cylindrical quartz housing concentric to the reactor cylinder, leaving an effective light pathlength of 9 cm through the solution. The two lamps give essentially the same spectral distribution as the single 6-in., 3675-W lamps used in the gas phase system. The lamp housing extended nearly to the bottom of the reactor, leaving space for a large magnetic stir bar that was used to stir the water sample. The concentration of trichloroethene (TCE) was determined by head space analysis (HP19395A), followed by capillary gas chromatography with electron capture detection (HP 5890).

RESULTS AND DISCUSSION

Pulsed Xenon Lamps

Figure 1 is a schematic diagram of the Purus 3.7-kW, 6-in. flashlamp. The lamp contains a xenon fill gas in a quartz tube, which is itself held in a 2×9 in. outer quartz tube that contains cooling water. The water is deionized to prevent current leakage through it.

Flashlamps operate in the pulsed mode with peak intensities much greater than those occurring with continuous sources of the same average power. The pulse duration is typically in the microsecond time scale, while the interval between pulses is on the order of milliseconds. The electrical discharge quickly heats the fill gas to a high enough temperature ($\geq 13,000$ K) to create a plasma that emits black body light characteristic of its temperature. Thus, while continuous lamps primarily emit lines associated with the electronic excitation levels of the un-ionized fill gas, pulsed lamps emit light consisting mostly of black body radiation upon which the electronic lines are superimposed. Commonly used fill gases for flashlamps include xenon, argon, krypton, and other inert gases or mixtures. Xenon plasmas generally have the greatest efficiency for photon production.

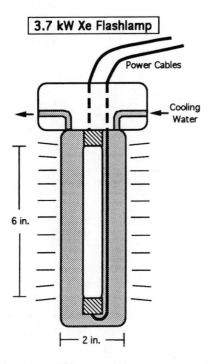

Figure 1.   Schematic diagram of the Purus 3.7-kW pulsed xenon lamp.

Unique features of pulsed lamps include the ability to come to full power immediately (without a warmup period) and the ability to shift the spectrum of a single lamp simply by changing the peak pulse power:

$$\text{Peak Power} = \frac{1}{2}\,CV^2\big/\Delta t = 12\ \text{MW} \tag{1}$$

$$\text{Average Power} = \frac{1}{2}\,CV^2 \cdot f = 3.7\ \text{kW} \tag{2}$$

where C is the capacitance in farads, V is the voltage in volts, Δt is the pulse width in seconds, f is the frequency in hertz, and the numerical values are the currently used conditions of the Purus lamps. The quantity $\frac{1}{2}CV^2$ is the power per pulse in joules. Figure 2 shows the effect of varying peak power on the emission spectra of a pulsed xenon lamp.[5] The pulse width is controlled by an inductor, and therefore Figure 2a corresponds to about 16 times the peak power as in Figure 2b. Figure 2b gives a spectrum similar to that of a continuous xenon source, while the higher peak power in Figure 2a gives a much higher plasma temperature, which lowers the wavelength of its emission maximum. Excessively high pulse energies increase thermal stresses, thus shortening lamp life.

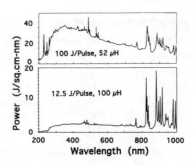

Figure 2.   Effect of peak power on the emission spectrum of a pulsed xenon lamp (courtesy of ILC Technology, Sunnyvale, CA,[5] With permission).

Comparison with Continuous Light Sources

Figure 3 compares the emission spectrum of a 6-in. Purus 3675-W xenon flashlamp with that of a conventional medium-pressure mercury lamp. The data are presented as the output power integrated over the total area of the lamp and normalized to the same input power. The xenon flashlamp has a maximum at 230 nm and significant output at wavelengths as low as 200 nm, whereas the conventional mercury lamp has most of its output at wavelengths above 250 nm. The efficiency of light output at ≤300 nm per energy input is 18.6% for the pulsed xenon lamp, which is about 50% greater than for the conventional mercury lamp. Although the mercury lamp has strong lines at 313 nm and especially 366 nm, it is difficult to have sufficient pathlength in a practical reactor for H_2O_2 to absorb a significant fraction of this light unless reflectors are used and maintained free of corrosion and deposits. Peroxidation Systems, Inc. (PSI) reportedly have newer proprietary medium-pressure mercury lamps with efficiencies at ≤300 nm ranging from 20 to

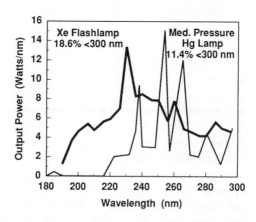

Figure 3.   Comparison of the emission spectra of a Purus xenon lamp operated at its standard conditions with that of a PSI medium-pressure mercury lamp normalized to 3675 W.

35%, depending on the power.[6] At least half of this light is reported to be below 250 nm,[6] but the actual spectral output is not available at present. If these efficiencies can be verified, the proprietary lamps approach the theoretical efficiency limit for output between 200 and 300 nm by a black body radiator (approximately 35%). The proprietary PSI sources have average lamp lives given at 3000 hr.[6] The Purus xenon flashlamp has demonstrated a life of 1000 hr and improvements are to be expected with further development.

Advantages of low-pressure mercury lamps for low-power applications are that they are very efficient and have long life. Low-pressure mercury lamps typically have 30–60% of the input energy converted to photons below 300 nm, medium-pressure mercury lamps about 10–15%, and the Purus flashlamps about 15–20%. However, a disadvantage of continuous sources is that to increase intensity the fill gas pressure needs to be increased, which in turn causes the emitted light to be reabsorbed, because the photons have energies equal to the electronic transitions of the ground state fill gas. Thus, as the mercury pressure is increased, the 254- and 185-nm lines are reabsorbed and their energy reemitted at longer wavelengths. By contrast, flashlamp plasma generation allows (and requires) higher power for a given fill gas pressure and emits nonresonant radiation that is not as readily reabsorbed. Flashlamp intensities of several hundred to 1000 W/in. (discharge length) are possible compared to a few to a few hundred watts per inch for low- and medium-pressure mercury lamps, respectively. For example, a low-pressure mercury lamp would need to be on the order of 300 m long to deliver the same power as a 6-in. 3675-W xenon flashlamp. The greater power density allows smaller reactor dimensions and greater ease of lamp cleaning and replacement, but more difficulty in achieving plug flow in continuous process reactors. The small size is particularly advantageous for sites where space is at a premium, such as offshore oil rigs. The development of flashlamps for remediation is still at an early stage, and some improvements in efficiency may be expected.

Excimer lamps are another source of low-wavelength UV light, among the most efficient of which is the 172-nm emission from the excited Xe_2 dimer. These sources, like low-pressure mercury lamps, operate in glow discharge mode, where the fill gas is near room temperature and thus gives emissions associated with specific electronic transitions and almost no black body radiation. By contrast, high-pressure mercury and pulsed xenon lamps operate in arc discharge mode, where the entire fill gas is heated and black body radiation also becomes important. Excimer lamps typically use rf or microwave excitation to ionize and excite the fill gas. The excited molecule associates with a ground state molecule to form a dimer or trimer (excimer). When the excimer emits a photon, it dissociates, and thus the light cannot be reabsorbed because no ground state of the excimer exists. Excimer sources for large-scale treatment systems are not yet commercially available, but offer potential benefits of good inherent efficiency (5–30%), unique lamp geometries, and a range of wavelength choices including sub-200-nm wavelengths suitable for direct photolysis of many organic compounds.[7] Competitive light absorption by oxygen and quartz envelopes may limit the usefulness of the 172-nm xenon excimer source in commercial systems.

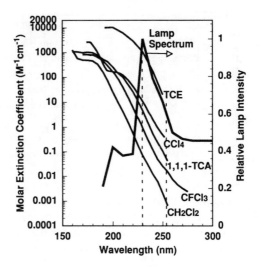

Figure 4. Comparison of pulsed xenon lamp spectrum with the gas-phase absorption spectra of some VOCs.

Direct Photolysis of VOCs

Pulsed xenon lamps are capable of direct photolysis of many VOCs of concern to the Environmental Protection Agency. Figure 4 compares the emission spectrum of the flashlamp with the gas-phase absorption spectra of several VOCs. The halomethanes and 1,1,1-trichloroethane (TCA) are weak absorbers, whereas TCE and other chloroolefins absorb strongly in the deep UV region. A shift in peak output from 254 to 230 nm is significant because it corresponds to a one to two order of magnitude increase in absorptivity of many VOCs, thereby greatly enhancing the rates of direct photolysis. Quantum yields for chloroalkanes are high (see below), so light absorption controls the rate. A shift to lower wavelength also enhances the rate of H_2O_2 photolysis, although to a lesser extent because the peroxide absorption band increases less steeply with decreasing wavelength (see Figure 5). The 185-nm peak of low-pressure mercury lamps causes some direct photolysis, but this line comprises ≤15% of the total light output. The percentage output of the 185-nm line can be increased by using a lower-pressure mercury lamp, albeit at the expense of lower overall intensity.

Cost of Photons and Other ROP Reagents

Table 1 lists the operating cost of generating photons from various sources and compares them to the costs of chemicals used in ROPs and other treatment processes. These estimates are based on vendor or user quotes for bulk quantities and/or electricity costs of $0.08/kWh. They do not include the costs of delivering the chemicals, drying the air used to generate ozone, or lamp replacement and

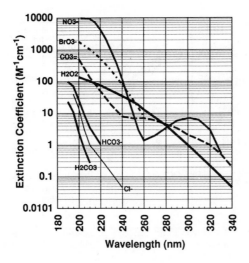

Figure 5. Aqueous absorption spectra of H_2O_2 and some common natural water solutes.

Table 1. 1992 Operating Costs of ROP Treatment Chemicals

Chemical	Bulk Price (Pure)	
	$/Pound	$/Mole
NaOH (Na$_2$CO$_3$ impurity)	0.14[a]	0.01
H$_2$SO$_4$	0.10[a], 0.40[d]	0.02, 0.08
Cl$_2$	1.05[a]	0.16
Cl$_2$	0.14[b]	0.02
H$_2$O$_2$	0.68[a], 0.99[d]	0.05, 0.07
O$_3$	0.95[c]	0.10
O$_3$	1.50[b]	0.16
ClO$_2$	2.00[a]	0.30
FeSO$_4$ · H$_2$O	2.54[a]	0.95
FeSO$_4$ · 7H$_2$O	0.80[d]	0.45

Photons	Source and Efficiency	$/Mole
172 nm	Xenon excimer, 20%	0.077
230 nm, broad	Pulsed xenon, 19%	0.061
240–303 nm	Med-pressure mercury, 11%	0.094
185, 254 nm	Low-pressure mercury, 40%	0.025

[a] Vendor quote.
[b] Municipal water plant cost quote.
[c] Calculated based on costs of O$_2$ and electricity but not air drying.
[d] Chemical Marketing Reporter, July 22, 1992.

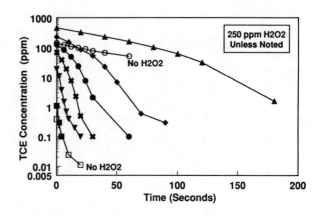

Figure 6.   Effect of initial concentration on TCE photooxidation rate in water. All reactions at 300 K with a xenon flashlamp in the presence of 250 ppm H_2O_2 in a stirred, 9-L reactor.

cleaning costs. Clearly, cost issues are more complex than indicated in Table 1; nevertheless, the data show that photons can be obtained commercially in a price range similar to that of chemical reagents. The efficiency chosen for the xenon excimer lamp (20%) is midway between the measured efficiency of 10% and the stated inherent efficiency of ≥30%.

Photooxidation in Water

Trichloroethene in Pure Water

Figure 6 shows the photolysis of TCE in the absence and presence of 250 ppm (7 mM) H_2O_2. Trichloroethene is one of the most rapidly treated compounds studied; the direct photolysis is so rapid that 250 ppm peroxide only increases the rate constant by a factor of 2 to 3. Direct photolysis contributes to the loss of many other VOCs, especially highly halogenated ones, but generally it is more cost effective to add H_2O_2, which results in relative rates that agree reasonably with expectations from their HO˙ radical rate constants.[8,9]

In Figure 6, as the initial concentration of TCE increases its efficiencies for trapping HO˙ radicals and light increase. Thus, despite the higher rate constants at lower concentration, the mass of TCE destroyed per unit time is greater at higher initial concentration and reaches an upper limit of about 6.5 ppm sec^{-1} above about 60 ppm. At this point, most of the light and HO˙ are trapped by TCE. In a given experiment, the slopes tend to increase with time as the solutions become optically thinner and the HO˙ concentration increases. However, products are clearly affecting the rates at later times because, for example, when the initially 400 ppm TCE solution reached 100 ppm, the slope is noticeably lower than in the solution of 100 ppm initially.

Efficiency of Photon Use

In practice, we have found that it is rarely possible to oxidize one organic molecule per photon emitted in aqueous peroxide systems, as would be expected from the unit quantum yield of $HO^{\cdot}$ formation by H_2O_2 photolysis.[2] At low conversions of optically thick solutions of CH_2Cl_2 or TCE (Figure 6) containing approximately 4:1 molar ratio of H_2O_2, we have found apparent quantum efficiencies (parent compound lost per sub-300-nm photon input) of 0.31 and 0.37, respectively. For CH_2Cl_2, this yield reflects competition for $HO^{\cdot}$ by the H_2O_2, as the respective $HO^{\cdot}$ reaction rate constants[8,9] of 6×10^7 and 2.7×10^7 $M^{-1} s^{-1}$ are similar. For TCE on the other hand, the efficiency is probably limited by the direct absorption of light by this compound and its inherent direct photolysis quantum yield in water. These efficiencies are for low conversion and are lower when complete mineralization is required. In the case of a cutting oil emulsion in water, we have found quantum efficiencies of 0.02–0.03 for complete mineralization, with a consumption of 10–20 mol of H_2O_2 per mole of organic carbon.

These efficiencies, obtained at high concentrations in the absence of competing natural water absorbers, are probably similar to the results that can be obtained with other ROPs not employing light. The efficiency in dark ROPs drops as the concentration of reactant decreases, because the oxidant source or the matrix components become important radical scavengers. This is also the case in photolytic ROPs when photooxidation involves $HO^{\cdot}$ generation from H_2O_2. On the other hand, when the mechanism involves direct photolysis, the rate is affected little by radical scavenging, but the efficiency still drops as the reactant concentration decreases and light is lost to the reactor walls. With either mechanism, photolytic ROPs are sensitive to competing light absorbers, whereas dark ROPs are not.

Effect of Natural Solutes

Figure 7 shows that nitrate, and to a lesser extent bicarbonate, inhibit the direct photolysis of TCE by the Purus lamp. Bicarbonate acts primarily as a radical scavenger, while nitrate is a competitive light absorber with extinction coefficients similar to those of TCE (Figure 5). Although nitrate photolyzes to yield $HO^{\cdot}$ radicals, the quantum yield of approximately 2%[10] is too low for effective treatment. For waters with high nitrate concentrations (greater than a few ppm), feasible solutions include nitrate removal by ion exchange or addition of peroxide with use of a longer-wavelength lamp. High levels of iron, manganese, or organic color show similar inhibitory effects and need to be removed or reduced before treatment is cost effective.

Products

Products of aqueous photolyses depend, among other things, on compound type and concentration. In simple cases like carbon tetrachloride or one-carbon

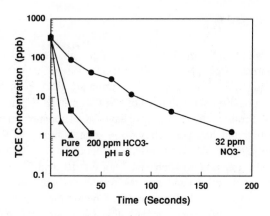

Figure 7. Effect of nitrate and bicarbonate on TCE photooxidation in water in the absence of H_2O_2 in the 9-L reactor.

freons, a single carbon-halogen bond photolysis is enough to mineralize the compound because subsequent dark oxidation and hydrolysis steps lead all the way to CO_2 and hydrohalic acids. Traces of haloforms and perhaloethanes can be found at high initial concentrations. For chloroform under conditions similar to those in Figure 6, we have observed quantitative chloride yields at initial chloroform concentrations up to at least 500 ppm. With larger or less halogenated compounds, more oxidation steps are needed, and many partially oxidized products are present as intermediates. For the example of TCE as in Figure 6, quantitative chloride was found at the lowest initial TCE concentrations, whereas at higher concentrations (160 ppm), we observed transient formation of 10–20% dichloroethenes and traces of chlorinated methanes and ethanes. At very high concentrations near the saturation limit of 1100 ppm, the solutions turned milky, suggesting polymerization. The initial products can almost always be mineralized at the cost of further photooxidation, and the desirability of doing so depends on the product concentrations and toxicity and the subsequent use of the water.

Photooxidation in Air

Kinetics

Figure 8 and Table 2 show that photolysis of chloroolefins in the gas phase is especially rapid, a result which has been attributed to chlorine atom chain reactions.[3] For these compounds, as well as for CH_2Cl_2, $CHCl_3$, and possibly 1,2-dichloroethane, chain mechanisms are supported by four lines of evidence: (1) apparent quantum yields greater than one (Table 2); (2) nonfirst-order kinetic behavior; (3) inhibition by the Cl^- scavenger ethylene (Table 2); and (4) plausible mechanisms for chlorine attack on these compounds at such positions that, after

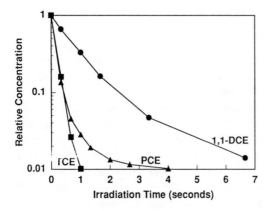

Figure 8.   Photolysis of chloroolefins in air in a 208-L reactor. Initial concentrations were 218 ppmv 1,1-DCE (vinylidene chloride), 144 ppmv TCE, and 165 ppmv PCE (perchloroethene).

oxidation, yield alkoxy radicals that can cleave to regenerate a Cl⋅ atom.[3] With the current Purus lamps, only the rate constants for the chloroolefins are high enough for a commercially viable process. Despite the high quantum yields for CH_2Cl_2 and $CHCl_3$, their absorption cross sections are too low to cause enough initiation. Nevertheless, the chain reactions for these two compounds suggest that sensitization with Cl_2 gas is a useful approach. Although the chloroolefins rapidly photogenerate Cl⋅, they do not sensitize the photooxidation of CH_2Cl_2 or $CHCl_3$ because the olefins themselves scavenge the Cl⋅ very efficiently.

Table 2.   First-Order Decay Coefficients and Wavelength-Averaged Disappearance Quantum Yields in Air

Compound	k (sec^{-1})[a]	$\dfrac{\sum I_\lambda \epsilon_\lambda^{CCl_4}}{\sum I_\lambda \epsilon_\lambda^{VOC}}$	×	$\dfrac{k_{VOC}}{k_{CCl_4}}$ =	Apparent Φ
CCl_4	0.00432	1.0		1.0	1.0
CCl_2FCClF_2	0.00093	5.09		0.22	1.1
$CFCl_3$	0.0036	1.18		0.84	0.99
CCl_3CH_3	0.0041	0.79		0.94	0.74
CH_2Cl_2	0.0070	4.60		1.62	7.5
$CHCl_3$	0.0366	1.79		8.47	15
CH_2ClCH_2Cl	0.0024	N.D.[b]		N.D.	N.D.
1,1-DCE	1.24	0.0389		287	11
PCE	1.7	0.0134		394	5.3
TCE	5.5	0.0236		1300	31
TCE + ethylene	0.075	0.0236		17	0.4

[a] Initial rate constants are taken for nonlog-linear curves.
[b] N.D. = not determined.

Products

The products of chloroolefin photolyses include chloroacetyl chlorides and phosgene, which need further treatment.[11] For example, photolysis of TCE results initially in nearly quantitative formation of dichloroacetyl chloride (DCAC), which is gradually converted to phosgene, HCl, and unidentified nonchlorinated products. Dichloroacetyl chloride and phosgene are acutely toxic, and DCAC hydrolyzes to dichloroacetic acid, which is believed to be carcinogenic. Compared to TCE removal, about 100 times more light exposure is needed to reduce the DCAC concentration by 90%.[11] Thus, although chloroolefins are eliminated rapidly by virtue of chain reactions, the overall photolysis is limited by the rate of destruction of partial oxidation products, which occurs on a time scale not much greater than other compounds that do not involve chain reactions (Table 2). Therefore, taking advantage of the rapid chain reactions depends on the ability to detoxify the initial products by methods other than direct photolysis. Hydrolysis in a scrubber removes the acute toxicity of these compounds, and we are currently investigating other processes for oxidizing or adsorbing any chloroacetic acids formed.

Comparison of Photooxidation in Air and Water

In the case of compounds that can readily be stripped from water to air, one has the choice of medium in which to perform the oxidation. Because gas densities are so much lower than liquid densities, much larger air reactors are needed to efficiently trap the light to the extent that at least some light is usually lost to the walls. Thus, although rate constants are similar, the mass of compound treated in air is usually lower for a given energy input. Notable exceptions are the chloroolefins, because chain reactions occur in air much more efficiently than in water. This observation may be a result of Cl⁻ conversion to Cl_2^- or slower bimolecular reaction of peroxyradicals in water.

To summarize, the advantages of carrying out the reaction in air include less interference from competing absorbers like nitrate and humic acid, the availability of copious amounts (usually a large excess) of O_2, the occurrence of chain reactions for some compounds, fewer corrosion and lamp cleaning problems, and higher discharge limits (no endorsement implied). The advantages of carrying out the oxidation in water include the greater density, allowing much smaller reactors to efficiently utilize the light, and the greater ease of adding H_2O_2. This oxidant is also a more efficient HO· source in water because it scavenges HO· with a rate constant of only $2.7 \times 10^7 \, M^{-1} s^{-1}$ compared to a value of $1.0 \times 10^9 \, M^{-1} s^{-1}$ in air.[9,12] In both air and water, products need to be considered in evaluating the treatment effectiveness.

CONCLUSIONS

Photons can be obtained commercially at a cost similar to other reagents commonly used for ROPs. Pulsed light sources have advantages over conventional (nonpulsed) mercury or xenon lamps for water and air treatment in that (1) a much greater fraction of their emission occurs in the far UV for the same average power usage; (2) they are considerably smaller for a given power output, allowing easier lamp cleaning, e.g., wiper system, and smaller reactor dimensions and thus smaller overall footprint; (3) their wavelength maximum and pulse rate can be changed electronically, allowing greater flexibility for a variety of treatment problems. The low-wavelength light is especially useful in photolyzing some compounds like TCE and freons that react slowly or not at all with HO˙ radicals, provided that competing absorbers like NO_3^- are absent. Reports of commercial mercury lamps with efficiencies of 20–35%, if verified, suggest that UV technology is approaching the theoretical efficiency limit for a black body radiator between 200 and 300 nm. Photolysis in air avoids competing absorbers, but much larger reactors are needed to efficiently absorb the light than are needed in water. For the special case of chloroolefins in air, efficient chain reactions allow especially rapid treatment, provided that undesirable products can be destroyed or removed by a subsequent process like hydrolysis or adsorption.

ACKNOWLEDGMENTS

The author thanks Marc van den Berg, Mark Johnson, Bill Maxfield, and Minggong Su for their careful experimental work and Emery Froelich, Bart Mass, and Paul Blystone for their insightful comments.

REFERENCES

1. Peyton, G. R., Oxidative methods for removal of organic compounds from drinking water supplies, in *Significance and Treatment of Volatile Compounds in Water Supplies*, Ram, N. M., Christman, R. F., and Cantor, K. P., Eds., Lewis Publishers, Chelsea, MI, 1988, chap. 14.
2. Nicole, I., De Laat, J., Dore, M., Duguet, J. P., and Bonnel, C., Utilisation du rayonnement ultraviolet dans le traitement des eaux: mesure du flux photonique par actinometrie chimique au peroxyde d'hydrogene, *Water Res.*, 24, 157, 1990.
3. Blystone, P. G., Johnson, M. D., Haag, W. R., and Daley, P. F., Advanced ultraviolet flashlamps for the destruction of organic contaminants in air in Proceedings of Industrial & Engineering Division Symposium Emerging Technology of Hazardous Waste Management, Atlanta, GA, October 1–3, 1991, American Chemical Society.
4. Rebbert, R. E. and Ausloos, P. J., Gas-phase photodecomposition of carbon tetrachloride, *J. Photochem.*, 6, 265, 1976/77.
5. Smith, B., Overview of flashlamps and arc lamps, *J. Soc. Photo-Optic. Instr. Eng.*, 609, 1, 1986.

6. Froelich, E. M., Peroxidation Systems, Inc., personal communication, 1992.

7. Eliasson, B. and Kogelschatz, U., UV excimer radiation from dielectric barrier discharges, *Appl. Phys.*, B46, 299, 1988.

8. Haag, W. R. and Yao, C.C.D., Rate constants for reaction of hydroxyl radical with several drinking water contaminants, *Environ. Sci. Technol.*, 26, 1005, 1992.

9. Buxton, G. V., Greenstock, C. L., Helman, W. P., and Ross, A. B., Critical review of rate constants for reactions of hydrated electrons, hydrogen atoms and hydroxyl radicals ($\cdot$OH/O$^-$) in aqueous solution, *J. Phys. Chem. Ref. Data,* 17, 513, 1988.

10. Zellner, R., Exner, M., and Herrmann, H., Absolute OH quantum yields in the laser photolysis of nitrate, nitrite, and dissolved H_2O_2 at 308 and 315 nm in the temperature range 278–353 K, *J. Atmos. Chem.*, 10, 411, 1990.

11. Johnson, M. D., Haag, W. R., Blystone, P. G., and Daley, P. F., Destruction of organic contaminants in air using advanced ultraviolet flashlamps, Final Report, EPA Contract No. CR 818209-01-0, Cincinnati, OH, July, 1992.

12. Baulch, D. L., Cox, R. A., Hampson, R. F., Kerr, J. A., Troe, J., and Watson, R. T., Evaluated kinetic and photochemical data for atmospheric chemistry, *J. Phys. Chem. Ref. Data,* 9, 295, 1980.

INDEX

A

Aberchrome 540, 351

Absorbance spectra, 112

Absorption bands
  hyposochromically shifted (H-band), 132
  micropolarity in, 135

Absorption coefficient, 470

Absorption cross section, see Molar
  absorptivity

Absorption rate, of sunlight, 13

Absorption spectroscopy, 129

Absorption spectrum
  of cloud waters, 225, 233-234
  of Fe^{3+}, 12
  of $FeBr^{2+}$, 12
  $FeCl^{2+}$, 12
  of H_2O, 470-472
  of H_2O_2, 470-473
  hypochromically-shifted, 363
  of O_2, 470-472
  of O_3, 470-473
  of S7, 130-132

ABZA, see 4-Aminobenzoic acid

Accessible surface sites, see Surface O-T-O
  sites

Acetaldehyde, see Ethanal

Acetate, 382-383
  ˙OH radicals reaction with, 86-87, 89-94

Acetic acid, 382-383
  photocatalytic degradation of, 291
  photo-Kolbe decarboxylation of, 282

Acetone, 165, 169

Acetonitrile, 330

Acetophenone, 165, 169

Acid deposition, 4, 223, see also Sulfate
  deposition

Acid rain, components of, 223

Actinometry, 351

Activation energy, of DCA degradation, 355,
  364

Additional solute concentration decrease
  (ΔC^*), 334, 342

Adsorption, see also Dark adsorption
  competitive solvent vs solute, 328
  LH kinetic model of, 268-273, 277
  LH-type interpretation of, 317
  within mixed multilayers, 319
  within mixed-solution monolayer, 319
  onto semiconductor surfaces, 317
  onto TiO_2, 276-281
    adsorption isotherm, 273-275
    of model pollutants, 317-346
    pH effect on, 271-273

Adsorption constant determined directly
  (K_{ads}), 318, 326-328, 335, 340, 342-
  343

Adsorption isotherms, see also Curved
    adsorption isotherm-type plot
  for Group I pollutants, 320
  for Group II pollutants, 321
  Langmuir-type, 318
  $n_{2(max)}^s$, 325-326
  for substituted benzoic acids, 324-326
  Langmuir-type model, 327-328

Adsorption sites, 317-318, see also Extent of
  adsorption

Adsorption studies
  dark adsorption
    for group II solutes, 328-330
    pH dependence of, 332
    solvent polarity effects upon, 330-331
  dark adsorption-desorption equilibria, 330
  extent of adsorption, 317-319, 322-323,
    326, 328, 335-341, 343
  Langmuir-type isotherms, 327
  monolayer surface-solution phase, 323
  solution-multilayer regime, 326
  for substituted benzoic acids, 323-325
  K value, 325-328

Advanced oxidation processes (AOPs), 261,
    451, 468, see also Heterogeneous
    photolysis; Homogeneous photolysis;
    Radical oxidation processes
  chemical efficiencies of, 508-509
  combinations of, 452
  dark homogeneous oxidation, 468
  factors in selection/optimization of, 509
    chemical efficiency/UV light, 510-511
    oxidant chemical efficiency, 510
    residence time considerations, 510
    UV lamps, 509
  H_2O_2-UV kinetic model, 502-503
  H_2O_2-UV system, 500-501, 503-504
  major oxidants in, 453, 468, 499-500, 511-
    512
  mechanism of, 480-482
  ˙OH in, 453, 468, 511-512

C

I